Springer Monographs in Mathematics

Springer
New York
Berlin
Heidelberg
Barcelona
Hong Kong
London
Milan
Paris
Singapore
Tokyo

Books of Related Interest by Serge Lang

Linear Algebra, Third Edition
1987, ISBN 0-387-96412-6

Undergraduate Algebra, Second Edition
1990, ISBN 0-387-97279-X

Undergraduate Analysis, Second Edition
1997, ISBN 0-387-94841-4

Complex Analysis, Third Edition
1993, ISBN 0-387-97886-0

Real and Functional Analysis, Third Edition
1993, ISBN 0-387-94001-4

Algebraic Number Theory, Second Edition
1994, ISBN 0-387-94225-4

Introduction to Complex Hyperbolic Spaces
1987, ISBN 0-387-96447-9

Other Books by Lang Published by Springer-Verlag

Introduction to Arakelov Theory • Riemann-Roch Algebra (with William Fulton) • Complex Multiplication • Introduction to Modular Forms • Modular Units (with Daniel Kubert) • Fundamentals of Diophantine Geometry • Elliptic Functions • Number Theory III • Survey of Diophantine Geometry • Cyclotomic Fields I and II • $SL_2(\mathbf{R})$ • Abelian Varieties • Introduction to Algebraic and Abelian Functions • Introduction to Diophantine Approximations • Elliptic Curves: Diophantine Analysis • Math Talks for Undergraduates • Introduction to Linear Algebra • Calculus of Several Variables • First Course in Calculus • Basic Mathematics • Geometry: A High School Course (with Gene Murrow) • Math! Encounters with High School Students • The Beauty of Doing Mathematics • THE FILE • CHALLENGES

Jay Jorgenson   Serge Lang

Spherical Inversion on $SL_n(\mathbf{R})$

Springer

Jay Jorgenson
Department of Mathematics
City College of New York, CUNY
138 Convent Avenue
New York, NY 10031
USA

Serge Lang
Department of Mathematics
Yale University
New Haven, CT 06520-8283
USA

With 2 Illustrations

Mathematics Subject Classification (2000): 22E46

Library of Congress Cataloging-in-Publication Data
Jorgenson, Jay.
Spherical inversion on SLn(R) / Jay Jorgenson, Serge Lang.
p. cm.
Includes bibliographical references and index.

ISBN 978-1-4419-2883-2 ISBN 978-1-4684-9302-3 (eBook)
DOI 10.1007/978-1-4684-9302-3

1. Spherical functions. 2. Decomposition (Mathematics) I. Lang, Serge, 1927–
II. Title.
QA406.J67 2001
515′.53—dc21 00-058307

Printed on acid-free paper.

Production coordinated by Brian Howe and managed by Terry Kornak; manufacturing supervised by Joe Quatela.
Typeset by Archetype Publishing Inc., Monticello, IL.
9 8 7 6 5 4 3 2 1

A member of BertelsmannSpringer Science + Business Media GmbH.

Contents

Acknowledgments

Jorgenson acknowledges support from NSF grants, and thanks Anthony Petrello and Adreas Typaldos for general financial assistance. Lang thanks the Max-Planck-Institut for productive yearly visits.

We thank several people with whom we had informative conversations in learning the subject, notably Cogdell, Coifman, Gangolli, Helgason, Howe, and Wallach.

Jay Jorgenson
Serge Lang

Overview

For the most part we shall be concerned with $SL_n(\mathbf{R})$ and with invariant differential operators, the invariance being with respect to various subgroups. To a large extent, this book carries out the general results of Harish-Chandra [Har 58a] on $SL_n(\mathbf{R})$. We are striving to the analogue of Fourier inversion for functions on $SL_n(\mathbf{R})$ which are bi-invariant under the action of the real unitary group. We recommend immediate reading of Chapter IX, which has been written in a self-contained way and could be a Chapter 0. Readers can then see where we are going.

Fourier analysis on ordinary euclidean space is based on the decomposition of a function with respect to e^{ixy}, which is viewed as a kernel for the Fourier integral operator. A euclidean space is usually identified with its dual space in this Fourier decomposition. The group $G = SL_n(\mathbf{R})$ has a polar decomposition $G = KAK$, with the real unitary group K and a group A which is isomorphic under the exponential map with a euclidean space. Thus the Fourier analysis of K-bi-invariant functions can be reduced to the euclidean space, but the reduction is somewhat complicated, and we seek the kernel function $\phi(\mathbf{i}\lambda, x)$ with $x \in G$ and λ in the euclidean space to play the same role as the above Fourier kernel function. There are two essential features to this kernel: it comes from characters on A, and it is an eigenfunction of a certain algebra of differential operators.

Not so roughly speaking, there is an integral transform called the (Harish-Chandra) spherical transform, coming from the kernel function called the (Harish-Chandra) spherical kernel, which transforms objects associated with Fourier–Mellin euclidean inversion to similar objects on certain non-commutative Lie groups G. Then analysis on G can be carried out in terms of a euclidean model.

For instance, the usual Gauss function with square exponential decay corresponds under the inverse spherical transform to a function on the group. This is significant, because it immediately shows in a vast structural context how to get the heat kernel for spherical functions (Gangolli [Gan 68]), and we are after the heat kernel—cf. the end of this introduction. In Chapter IV, §7, and Chapter X, Theorem 3.1, we recall extensively inversion results from standard euclidean space analysis. These results provide the springboard from which we take off for the non-abelian extensions.

Harish-Chandra's work, culminating for our purposes with [Har 66], was complemented and simplified in some respects over three decades by Helgason, Gangolli, Rosenberg, and Anker whose contributions have been taken into account in the present book.

The bi-invariant functions have a spectral decomposition in terms of elementary spherical functions, arising from characters in a suitable sense. These spherical functions are eigenfunctions of certain differential operators, one of which is especially important (Casimir, see below). The study of these differential operators is carried out by using various decompositions of the group. We now go somewhat more extensively into this development.

Decompositions. We start with the most basic decomposition, the Iwasawa decomposition and the corresponding decomposition of Haar measure. Chapter I, §4 and §5, provide a piece of self-contained linear algebra concerning positivity and convexity, with an application in the context of Iwasawa decompositions. Then we get on the first main theme of the book, namely the invariant differential operators. They enter the inversion picture because the basic spherical functions are those which are not only K-bi-invariant (K = unitary group) but are eigenfunctions of the invariant differential operators, just like the exponential characters e^{ixy} in Fourier inversion.

The study of invariant differential operators will be made in conjunction with various additional structures on the group or homogeneous space, systematically. These include:

Left and right translation, and conjugation.
The Iwasawa decomposition $G = UAK$.
The Cartan Lie decomposition $\mathfrak{g} = \mathfrak{p} + \mathfrak{k}$.
The Gelfand–Naimark decomposition.
The global Cartan decomposition $G = \mathbf{P}K$.
The polar decomposition $G = KAK$.
The Bruhat decomposition $G = \bigcup BwB$ $(w \in W)$.
The semisimple decomposition of the Lie algebra under the A-conjugation action, or under the regular representation of $\mathfrak{a}$ on the Lie algebra of G.
A G-invariant non-singular symmetric bilinear form.

The systematic analysis associated with these structures includes:

The computations of the Jacobian determinant for each one of the various decompositions.

Computation of the direct images under various projections, notably the direct image on the A-component in the various decompositions.

Casimir operator. Chapter VII gives an example, with a special differential operator, the Casimir operator, which can be defined with no knowledge of differential geometry. It has the expository advantage of allowing a more self-contained exposition with no background in Riemannian geometry. Only general manifold theory and the Lie group structure are needed to handle this operator. Although the Laplacian has a more familiar name, actually the Casimir operator is much easier to introduce and to handle. Harish-Chandra used it systematically. Readers should not let themselves be swayed by a false impression that Casimir is somehow exotic. Anyhow, it's equal to the Laplacian whenever one wants it to be.

Chapter VIII complements the study of invariant differential operators by using the Casimir operator to define the Harish-Chandra series expression for the spherical functions, as eigenfunctions of the Casimir operator. Gangolli made a substantial contribution in determining the convergence of this series. The series is used among other things for the main argument in Helgason's support theorem.

Inversion theorem. Chapter IX gives Rosenberg's proof for the Harish-Chandra inversion and Helgason's contribution concerning the Paley–Wiener space. This chapter is written in a self-contained way in a euclidean, abelian situation to which the non-commutative theory has been reduced. It may thus be more useful to a much larger audience than the rest of the book, and may be useful to analysts at large in pursuing various generalizations of Fourier inversion, especially in connection with the spectral theory of differential equations, irrespective of the context of Lie groups.

Chapters X and XI follow Anker's proof of Harish-Chandra inversion on the Schwartz spaces, reducing the question to the $C_c^\infty(K\backslash G/K)$-Paley–Wiener case by an appropriate continuity argument. Since the 1950s, the space of C^∞ functions with compact support has been regarded as the basic space of test functions, and other spaces arising naturally are viewed as contained in its completions with respect to various topologies defined by norms or families of seminorms. Anker saw how to fit the Schwartz spaces inversion simply into this pattern.

Chapter XII shows how the theory simplifies on $SL_n(\mathbf{C})$, and we shall make comments on this phenomenon later.

Parabolic induction. Before Anker's proof, Harish-Chandra's method of induction from parabolic subgroups was central to the arguments for inversion on the Schwartz spaces. This method has now disappeared from the inversion proofs. It reappears in the study of $\Gamma\backslash G$ with a discrete group Γ, as in [Har 68] and his papers from 1970–1976; or also in his works of Langlands [Lgl 76]. It will be central to our further work on heat Eisenstein series.

Partial axiomatizations. Another main item contributing to an easier access is the partial axiomatization, which allows for local logical independence of certain items, making proofs easier to read, besides extending the domain of validity of some theorems. Such axiomatization also substitutes for hundreds of pages of background prerequisites, say on Lie algebras. On concrete special groups, the axioms can usually be verified at once. However, we give priority to keeping things simple. If the axiomatization threatens to become too heavy, we concentrate on the concrete case of SL_n. Hence there remains to be written a book striving for generality, but axiomatized along the lines we have indicated, i.e. according to the various decompositions. Such a book would be organized quite differently from those which start under the influence of Harish-Chandra: "Let G be a semisimple Lie group," or its softer version, "Let G be a reductive group" or a group in Harish-Chandra's class. Both general versions of the theory, and the working out of specific cases, are needed for different connections with other parts of mathematics. For an introduction to the subject, however, focusing on $\mathrm{SL}_n(\mathbf{R})$ seemed to us to be the most effective approach.

Internal or external use? Over the past decades, since the 1950s, the subject developed essentially internally, for use by people in the specific field of representation theory and (generalized) Fourier analysis on semisimple Lie groups. It developed certain habits which make its access difficult if not impossible by outsiders. One of these has to do with idiosynchratic terminology. As far as we are concerned, eigenvalues, eigenfunctions and eigenspaces deserve being called by these names. We also see no reason to use the expression "wave packet" for an element in the image of the transpose spherical transform, just because it was once upon a time considered in the context of physics. *Ad lib.*

Even more importantly, certain general attitudes have resulted in some excesses, creating obstacles for newcomers to the subject. On the one hand, as Knapp writes [Kna 86]: "The early works ... established the explicit determination of the Plancherel formula and the explicit description of the unitary dual as important initial goals. This attitude of requiring explicit results ultimately forced a more concrete approach to the subject than was possible with abstract functional analysis, and the same attitude continues today." This attitude stems mostly from the works of Harish-Chandra, and we fully subscribe to it. The present book is in fact motivated by the explicit formulas for the heat kernel in connection with inversion theory.

However, Knapp goes on to describe a further development as follows: "More recently, this attitude has been refined to insist that significant results not only be explicit but also be applicable to all semisimple groups. A group-by-group analysis is rarely sufficient now: It usually does not give the required amount of insight into the subject. To be true to the field, this book attempts to communicate such attitudes and approaches, along with the results."

We definitely do not regard this last attitude as a "refinement" of the first. As to the phrase "the required amount of insight into the subject," we ask: Required by whom? Under what authority? When? For what purpose?

The condition "to be true to the field" is sectarian. "The field" is in any case not well-defined. Even internally, there arises a serious question how much one places everything in the context of representation theory, and how much one develops certain parts of analysis independently of representation theory, for instance, the Harish-Chandra inversion theory as carried out for SL_n in the present book. In different ways, representation theory both illuminates and interferes with the analysis.

As described by Knapp, the insistence on the last-mentioned attitude had the effect of making "the field" more and more directed at insiders and impenetrable to outsiders, who may wish to connect some aspects of "the field" (whatever it is) with several if not all other parts of mathematics, and do so in specific cases of intrinsic interest. We are among the outsiders.

We got into the Harish-Chandra inversion theory because as was shown by Gangolli, Harish-Chandra's inversion measure is also the measure which defines the heat kernel on G/K as the inverse spherical transform of the Gauss function, and the heat kernel is the fundamental object from which we are building up our ladders of zeta functions. We found it rather hard to get into this inversion theory, because current expositions are geared to insiders. As a result we have tried to make the exposition in this book reader friendly, starting with ourselves. Two major factors contribute to making access easier: first the choice of the special case $\mathrm{SL}_n(\mathbf{R})$ in a wide class of groups which include the semisimple Lie groups, but definitely more; and second partial axiomatizations as in our list of decompositions. Although it was of course useful to have the theory on $\mathrm{SL}_2(\mathbf{R})$ tabulated about 25 years ago, still $\mathrm{SL}_2(\mathbf{R})$ is not a representative sample of the general case, whereas $\mathrm{SL}_n(\mathbf{R})$ is. All the essential features of the Harish-Chandra theory are exhibited on $\mathrm{SL}_n(\mathbf{R})$, but hundreds of pages of background on Lie algebras can be replaced by short direct verifications. Thus we find $\mathrm{SL}_n(\mathbf{R})$ very appropriate to serve as a model for a first introduction to Harish-Chandra inversion.

Other groups, starting with $SL_n(\mathbf{C})$. As will become apparent in the last chapter, however, we give immediately an example why one might prefer to deal with $SL_n(\mathbf{C})$. One major feature is that the objects entering into inversion theory such as spherical functions and the heat kernel are "split" in various ways, essentially amounting to combinations of trigonometric or ordinary polynomials and simple exponential factors. The case of SL_n (especially $SL_n(\mathbf{C})$) seemed to us to be the simplest case on which we could investigate our ladders of zeta functions, which arise as follows. On a suitable universal covering space (Cartan–Hadamard manifold in differential geometric terms), one periodizes the heat kernel with respect to the action of a discrete group. On the quotient space the heat kernel has a spectral decomposition (Fourier expansion of some sort), and the equality between the two expressions is a theta relation. For compact quotients, Gangolli's pioneering paper [Gan 68] took the trace of the heat kernel and called its Fourier expansion a theta relation. However, in our series of articles we deal with very general types of theta series, and there is no need to take the trace to call the expansion a theta relation. Furthermore, we are especially interested in the case of non-compact quotients, arising naturally in analytic number theory, algebraic geometry, differential geometry, and spectral theory. We then apply our Gauss transform (rather than the Mellin transform) as in [JoL 94] to get a certain type of zeta function (Dirichlet series, Bessel series, ...) which form our main objects of study. Perhaps the easiest discrete groups which arise are $SL_n(\mathbf{Z})$ in $SL_n(\mathbf{R})$ and $SL_n(\mathbf{Z}[\mathbf{i}])$ in $SL_n(\mathbf{C})$.

Ultimately, investigations will involve many other groups, and will exhibit more complicated features of algebraic geometry and differential geometry. The general theorems of inversion theory then get tabulated in the specific instances to determine precisely how they affect the differential geometric and algebraic geometric properties of these objects. Books developing the general inversion theory and books describing the specific behavior in each important specific case will thus complement each other. They need each other. For example, the general inversion theory could be made explicit in cases involving not only the Siegel modular space and the Hilbert–Asai modular case, but moduli spaces of many other kinds, such as moduli spaces for K3 surfaces and Calabi–Yau manifolds, moduli spaces of curves, and moduli spaces of forms of arbitrary degree as in a paper of Jordan [Jor 1880]. In this way, both differential and algebraic geometry come to the fore, connecting both with inversion theory and our theory of zeta functions and regularized products or regularized harmonic series. Thus, both the general theory of the Lie industry, and the special aspects of each particular case, would merge and make both aspects useful simultaneously. The situation is completely open ended.

Table of the Decompositions

Iwasawa decomposition

Let G be a Lie group. By a **weak Iwasawa decomposition** of G we mean the data of three closed Lie subgroups U, A, K satisfying the conditions:

IW 1. The product map $U \times A \times K \to UAK = G$ is a differential isomorphism.

IW 2. The group A is abelian, contained in the normalizer of U.

We say that the decomposition is an **Iwasawa decomposition** if in addition it satisfies:

IW 3. The group A has a Lie group isomorphism $A \approx \mathbf{R}^+ \times \cdots \times \mathbf{R}^+$.

The factor group of the normalizer of A in K modulo the centralizer of A in K is called the **Weyl group**, denoted by W. We assume W finite.

We let Iw_U, Iw_A, Iw_K be the projection of G on U, A, K respectively.

Cartan Lie decomposition

Let G have a weak Iwasawa decomposition. Let $\mathfrak{g} = \mathrm{Lie}(G)$. For $x \in G$, let $\mathbf{c}_{\mathrm{Lie}}(x)$ be the action on $\mathfrak{g}$ induced by conjugation with x. We say that $\mathfrak{g}$ admits a **Cartan** or **Cartan Lie decomposition** under the following three conditions:

CA 1. Let $\mathfrak{p} = \mathbf{c}_{\text{Lie}}(K)\mathfrak{a}$. Then $\mathfrak{g} = \mathfrak{p} \oplus \mathfrak{k}$, where $\mathfrak{k} = \text{Lie}(K)$.

CA 2. $\mathfrak{p} = \mathfrak{a} \oplus \mathfrak{q}$ with $\mathfrak{q} \subset \mathfrak{n} \oplus \mathfrak{k}$, where $\mathfrak{n} = \text{Lie}(U)$ and $\mathfrak{a} = \text{Lie}(A)$.

CA 3. In the symmetric algebra, let $P \to P_{\mathfrak{a}}$ be the projection of $S(\mathfrak{p})$ to $S(\mathfrak{a})$. Then this projection is a linear isomorphism

$$S(\mathfrak{p})^{c(K)} \xrightarrow{\approx} S(\mathfrak{a})^{W}.$$

This projection is called the **Chevalley isomorphism**.

Global Cartan decomposition

Given a Cartan Lie decomposition, let $\mathbf{P} = \exp \mathfrak{p}$. We say that the **Cartan decomposition** is **global** if $G = \mathbf{P}K$ and the product map $\mathbf{P} \times K \to \mathbf{P}K = G$ is a differential isomorphism.

Lie semisimplicity

Let G be given with an Iwasawa decomposition. We call it **Lie semisimple** if $\mathfrak{n}$ is a direct sum

$$\mathfrak{n} = \bigoplus \mathfrak{n}_{\alpha},$$

where each $\mathfrak{n}_{\alpha}$ is a non-zero eigenspace for $\mathfrak{a}$, with eigencharacter $\alpha \neq 0$. We let $\mathcal{R}(\mathfrak{n})$ be the set of such characters, which we call **relevant characters** or $(\mathfrak{a}, \mathfrak{n})$**-characters**. The trace of the representation is denoted by τ, and $\rho = \tau/2$. The corresponding multiplicative character $a \mapsto a^{\tau}$ on A is denoted by $\delta = \delta_{\text{Iw}}$, and is called the **Iwasawa character**.

Gelfand–Naimark decomposition

Let $G = PK = UAK$, $P = UA$ be a weak Iwasawa decomposition with K compact, G, U, A unimodular. By a **Gelfand–Naimark decomposition** of such G, we mean the additional data of a closed subgroup M of K and a closed unimodular subgroup V of G such that, if we put $B = PM$, then:

GN 1. $P \times M \to PM = B$ is a differential isomorphism.

GN 2. $B \times V \to BV$ is a differential isomorphism of $B \times V$ with an open subset of G, whose complement has measure 0.

GN 3. M normalizes U, V and centralizes A.

GN 4. The map $\psi_{M/K}: V \to M\backslash K$ defined by $\psi_{M\backslash K}(v) = M\,\mathrm{Iw}_K(v)$ gives a differential isomorphism of V with an open subset of $M\backslash K$ whose complement has measure 0. We call $\psi_{M\backslash K}$ the **Harish-Chandra mapping.**

Remark. Condition **GN 1** is automatically satisfied because of the direct product decomposition $G = PK$ and the fact that a Lie subgroup is locally split.

Polar decomposition

Let G have an Iwasawa decomposition $G = UAK$ with K compact. We suppose that $\mathfrak{g}$ is Lie semisimple. We then define:

$$A' = \text{set of regular elements}$$
$$= \{a \in A \text{ such that } a^\alpha \neq 1 \text{ for all } \alpha \in \mathcal{R}(\mathfrak{n})\}.$$
$$A^+ = \text{subset of } a \in A' \text{ such that } a^\alpha > 1 \text{ for all } \alpha \in \mathcal{R}(\mathfrak{n}).$$

We say G has a **polar decomposition** associated with the Iwasawa decomposition if the product map

$$\mathbf{p}\colon\ K \times A \times K \to KAK = G$$

is a surjection, and satisfies the following two conditions:

POL 1. The set A^+ is a fundamental domain for the action of W on A'.

POL 2. The polar map

$$K/M \times A' \times K \to KA'K$$

induces a covering of degree $|W|$ over the set of regular elements, and the complement of the image has measure 0. Thus we get a differential isomorphism

$$K/M \times A^+ \times K \to KA^+K.$$

The (anti-)involution

Let θ be an **involution** of G (Lie group automorphism of order 2, $\theta \neq \mathrm{id}$) and let $\mathbf{t}$ be defined by $\mathbf{t}x = \theta x^{-1}$. Equivalently, $\mathbf{t}$ could be given as an anti-involution, then defining θ by $\theta x = \mathbf{t}x^{-1}$ and assuming $\theta \neq \mathrm{id}$. By the chain rule, $T_e\mathbf{t} = -T_e\theta$. We may write $\mathbf{t}_{\mathrm{Lie}}$ or θ_{Lie} for the action of $\mathbf{t}$ (resp. θ) on $\mathfrak{g} = T_eG$. We suppose G is given with a Lie semisimple Iwasawa decomposition. We say that θ (or $\mathbf{t}$) is **adapted** if it satisfies the following five conditions. We could also say that θ is a **Cartan involution**.

TR 1. The Lie algebra $\mathfrak{g}$ has the direct sum decomposition

$$\mathfrak{g} = \mathfrak{a} \oplus \mathfrak{n} \oplus \mathbf{t}\mathfrak{n},$$

and $T_e\mathbf{t}$ preserves the eigenspace decomposition, that is, $\mathbf{t}$ induces an isomorphism

$$\mathbf{t}_\alpha : \mathfrak{n}_\alpha \to (\mathbf{t}\mathfrak{n})_{-\alpha} \qquad \text{for each } \alpha \in \mathcal{R}(\mathfrak{n}).$$

Since $\mathfrak{n} + \mathbf{t}\mathfrak{n}$ is stable under $\mathbf{t}$, we may then define the **skew-symmetric** (resp. **symmetric**) subspaces of $\mathfrak{n} + \mathbf{t}\mathfrak{n}$ in the natural way with respect to $\mathbf{t}$, and denote them by Sk (resp. $\mathrm{Sym}^{(0)}$).

TR 2. We have $T_e\mathbf{t} = \mathrm{id}$ on $\mathfrak{a}$.

Then we define

$$\mathrm{Sym} = \mathfrak{a} \oplus \mathrm{Sym}^{(0)}.$$

The direct sum decomposition of the Lie algebra can then be expressed as

$$\mathfrak{g} = \mathrm{Sym} \oplus \mathrm{Sk}.$$

TR 3. $\mathrm{Sym} = \mathbf{c}(K)\mathfrak{a}$.

TR 4. $\mathrm{Sk} = \mathfrak{k}$.

Letting $\mathfrak{p} = \mathrm{Sym}$, we recover a Cartan Lie decomposition. By **Tr 2** and the fact that A is connected we conclude that A is pointwise fixed by $\mathbf{t}$, and the connected component of K is pointwise fixed by θ. We require the global condition:

TR 5. For all $k \in K$ we have $\mathbf{t}k = k^{-1}$ or $\theta k = k$.

Depending on one's purposes and context, one may wish to add the following global axiom for a Cartan involution.

POL SSk or Global Cartan Decomposition. Let $\mathbf{P} = \exp \mathfrak{p}$. Then $G = \mathbf{P}K$, and

$$\mathbf{p}_{\mathrm{SSk}} \text{ or } \mathbf{p}_{\mathrm{CA}} : \mathbf{P} \times K \to \mathbf{P}K = G$$

is a differential isomorphism.

CHAPTER I

Iwasawa Decomposition and Positivity

The most basic of all the decompositions is the Iwasawa decomposition, which we introduce in the first section. The section computes appropriate Haar measures and Jacobians for the Iwasawa decomposition. For more similar Haar measure computations, see Chapter V, §3. In §3 we consider the Cartan Lie decomposition in connection with polynomial invariants. The last three sections are devoted to some linear algebra in connection with the notion of positivity (partial ordering) which will play an important role later. The polar decomposition is introduced in connection with a theorem of Harish-Chandra concerning this partial ordering. The theorem compares the size of the A-component in the Iwasawa and polar decompositions. Jacobian computations and basic effects of polar decomposition on differential operators will be given in Chapter VI.

I, §1. THE IWASAWA DECOMPOSITION

Let:

$G = G_n = \mathrm{GL}_n(\mathbf{R})$.

$\mathrm{Pos}_n = \mathrm{Pos}_n(\mathbf{R})$ = space of symmetric positive definite real matrices.

$K = O(n) = \mathrm{Uni}_n(\mathbf{R})$ = group of real unitary $n \times n$ matrices.

U = groups of real unipotent upper triangular matrices, i.e. of the form

$$u(X) = u = \begin{pmatrix} 1 & & & x_{ij} \\ 0 & 1 & & \\ \vdots & & \ddots & \\ 0 & 0 & \cdots & 1 \end{pmatrix} \qquad \text{so}$$

$$u(X) = I + X, \qquad X = (x_{ij}), \qquad 1 \leqq i < j \leqq n.$$

A = group of diagonal matrices with positive components,

$$a = \begin{pmatrix} a_1 & & & 0 \\ & a_2 & & \\ & & \ddots & \\ 0 & & & a_n \end{pmatrix} \qquad a_i > 0 \text{ all } i.$$

Theorem 1.1. *The product mapping*

$$U \times A \times K \to UAK = G$$

is a differential isomorphism. Actually, the map

$$U \times A \to \mathrm{Pos}_n(\mathbf{R}) \qquad \textit{given by} \qquad (u, a) \mapsto ua{}^t u$$

is a differential isomorphism.

Proof. Let $\{e_1, \ldots, e_n\}$ be the standard unit vectors of $\mathbf{R}^n$. Let $x \in \mathrm{GL}_n(\mathbf{R})$. Let $v_i = xe_i$. We orthogonalize $\{v_1, \ldots, v_n\}$ by the standard Gram–Schmidt process, so we use a transformation by a matrix $u \in U$, namely we let

$$w_1 = v_1, \qquad w_2 = v_2 - c_{21}w_1 \perp w_1,$$
$$w_3 = v_3 - c_{32}w_2 - c_{31}w_1 \perp w_1 \text{ and } w_2, \qquad \text{and so on.}$$

Then $e_i' = w_i/\|w_i\|$ is a unit vector, and the matrix a having $\|w_i\|^{-1}$ for its diagonal elements is in A. Let $k = aux$ so $x = u^{-1}a^{-1}k$. Then k is unitary, which proves that $G = UAK$. To show uniqueness, suppose that

$$u_1 a {}^t u_1 = u_2 b {}^t u_2 \qquad \text{with} \quad u_1, u_2 \in U \quad \text{and} \quad a, b \in A,$$

then putting $u = u_2^{-1}u_1$ we find

$$ua = b{}^t u.$$

Since u and ${}^t u$ are triangular in opposite direction, they must be diagonal, and finally $a = b$. That the decomposition is differentially a product is proved by computing the Jacobian of the product map, done in §2.

The group K is the subset of elements of G fixed under the involution

$$g \mapsto {}^t g^{-1}.$$

We write the transpose on the left to balance the inverse on the right. We have a surjective mapping

$$G \to \mathrm{Pos}_n \qquad \text{given by } g \mapsto {}^t g g.$$

This mapping gives a bijection of the coset space

$$\varphi: G/K \to \mathrm{Pos}_n,$$

and this bijection is a real analytic isomorphism. Furthermore, the group G acts on Pos_n by a homomorphism $g \mapsto [g] \in \mathrm{Aut}(\mathrm{Pos}_n)$, where $[g]$ is given by the formula

$$[g]p = gp\,{}^t g.$$

This action is on the left, contrary to right wing action by some people. On the other hand, there is an action of G on the coset space G/K by translation

$$\tau: G \to \mathrm{Aut}(G/K) \qquad \text{such that} \qquad \tau(g)g_1K = gg_1K.$$

Under the bijection φ, a translation $\tau(g)$ corresponds precisely to the action $[g]$.

Thus $\mathrm{Pos}_n(\mathbf{R})$ is differentially isomorphic to G/K for $G = \mathrm{GL}_n(\mathbf{R})$, and we call it the **quadratic model** for G/K. Cf. [Mos 53] and [Lan 99], Chapter XII, for differential geometric properties of this model. Similarly, we could let $G = \mathrm{SL}_n(\mathbf{R})$ be the subgroup of $\mathrm{GL}_n(\mathbf{R})$ consisting of the elements of determinant 1, and K the subgroup of unitary matrices of determinant 1. Then we write

$$\mathrm{SPos}_n(\mathbf{R}) = G/K$$

for the quadratic model of elements of $\mathrm{Pos}_n(\mathbf{R})$ having determinant 1.

Our point of view is that $\mathrm{SL}_n(\mathbf{R})$ is a prototype for much more general Lie groups for which the theory we are developing is valid. Thus we shall systematically propose local axiomatizations which indicate more general conditions under which certain theorems are valid. We start with an axiomatization of the Iwasawa decomposition.

Let G be a Lie group. By a **weak Iwasawa decomposition** of G we mean the data of three closed Lie subgroups U, A, K satisfying the following conditions:

IW 1. The product map

$$U \times A \times K \to UAK = G$$

is a differential isomorphism.

IW 2. The group A is abelian, contained in the normalizer of U.

We say that the decomposition is an **Iwasawa decomposition** if in addition A satisfies:

IW 3. The group A has a Lie group isomorphism

$$A \xrightarrow{\approx} \mathbf{R}^+ \times \cdots \times \mathbf{R}^+.$$

A given isomorphism then constitutes an added structure, which allows a coordinatization of A, as we shall see in Chapter III. One reason for splitting off the third condition is that in a number of abstract nonsense measure theoretic considerations, only **IW 1** and **IW 2** are relevant, and they occur often enough to warrant a name. In particular, with only **IW 1**, **IW 2** we have what we call the **Iwasawa projection** on A,

$$\mathrm{Iw}_A : G \to A \qquad \text{defined by} \qquad \mathrm{Iw}_A(uak) = a.$$

Remark. Unless we integrate over K, it is irrelevant whether K is compact or not, and we do *not* assume this additional property, unless otherwise specified.

Examples. Of course, the standard example occurs when A is the group of positive diagonal matrices, U is the group of unipotent upper triangular matrices, and K is the real unitary group. However, what is usually called the Langlands decomposition for certain subgroups also falls under the above axioms. Specifically, let

$$n = n_1 + \cdots + n_{r+1}$$

be a partition P of n into a sum of positive integers. We consider blocks of $n_i \times n_i$ matrices along the diagonal ($i = 1, \ldots, r+1$). We let K_P be the group of such blocks such that each block has determinant ± 1. We let U_P be the group of unipotent matrices (upper triangular, 1's on the diagonal) which have otherwise non-zero components only above the blocks. We let A_P be the group of positive diagonal matrices

$$\begin{pmatrix} \boxed{a_1 I_1} & & & 0 \\ & \boxed{a_2 I_2} & & \\ & & \ddots & \\ 0 & & & \boxed{a_{r+1} I_{r+1}} \end{pmatrix} \qquad \text{with} \quad I_i = n_i \times n_i \text{ unit matrix,}$$

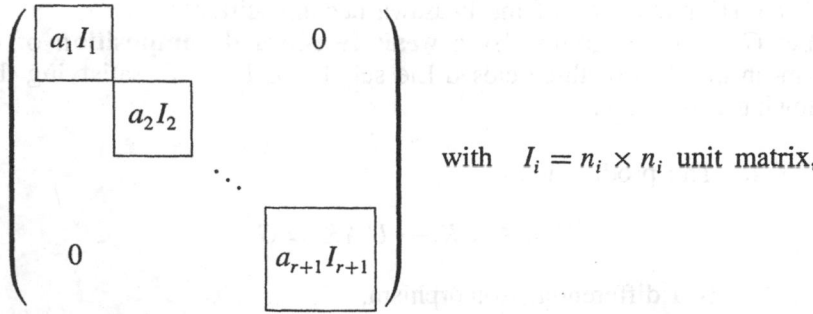

so the diagonal components are constant within each block, and $a_i > 0$ all i. The group $U_P A_P K_P$ gives an example satisfying the axioms. Observe that in this case, the group K_P is not compact.

I, §2. HAAR MEASURE AND IWASAWA DECOMPOSITION

For a much more general discussion, see [Rag 72], Chapter I. We use notation to fit immediate applications, but the first proposition is very general. At first, we do not assume A abelian.

Proposition 2.1. *Let P be a locally compact group with two closed subgroups A, U such that A normalizes U, and such that the product*

$$A \times U \to AU = P$$

is a topological isomorphism. Then for $f \in C_c(P)$ the functional

$$f \mapsto \int_U \int_A f(au)\, da\, du = \int_A \int_U f(au)\, du\, da$$

is a Haar (left invariant) functional on P.

Proof. Left invariance by A is immediate. Let $u_1 \in U$. Then

$$\begin{aligned}
\int_A \int_U & f(u_1 au)\, du\, da \\
&= \int_A \int_U f(aa^{-1}u_1 au)\, du\, da \\
&= \int_A \int_U (f \circ a)\left(u_1^a u\right)\, du\, da \quad \text{where } u_1^a = a^{-1}u_1 a \\
&= \int_A \int_U f(au)\, du\, da \quad \text{by Haar measure property}
\end{aligned}$$

thus proving our assertion.

Let G be a locally compact group with Haar measure dx. We recall that the **modular function** $\Delta = \Delta_G$ on G is the function (actually continuous homomorphism into $\mathbf{R}^+$) such that for all $f \in C_c(G)$,

$$\int_G f(xx_1)\, dx = \Delta(x_1) \int_G f(x)\, dx.$$

Proposition 2.2. *Notation as in Proposition 2.1, there exists a unique continuous homomorphism* $\delta: A \to \mathbf{R}^+$ *such that for* $f \in C_c(U)$,

$$\int_U f(a^{-1}ua)\,du = \delta(a) \int_U f(u)\,du,$$

or in other words, for $f \in C_c(P)$,

$$\int_U f(ua)\,du = \delta(a) \int_U f(au)\,du,$$

If U *is unimodular, then* δ *is the modular function on* P, *that is*

$$\Delta_P(p) = \Delta_P(au) = \delta(a).$$

Proof. The first statement is immediate because the map

$$u \mapsto u^a = a^{-1}ua$$

is a topological group automorphism of U, which preserves Haar measure up to a constant factor, by uniqueness of Haar measure. For the second statement, we first note that the functional

$$f \mapsto \int_U \int_A f(au)\,da\,du$$

is right invariant under U. Let $a_1 \in A$. Then

$$\begin{aligned}\iint f(aua_1)\,da\,du &= \iint f(aa_1a_1^{-1}ua_1)\,da\,du \\ &= \delta(a_1) \iint f(au)\,da\,du,\end{aligned}$$

which proves the formula. The proposition will be complemented in Chapter V, §3.

Proposition 2.3. *Let* G *be a locally compact group with two closed subgroups,* P, K *such that*

$$P \times K \to PK = G$$

is a topological isomorphism (not group isomorphism). Assume that G, K *are unimodular. Let* dx, dp, dk *be given Haar measures on* G, P, K *respectively. Then there is a constant* c *such that for all* $f \in C_c(G)$,

$$\int_G f(x)\,dx = c \int_P \int_K f(pk)\,dp\,dk.$$

If in addition $P = AU$ as in Proposition 2.1, *with U unimodular, so we have the product decomposition*

$$U \times A \times K \to G,$$

then

$$\int_G f(x)\,dx = c \int_U \int_A \int_K f(uak)\delta(a)^{-1}\,du\,da\,dk.$$

Proof. The first assertion comes from a standard fact of homogeneous spaces, that there exists a left invariant Haar measure on $G/K = P$. See for instance [Lan 99], Chapter XVI, Theorem 5.1. It also follows from Lemma 3.3 of Chapter V. The second integral formula simply comes from plugging in Proposition 2.2.

We call $\delta = \delta_{\mathrm{Iw}}$ the **Iwasawa character** of the product decomposition $G = UAK$, defining δ on G by

$$\delta(uak) = \delta(auk) = \delta(a),$$

so δ is U-invariant on the left and K-invariant on the right. We see that δ may then be viewed as the modular function on G.

Standard example. Let $G = \mathrm{GL}_n(\mathbf{R})$ or $\mathrm{SL}_n(\mathbf{R})$. Let:

$U =$ upper triangular unipotent subgroup of G with 1 on the diagonal;
$A =$ diagonal subgroup with positive diagonal components;
$K =$ real unitary subgroup.

Then as in §1 we have the Iwasawa product decomposition $G = UAK$. We let $\mathfrak{n} = \mathrm{Lie}(U)$ be the space of strictly upper triangular matrices.

Proposition 2.4.

(i) *The subgroup U is unimodular. Writing $u \in U$ as $u = I + X$ with $X \in \mathfrak{n}$, and coordinates x_{ij} $(i < j)$, the Haar measure is given by the euclidean measure*

$$\prod_{i<j} dx_{ij}.$$

(ii) *For each $i < j$, let χ_{ij} be the character on A such that*

$$\chi_{ij}(a) = a_i/a_j.$$

Then the Iwasawa character in Proposition 2.3 *is given by*

$$\delta = \prod_{i<j} \chi_{ij}.$$

Proof. We first verify that U is unimodular and that a Haar measure on U is given by the euclidean measure, namely writing $u = I + X$ with X being strictly upper triangular, with coordinates x_{ij} $(i < j)$, Haar measure is given by

$$\prod_{i<j} dx_{ij}.$$

The group U has a global chart, coordinatized by the variables x_{ij}. Let $v \in U$, and write $v = I + Y$ with coordinates y_{ij} $(i < j)$ for Y. Left translation L_v is represented in the chart by the map f_Y such that

$$f_Y(X) = X + Y + YX.$$

Then the tangent linear map $f'_Y(X)$ mapping the vector space of strictly upper triangular matrices into itself is again given by multiplication by $I + Y$, and for Z strictly upper triangular,

$$(I + Y)Z = Z + YZ.$$

But the above vector space is filtered by the natural subspaces (having zeros on successive shrinking diagonals above the main diagonal). Multiplication $Z \mapsto YZ$ strictly decreases the filtration, and so the determinant of $f'_Y(X)$ is 1. This proves that the euclidean measure on U is invariant under left translation. The similar argument shows that it is invariant under right translation, so U is unimodular with the Haar measure as stated.

To compute $\delta(a)$, let $a = \mathrm{diag}(a_1, \ldots, a_n)$ be the matrix with positive diagonal components $a_1, \ldots, a_n$. Let $u = (u_{ij})$. Then

$$(aua^{-1})_{ij} = a_i u_{ij} a_j^{-1}. \tag{1}$$

Denote by $\mathbf{c}_U(a)$ the conjugation automorphism $u \mapsto aua^{-1}$ of U. Then

$$\delta(a) = |\det \mathbf{c}_U(a)|. \tag{2}$$

Hence we obtain

$$\boxed{\delta(a) = \prod_{i<j} (a_i/a_j) = \prod_{i=1}^{n} a_i^{n-2i+1}.} \tag{3}$$

We shall use the exponential notation with α_{ij} instead of χ_{ij}, namely

$$a^{\alpha_{ij}} = a_i/a_j \qquad \text{so that} \qquad \log \delta(a) = \sum_{i<j} \alpha_{ij}(\log a)$$

On $\mathrm{SL}_2(\mathbf{R})$, for a diagonal matrix $\mathrm{diag}(a_1, a_1^{-1})$ we get $\delta(a) = a_1^2$.

Remark. The essential aspect of the above argument is that the Lie algebra $\mathfrak{n}$ of U is semisimple for the conjugation action of A on $\mathfrak{n}$. Indeed,

$$\mathfrak{n} = \sum_{\alpha} \mathfrak{n}_\alpha,$$

where $\mathfrak{n}_\alpha$ is the α-eigenspace, and α ranges over the characters α_{ij}, $i < j$. Let E_{ij} be the matrix with 1 as ij-component and 0 elsewhere. Then E_{ij} is a basis for the α_{ij}-eigenspace. Writing the formula in the form

$$\log \delta(a) = \sum_{\alpha} \alpha(\log a)$$

then generalizes to any Iwasawa decomposition such that $\mathfrak{n}$ is semisimple over the conjugation action of A, except that in general, the eigenspaces may have dimension > 1, so one has to multiply each term in the sum by the multiplicity $m(\alpha)$, equal to the dimension of the eigenspace. For $\mathrm{SL}_n(\mathbf{C})$, this multiplicity is 2.

In general, the formula reads

$$\delta = \prod_{\alpha} \chi_\alpha^{m(\alpha)}.$$

We shall come back to all this in Chapter III, §2.

Here we have been interested in computing a Jacobian. But for future reference, we note that we have proved the following important relation, of interest by itself. Let $p_i(a) = a_1 \cdots a_i$ be the product of the first i components. Then (3) on $A \cap \mathrm{SL}_n(\mathbf{R})$, i.e. on diagonal matrices with determinant 1, yields

$$\boxed{p_1 \cdots p_{n-1} = \delta^{1/2}.} \tag{4}$$

The characters p_d $(d = 1, \ldots, n-1)$ will appear in §5. We shall consider formula (4) from another point of view in Chapter III, §2.

We may also describe the Haar measure in terms of the euclidean coordinates on $\mathrm{GL}_n(\mathbf{R})$ as follows. Define

$$\beta(a) = \beta(au) = \prod_{i=1}^{n} a_i^i$$

Note that $AU = UA$ is the upper triangular group with positive diagonal elements. We denote the variable in this group by $T = (t_{ij})$, with $1 \leqq i \leqq j \leqq n$. Then we have the functions δ and β defined on UA by

$$(5) \qquad \delta(T) = \prod_{i=1}^{n} t_{ii}^{n-2i+1} \quad \text{and} \quad \beta(T) = \prod_{i=1}^{n} t_{ii}^{i}.$$

Proposition 2.5. *Let UA be as above on $\mathrm{GL}_n(\mathbf{R})$. A Haar measure μ on UA is given in terms of the coordinates t_{ij} $(i \leqq j)$ of a matrix T by*

$$d\mu(T) = \delta(T)^{-1}\beta(T)^{-1}d\mu_{\mathrm{euc}}(T) \text{ where } d\mu_{\mathrm{euc}}(T) = \prod_{i \leqq j} dt_{ij}.$$

This follows easily but we won't use it.

I, §3. THE CARTAN LIE DECOMPOSITION, POLYNOMIAL ALGEBRA AND CHEVALLEY'S THEOREM

For use in the next chapter, we make some comments on polynomial algebras especially as they relate subspaces of the Lie algebra of $\mathrm{GL}_n(\mathbf{R})$ or $\mathrm{SL}_n(\mathbf{R})$.

Let V be a finite dimensional vector space over the reals. We let:

$\mathrm{Pol}(V)$ = algebra of polynomial functions on V;

$S(V) = \mathrm{Pol}(V^\vee)$ = symmetric algebra of V, where $V^\vee$ is the dual space.

In non-invariant terms, if $\{\lambda_1, \dots, \lambda_N\}$ is a basis of $V^\vee$, the monomials $\{\lambda_1^{m_1} \cdots \lambda_N^{m_N}\}$ form a basis of $\mathrm{Pol}(V)$. We apply this construction to two vector spaces as follows. First, let:

$\mathfrak{a}$ = vector space of $n \times n$ diagonal matrices;

W = group of permutations of the diagonal elements of a diagonal matrix.

Because of the way W generalizes in the theory of Lie algebras, we call W the **Weyl group**. Let E_{ii} be the diagonal matrix with 1 in the i-th component and 0 elsewhere. Then every element $v \in \mathfrak{a}$ can be expressed as a linear combination

$$v = \sum_{i=1}^{n} h_i E_{ii} \text{ with coordinate functions } h_i.$$

Let

$\mathrm{Pol}(\mathfrak{a})^W$ = subalgebra of $\mathrm{Pol}(\mathfrak{a})$ consisting of elements invariant under W.

Thus $\mathrm{Pol}(\mathfrak{a})^W$ consists of the symmetric polynomials in the algebraically independent elements (variables) $h_1, \ldots, h_n$. Next we let:

$V = \mathrm{Sym}_n(\mathbf{R}) = \mathrm{Sym}$ = vector space of $n \times n$ real symmetric matrices.

Let E_{ij} be the matrix with ij-component equal to 1, and all other components equal to 0. Then Sym has a basis (actually orthogonal) consisting of the elements

$$v_{ii} = E_{ii} \qquad \text{and} \qquad v_{ij} = \tfrac{1}{2}(E_{ij} + E_{ji}) \qquad \text{for } i < j.$$

Then the algebra Pol(Sym) can be viewed as the algebra of polynomials

$$P(X) = P(\ldots, x_{ij}, \ldots)_{i \leqq j}$$

where X is the coordinate matrix of a vector $v = \sum_{i \leqq j} x_{ij} v_{ij} = v_X$. Let K be the usual compact group of real unitary matrices. We let:

$\mathrm{Pol}(\mathrm{Sym})^K$ = subalgebra consisting of the elements invariant under the conjugation action by K.

Theorem 3.1. *The restriction* $\mathrm{Pol}(\mathrm{Sym}) \to \mathrm{Pol}(\mathfrak{a})$ *induces an algebra isomorphism*

$$\mathrm{Pol}(\mathrm{Sym})^K \xrightarrow{\approx} \mathrm{Pol}(\mathfrak{a})^W.$$

In other words, every W-invariant polynomial on $\mathfrak{a}$ can be uniquely extended to a K-invariant polynomial on Sym.

Proof. Every element of Sym can be diagonalized with respect to some orthonormal basis. This means that

$$\mathrm{Sym} = [K]\mathfrak{a},$$

or that every element of Sym is of the form $kv\,{}^tk = kvk^{-1}$ for some $v \in \mathfrak{a}$ and $k \in K$. Thus the restriction map is injective. We have to prove that it is surjective. For this we recall that a symmetric polynomial in variables $h_1, \ldots, h_n$ can be expressed uniquely as a polynomial in the elementary symmetric functions $s_1, \ldots, s_n$. Furthermore, these symmetric functions are the coefficients of the characteristic polynomial of elements $v \in \mathfrak{a}$:

$$\det(tI + v) = t^n + s_1 t^{n-1} + \cdots + s_n.$$

But then a polynomial $Q(s_1, \ldots, s_n)$ can be viewed as an element of the symmetric algebra Pol(Sym), by taking $v \in \text{Sym}$, extending the polynomial in Pol($\mathfrak{a}$), and obviously K-invariant, thus proving the theorem.

Remark. The above result was proved by Chevalley for semisimple Lie algebras. Cf. Wallach [Wal 88], Theorem 3.1.2, and Helgason [Hel 84], Chapter II, Corollary 5.12, for a proof as a consequence of a much more analytic theorem. A more direct proof was given by Harish-Chandra essentially along the same lines as the proof we gave above for Theorem 1.1, but with technical complications. Cf. [Hel 62], Chapter X, Theorem 6.16, which gives a complete exposition of this proof, not kept in [Hel 84], but only mentioned in Exercise D1, p. 340, following Harish-Chandra.

Let $\mathfrak{g} = \text{Mat}_n(\mathbf{R})$. Then of course we have the direct sum decomposition

$$\mathfrak{g} = \mathfrak{n} \oplus \mathfrak{a} \oplus \text{Sk}$$

where Sk is the vector space of skew-symmetric matrices and $\mathfrak{n} = \text{Lie}(U)$. But we also have the direct sum decomposition

$$\mathfrak{g} = \text{Sym} \oplus \text{Sk}\,.$$

The second decomposition will appear in the next chapter as a complement to the Iwasawa decomposition. Note that we also use the notation

$$\text{Sk} = \mathfrak{k}.$$

We now give a general axiomatization of the above decomposition. We start with a brief discussion of conjugation in the general context of Lie groups.

Let G be a Lie group.

We let $\mathbf{c}_g$ or $\mathbf{c}(g)$ denote conjugation by an element $g \in G$ acting on G, so $g \mapsto \mathbf{c}(g)$ is a representation of G in Aut(G) (Lie group automorphisms of G). For $g \in G$, we let

$$(1) \qquad T_e\mathbf{c}(g) = \mathbf{c}_{\mathfrak{g}}(g) \qquad \text{or also } \mathbf{c}_{g,\mathfrak{g}} \qquad \text{or also } \mathbf{c}_{\mathrm{Lie}}(g)$$

be the corresponding representation on $T_eG = \mathfrak{g}$. So $\mathbf{c}_{\mathfrak{g}}(g) = \mathbf{c}_{\mathrm{Lie}}(g)$ is the tangent map of $\mathbf{c}(g)$ at the origin. Classical literature writes Ad for the **conjugation representation on** $\mathfrak{g}$.

The Lie algebra $\mathfrak{g}$ is only one possible functor (vis-à-vis Lie group isomorphisms). In the next chapter, we consider the algebras of C^∞ functions, or algebras of differential operators, or algebras of invariant differential operators. By functoriality, conjugation by an element of G induces an automorphism on all objects associated functorially to G. It will be useful to use various notations for the functorial action. One notation is to write $[\mathbf{c}(g)]$, without subscript, and let the context determine which functor is involved. Another notation is to index $\mathbf{c}(g)$ as we have done above, in the present instance by the Lie algebra $\mathfrak{g}$. A third notation is simply to use $\mathbf{c}(g)$ without any subscript, and let the context prescribe which functor it is acting on. We shall use all three notations.

Remark. If G is given as a Lie subgroup of $\mathrm{GL}_n(\mathbf{R})$ for some n, so we have what is called a representation of G, then $\mathbf{c}_{\mathfrak{g}}(g)$ is actual conjugation of a tangent vector (a matrix) by g. It is then no surprise that we have the general rule

$$(2) \qquad \exp \mathbf{c}_{\mathfrak{g},g}(v) = \mathbf{c}_g(\exp v) \qquad \text{for } v \in \mathfrak{g} = T_eG.$$

We assume given a weak Iwasawa decomposition of G. What will now be used is actually the Iwasawa decomposition of its Lie algebra, that is,

$$\mathfrak{g} = \mathfrak{n} \oplus \mathfrak{a} \oplus \mathfrak{k}.$$

Let

$$\mathfrak{p} = \mathbf{c}_{\mathrm{Lie}}(K)\mathfrak{a}$$

be the subspace of $\mathfrak{g}$ consisting of all conjugates of elements in $\mathfrak{a}$ under the conjugation action of K. Then $\mathfrak{p}$ is stable under $\mathbf{c}_{\mathrm{Lie}}(K)$. Note that $\mathbf{c}_{\mathrm{Lie}}(K)$ contains the subgroup leaving $\mathfrak{a}$ stable, and this subgroup contains the further subgroup of elements leaving $\mathfrak{a}$ elementwise fixed. The factor group is called the **Weyl group**, denoted by W.

For a more systematic discussion of the Weyl group, emphasizing one additional feature which is seen already on $\mathrm{SL}_n(\mathbf{R})$, see Chapter III, §3. For relations between the Cartan Lie decomposition, the transpose operation, and polar decompositions, see Chapters V, VI, §1.

We shall say that $\mathfrak{g}$ admits a **Cartan** or **Cartan Lie decomposition** if the following three conditions are satisfied:

CA 1. Let $\mathfrak{p} = \mathbf{c}_{\mathrm{Lie}}(K)\mathfrak{a}$. Then

$$\mathfrak{g} = \mathfrak{p} \oplus \mathfrak{k}.$$

CA 2. The space $\mathfrak{p}$ has a direct sum decomposition

$$\mathfrak{p} = \mathfrak{a} \oplus \mathfrak{q} \qquad \text{such that} \qquad \mathfrak{q} \subset \mathfrak{n} \oplus \mathfrak{k}.$$

Note that condition **CA 2** is the point where the Iwasawa decomposition meets the Cartan decomposition.

Example. In the case $G = \mathrm{GL}_n(\mathbf{R})$, $\mathfrak{p} = \mathrm{Sym}$, we let $\mathfrak{q} = \mathrm{Sym}^{(0)}$ be the space of symmetric matrices with 0 diagonal components. Such a matrix can be written in the form

$$\begin{pmatrix} 0 & & X \\ & \ddots \; 0 \; \ddots & \\ {}^tX & & 0 \end{pmatrix} = \begin{pmatrix} 0 & & 2X \\ & \ddots \; 0 \; \ddots & \\ 0 & & 0 \end{pmatrix} - \begin{pmatrix} 0 & & X \\ & \ddots \; 0 \; \ddots & \\ -{}^tX & & 0 \end{pmatrix}$$

so Condition **CA 2** is trivially satisfied. For $\mathrm{SL}_n(\mathbf{R})$, we merely require in addition that the matrices have trace 0.

Remark. Here we are only concerned with the Lie algebra. For a global condition arising both in the context of a Cartan decomposition and the group G, see Chapters V, VI, §1.

A third condition will be provided in a moment, after we make relevant remarks. Suppose that G satisfies **IW 1**, **IW 2**, and $\mathfrak{g}$ satisfies **CA 1** and **CA 2**. Then first we obtain a direct sum decomposition

$$S(\mathfrak{p}) = S(\mathfrak{a}) \oplus S(\mathfrak{p})\mathfrak{q}. \tag{3}$$

We let

$$\mathrm{CA}_{\mathfrak{a}} : S(\mathfrak{p}) \to S(\mathfrak{a})$$

be the projection on $S(\mathfrak{a})$, which we call the **Cartan projection**. Similarly we have the Iwasawa direct sum decomposition

$$S(\mathfrak{g}) = S(\mathfrak{a}) \oplus (\mathfrak{n}S(\mathfrak{g}) + S(\mathfrak{g})\mathfrak{k}).$$

Of course, $S(\mathfrak{g})$ is commutative so it doesn't make any mathematical difference which side we write $\mathfrak{n}$ and $\mathfrak{k}$, but it makes a psychological difference when we shall map $S(\mathfrak{g})$ into a non-commutative algebra in the next chapter. Let

$$\mathrm{IW}_{\mathfrak{a}}: S(\mathfrak{g}) \to S(\mathfrak{a})$$

be the projection on $S(\mathfrak{a})$, which we call the **Iwasawa projection**. We shall apply both these projections to the $\mathbf{c}(K)$-invariants $S(\mathfrak{p})^{\mathbf{c}(K)}$. In any case, since $\mathfrak{q} \subset \mathfrak{n} + \mathfrak{k}$ by **CA 2**, it follows that

(4) *On* $S(\mathfrak{p})$, *we have* $\mathrm{CA}_{\mathfrak{a}} = \mathrm{IW}_{\mathfrak{a}}$.

Later we shall consider a **global Cartan–polar decomposition**, which is such that if $\mathbf{P} = \exp \mathfrak{p}$, then G has the direct product decomposition

$$G = \mathbf{P}K.$$

Cf. Chapter VI, conditions **TR 1** through **TR 5**, and the subsequent decomposition. On the whole, our treatment will emphasize more the $G = KAK$ decomposition, as explained at the beginning of Chapter VI.

We shall need a third axiom, which axiomatizes the Chevalley theorem. Given a polynomial $P \in S(\mathfrak{p})$, we let $P_{\mathfrak{a}}$ be its projection in $S(\mathfrak{a})$, so $\mathrm{CA}_{\mathfrak{a}}(P)$ or $\mathrm{IW}_{\mathfrak{a}}(P)$, which amount to the same thing. We note that if $P \in S(\mathfrak{p})^{\mathbf{c}(K)}$, that is, P is invariant under the conjugation action by K, then $P_{\mathfrak{a}} \in S(\mathfrak{a})^W$, that is, the projection $P_{\mathfrak{a}}$ is invariant under the action of W. We can now state our third Cartan condition.

CA 3. The projection $P \mapsto P_{\mathfrak{a}}$ is a linear isomorphism

$$S(\mathfrak{p})^{\mathbf{c}(K)} \xrightarrow{\approx} S(\mathfrak{a})^W.$$

The projection will be called the **Chevalley isomorphism**.

Let $P \in S(\mathfrak{p})^{\mathbf{c}(K)}$, $P \neq 0$. Since the action of $\mathbf{c}_{\mathrm{Lie}}(K)$ preserves degrees in $S(\mathfrak{g})$, it follows that the homogeneous component of maximal degree of P is also in $S(\mathfrak{p})^{\mathbf{c}(K)}$. Then **CA 3** implies:

The projection $P \mapsto P_{\mathfrak{a}}$ *preserves the degree, that is,* $\deg P = \deg P_{\mathfrak{a}}$.

For the reader's convenience, we recall specific properties of the duality for polynomial functions. We do so in a general context. Let V be a finite dimensional vector space over a field of characteristic 0. Let d be a positive integer, and let $\mathrm{Pol}^d(V)$ be the vector space of homogeneous polynomials of degree d on V. The usual "variables" are the coordinate functions with respect to a basis, and such polynomials are therefore polynomial functions on V. If $V^\vee$ is the dual space, then $V = V^{\vee\vee}$, and V, $V^\vee$ play a symmetric role with respect to each other. We denote the elements of V by v and elements of $V^\vee$ by λ.

The vector spaces $\mathrm{Pol}^d(V)$ and $\mathrm{Pol}^d(V^\vee)$ are dual to each other, under the pairing whose value on monomials is given by

$$\langle \lambda_1 \cdots \lambda_d, v_1 \cdots v_d \rangle = \sum_\sigma \langle \lambda_1, v_{\sigma(1)} \rangle \cdots \langle \lambda_d, v_{\sigma(d)} \rangle.$$

The sum is here taken over all permutations σ of $\{1, \ldots, d\}$. Given a non-degenerate bilinear map between V and another vector space $V^\vee$, the same formula defines a duality on their algebras of polynomial functions. In practice, one is usually given some non-degenerate symmetric bilinear form on V itself, identifying V with its dual space. Note that the sum defining the scalar product on monomials is the same as the sum defining determinants, except that the alternating signs are replaced by all plus signs, thus making the sum symmetric rather than skew symmetric in the two sets of variables $(\lambda_1, \ldots, \lambda_d)$ and $(v_1, \ldots, v_d)$. If $\{v_1, \ldots, v_n\}$ is a basis for V and $\{\lambda_1, \ldots, \lambda_n\}$ is the dual basis, then the value of the above pairing on their monomials is 1 or 0. Thus the distinct monomials of given degree d form dual bases for $\mathrm{Pol}^d(V)$ and $\mathrm{Pol}^d(V^\vee)$.

Let K be a group acting on V. Then K also acts functorially on the dual space $V^\vee$. For a functional $\lambda \in V^\vee$, and $v \in V$, we have by definition

$$([k]\lambda)(v) = \lambda([k^{-1}]v).$$

Proposition 3.2. *The above pairing between* $\mathrm{Pol}(V)$ *and* $\mathrm{Pol}(V^\vee)$ *is K-invariant, in the sense that for* $P \in \mathrm{Pol}(V)$ *and* $Q \in \mathrm{Pol}(V^\vee)$, *we have*

$$\langle [k]P, [k]Q \rangle = \langle P, Q \rangle.$$

This is an immediate consequence of the definitions.

Let $\mathfrak{a}$ be a subspace of V. The exact sequence $0 \to \mathfrak{a} \to V$ has the dual exact sequence

$$V^\vee \to \mathfrak{a}^\vee \to 0.$$

The restriction map

$$\mathrm{Pol}(V) \to \mathrm{Pol}(\mathfrak{a}) \to 0$$

corresponds to the dual sequence

$$0 \to \mathrm{Pol}(\mathfrak{a}^\vee) \to \mathrm{Pol}(V^\vee).$$

Let W be the subgroup of K leaving $\mathfrak{a}$ stable, modulo the subgroup leaving $\mathfrak{a}$ elementwise fixed. We have

$$\mathrm{Pol}(\mathfrak{a}^\vee)^W = S(\mathfrak{a})^W.$$

Immediately from the definitions, we get:

Proposition 3.3. *The restriction*

$$\mathrm{Pol}(V)^K \to \mathrm{Pol}(\mathfrak{a})^W$$

is an isomorphism if and only if the dual sequence

$$\mathrm{Pol}(\mathfrak{a}^\vee)^W \to \mathrm{Pol}(V^\vee)^K$$

is an isomorphism.

I, §4. POSITIVITY

This section and the next two are put here for convenience, in order not to interrupt the flow of the inversion theory. Most of this section introduces the basic notion of positivity and partial ordering on $\mathfrak{a}$ and $\mathfrak{a}^\vee$, and tabulates some properties from basic linear algebra in the context of an Iwasawa decomposition. In §6 we compare the A–Iwasawa component and the polar component. We are essentially carrying out some results of Harish-Chandra [Har 58a], Lemma 35, and its corollaries. We shall make further historical comments as we go along.

To start, we let:

$\mathfrak{a} =$ finite dimensional real vector space of dimension r, and

$\mathfrak{a}^\vee =$ dual space.

We suppose given a positive definite scalar product

$$(u, v) \mapsto \langle u, v\rangle = B(u, v)$$

on $\mathfrak{a}$.

Given $H \in \mathfrak{a}$, we let λ_H be the element of $\mathfrak{a}^\vee$ such that

$$\langle H, u\rangle = \lambda_H(u) \qquad \text{for all } u \in \mathfrak{a}.$$

The full correct notation should have the scalar product visible, so $\lambda_H = \lambda_{H,B}$. The form B will be fixed throughout so we omit the index B for simplicity. Similarly, given $\lambda \in \mathfrak{a}^\vee$, we let H_λ be the element of $\mathfrak{a}$ such that

$$\langle H_\lambda, u\rangle = \lambda(u) \qquad \text{for all } u \in \mathfrak{a}.$$

The form B induces an isomorphism of $\mathfrak{a}$ with its dual $\mathfrak{a}^\vee$. It thereby induces the scalar product on $\mathfrak{a}^\vee$ corresponding to B on $\mathfrak{a}$. This scalar product is determined by the formula

$$\langle \lambda, \mu \rangle = \langle H_\lambda, H_\mu \rangle \qquad \text{for all } \lambda, \mu \in \mathfrak{a}^\vee.$$

Let $\mathcal{A} = \{\alpha_1, \ldots, \alpha_r\}$ be a basis of $\mathfrak{a}^\vee$. We let $\{\alpha'_1, \ldots, \alpha'_r\} = \mathcal{A}'$ be the dual basis of $\mathfrak{a}^\vee$, characterized by the property

$$\langle \alpha'_i, \alpha_j \rangle = \delta_{ij}.$$

We note that $\{\alpha'_1, \ldots, \alpha'_r\}$ is also the dual basis of $\{H_{\alpha_1}, \ldots, H_{\alpha_r}\}$, that is

$$\alpha'_i(H_{\alpha_j}) = \delta_{ij} \qquad \text{as well as} \qquad \langle \alpha'_i, \alpha_j \rangle = \delta_{ij}.$$

We define the notion of positivity with respect to the basis $\{\alpha_1, \ldots, \alpha_r\}$ as follows. We say that an element $H \in \mathfrak{a}$ is **$\mathcal{A}$-positive** and write

$$H > 0 \qquad \text{or} \qquad H >_{\mathcal{A}} 0$$

if and only if any one of the following equivalent conditions is satisfied:

Pos 1. $\alpha_i(H) > 0$ for $i = 1, \ldots, r$.

Pos 2. $\langle H_{\alpha_i}, H \rangle > 0$ for $i = 1, \ldots, r$.

Pos 3. If H is written as a linear combination

$$H = \sum_{i=1}^{r} s_i H_{\alpha'_i},$$

of the dual basis, then $s_i > 0$ for all i.

The proof that the above conditions are equivalent is immediate. These conditions define a **partial ordering** on $\mathfrak{a}$. We write as usual for $H_1, H_2 \in \mathfrak{a}$,

$$H_1 > H_2 \qquad \text{or} \qquad H_2 < H_1 \qquad \text{if and only if} \qquad H_1 - H_2 > 0.$$

We use the notation

$$\mathfrak{a}_{\mathcal{A}>0} = \mathfrak{a}^+ = \text{positive cone of } \mathcal{A}\text{-positive elements in } \mathfrak{a}.$$

If we replace the strict inequality > 0 by $\geqq 0$ in the three conditions, we define what we call the **$\mathcal{A}$-semipositive** elements, and the set of these elements is denoted by the **closure sign**, so

$$\mathfrak{a}_{\mathcal{A}\geqq 0} = \mathrm{Cl}(\mathfrak{a}^+) = \text{positive cone of } \mathcal{A}\text{-semipositive elements in } \mathfrak{a}.$$

We define positivity in $\mathfrak{a}^\vee$ by duality, in terms of positivity in $\mathfrak{a}$, namely for an element $\lambda \in \mathfrak{a}^\vee$, we define

$$\lambda \geqq 0 \quad \text{if and only if} \quad H_\lambda \geqq 0.$$

We denote $\mathfrak{a}^\vee_{\mathcal{A}\geqq 0}$ the set of semipositive elements in $\mathfrak{a}^\vee$, and similarly for the set of positive elements. The definition is such that:

The map

$$\lambda \mapsto H_\lambda \quad \textit{of} \quad \mathfrak{a}^\vee_{\mathcal{A}\geqq 0} \to \mathfrak{a}_{\mathcal{A}\geqq 0}$$

is a bijection.

Furthermore, the following conditions are equivalent:

$\lambda \geqq 0$ (that is, $H_\lambda \geqq 0$).

$\langle \alpha_i, \lambda \rangle \geqq 0$ for $i = 1, \ldots, r$, or alternatively, $\lambda(H_{\alpha_i}) \geqq 0$ for all i.

$\lambda = s_1\alpha'_1 + \cdots + s_r\alpha'_r$ with $s_i \geqq 0$ for all $i = 1, \ldots, r$.

The last equivalence comes from **Pos 3**. The other is clear.

We define $\lambda > 0$ similarly, replacing the inequalities $\geqq 0$ by > 0 throughout. For $\lambda, \mu \in \mathfrak{a}^\vee$ we write

$$\lambda \leqq \mu \quad \text{if and only if} \quad \mu - \lambda \geqq 0.$$

We may also define:

$\mathfrak{a}_{\mathcal{A}'>0} =$ set of **$\mathcal{A}'$-positive elements**

to be the set of elements $H \in \mathfrak{a}$ satisfying any one of the following conditions, whose equivalence is immediate:

Pos 1′. $\alpha'_i(H) > 0$ for $i = 1, \ldots, r$.

Pos 2′. $\langle H_{\alpha'_i}, H \rangle > 0$ for $i = 1, \ldots, r$.

Pos 3′. If H is written as a linear combination

$$H = \sum t_i H_{\alpha_i}$$

then $t_i > 0$ for all i.

Of course, the set of **$\mathcal{A}'$-semipositive elements** $\mathfrak{a}_{\mathcal{A}'\geqq 0}$ is defined by replacing $>$ by $\geqq$ in the above conditions. We define $\mathcal{A}'$-positivity in $\mathfrak{a}^\vee$ by:

$$\lambda \mapsto H_\lambda \text{ is a bijection } \mathfrak{a}^\vee_{\mathcal{A}'>0} \to \mathfrak{a}_{\mathcal{A}'>0}.$$

We get another partial ordering on $\mathfrak{a}$ or $\mathfrak{a}^\vee$, and thereby for $H_1, H_2 \in \mathfrak{a}$ we write

$$H \geqq_{\mathcal{A}'} H_2 \text{ or } H_2 \leqq_{\mathcal{A}'} H_1 \text{ if and only if } H_1 - H_2 \geqq_{\mathcal{A}'} 0,$$
$$\text{that is, } H_1 - H_2 \in \mathfrak{a}_{\mathcal{A}'\geqq 0}.$$

The above conditions of $\mathcal{A}'$-positivity are equivalent to still another one, as in [Har 58a], Lemma 35, which we now give.

Proposition 4.1. *Let $H \in \mathfrak{a}$. Let $\mathcal{A}$ be the basis of $\mathfrak{a}^\vee$ as above, and $\mathcal{A}'$ the dual basis. Then $H \in \mathfrak{a}_{\mathcal{A}'\geqq 0}$ if and only if*

$$\langle H, v\rangle \geqq 0 \quad \textit{for all } v \in \mathfrak{a}_{\mathcal{A}\geqq 0} \quad \textit{or symbolically} \quad \langle H, \mathfrak{a}_{\mathcal{A}\geqq 0}\rangle \geqq 0.$$

Proof. We have the equivalences:

$$\begin{aligned}\alpha_i'(H) \geqq 0 \quad \text{for all } i &\iff \langle H, H_{\alpha_i'}\rangle \geqq 0 \quad \text{for all } i\\ &\iff \left\langle H, \sum s_i H_{\alpha_i'}\right\rangle \geqq 0 \quad \text{for all } s_i \geqq 0\\ &\iff \langle H, v\rangle \geqq 0 \quad \text{for all } v \in \mathfrak{a}_{\mathcal{A}\geqq 0}\end{aligned}$$

by condition **Pos 3**. This concludes the proof.

The standard example

For the rest of this section, we let:

$$\mathfrak{a} = \text{real vector space of real } n \times n \text{ diagonal matrices with trace } 0.$$

An element $H \in \mathfrak{a}$ is therefore a matrix

$$H = \begin{pmatrix} h_1 & & 0 \\ & \ddots & \\ 0 & & h_n \end{pmatrix} \qquad \text{with} \quad \sum_{i=1}^n h_i = 0.$$

We define the **trace form** (the ordinary dot product in the diagonal matrix context)

$$\langle H, H'\rangle = \operatorname{tr}(HH')$$

where tr is the trace (sum of the diagonal elements). This trace form is positive definite on $\mathfrak{a}$. If $H \in \mathfrak{a}$ we let λ_H be the element of the dual space $\mathfrak{a}^\vee$ such that

$$\lambda_H(H') = \langle H, H' \rangle \qquad \text{for all } H' \in \mathfrak{a}.$$

For $i = 1, \ldots, r = n - 1$ we let

$$H_{i,i+1} = \begin{pmatrix} 0 & & & & & & \\ & \ddots & & & & & \\ & & 0 & & & & \\ & & & 1 & & & \\ & & & & -1 & & \\ & & & & & 0 & \\ & & & & & & \ddots \\ & & & & & & & 0 \end{pmatrix} \qquad \begin{matrix} \text{with } h_i = 1, h_{i+1} = -1, \\ h_j = 0 \text{ for } j \neq i, i+1. \end{matrix}$$

We let $\alpha_i \in \mathfrak{a}^\vee$ be the functional defined by

$$\alpha_i(H) = h_i - h_{i+1}.$$

Thus

$$H_{\alpha_i} = H_{i,i+1}.$$

The elements $\alpha_1, \ldots, \alpha_r$ obviously form a basis of $\mathfrak{a}^\vee$. These elements are called the **simple characters** of $\mathfrak{a}$.

We let $\lambda_i \in \mathfrak{a}^\vee$ be the functional such that

$$\lambda_i(H) = h_1 + \cdots + h_i.$$

It is immediately verified that $\lambda_i = \alpha_i'$, namely **$\{\lambda_1, \ldots, \lambda_r\}$ is the dual basis of $\{\alpha_1, \ldots, \alpha_r\}$ as well as the dual basis of** $\{H_{\alpha_1}, \ldots, H_{\alpha_r}\}$. The form on $\mathfrak{a}^\vee$ defined by

$$\langle \lambda, \mu \rangle = \langle H_\lambda, H_\mu \rangle \qquad \text{for } \lambda, \mu \in \mathfrak{a}^\vee$$

will be called the **dual trace form.**

Note that $\{H_{12}, \ldots, H_{r,r+1}\}$ is not orthogonal nor do the vectors have norm 1. Indeed, applying the definition shows that

$$\langle H_{i,i+1}, H_{i,i+1} \rangle = 2 \qquad \text{for } i = 1, \ldots, r.$$

In practice, it will be useful to use the notation

$H > 0$ (resp. $H \geqq 0$) *if and only if* $H \in \mathfrak{a}_{A>0}$ (resp. $H \in \mathfrak{a}_{A\geqq 0}$).

In other words, unless otherwise specified, positivity refers to the effect of $\alpha_1, \ldots, \alpha_r$. Condition **Pos 3**, however, gives an equivalent condition in terms of the dual basis. We also obtain an estimate showing how some functionals behave like a norm. *Let* $\lambda = s_1\lambda_1 + \cdots + s_r\lambda_r$ *with* $s_i > 0$ *for all* i. *Then there exist constants* $c_1, c_2 > 0$ *such that for all* $H \in \mathrm{Cl}(\mathfrak{a}^+)$ *we have*

$$(1) \qquad c_1|H| \leqq \lambda(H) \leqq c_2|H|.$$

The right inequality just expresses the continuity of a functional. The left one is proved in the same way one proves two norms are equivalent. Let S be the unit sphere in $\mathfrak{a}$, i.e. the set of $H \in \mathfrak{a}$ such that $|H| = 1 = \langle H, H \rangle$. Then λ has a minimum at a point H in the compact set $S \cap \mathrm{Cl}(\mathfrak{a}^+)$, and this minimum cannot be 0, otherwise $\lambda_i(H) = 0$ for all i so $H = 0$. The existence of c_1 then follows by homogeneity.

The next proposition is more serious (Lemma 35 of [Har 58a]). We give an ad hoc proof. The general case takes more machinery.

Proposition 4.2.

(i) *For each* i,

$$(r+1)\lambda_i \in \sum_{j=1}^{r} \mathbf{Z}_{>0}\alpha_j,$$

so λ_i *is a linear combination with positive rational coefficients of* $\alpha_1, \ldots, \alpha_r$. *In particular,*

$$\mathfrak{a}_{\mathcal{A}>0} \subset \mathfrak{a}_{\mathcal{A}'>0} \quad \textit{and} \quad \mathfrak{a}_{\mathcal{A}\geqq 0} \subset \mathfrak{a}_{\mathcal{A}'\geqq 0}.$$

Hence we also have

$$\mathfrak{a}^\vee_{\mathcal{A}>0} \subset \mathfrak{a}^\vee_{\mathcal{A}'>0} \quad \textit{and} \quad \mathfrak{a}^\vee_{\mathcal{A}\geqq 0} \subset \mathfrak{a}^\vee_{\mathcal{A}'\geqq 0}.$$

(ii) *Let* $H \in \mathfrak{a}$ *and write* H *as a linear combination*

$$H = t_1 H_{\alpha_1} + \cdots + t_r H_{\alpha_r}.$$

If $H > 0$ (resp. $\geqq 0$) *then* $t_i > 0$ (resp. $t_i \geqq 0$) *for* $i = 1, \ldots, r$.

Proof. Write

$$\begin{aligned} h_i &= h_i - h_{i+1} + \cdots + h_r - h_{r+1} + h_{r+1} \\ &= h_i - h_{i+1} + \cdots + h_r - h_{r+1} - (h_1 + \cdots + h_r). \end{aligned}$$

We add $h_1 + \cdots + h_{i-1}$ to both sides, giving

$$h_1 + \cdots + h_i = L_1 - (h_i + \cdots + h_r),$$

where $L_1 = \alpha_i + \cdots + \alpha_r$. We now add

$$\frac{r-i+1}{i}(h_1 + \cdots + h_i)$$

to both sides. We clear denominators and get

$$\begin{aligned}(r+1)(h_1 + \cdots + h_i) &= iL_1 - i(h_i + \cdots + h_r) \\ &\quad + (r-i+1)(h_1 + \cdots + h_i) = iL_1 + L_2\end{aligned}$$

with a positive integral linear combinations $iL_1 + L_2$ of $\alpha_1, \ldots, \alpha_r$. This concludes the proof of the first inclusion in (i). The dual inclusion comes from the fact that the map $\lambda \mapsto H_\lambda$ is positivity preserving. Finally (ii) follows immediately from **Pos 3′**.

The Weyl group

Next we consider positivity in connection with the group W consisting of the permutations of the coordinates $(h_1, \ldots, h_n)$. This group is called the **Weyl group** for historical reasons. Also in extending the theory, one does not have such a direct definition for it, but the present section is being kept as simple and naive as possible. For a more general discussion of the Weyl group, see Chapter III, §3.

The Weyl group acts on everything in sight, including functionals, by functoriality. If $H \in \mathfrak{a}$ or $\lambda \in \mathfrak{a}^\vee$ we write wH or $w\lambda$ for the effect of a Weyl group element w on H or λ. An orbit of W is denoted by WH or $W\lambda$.

Proposition 4.3. *The set* $\mathrm{Cl}(\mathfrak{a}^+) = \mathfrak{a}_{\mathcal{A} \geqq 0}$ *of semipositive elements in* $\mathfrak{a}$ *with respect to the basis* $\mathcal{A} = \{\alpha_1, \ldots, \alpha_r\}$ *is a fundamental domain for the Weyl group on* $\mathfrak{a}$. *In other words, given an element* $H \in \mathfrak{a}$, *there exists a unique element* $H^+ \in WH$ *such that* $H^+ \geqq 0$.

Proof. In the present case dealing with the diagonal matrices, this is obvious because we can always reorder a sequence $(h_1, \ldots, h_n)$ (permute the elements) so that

$$h_1 \geqq h_2 \geqq \cdots \geqq h_n = h_{r+1}.$$

Such an element is uniquely determined. Indeed, if all the components are distinct, only the identity permutation preserves the semipositivity. If some components are equal (semipositivity), they have to be adjacent, and only a permutation permuting equal elements will preserve the semipositivity property of the sequence.

In light of Proposition 4.3, we call $\mathrm{Cl}(\mathfrak{a}^+) = \mathfrak{a}_{\mathcal{A} \geqq 0}$ the **semipositive fundamental domain**, and $\mathfrak{a}^+$ the **positive fundamental domain** for W.

Similarly, letting $A^+ = \exp(\mathfrak{a}^+)$, we call A^+ the **positive fundamental domain** (in A).

We note that the basic scalar product is W-invariant, that is

$$\begin{aligned} \langle wH_1, wH_2 \rangle &= \langle H_1, H_2 \rangle \qquad \text{for all } H_1, H_2 \in \mathfrak{a}, \\ \langle w\lambda, w\mu \rangle &= \langle \lambda, \mu \rangle \qquad \text{for all } \lambda, \mu \in \mathfrak{a}^\vee. \end{aligned}$$

We immediately get the formulas for $H \in \mathfrak{a}$ and $\lambda \in \mathfrak{a}^\vee$:

$$\begin{aligned} (w\lambda)(H) &= \lambda(w^{-1}H), \\ wH_\lambda &= H_{w\lambda}. \end{aligned}$$

Thus the correspondence $\lambda \mapsto H_\lambda$ preserves not only positivity but the Weyl group action.

We may then transmute Proposition 4.3 to the dual situation.

Proposition 4.3$^\vee$. *The set $\mathfrak{a}^\vee_{\mathcal{A}\geqq 0}$ of semipositive elements λ in $\mathfrak{a}^\vee$ (with respect to the basis $\mathcal{A} = \{\alpha_1, \ldots, \alpha_r\}$), i.e. $\lambda \geqq 0$, is a fundamental domain for W acting on $\mathfrak{a}^\vee$. In other words, given $\lambda \in \mathfrak{a}^\vee$ there exists a unique element $\lambda^+ \in W\lambda$ such that $\lambda^+ \geqq 0$.*

Proof. Direct consequences of the above formulas and Proposition 4.3.

The next proposition is of a different kind. It will be essentially dealt with ad hoc in Lemma 3.5 of Chapter VIII. Making the definition $\lambda_i = \alpha_i'$ explicit, we see that for $H \in \mathfrak{a}$,

$H \geqq_{\mathcal{A}'} wH$ for all $w \in W$ if and only if for all $i = 1, \ldots, r$ and all $w \in W$ (viewed as permutation of the indices), we have

$$h_1 + \cdots + h_i \geqq h_{w(1)} + \cdots + h_{w(i)} \quad \text{that is} \quad \alpha_i'(H) \geqq \alpha_i'(wH).$$

Proposition 4.4.

(i) *Let $H \in \mathfrak{a}$ and write $H = \operatorname{diag}(h_1, \ldots, h_n)$. Then $H \geqq 0$ if and only if $H \geqq_{\mathcal{A}'} wH$ for all $w \in W$. In particular, $H^+ \geqq_{\mathcal{A}'} H$.*

(ii) *Let $\lambda \in \mathfrak{a}^\vee$. Then $\lambda^+(H^+) = \max_w \lambda(wH)$.*

Proof. Suppose $H \geqq 0$. For each i, the i-tuple $(w(1), \ldots, w(i))$ can be reordered in decreasing order, and the hypothesis that $H \geqq 0$ yields the inequality stated in (i). Conversely, assume this inequality. If H is not $\geqq 0$, then for some j we have

$$h_1 \geqq \cdots \geqq h_j \quad \text{but} \quad h_j < h_{j+1}.$$

Let w be a permutation which reorders the elements $h_1, \dots, h_n$ in decreasing order. Then

$$\alpha'_j(wH) > \alpha'_j(H),$$

which contradicts the inequality, thus proving (i). As for (ii), from Proposition 4.3$^\vee$ we have $\lambda^+ = w_0\lambda$ (some $w_0 \in W$), and from **Pos 3** we can write λ^+ as a linear combination

$$\lambda^+ = s_1\alpha'_1 + \cdots + s_r\alpha'_r \qquad \text{with } s_i \geqq 0.$$

Thus (ii) follows from (i), thus concluding the proof.

Corollary 4.5. *Let $\lambda \in \mathfrak{a}^\vee$, $\lambda \geqq 0$. Then $\lambda \geqq_{\mathcal{A}'} w\lambda$ for all $w \in W$.*

Proof. Immediate from Proposition 4.4 and the correspondence $\lambda \mapsto H_\lambda$ which preserves $\mathcal{A}'$-positivity as well as the Weyl group structure.

I, §5. CONVEXITY

Next we relate the inequalities of Proposition 4.4 and Corollary 4.5 to convexity. We do this in a spirit of completeness, but we are really after the Harish-Chandra estimate in relation to the Iwasawa decomposition which follows in §6. The proofs of Proposition 5.1 and Theorem 6.1 are independent of each other. One can put the two results together to get the estimate of Theorem 6.2, which will be relevant in Chapter IV, Theorem 8.7.

For Proposition 5.1, readers may find it profitable to look at both [Har 58a] and [GaV 88], Propositions 3.5.2, 3.5.4, and 6.2.7, which we found useful. The latter deal with part of a very general theorem of Kostant [Kos 73], see the comments below.

Let S be a subset of a real vector space. We denote by $\text{Co}(S)$ the convex closure of S. In our application S consists of a finite number of points $P_1, \dots, P_N$. Then $\text{Co}_N(S)$ is the set of all vectors

$$\sum_{k=1}^{N} t_k P_k \quad \text{with } t_k \geqq 0 \text{ and } \sum_{k=1}^{N} t_k = 1.$$

In the application, the zero element 0 is in S, in which case the convex closure of S is also defined by replacing the equality sign on the right by the inequality $\leqq 1$.

Let us consider our vector space $\mathfrak{a}$ of diagonal real matrices with trace 0. Let $H \in \mathfrak{a}$. Then $\mathrm{Co}(WH)$ contains 0 because the element

$$X = |W|^{-1} \sum_{w \in W} wH$$

is fixed under W, so all its components are equal, and the trace is 0, so the element is equal to 0. (In more general situations, one would argue that $\alpha_i(X) = 0$ for all i, so $X = 0$.)

We shall use convexity to estimate certain vectors. Let S consist of a finite number of points P_k, $k = 1, \ldots, N$, again. Let $X \in \mathrm{Co}(S)$. Then for any norm on the vector space,

$$|X| \leqq \max_k |P_k|.$$

This is immediate from the definition of the convex closure. One has a similar inequality using a functional instead of a norm, but even more importantly, by elementary linear algebra (cf. for instance [Lan 87], Chapter XII, Theorems 2.1 and 4.1):

Convexity criterion. *An element $X \in \mathfrak{a}$ is in the convex closure* $\mathrm{Co}(P_1, \ldots, P_N)$ *if and only if, for every $\lambda \in \mathfrak{a}^\vee$, we have*

$$\lambda(X) \leqq \max_k \lambda(P_k).$$

Proposition 5.1.

(i) *Let $H_0, H \in \mathfrak{a}$. Then*

$$H_0 \in \mathrm{Co}(WH) \Longleftrightarrow H^+ \geqq_{\mathcal{A}'} wH_0 \textit{ for all } w \in W.$$

(ii) *Let H_0 and $H \geqq 0$. Then*

$$H_0 \in \mathrm{Co}(WH) \Longleftrightarrow H \geqq_{\mathcal{A}'} H_0.$$

Proof. Part (ii) is an immediate consequence of (i) and Proposition 4.4(i). Let us show the implication $\Rightarrow$ of (i). Suppose $H_0 \in \mathrm{Co}(WH)$. Since the convex closure $\mathrm{Co}(WH)$ is stable under the action of W, it suffices to prove that $H^+ \geqq_{\mathcal{A}'} H_0$. But using the above convex closure property,

$$\alpha_i'(H_0) \leqq \max_w \alpha_i'(wH) \leqq \alpha_i'(H^+)$$

by Proposition 4.4(ii). This proves that $H^+ \geqq_{\mathcal{A}'} H_0$, which is the first implication.

Conversely, suppose $H \geqq_{\mathcal{A}'} wH_0$ for all $w \in W$. By Proposition 4.3, there exists $w_0 \in W$ such that $w_0 H_0 \geqq 0$. Since $w \mapsto ww_0$ is a permutation of W, and $\mathrm{Co}(WH)$ is stable under the action of W, we may assume without loss of generality that $H_0 \geqq 0$.

So assume $H \geqq_{A'} H_0$, that is, $\alpha_i'(H) \geqq \alpha_i'(H_0)$ for $i = 1, \ldots, r$. We must show that $H_0 \in \mathrm{Co}(WH)$. By elementary algebra mentioned above, we are reduced to proving that for every functional $\lambda \in \mathfrak{a}^\vee$, we have

$$\lambda(H_0) \leqq \max_w \lambda(wH).$$

By Proposition 4.3$^\vee$ there is an element $\lambda^+ \geqq 0$ in the orbit $W\lambda$. Then

$$\begin{aligned}\lambda(H_0) &\leqq \lambda^+(H_0) && \text{because } H_0 \geqq 0 \text{ and we apply Proposition 4.4(ii)}\\ &\leqq \lambda^+(H) && \text{because } H \geqq_{A'} H_0 \text{ and } \lambda^+ = \textstyle\sum s_i\alpha_i' \text{ with } s_i \geqq 0\\ &= \max_w \lambda(wH) && \text{by Proposition 4.4(ii),}\end{aligned}$$

which concludes the proof of the proposition.

For convenience of reference, we reformulate the proposition on the dual space.

Proposition 5.1$^\vee$. *Let* $\lambda \in \mathfrak{a}^\vee$, $\lambda > 0$.

(i) $\lambda_0 \in \mathfrak{a}^\vee$. *Then*

$$\lambda_0 \in \mathrm{Co}(W\lambda) \iff \lambda \geqq_{A'} w\lambda_0 \textit{ for all } w \in W.$$

(ii) *Let* $\lambda_0 \geqq 0$. *Then* $\lambda_0 \in \mathrm{Co}(W\lambda)$ *if and only if* $\lambda \geqq_{A'} \lambda_0$.

Proof. Use the fact that $\lambda \mapsto H_\lambda$ is a linear isomorphism of $\mathfrak{a}^\vee$ with $\mathfrak{a}$, preserving positivity, and commuting with the action of W.

Corollary 5.2. *Let* $\xi, \lambda \in \mathfrak{a}^\vee$ *and* $\lambda > 0$. *Assume that* $-\lambda \in W\lambda$, *so* $\mathrm{Co}(W\lambda)$ *is symmetric. Then* $\xi \in \mathrm{Co}(W\lambda)$ *if and only if*

$$|w\xi(H)| \leqq \lambda(H) \textit{ for all } w \in W \textit{ and } H \in \mathrm{Cl}(\mathfrak{a}^+).$$

Proof. Immediate from Proposition 5.1$^\vee$. The symmetry of $\mathrm{Co}(W\lambda)$ is of course used here to put the absolute value sign on the left of the above inequality.

Example. The character ρ. The most important character to arise is the character

$$\rho = \lambda_1 + \cdots + \lambda_r = \frac{1}{2} \sum_{\alpha \in \mathcal{R}(\mathfrak{n})} \alpha,$$

where $\mathcal{R}(n) = \{\alpha_{ij}\}$ $(1 \leqq i < j \leqq n)$ as in §4. Sometimes the natural character to consider will be 2ρ, as in Chapter III, §1. See also Chapter X, §1, (5) used in Chapter XI. Note that we have both

$$\rho >_{\mathcal{A}} 0 \quad \text{and} \quad \rho >_{\mathcal{A}'} 0.$$

In Chapter X, §1, (5), we shall see that there exists an element $w_0 \in W$ such that $w_0\rho = -\rho$, so the convex set $\mathrm{Co}(W\rho)$ is symmetric.

I, §6. THE HARISH-CHANDRA *U*-POLAR INEQUALITY; CONNECTION WITH THE IWASAWA AND POLAR DECOMPOSITIONS

We now return to a mixture of multiplicative and additive structures. We let $G = \mathrm{SL}_n(\mathbf{R})$ with its Iwasawa decomposition

$$G = UAK.$$

We suppose $A = \exp \mathfrak{a}$ with the standard vector space of real diagonal matrices $\mathfrak{a}$. We let:

$$A^+ = \exp \mathfrak{a}^+ \quad \text{and} \quad \mathrm{Cl}(A^+) = \exp \mathrm{Cl}(\mathfrak{a}^+).$$

We define the **partial products** corresponding to the functionals λ_d of §4 by

$$p_d(a) = a_1 \cdots a_d \quad \text{for} \quad d = 1, \ldots, r.$$

We may write

$$p_d(a) = a^{\lambda_d} \quad \text{or} \quad \lambda_d(\log a) = \log p_d(a).$$

If $a = \exp H$ we also write $H = \log a$ and we have the formula

$$p_d(\exp H) = e^{\lambda_d(H)}.$$

For $g \in G$, the action of G on $\mathbf{R}^n$ extends to the alternating products $\bigwedge^d g$ on $\bigwedge^d \mathbf{R}^n$. For $x, y \in G$ we have

$$\bigwedge^d(xy) = \bigwedge^d(x)\bigwedge^d(y).$$

We let $\{e_1, \ldots, e_n\}$ be the standard unit column vectors. We note that

$$(1) \qquad \left(\bigwedge^d u\right)(e_1 \wedge \cdots \wedge e_d) = ue_1 \wedge \cdots \wedge ue_d = e_1 \wedge \cdots \wedge e_d,$$

because for each i, $ue_i = e_i$ + linear combination of e_j with $j < i$, and the extra linear combination is killed in the wedge product.

We take the natural positive definite scalar product on $\bigwedge^d \mathbf{R}^n$, for which the elements $e_{j_1} \wedge \cdots \wedge e_{j_d}$ with $j_1 \wedge \cdots \wedge j_d$ form an orthonormal basis. Then for the norm of this scalar product, using the fact that k is unitary, we conclude that for $k \in K$,

(2a) $$\left\| \bigwedge^d (k)(e_1 \wedge \cdots \wedge e_d) \right\| = \|ke_1 \wedge \cdots \wedge ke_d\| = 1$$

(2b) $$\left\| \bigwedge^d (k) Z \right\| = \|Z\| \qquad \text{for } Z \in \bigwedge^d \mathbf{R}^n.$$

Thirdly, for the A-component,

(3) $$\left\| \bigwedge^d (a)(e_1 \wedge \cdots \wedge e_d) \right\| = \|ae_1 \wedge \cdots \wedge ae_d\| = p_d(a).$$

Theorem 6.1 ([Har 58a], Lemma 35 et seq.). *Let*

$$au = k_1 bk \qquad \textit{with } a, b \in A, u \in U \textit{ and } k_1, k \in K.$$

Suppose $a, b \in \mathrm{Cl}(A^+)$ (*so* $\log a, \log b \in \mathrm{Cl}(\mathfrak{a}^+)$). *Then for all* $i = 1, \ldots, r$,

$$p_i(a) \leqq p_i(b) \qquad \textit{for } i = 1, \ldots, r.$$

Or, written additively,

$$\log a \leqq_{A'} \log b.$$

Proof. First, we have

$$\left\| \bigwedge^d (au) e_1 \wedge \cdots \wedge e_d \right\| = \left\| \left(\bigwedge^d a\right) ue_1 \wedge \cdots \wedge ue_d \right\| = a_1 \cdots a_d = p_d(a).$$

On the other hand, computing the norm with the expression $k_1 bk$ yields

$$\left\| \bigwedge^d (k_1 bk) e_1 \wedge \cdots \wedge e_d \right\| = \left\| \left(\bigwedge^d b\right) ke_1 \wedge \cdots \wedge ke_d \right\|.$$

Let

$$ke_1 \wedge \cdots \wedge ke_d = \sum c_{(j)} e_{j_1} \wedge \cdots \wedge e_{j_d},$$

with $(j) = (j_1, \ldots, j_d)$ and $j_1 < \cdots < j_d$. Then we get

$$bke_1 \wedge \cdots \wedge bke_d = \sum c_{(j)} b_{j_1} \cdots b_{j_d} e_{j_1} \wedge \cdots \wedge e_{j_d}.$$

By Pythagoras, it follows that

$$p_d(a)^2 = \sum c_{(j)}^2 (b_{j_1} \cdots b_{j_d})^2.$$

Since $b \in A^+$ by assumption, it follows that

$$b_{j_1} \cdots b_{j_d} \leqq b_1 \cdots b_d \qquad \text{for all } (j).$$

Furthermore, Pythagoras also gives

$$\sum c_{(j)}^2 = 1.$$

Hence

$$p_d(a)^2 \leqq \sum c_{(j)}^2 p_d(b)^2 = p_d(b)^2,$$

which proves the theorem.

Remarks. Actually Harish-Chandra formulated the inequality in the form

$$\log b - H(bk) \in \mathfrak{a}_{\mathcal{A}' \geqq 0}.$$

where $H(x)$ is the A-projection in the anti-Iwasawa decomposition $x = kbu$, with $k \in K, b \in A$, and $u \in U$. So $H(x) = \log b$ by definition. This formulation and the one we have given in Theorem 6.1 are obviously equivalent.

The map $k \mapsto H(ak)$ for $a \in A^+$ has been considered extensively in the literature. Things start independently of Harish-Chandra, in papers of Horn [Hor 54a], [Hor 54b] concerning eigenvalues of hermitian matrices. These got extended by Lenard [Len 71] and Thompson [Tho 71]. Finally Kostant [Kos 73] formulated the previous results from these authors in a theorem on the image of the above map on arbitrary semisimple Lie groups, namely:

Kostant's convexity theorem. *Let $a \in A$. The image of the map $k \mapsto H(ak)$ is the convex closure of the points $\log a^w$ $(w \in W)$.*

We shall not use this theorem.

We let $|v|$ $(v \in \mathfrak{a})$ be the norm associated with the trace form on $\mathfrak{a}$. We then get a corollary of Theorem 6.1. (Cf. Chapter IV, Proposition 8.10.)

Theorem 6.2. *Let $au = k_1 b k_2$ with $a, b \in A$, $u \in U$ and $k_1, k_2 \in K$. Then*

$$|\log a| \leqq |\log b|.$$

Proof. By Proposition 4.3, we can suppose a and $b \in \mathrm{Cl}(A^+)$, since the norm associated with our scalar product on $\mathfrak{a}$ is W-invariant. Theorem 6.1 states that $\log a \leqq_{\mathcal{A}'} \log b$. Proposition 5.1(ii) then implies $\log a \in \mathrm{Co}(W \log b)$, whence the inequality of Theorem 6.2 follows because of the W-invariance of the norm.

Remark. The above inequality can also be viewed from a differential geometric point of view. The space G/K is a Cartan–Hadamard space, whose exponential map is metric semi-increasing. The U-fibers are perpendicular to A. The metric increasing property shows that the law of cosines with the inequality in the right direction is valid on G/K, and the desired inequality $|\log a| \leqq |\log b|$ then follows at once. For the law of cosines, see Mostow's exposition of Cartan's work [Mos 53], Lemma 4, reproduced in [Lan 99], Chapter IX, Theorem 4.8 (the reference to Helgason should be replaced by a reference to Mostow and Cartan).

Theorems 6.1 and 6.2 have been formulated with a minimum of terminology. However, it may be helpful to the reader to see a larger context for the inequalities, which we now explain in light of later developments. We recall the following theorem from basic linear algebra:

Theorem 6.3. *Let $G = \mathrm{SL}_n(\mathbf{R})$ or $\mathrm{GL}_n(\mathbf{R})$, and let K be the real unitary subgroup of G. Then every element $x \in G$ has a decomposition (called* **a polar decomposition***)*

$$x = k_1 b k_2 \qquad \text{with } k_1, k_2 \in K \text{ and } b \in A.$$

The element b is uniquely determined up to the action of the Weyl group, i.e. up to a permutation of the diagonal elements of b.

Proof. Let $\mathrm{Pos}_n(\mathbf{R})$ be the space of symmetric positive definite $n \times n$ real matrices, and $\mathrm{SPos}_n(\mathbf{R})$ the subspace of those elements with determinant 1. Say $G = \mathrm{SL}_n(\mathbf{R})$. The quadratic map $x \mapsto x\,{}^t x$ is a surjective map of G onto $\mathrm{SPos}_n(\mathbf{R})$, for instance, because every element of $\mathrm{SPos}_n(\mathbf{R})$ has a square root in $\mathrm{SPos}_n(\mathbf{R})$ (diagonalize the matrix acting on $\mathbf{R}^n$ and take the natural positive square root of the diagonal elements). The subgroup K is the inverse image of the identity under the quadratic map. Thus we have a bijection of the coset space $G/K \to \mathrm{SPos}_n(\mathbf{R})$. Given $x \in G$, there exists $k_1 \in K$ such that $x\,{}^t x = k_1 b^2 k_1^{-1}$ with $b \in A$ (again by the theorem of linear algebra that there is an orthonormal basis of $\mathbf{R}^n$ consisting of eigenvectors for $x\,{}^t x$). Then by the above bijection, there exists $k_2 \in K$ such that $x = k_1 b k_2$. The element b is uniquely determined up to a permutation (action of W) because b^2 is the matrix of eigenvalues of $x\,{}^t x$ (roots of the characteristic polynomial). This proves the theorem.

Now we see that Theorem 6.2 relates the estimate for the Iwasawa decomposition and the polar decomposition. The unique element $\geqq 0$ in the orbit Wb is denoted by b^+, or also

$$b^+ = \mathrm{pol}^+(x).$$

If $b^+ = \exp H^+$ we use the notation

$$\sigma(x) = |H| = |\log b|.$$

The inequality of Theorem 6.2 may then be written

$$\sigma(a) \leqq \sigma(au) \qquad \text{for all } a \in A \text{ and } u \in U.$$

The function σ will reappear in a very significant fashion in Chapter X, §1. We call it the **polar height**.

The technique of using the wedge product representation will recur in Chapter V and also in Chapter X, where we use (1), (2), (3) systematically. In these later uses, we deal with the inverse of an element $x = uak$, that is

$$x^{-1} = k^{-1}a^{-1}u^{-1}.$$

As before, let $a = x_A$ be the A-Iwasawa projection of x. Let $\{\lambda_1, \ldots, \lambda_r\}$ be the dual basis of the positive simple $(\mathfrak{a}, \mathfrak{n})$-characters as above. Then we shall deal with the fundamental formula

$$x_A^{-\lambda_i} = a^{-\lambda_i} = \left\| \bigwedge^i (x^{-1})(e_1 \wedge \cdots \wedge e_i) \right\|. \tag{4}$$

This formula falls out of the successive use of (1), (2), (3) and the functoriality

$$\bigwedge^i (x^{-1}) = \bigwedge^i (k^{-1}) \bigwedge^i (a^{-1}) \bigwedge^i (u^{-1}).$$

For applications, see Chapter V, Lemma 5.2, and Chapter X, Theorem 1.3, et seq. For a formulation in the context of semisimple Lie groups, cf. [GaV 88], 3.4.4, and the subsequent example tying up with (4) above.

CHAPTER II

Invariant Differential Operators and the Iwasawa Direct Image

We start our main business, the study of invariant differential operators, by using certain invariance properties with respect to certain subgroups arising from basic decompositions. This chapter deals with the Iwasawa and Cartan Lie decompositions, which are immediate on $\mathrm{GL}_n(\mathbf{R})$ or $\mathrm{SL}_n(\mathbf{R})$. Further chapters will deal with the properties which arise from other decompositions.

Except as noted, the essential theorems of this chapter are due to Harish-Chandra, who studied systematically the relations between differential operators on the Lie algebra and those on the group. Cf. [Har 57a], [Har 57b], [Har 58a]. We found the exposition in [Wal 73], [Hel 84], and [GaV 88] useful. Harish-Chandra's results were complemented in one important respect by Helgason [Hel 59] on the symmetric space, see Theorem 2.5. Roughly speaking, because of the Iwasawa decomposition, we have two natural submersions

$$G \to X \to A \quad (\text{with } X \approx G/K),$$

and we take the direct image of differential operators at each step. In key situations, this procedure reduces certain theorems on G or X to simple facts on A. The kernel of the direct image was determined by Harish-Chandra.

In this chapter we have carried out whatever we could without using characters. These are needed to complete Harish-Chandra's reduction theorem to A, which will be carried out in the next chapter. Only the direct image with respect to the Iwasawa decomposition will be needed. We shall treat the direct image for the polar decomposition in Chapter VI, §6.

By a **differential operator** on a manifold of dimension n, we shall always mean a C^∞ differential operator, which can be expressed locally in terms of coordinates $x_1, \ldots, x_n$ in the form

$$D = \sum_{(j)} \varphi_{(j)}(x_1, \ldots, x_n)\partial_{j_1} \ldots \partial_{j_m}$$

where $\partial_j = \partial/\partial x_j$, and the coefficients $\varphi_{(j)}$ are C^∞ functions. If such an expression exists in one chart, then it exists in another chart, after applying a C^∞ isomorphism and the chain rule. We may write the operator as

$$D = \sum_{(m)} \varphi_{(m)}(x_1, \ldots, x_n)\partial_1^{m_1} \ldots \partial_n^{m_n}.$$

The **degree** of the operator at a point (x) is the maximum of $m_1 + \cdots + m_n$ for which there is a non-zero coefficient in the above polynomial expression (with functions as coefficients). Under a change of charts (differential isomorphism) the degree cannot increase (by the chain rule) so the degree is preserved.

Suppose a Lie group G acts on a manifold X. Then the action of an element is a differential automorphism of the manifold, and locally can be viewed as a change of charts by translation, as one says. In some examples, the Lie group acts homogeneously. For instance, we may consider G acting on itself by left translation. On a homogeneous space, if D is a differential operator invariant under the action of G, then the degree of the differential operator is defined globally, and is the degree at each point, independently of the point.

Functoriality

Let θ be an automorphism (differential) of the manifold X. Then θ induces a covariant action on all the standard functors of X, for instance the vector spaces of functions, differential operators, the tangent bundle, ad lib. We may index θ by the corresponding functor, so we get θ_{Fu}, θ_{DO}, θ_T, ad lib. For any function $f \in \mathrm{Fu}(X)$ (C^∞ functions on X), we want the commutative diagram relation

$$(\theta_{\mathrm{Fu}} f)(\theta x) = f(x) \text{ for all } x \in X.$$

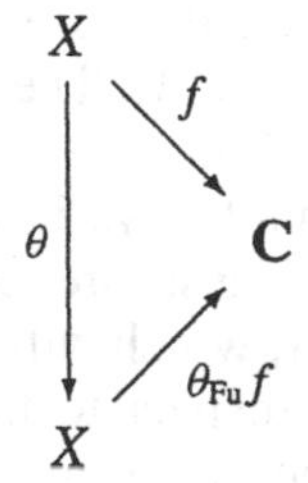

Thus one defines the action θ_{Fu} by

FUNCT 1. $(\theta_{\mathrm{Fu}} f)(x) = f(\theta^{-1}x)$, also written $\theta_{\mathrm{Fu}} f = f \circ \theta^{-1}$.

If $D: \mathrm{Fu}(X) \to \mathrm{Fu}(X)$ is a linear map, then writing θ_* for the induced automorphism of $\mathrm{End}(\mathrm{Fu}(X))$, we have

FUNCT 2. $\theta_* D = \theta_{\mathrm{Fu}} \circ D \circ \theta_{\mathrm{Fu}}^{-1}$ or also $\theta_{\mathrm{Fu}}^{-1} \circ \theta_* D = D \circ \theta_{\mathrm{Fu}}^{-1}$.

Thus the functorial action on $\mathrm{End}(\mathrm{Fu}(X))$ is conjugation by the functorial action on functions, and makes the following diagram commutative:

$$\begin{array}{ccc} \mathrm{Fu} & \xrightarrow{\;D\;} & \mathrm{Fu} \\ {\scriptstyle \theta_{\mathrm{Fu}}}\big\downarrow & & \big\downarrow{\scriptstyle \theta_{\mathrm{Fu}}} \\ \mathrm{Fu} & \xrightarrow[\;\theta_* D\;]{} & \mathrm{Fu} \end{array}$$

In particular, for a function $f \in \mathrm{Fu}$, we have

FUNCT 3. $(\theta_* D)(\theta_{\mathrm{Fu}} f) = \theta_{\mathrm{Fu}}(Df) = (Df) \circ \theta^{-1}$.

Putting **FUNCT 2** and **FUNCT 3** together yields for $x \in X$, $f \in \mathrm{Fu}(X)$:

FUNCT 4. $(D(\theta_{\mathrm{Fu}}^{-1} f))(x) = (\theta_{\mathrm{Fu}}^{-1}((\theta_* D) f))(x) = ((\theta_* D) f)(\theta x)$.

Warning. Let G be a Lie group. There are two natural structures on G: the Lie group structure, and the differential manifold structure, thus giving rise to two different kinds of automorphisms. For instance, given $g \in G$, **left translation** by g (that is, $x \mapsto gx$ for $x \in G$) is a differential automorphism but not a Lie group automorphism. **Conjugation** $x \mapsto gxg^{-1}$ is a Lie group automorphism. Both will occur in the sequel.

We say that D is **θ-invariant** if $\theta_* D = D$. Thus a linear operator D on $\mathrm{Fu}(X)$ is θ-invariant if and only if it commutes with θ_{Fu}; that is, for all $f \in \mathrm{Fu}(X)$, we have

$$D(f \circ \theta)) = (Df) \circ \theta.$$

Let G be a Lie group acting on X, and for $g \in G$, let $\tau(g): X \to X$ denote the action. Let $D: \mathrm{Fu}(X) \to \mathrm{Fu}(X)$ be a linear map, for example a differential operator. We say that D is **$\tau(G)$-invariant** or simply G-invariant if D is $\tau(g)$-invariant for every $g \in G$. Thus D is G-invariant if and only if D commutes with $\tau_{\mathrm{DO}}(g)$, that is

$$D(f \circ \tau(g)) = (Df) \circ \tau(g) \quad \text{for all } g \in G.$$

Let G act linearly on some vector space, or a module M over a ring, so we have what is called a representation of G on M. We use the standard notation

$$M^G$$

for the submodule of **invariant elements**, i.e. the elements fixed by the action of G. Let $L: G \to \mathrm{Aut}(M)$ denote the representation, if the representation needs to be included in the notation. We may then write more precisely

$$M^G = M^{L(G)}.$$

Let $\tau: G \to \mathrm{DiffAut}(X)$ be an action of G on a manifold X. We write

$$\mathrm{DO}(X)^{\tau(G)}$$

for the subalgebra of $\tau_{\mathrm{DO}}(G)$-invariant differential operators on X.

We shall start by studying the action of left translation $L_g: G \to G$ defined by $L_g(x) = gx$ for $x \in G$. In this case, we abbreviate

$$\mathrm{DO}(G)^{L(G)} = \mathrm{IDO}(G).$$

We call $\mathrm{IDO}(G)$ the algebra of **invariant differential operators**, meaning left invariant, or invariant under left translation.

We shall also deal with other types of homogeneous spaces, which are isomorphic to G/K with a closed Lie subgroup K.

Quite generally, let G be a Lie group acting on a manifold X. Let D be a G-invariant linear operator on the space of functions $\mathrm{Fu}(X)$. Let $f \in \mathrm{Fu}(X)$ be G-invariant. Then immediately from the definitions, we see that Df is also G-invariant. We shall study cases when the quotient space $G\backslash X$ is a manifold, and G-invariant smooth functions on X are in bijection with smooth functions on the quotient space $G\backslash X$.

We shall give explicit generators for the algebras of invariant differential operators in the concrete cases of interest to us. We are also interested in the characters of the algebras of invariant differential and integral operators, i.e. eigenfunctions and their eigenvalues, which will also be explicitly exhibited.

II, §1. INVARIANT DIFFERENTIAL OPERATORS ON A LIE GROUP

Let G be a Lie group with Lie algebra $\mathfrak{g} = T_e G$. For $v \in \mathfrak{g}$ we let ξ_v be the left invariant vector field obtained by translating v over G. We let $\mathcal{D}(v)$ be the left invariant differential operator defined on a function $f \in C^\infty(G)$ by the Lie derivative

$$(1) \qquad (\mathcal{D}(v)f)(g) = \mathcal{L}_{\xi_v}(f \circ L_g)(e) = \frac{d}{dt} f(g \exp tv)\bigg|_{t=0}.$$

If we let $F_{v,g}(t) = f(g \exp tv)$, then it follows at once by induction that

$$(1_m) \qquad (\mathcal{D}(v)^m f)(g) = F_{v,g}^{(m)}(0) = \left(\frac{d}{dt}\right)^m F_{v,g}(t)\bigg|_{t=0}.$$

Let $\mathrm{IDO}(G)$ denote the algebra (over $\mathbf{R}$) of left invariant differential operators (invariance under left translations). Then immediately from the defining formula (1), it is clear that $\mathcal{D}(v) \in \mathrm{IDO}(G)$. Furthermore, $\mathcal{D}(v)$ is a differential operator of degree 1. For any vector fields ξ, η on a manifold, we have

$$\mathcal{L}_{[\xi,\eta]} = \mathcal{L}_\xi \circ \mathcal{L}_\eta - \mathcal{L}_\eta \circ \mathcal{L}_\xi,$$

whence for $v, w \in \mathfrak{g}$ we see that $\mathcal{D}$ is a representation of the Lie algebra, that is

$$(2) \qquad \mathcal{D}([v, w]) = \mathcal{D}(v)\mathcal{D}(w) - \mathcal{D}(w)\mathcal{D}(v).$$

We shall deal with the **symmetric algebra**, but first make general remarks in a more abstract context going back to Bourbaki's *Algebra*. Let V be a finite dimensional vector space over $\mathbf{R}$. Let $\{v_1, \ldots, v_N\}$ be a basis of V. An element $P \in S(V)$ is a polynomial

$$P = \sum c_{(j)} v_1^{j_1} \ldots v_N^{j_N}, \qquad c_{(j)} \in \mathbf{R},$$

that is, a polynomial in the algebraically independent commutative variables $v_1, \ldots, v_N$. Hence this polynomial can be evaluated at N elements in any commutative algebra over $\mathbf{R}$.

Let $L: V \to \mathcal{A}$ be a linear map of V into an algebra (associative, but not necessarily commutative). This linear map extends to a linear map of the tensor algebra of V, which on the homogeneous subspace of degree p maps

$$w_1 \otimes \cdots \otimes w_p \mapsto L(w_1) \cdots L(w_p) \qquad \text{for all } w_1, \ldots, w_p \in V.$$

As a vector space, $S(V)$ is the quotient of the tensor algebra by the subspace generated by all elements

$$w_1 \otimes \cdots \otimes w_p - w_{\sigma(1)} \otimes \cdots \otimes w_{\sigma(p)},$$

with all permutations σ of $\{1, \ldots, p\}$. Hence, there exists a unique linear map

$$S(L): S(V) \to \mathcal{A}$$

called the **symmetrization**, such that for all $w_1, \dots, w_p \in V$ we have

$$S(L): w_1 \cdots w_p \mapsto \frac{1}{p!} \sum_\sigma L(w_{\sigma(1)}) \cdots L(w_{\sigma(p)}),$$

and this homomorphism coincides with $v \mapsto L(v)$ on V. Hence we often write L instead of $S(L)$. We make two remarks.

If $L(w_1), \dots, L(w_p)$ commute, then $L(w_1 \cdots w_p) = L(w_1) \cdots L(w_p)$.

Let $L_1, L_2: S(V) \to \mathcal{A}$ be linear maps such that $L_1(v^m) = L_2(v^m)$ for all $v \in V$ and all positive integers m. Then $L_1 = L_2$.

This is a standard fact of ordinary polynomials, depending on the fact that the pure powers v^m $(v \in V)$ generate $S(V)$ as vector space over $\mathbf{R}$.

We apply the above basic facts to the linear map

$$\mathcal{D}: \mathfrak{g} \to \mathrm{IDO}(G),$$

in which case the extension to the symmetric algebra $S(\mathfrak{g})$ will be called the **Harish symmetrization**. This map occurred previously in the context of what the Lie industry calls the Poincaré–Birkhoff–Witt theorem, but we bypass this sort of general treatment to go directly where we want. Note that if $\mathcal{D}(w_1), \dots, \mathcal{D}(w_p)$ commute, then

$$S(\mathcal{D})(w_1 \cdots w_p) = \mathcal{D}(w_1) \cdots \mathcal{D}(w_p).$$

In particular, $S(\mathcal{D})(v^m) = \mathcal{D}(v)^m$ for all positive integers m and all $v \in \mathfrak{g}$.

The following theorem comes from Harish-Chandra [Har 49] and also [Har 53], p. 192. We write $\mathcal{D}(P)$ instead of $S(\mathcal{D})P$.

Theorem 1.1. *The Harish linear homomorphism*

$$S(\mathcal{D}) = \mathcal{D}: S(\mathfrak{g}) \to \mathrm{IDO}(G)$$

is an isomorphism. If $\{v_1, \dots, v_N\}$ is a basis of $\mathfrak{g}$, and $P \in S(\mathfrak{g})$, then for $f \in C^\infty(G) = \mathrm{Fu}(G)$ we have

(3) $$(\mathcal{D}(P)f)(g) = P(\partial_1, \dots, \partial_N) f(g \exp(t_1 v_1 + \cdots + t_N v_N))|_{(t)=0},$$

where $\partial_i = \partial / \partial_{t_i}$ and $(t) = (t_1, \dots, t_N)$.

Proof. Let $\tilde{P}$ be defined by

$$(\tilde{P} f)(g) = P(\partial_1, \dots, \partial_N) f(g \exp(t_1 v_1 + \cdots + t_N v_N))|_{(t)=0}.$$

Since $(t_1, \dots, t_N) \mapsto g\exp(t_1v_1+\dots+t_Nv_N)$ is a chart in a neighborhood of g, it is immediate from the definition that $\tilde{P}$ is an invariant differential operator. The map $P \mapsto \tilde{P}$ is linear. It is injective on $S(\mathfrak{g})$, as one sees by applying $\tilde{P}$ to monomials $t_1^{j_1} \cdots t_N^{j_N}$ in the above chart. Furthermore, let $v \in \mathfrak{g}$, $v = c_1v_1 + \dots + c_Nv_N$ with coefficients $c_i \in \mathbf{R}$. By definition, for $P = v$, we have

$$(\tilde{v}f)(g) = (c_1\partial_1 + \dots + c_N\partial_N)f(g\exp(t_1v_1 + \dots + t_Nv_N))|_{(t)=0}.$$

Letting $F(w) = f(g\exp w)$ for $w \in \mathfrak{g}$, the chain rule shows immediately that

$$(\tilde{v}f)(g) = F'(0)v = (\mathcal{D}(v)f)(g),$$

so $\tilde{v} = \mathcal{D}(v)$. It follows that $\tilde{P} = \mathcal{D}(P)$ for all $P \in S(\mathfrak{g})$, thus proving (3). There remains only to prove that $P \mapsto \mathcal{D}(P)$ is surjective. Let D be a given invariant differential operator. There is a unique polynomial P such that for all functions f,

$$(Df)(e) = P(\partial_1, \dots, \partial_N)f(\exp(t_1v_1 + \dots + t_Nv_N))|_{(t)=0}.$$

This P is what we call $P_{D,e}$, the **polynomial associated to D at the origin in the chart.** By left invariance, we get $D = \mathcal{D}(P)$, which concludes the proof.

Remark on notation. We use $\mathcal{D}(P)$ instead of the Harish-Chandra notation $\lambda(P)$ to denote the map from polynomials to differential operators, because λ is occupied in several other places, e.g. for functionals or eigenvalues.

The filtration and graded structure

Quite generally, let R be an algebra with a filtration, i.e. vector subspaces R_i (i ranging over the integers $\geqq 0$) such that

$$R_0 \subset R_1 \subset \dots \subset R_j \subset \cdots \quad \text{and} \quad \bigcup R_j = R,$$

and $R_iR_j \subset R_{i+j}$ for all $i, j \geqq 0$. In particular, R is an R_0-algebra. We then call R a **filtered algebra**. We have two examples in mind, namely the two algebras $S(\mathfrak{g})$ and $\mathrm{IDO}(G)$ which are filtered by the degree. Note that a change of charts preserves the degree of a differential operator, whence the notion of degree made sense geometrically, i.e. independently of charts. We now observe additionally that the Harish linear isomorphism preserves the filtration. In other words, denote by $S(\mathfrak{g})_d$ the space of polynomials of degree $\leqq d$, and $\mathrm{IDO}(G)_d$ the space of invariant differential operators of degree $\leqq d$. Then

$$\mathcal{D}_d : S(\mathfrak{g})_d \to \mathrm{IDO}(G)_d$$

maps $S(\mathfrak{g})_d$ into $\mathrm{IDO}(G)_d$. In fact, we have

$$\deg P = \deg \mathcal{D}(P) = \deg P_{D,e},$$

where $\deg \mathcal{D}(P)$ is the degree of the differential operator. In particular, for $v \in \mathfrak{g}$, the differential operator $\tilde{v} = \mathcal{D}(v)$ has degree 1.

Note, however, that even if P is a homogeneous polynomial of a given degree in $S(\mathfrak{g})$, then $\mathcal{D}(P)$ as a differential operator in a chart may contain terms of lower degree, and in fact usually does, because a change of charts does not preserve the property of being homogeneous of given degree. We thus introduce another notion. Let R be a filtered algebra. We define the **associated graded algebra** by letting

$$\mathrm{gr}_d(R) = R_d/R_{d-1} \quad \text{and} \quad \mathrm{gr}(R) = \bigoplus \mathrm{gr}_d(R).$$

Of course, we put $R_{-1} = \{0\}$. We may complement Theorem 1.1 by taking into account the multiplicative structure as follows.

Proposition 1.2. *The Harish linear isomorphism $\mathcal{D}$ is in fact multiplicative on the graded algebras, and induces a graded algebra isomorphism*

$$\mathrm{gr}(\mathcal{D}) : \mathrm{gr}\, S(\mathfrak{g}) \to \mathrm{gr}\ \mathrm{IDO}(G).$$

Proof. The multiplicative statement is immediate, because the commutator of two differential operators of degrees i, j has degree $\leqq i + j - 1$. Thus if $\deg P \leqq i$ and $\deg Q \leqq j$, then

$$\mathcal{D}(PQ) \equiv \mathcal{D}(P)\mathcal{D}(Q) \mod \mathrm{IDO}(G)_{i+j-1}.$$

The fact that $\mathcal{D}$ is a linear isomorphism from Theorem 1.1 then implies that $\mathrm{gr}(\mathcal{D})$ is a graded algebra isomorphism.

Corollary 1.3. *Let $\{v_1, \ldots, v_N\}$ be a basis of $\mathfrak{g}$, and $D \in \mathrm{IDO}(G)$. Then D has a unique expression as a (non-commutative) polynomial*

$$D = \sum c_{(j)} \tilde{v}_1^{j_1} \cdots \tilde{v}_N^{j_N}$$

with constants $c_{(j)}$.

Proof. This is an immediate consequence of Proposition 1.2, making this proposition more explicit in terms of the basis. Indeed, for $v, w \in \mathfrak{g}$ by (2) we have $[\tilde{v}, \tilde{w}] = [v, w]^{\sim}$, and $[v, w]^{\sim}$ has degree 1, so $\tilde{v}$ and

$\tilde{w}$ commute modulo lower degrees, and inductively we get the stated polynomial expression. The monomials

$$\tilde{v}_1^{j_1} \cdots \tilde{v}_N^{j_N}$$

are linearly independent, as one sees by induction, first by considering those of given degree $j_1 + \cdots + j_N = d$, modulo differential operators of degree $< d$, i.e. the image of the monomials in $\mathrm{gr}_d(\mathrm{IDO}(G))$.

Remark. One can define a filtration on $\mathrm{IDO}(G)$ by letting a product $\tilde{v}_{i_1} \cdots \tilde{v}_{i_d}$ have filtration $\leqq d$. Corollary 1.3 then implies that this filtration is the same as the previously used degree filtration.

In some applications, one wants a converse, deducing a linear isomorphism from a graded one. We include the relevant lemma here for convenience.

Lemma 1.4. *Let R, S be filtered algebras, $R = \bigcup R_j$ and $S = \bigcup S_j$. Let*

$$L: S \to R$$

be an (R_0, S_0)-linear map preserving the filtration, and inducing an (R_0, S_0)-linear isomorphism

$$\mathrm{gr}_d(L): \mathrm{gr}_d(S) \to \mathrm{gr}_d(R)$$

for all $d \geqq 0$. Then L is an (R_0, S_0)-linear isomorphism.

Proof. Routine algebra, inductively on the degree for showing both the injectivity and surjectivity. The details are left to the reader.

Example. Consider the special case when $G = A$ is an r-fold product of positive multiplicative groups,

$$A = \mathbf{R}^+ \times \cdots \times \mathbf{R}^+.$$

This case will be important later. We can make the Harish isomorphism more explicit in this case, at the level of elementary calculus. Let $\{v_1, \ldots, v_r\}$ be a basis of the Lie algebra $\mathfrak{a} = \mathbf{R} + \cdots + \mathbf{R}$. Then $\mathcal{D}_A(S(\mathfrak{a}))$ consists of all polynomials with constant coefficients

$$\sum c_{(j)} \tilde{v}_1^{j_1} \cdots \tilde{v}_r^{j_r}$$

where for $v \in \mathfrak{a}$, now $\tilde{v}$ means the differential operator $\mathcal{D}_A(v)$.

Let $\{E_{11}, \ldots, E_{rr}\}$ be the standard basis for $\mathfrak{a}$, so E_{ii} has component 1 on the i-th diagonal element and 0 elsewhere. In terms of the multiplicative diagonal variables $a_1, \ldots, a_r$, let $\mathcal{D}_i = a_i \partial/\partial a_i$, so $\mathcal{D}_i \in \mathrm{IDO}(A)$. Let

$$v = \sum h_i E_{ii}.$$

Directly from the definitions, for a function $f \in \mathrm{Fu}(A)$, we have

$$(\tilde{v}f)(a) = \sum h_i a_i \partial_i f(a) \qquad \text{so} \qquad \tilde{v} = \sum h_i \mathcal{D}_i,$$

where $\partial_i f$ is the derivative of f with respect to the i-th variable. Thus $\mathrm{IDO}(A)$ is the commutative polynomial algebra

$$\mathrm{IDO}(A) = \mathbf{R}[\mathcal{D}_1, \ldots, \mathcal{D}_r].$$

If one pulls back one step further on $\mathbf{R}^r$ with coordinates $(h_1, \ldots, h_r)$, then

$$a_i \frac{\partial}{\partial a_i} = \frac{\partial}{\partial h_i}$$

becomes the ordinary partial derivative in euclidean space, and

$$\mathrm{IDO}(\mathbf{R}^r) = \mathbf{R}\left[\frac{\partial}{\partial h_1}, \ldots, \frac{\partial}{\partial h_r}\right].$$

We now go into the functoriality business. Let

$$\theta: G \to G'$$

be a *Lie group isomorphism.* Then θ induces an isomorphism on functors, for instance θ induces an isomorphism of the ring of differential operators on G, and also on the ring of invariant differential operators. The functoriality requires that for a differential operator D, a function f, and $g \in G$,

$$\theta_{\mathrm{DO}}(D)\theta_{\mathrm{Fu}}(f) = \theta_{\mathrm{Fu}}(Df) \qquad \text{and} \qquad \theta_{\mathrm{Fu}}(f)(\theta g) = f(g),$$

where we index θ by the category on which it induces a representation, so the algebra of differential operators and the algebra of functions respectively. The above notation can get heavy, and we often denote the functorial action by the subscript $*$, so we would rewrite the above relations in the form

(4) $$(\theta_* D)(\theta_* f) = \theta_*(Df) \qquad \text{and} \qquad (\theta_* f)(\theta g) = f(g).$$

Roughly speaking, any construction carried out using only the Lie group structure is "preserved" under Lie group isomorphisms. This is a pain to formalize into a theorem, but readers should realize the extent to which the next proposition is a piece of abstract nonsense. In practice, as we shall do, one actually gives the formal proof at least once to see what's going on.

Proposition 1.5. *Let $\theta: G \to G'$ be a Lie group isomorphism. Let $\theta_{\mathfrak{g}}$ and θ_* denote the induced maps on the Lie algebra and invariant differential operators respectively. Then the induced action of θ commutes with $\mathcal{D}$, that is, for $v \in \mathfrak{g}$,*

$$(\theta_{\mathfrak{g}} v)^{\sim} = \theta_*(v^{\sim}).$$

More generally, $\mathcal{D}$ commutes with the action of θ on polynomials, that is for $P \in S(\mathfrak{g})$,

$$\mathcal{D}_{G'}(\theta_{S(\mathfrak{g})} P) = \theta_* \mathcal{D}_G(P).$$

In other words, the following diagram is commutative:

$$\begin{array}{ccc} S(\mathfrak{g}) & \xrightarrow{\mathcal{D}_G} & \mathrm{IDO}(G) \\ \downarrow{\scriptstyle \theta_{S(\mathfrak{g})}} & & \downarrow{\scriptstyle \theta_*} \\ S(\mathfrak{g}') & \xrightarrow[\mathcal{D}_{G'}]{} & \mathrm{IDO}(G') \end{array}$$

Proof. Let f' be a function on G' and let $g' = \theta g$ with $g \in G$. Then

$$\begin{aligned} ((\theta_* \tilde{v}) f')(g') &= \theta_*(\tilde{v}(\theta_*^{-1} f'))(g') \\ &= \tilde{v}(f' \circ \theta)(\theta^{-1} g') = (\tilde{v}(f' \circ \theta))(g). \end{aligned}$$

On the other hand,

$$\begin{aligned} ((\theta_{\mathfrak{g}} v)^{\sim} f')(g') &= \frac{d}{dt} f'(g' \exp(t\theta_{\mathfrak{g}} v))\Big|_{t=0} \\ &= \frac{d}{dt} f'(\theta g \cdot \theta(\exp(tv)))\Big|_{t=0} \\ &= \tilde{v}(f' \circ \theta)(\theta^{-1} g') = \tilde{v}(f' \circ \theta)(g), \end{aligned}$$

which proves the first formula. Since θ is a Lie group isomorphism, its induced map θ_* is also an isomorphism for multiplication (composition) of differential operators and for the polynomial algebras $S(\mathfrak{g}) \to S(\mathfrak{g}')$. The commuting statement for arbitrary polynomials is then an immediate consequence of the statement for elements of $\mathfrak{g}$. This concludes the proof.

Conjugation

Given an element $g \in G$ conjugation $\mathbf{c}(g)$ (mapping $x \mapsto gxg^{-1}$) is a Lie group automorphism of G, to which we can apply Proposition 1.5. The first formula of Proposition 1.5 can be written

$$(\mathbf{c}_{\mathfrak{g}}(g)v)^{\sim} = \mathbf{c}_{\mathrm{DO}}(g)\tilde{v}.$$

We write the second with the star notation,

$$\mathbf{c}(g)_*\mathcal{D}(P) = \mathcal{D}(\mathbf{c}(g)_*P). \tag{5}$$

In terms of functions $f \in \mathrm{Fu}(G)$, we therefore have

$$\begin{aligned} \mathbf{c}(g)_*(\mathcal{D}(P)f) &= \mathcal{D}(\mathbf{c}(g)_*P)(\mathbf{c}(g)_*f) \\ &= (\mathbf{c}(g)_*\mathcal{D}(P))(\mathbf{c}(g)_*f). \end{aligned} \tag{6a}$$

Since $\mathbf{c}(g)e = e$, this yields

$$(\mathcal{D}(P)f)(e) = \mathbf{c}(g)_*\mathcal{D}(P)(\mathbf{c}(g)_*f)(e). \tag{6b}$$

Next, we note that for $v \in \mathfrak{g}$ and $P \in S(\mathfrak{g})$,

(7) $(\mathbf{c}(g)_*v)^{\sim} = R(g^{-1})_*\tilde{v}$ whence also $\mathcal{D}(\mathbf{c}(g)_*P) = R(g^{-1})_*\mathcal{D}(P)$.

Proof. We have $\mathbf{c}(g) = L(g)R(g^{-1})$ (composite of left and right translation), and

$$L(g)_*D = D$$

for every left invariant differential operator D, such as $\tilde{v}$ or $\mathcal{D}(P)$. Hence formula (7) follows.

II, §2. THE PROJECTION ON A HOMOGENEOUS SPACE

Invariant differential operators are by definition invariant under left translation by G. Let K be a subgroup of G. For the differential theory, one does not need to assume K compact, and we do not do so. Compactness will be assumed only when we have to integrate, starting with Theorem 4.5. Letting $R(K)$ denote right translation by K, these differential automorphisms of G as differential manifolds induce an action on $\mathrm{IDO}(G)$, as does conjugation by elements of K, and it follows that $\mathbf{c}(K) = R(K)$, so

$$\mathrm{IDO}(G)^{\mathbf{c}(K)} = \mathrm{IDO}(G)^{R(K)}. \tag{1}$$

Thus we shall denote by

$$\mathrm{IDO}(G)^K$$

the vector subspace of elements in IDO(G) invariant under the actions of conjugation or right translation by K. They are the same action, and thus have the same fixed subspace.

Theorem 2.1. *The Harish-Chandra linear isomorphism*

$$\mathcal{D}: S(\mathfrak{g}) \to \mathrm{IDO}(G)$$

induces a linear isomorphism on the $\mathbf{c}(K)$*-invariants*

$$\mathcal{D}: S(\mathfrak{g})^{\mathbf{c}(K)} \to \mathrm{IDO}(G)^K,$$

and similarly on the graded algebras.

Proof. This is immediate because a change of charts preserves the degree of a differential operator, and conjugation $\mathbf{c}_{\mathfrak{g}}(g)$ preserves the homogeneous components of a polynomial in $S(\mathfrak{g})$. We can then apply Proposition 1.5.

In [Har 58a] starting with Theorem 1, §4, Harish-Chandra carried out the direct image from G to A using the Iwasawa decomposition. He determines the kernel, especially using the space

$$\tilde{\mathfrak{n}}\,\mathrm{IDO}(G) + \mathrm{IDO}(G)\tilde{\mathfrak{k}},$$

and the K-invariant subspace. He works on the group. Helgason interpolated the homogeneous space $X = G/K$ between G and A, see [Hel 59] and [Hel 84], Chapter II, Theorems 4.8–4.10. As Helgason himself wrote [Hel 62], p. 455, Harish-Chandra's analysis "leads" to Helgason's formulation on X. In this section we develop first Helgason's formulation for a reductive space. The next sections give Harish-Chandra's direct image to A.

Let X be a manifold which is a homogeneous space for G, with the action denoted by $\tau: G \to \mathrm{Aut}(X)$, so for $g \in G$,

$$\tau_g = \tau(g): X \to X$$

is a differential isomorphism. Sometimes we denote this action by $[g]$. Cf. Chapter I, §1. We let I be a point of X, and K the isotropy group (**not necessarily compact**, although it will be compact in our applications). Then we get a morphism

$$\pi: G \to X \quad \text{given by} \quad \pi(g) = \tau(g)I.$$

This morphism commutes with translations, that is, for $g, g_1 \in G$,

$$\pi(gg_1) = \tau(g)\pi(g_1) \qquad \text{or also} \qquad \pi \circ L_g = \tau(g) \circ \pi.$$

We assume that K is a Lie subgroup (never mind that this can be proved). Then π gives rise to a differential isomorphism

$$\pi_{G/K}: G/K \to X.$$

We have the relation for $k \in K$,

$$\pi \circ L_k = \pi \circ \mathbf{c}_G(k) = \tau(k) \circ \pi \quad \text{on } G. \tag{2}$$

Taking the tangent map at the origin e, we get a commutative diagram

$$\begin{array}{ccc} T_eG & \xrightarrow{T_e\pi} & T_IX \\ {\scriptstyle \mathbf{c}_{\mathfrak{g}}(k) = T_e\mathbf{c}_G(k)}\downarrow & & \downarrow{\scriptstyle T_I(\tau(k))} \\ T_eG & \xrightarrow[T_e\pi]{} & T_IX \end{array} \tag{3}$$

which can be written out in the form

$$T_I(\tau(k)) \circ T_e\pi = T_e\pi \circ \mathbf{c}_{\mathfrak{g}}(k) \quad \text{on } \mathfrak{g}. \tag{4}$$

We take for granted the foundational fact that the map $\pi: G \to X$ is a submersion, or equivalently, $G \to G/K$ is a submersion.

Example. Keep in mind the quadratic map $\pi(g) = g^t g$ from $\mathrm{GL}_n(\mathbf{R})$ to Pos_n, or $\mathrm{SL}_n(\mathbf{R})$ to SPos_n. We have $\tau(g) = [g]$ in our standard notation.

The submersion can actually be made more explicit. Let $\mathfrak{m}$ be any complementary subspace of $\mathfrak{k}$, that is, $\mathfrak{g} = \mathfrak{k} \oplus \mathfrak{m}$. Let $g \in G$. Then the map

$$v \mapsto (g \exp v)$$

is a differential isomorphism of a neighborhood of 0 in $\mathfrak{m}$ with a neighborhood of $\pi(g)$ in X. The tangent map $T_e\pi$ restricted to $\mathfrak{m}$ gives a linear isomorphism

$$(T_e\pi)_{\mathfrak{m}}: \mathfrak{m} \to T_IX.$$

We shall now define a linear map, the **direct image** of π on the space of K-invariant differential operators.

Lemma 2.2. *There exists a unique linear map (the* **direct image***)*

$$\pi_*: \mathrm{IDO}(G)^K \to \mathrm{DO}(X)^G \tag{5}$$

such that for all functions f on X and $D \in \mathrm{IDO}(G)^K$, we have

$$(\pi_* D)f = (D(f \circ \pi))_X. \tag{6}$$

The subscript X means that we view $D(f \circ \pi)$ as a function on X via the map $\pi: G \to X$.

Proof. From (1) it follows that the function $D(f \circ \pi)$ is constant on cosets of K, i.e. is actually a function on G/K, and so a function on X via π, so that (6) holds, thus defining $(\pi_* D)f$ as a function on X. In this way, $\pi_* D$ is an operator on the space of functions, and we have to see that it is a differential operator. From the submersive property, one can choose locally a product decomposition, whereby a point g has local coordinates (x, y) with x being coordinates on X and y coordinates on K. The differential operator can then be expressed in terms of the partial derivatives with respect to the coordinates on x and the coordinates on K. Since $f \circ \pi$ is constant on fibers, it follows that partials with respect to the vertical coordinates annihilate $f \circ \pi$. That D is $R(K)$-invariant then shows that the coefficient functions depend only on x, so $\pi_* D$ is actually a differential operator on X. Thus (6) defines a linear map $\pi_*: \mathrm{IDO}(G)^K \to \mathrm{DO}(X)$.

For basic functoriality reasons, $\pi_* D$ is in $\mathrm{DO}(X)^G$. Indeed, for $f \in \mathrm{Fu}(X)$,

$$\begin{aligned}([\tau(g)](\pi_* D))f &= \tau_{\mathrm{Fu}}(g) \circ \pi_* D \circ \tau_{\mathrm{Fu}}(g)^{-1} f \\ &= \tau_{\mathrm{Fu}}(g)(D(f \circ \tau(g) \circ \pi))_X \\ &= \tau_{\mathrm{Fu}}(g)(D(f \circ \pi \circ L_g))_X \\ &= (L_g \circ D \circ L_g^{-1}(f \circ \pi))_X \\ &= (D(f \circ \pi))_X = (\pi_* D)f\end{aligned}$$

because D is assumed G-invariant. Thus $[\tau(g)]\pi_* D = \pi_* D$, which is therefore an element of $\mathrm{DO}(X)^G$. This concludes the proof of Lemma 2.2.

We shall call X or G/K **reductive** if the Lie algebra $\mathfrak{k} = \mathrm{Lie}(K)$ admits a complement $\mathfrak{m}$, that is a direct sum decomposition

$$\mathfrak{g} = \mathfrak{m} \oplus \mathfrak{k} \quad \text{(as vector space)},$$

such that $\mathbf{c}_{\mathfrak{g}}(k)\mathfrak{m} \subset \mathfrak{m}$ for all $k \in K$; in other words, $\mathfrak{m}$ is stable under the conjugation action by K. The main goal of this section is contained in Theorems 2.3 and 2.5. *From here on, we assume X reductive.*

We shall use the isomorphism of Theorem 2.1,

$$\mathcal{D}: S(\mathfrak{g})^{\mathbf{c}(K)} \xrightarrow{\approx} \mathrm{IDO}(G)^K, \tag{7}$$

which we also call the **Harish isomorphism**. From the stability of $\mathfrak{m}$ under $\mathbf{c}(K)$, the injection $S(\mathfrak{m}) \hookrightarrow S(\mathfrak{g})$ induces an injection

$$S(\mathfrak{m})^{\mathbf{c}(K)} \hookrightarrow S(\mathfrak{g})^{\mathbf{c}(K)}. \tag{8}$$

We define the map

$$\mathcal{D}_{\mathfrak{m},X}: S(\mathfrak{m})^{\mathbf{c}(K)} \to \mathrm{DO}(X)^G \quad \text{by} \quad \mathcal{D}_{\mathfrak{m},X} = \pi_* \circ \mathcal{D}, \tag{9}$$

so that the following diagram is commutative:

$$\begin{array}{ccc} \mathrm{IDO}(G)^K & \xrightarrow{\ \pi_*\ } & \mathrm{DO}(X)^G \\ {\scriptstyle \mathcal{D}_G = \mathcal{D}}\uparrow & & \uparrow{\scriptstyle \mathcal{D}_{\mathfrak{m},X}} \\ S(\mathfrak{g})^{\mathbf{c}(K)} & \xleftarrow[\text{inc}]{} & S(\mathfrak{m})^{\mathbf{c}(K)} \end{array}$$

The bottom map is the natural inclusion. We thus have by definition

(10) $\mathcal{D}_{\mathfrak{m},X}(P)f$

$$= (\mathcal{D}(P)(f \circ \pi))_X \qquad \text{for } P \in S(\mathfrak{m})^{\mathbf{c}(K)} \text{ and } f \in \mathrm{Fu}(X).$$

Let $\{v_1, \dots, v_r\}$ be a basis of $\mathfrak{m}$, and $t_1, \dots, t_r$ the coordinates of an element $\sum t_i v_i$ in $\mathfrak{m}$. Then (10) has the explicit expression

(11) $(\mathcal{D}_{\mathfrak{m},X}(P)f)(e)$

$$= P\left(\frac{\partial}{\partial t_1}, \dots, \frac{\partial}{\partial t_r}\right) f \circ \pi(\exp(t_1 v_1 + \cdots + t_r v_r))\big|_{(t)=0}.$$

Theorem 2.3 [Hel 59]. *If X is reductive, then $\mathcal{D}_{\mathfrak{m},X}$ is a linear isomorphism.*

Proof. First, the map is injective because $v \mapsto \pi(\exp v)$ is a chart from a neighborhood of 0 in $\mathfrak{m}$ to a neighborhood of I in X, so if $(\mathcal{D}_{\mathfrak{m},X}(P)f)(I) = 0$ for all f, then $P = 0$. For the surjectivity, let $E \in \mathrm{DO}(X)^G$. There exists a unique polynomial P in r variables such that for all functions $f \in \mathrm{Fu}(X)$ we have

$$(12)\quad (Ef)(I) = P\left(\frac{\partial}{\partial t_1}, \ldots, \frac{\partial}{\partial t_r}\right) f \circ \pi\, (\exp(t_1 v_1 + \cdots + t_r v_r)\Big|_{(t)=0}$$
$$= (\mathcal{D}(P)(f \circ \pi)(e).$$

We apply this formula to the function $f \circ \tau(g)$ instead of f. We use the relation $\tau(g) \circ \pi = \pi \circ L_g$. We also use the G-invariance of E, that is

$$E(f \circ \tau(g)) = (Ef) \circ \tau(g).$$

We then obtain

$$(Ef)(\tau(g)I) = (\mathcal{D}(P)(f \circ \pi \circ L_g)(e).$$

Now take $g = k \in K$. Then using $\pi \circ L_k = \pi \circ \mathbf{c}(k)$ we obtain:

$$\begin{aligned}(Ef)(I) = (Ef)(\tau(k)I) &= (\mathcal{D}(P))(f \circ \pi \circ L_k))(e)\\ &= (\mathcal{D}(P))(f \circ \pi \circ \mathbf{c}(k))(e)\\ &= (\mathcal{D}(P))(\mathbf{c}(k^{-1})_*(f \circ \pi))(e)\\ &= \mathcal{D}(\mathbf{c}(k)_* P)(f \circ \pi)(e).\end{aligned}$$

This formula is true for all f and all k. By the uniqueness of the polynomial P satisfying (11), it follows that P is $\mathbf{c}(K)$-invariant, which proves the theorem.

Next we are interested in splitting the two horizontal arrows in the diagram. The bottom arrow is trivially split since we are dealing with ordinary commutative polynomial algebras. In fact we have two splittings,

$$(13a)\qquad S(\mathfrak{g}) = S(\mathfrak{m}) \oplus S(\mathfrak{g})\mathfrak{k},$$

$$(13b)\qquad S(\mathfrak{g})^{\mathbf{c}(K)} = S(\mathfrak{m})^{\mathbf{c}(K)} \oplus (S(\mathfrak{g})\mathfrak{k})^{\mathbf{c}(K)}.$$

Note that (13b) follows from (13a) because each term of the direct sum in (13a) is $\mathbf{c}(K)$-stable.

From Proposition 1.5 we can formulate the corresponding splitting for the algebra of invariant differential operators. For convenience of notation, we define

$$(14a)\qquad \mathcal{J}(\mathfrak{k}) = \mathrm{IDO}(G)\tilde{\mathfrak{k}},$$

so

$$(14b)\qquad \mathcal{J}(\mathfrak{k})^K = \mathrm{IDO}(G)^K \cap \mathrm{IDO}(G)\tilde{\mathfrak{k}} = (\mathrm{IDO}(G)\tilde{\mathfrak{k}})^K.$$

In (14a), $\mathcal{J}(\mathfrak{k})$ is a left ideal of $\mathrm{IDO}(G)$, and in (14b), $\mathcal{J}(\mathfrak{k})^K$ is a left ideal of $\mathrm{IDO}(G)^K$.

Proposition 2.4. *If G/K is reductive, there is a direct sum decomposition of $(\mathbf{c}(K), \mathbf{R})$-vector spaces*

$$\mathrm{IDO}(G) = \mathcal{D}(S(\mathfrak{m})) \oplus \mathcal{J}(\mathfrak{k})$$

and so

$$\mathrm{IDO}(G)^K = \mathcal{D}(S(\mathfrak{m}))^K \oplus \mathcal{J}(\mathfrak{k})^K.$$

Proof. We apply $\mathcal{D}$ to the decomposition of (13), or more precisely to the space of elements of degree $\leqq d$. Thus we find

$$\begin{aligned} \mathcal{D}(S(\mathfrak{g})_d) &= \mathcal{D}(S(\mathfrak{m})_d) + \mathcal{D}(S(\mathfrak{g})_{d-1}\mathfrak{k}) \\ &\equiv (S(\mathfrak{m})_d) + \mathcal{D}(S(\mathfrak{g})_{d-1})\mathcal{D}(\mathfrak{k}) \quad \operatorname{mod} \mathrm{IDO}(G)_{d-1}. \end{aligned}$$

By induction on the degree, this implies the first decomposition as a sum. That the sum is actually direct is immediate. We then take $\mathbf{c}(K)$-invariants. Each summand in the first relation is $\mathbf{c}(K)$-stable, so the second decomposition is an immediate consequence of the first.

Remark. By Proposition 1.5, we have of course

$$\mathcal{D}(S(\mathfrak{m}))^K = \mathcal{D}(S(\mathfrak{m})^{\mathbf{c}(K)}).$$

Theorem 2.5 (Helgason [Hel 59]). *Let G be a Lie group, X a homogeneous space, $I \in X$, with isotropy group K and corresponding projection $\pi: G \to X$. We assume X (or G/K) reductive, with corresponding decomposition $\mathfrak{g} = \mathfrak{m} \oplus \mathfrak{k}$. Let $\pi_*: \mathrm{IDO}(G)^K \to \mathrm{DO}(X)^G$ be the direct image. Then*

$$\operatorname{Ker} \pi_* = \mathcal{J}(\mathfrak{k})^K.$$

Thus π_* induces a linear isomorphism

$$\pi_*: \mathrm{IDO}(G)^K / \mathcal{J}(\mathfrak{k})^K \to \mathrm{DO}(X)^G.$$

Furthermore, π_* induces a linear isomorphism

$$\pi_*: \mathcal{D}(S(\mathfrak{m}))^K \to \mathrm{DO}(X)^G.$$

The linear isomorphisms

$$S(\mathfrak{m})^{\mathbf{c}(K)} \to \mathcal{D}(S(\mathfrak{m})^{\mathbf{c}(K)}) = \mathcal{D}(S(\mathfrak{m}))^K \to \mathrm{DO}(X)^G$$

preserve the degree filtration.

Proof. We apply π_* to the direct sum decomposition of $\mathrm{IDO}(G)^K$ in Proposition 2.4. It is immediate from the definition that the summand

$$(\mathrm{IDO}(G)\tilde{\mathfrak{k}})^K = \mathcal{J}(\mathfrak{k})^K$$

annihilates the function $f \circ \pi$ for every function f on X. Then we apply Theorem 2.3 to conclude the proof of the first statement. The splitting of Proposition 2.4 gives rise to the second statement. The Remark preceding the theorem and Proposition 1.2 show that the map $\mathcal{D}$ applied to $S(\mathfrak{m})^{c(K)}$ preserves degrees, thus concluding the proof.

Remark. Cf. Harish-Chandra [Har 58a], §4, Theorem 1.

II, §3. THE IWASAWA PROJECTION ON *A*

The rest of this chapter is essentially due to Harish-Chandra [Har 58]. We are striving toward Theorems 4.1 and 4.2, diagram (4). These results will be completed in Chapter III, Theorem 5.3, using characters and the beginning of spherical functions.

Suppose given a submanifold Y of a manifold X, and an open subset S of X containing Y, together with a submersion

$$\pi : S \to Y$$

which is the identity on Y. Thus locally in a chart, π is a projection and Y itself is a section of the submersion. Given these data, we define the **direct image**

$$\pi_* : \mathrm{DO}(X) \to \mathrm{DO}(Y)$$

as follows. Given a function f on Y, we consider the composite function $f \circ \pi$, apply D, and restrict the resulting function to Y, so by definition

(1) $$(\pi_* D)f = D(f \circ \pi)_Y,$$

the subscript denoting restriction to Y. The operator $\pi_* D$ is a linear operator on functions. It is in fact a differential operator. One sees this by picking a chart such that in this chart, π is a projection

$$\pi : W \times V \to V,$$

with V a chart in Y. Let the coordinates be (w, y) with $y \in V$ and $w \in W$. Then D is represented in the chart $W \times V$ as a sum

$$D = \sum \varphi_{(j)}(w, y)\partial_1^{j_1} \cdots \partial_r^{j_r} + E(w, y),$$

where $r = \dim Y$, $\partial_i = \partial/\partial y_i$, and $E(y, w)$ is a differential operator in the left ideal generated by $\partial/\partial w_1, \dots, \partial/\partial w_s$ ($s = \dim W$). For any function $f = f(y)$ on V, the function $f \circ \pi$ given by $(f \circ \pi)(w, y) = f(y)$ is annihilated by $E(w, y)$.

An example of the above situation comes from a Lie group action as follows, when we can define a more global version of π_*.

Let H be a Lie group acting on a manifold X. Let

$$\pi : X \to Z$$

be a submersion, **homogeneously fibered** by H. For our purposes, we take this to mean that each point $z \in Z$ has an open neighborhood V with a section $\sigma : V \to X$ of π, such that the map

$$H \times \sigma(V) \to \pi^{-1}(V) \qquad \text{given by} \qquad (h, \sigma(z)) \mapsto h\sigma(z)$$

is a differential isomorphism. Thus $Z = H\backslash X$ is the quotient manifold, under the action of H on X. For a more general definition, see [Lan 99].

The map π then induces a **direct image** on H-invariant differential operators, namely there is a unique linear map

$$\pi_* : \mathrm{DO}(X)^H \to \mathrm{DO}(Z)$$

such that for every function f on Z, we have

$$(2) \qquad (\pi_* D) f = (D(f \circ \pi))_Z.$$

Indeed, the H-invariance of an element $D \in \mathrm{DO}(X)^H$ means that

$$D(F \circ L_h) = (DF) \circ L_h \qquad \text{for } F \in \mathrm{Fu}(X).$$

Putting $F = f \circ \pi$, it follows immediately that $D(f \circ \pi)$ is constant on each homogeneous fiber $X_z = \pi^{-1}(z)$, and therefore defines a function on Z, which we have denoted by $(D(f \circ \pi))_Z$.

Similarly, given two Lie groups H, K with H acting on the left and K on the right of X, we say that a submersion $\pi : X \to Z$ is **(H, K)-homogeneously fibered** if to each $z \in Z$ there is an open neighborhood V of z such that the map

$$(H, K) \times \sigma(V) \to \pi^{-1}(V) \qquad \text{given by} \qquad (h, k, \sigma(z)) \mapsto h\sigma(z)k$$

is a differential isomorphism. Thus Z is isomorphic to the quotient $H\backslash X/K$. Actually, the right action by K amounts to left action by the reverse group K' of K (defining the new product of two elements to be the old product in reverse order), so $H \times K'$ is acting on the left, and the present situation is just the same as the one discussed with only one group H previously. In any case, we can again define the **direct image**

$$\pi_*: \mathrm{DO}(X)^{(H,K)} \to \mathrm{DO}(Z)$$

by means of the same formula (1). Actually we are dealing with two submersions

$$X \xrightarrow{\pi_1} X/K \xrightarrow{\pi_2} H\backslash X/K$$

and directly from the definition of the direct image, we have

$$(\pi_2 \circ \pi_1)_* = (\pi_2)_* \circ (\pi_1)_*.$$

We shall now apply the direct image to a Lie group G with a weak Iwasawa decomposition as defined in Chapter I, §1, so

$$G = UAK.$$

In particular, we have the projection $\pi_A = \mathrm{Iw}_A$ on A.

Although we shall consider principally the direct image $(\pi_A)_*$, we make some comments in the context of lifted functions. Let $\mathrm{Fu}(G)^{U,K}$ be the space of C^∞ functions on G which are left U-invariant and right K-invariant. Then the restriction to A is an isomorphism

$$\mathrm{Fu}(G)^{(U,K)} \xrightarrow{\approx} \mathrm{Fu}(A).$$

Given $f \in \mathrm{Fu}(A)$, we define its **Iwasawa lift** f_G by $f_G = f \circ \mathrm{Iw}_A$. Given $F \in \mathrm{Fu}(G)^{U,K}$ we let F_A be its restriction to A. Directly from the definition of left U-invariance and right K-invariance, we get:

If $D \in \mathrm{IDO}(G)$ and $F \in \mathrm{Fu}(U\backslash G)$, then $DF \in \mathrm{Fu}(U\backslash G)$.

If $F \in \mathrm{Fu}(G)^{U,K}$ and $D \in \mathrm{IDO}(G)^K$, then $DF \in \mathrm{Fu}(G)^{U,K}$.

Thus if $F = f_G$ with some function $f \in \mathrm{Fu}(A)$, and $D \in \mathrm{IDO}(G)^K$, then $Df_G \in \mathrm{Fu}(G)^{U,K}$, so Df_G corresponds to a function in $\mathrm{Fu}(A)$. The above remarks allow us to pass from functions on G to functions on A in applying the next theorem to various contexts, especially in the next section.

From **IW 1** and **IW 2** in the weak Iwasawa decomposition, we have

$$\mathfrak{g} = \mathfrak{n} + \mathfrak{a} + \mathfrak{k},$$

where $\mathfrak{g} = \mathrm{Lie}(G)$, $\mathfrak{n} = \mathrm{Lie}(U)$, $\mathfrak{a} = \mathrm{Lie}(A)$, $\mathfrak{k} = \mathrm{Lie}(K)$.

Theorem 3.1. *Assume only the weak Iwasawa conditions* **IW 1** *and* **IW 2**. *Let*

$$(\mathrm{Iw}_A)_*: \mathrm{IDO}(G) \to \mathrm{IDO}(A)$$

be the direct image defined in (1). *Then the following diagram is commutative*:

$$(3)\qquad \begin{array}{ccc} \mathrm{IDO}(G) & \xrightarrow{(\mathrm{Iw}_A)_*} & \mathrm{IDO}(A) \\ \mathcal{D}_G \uparrow \approx & & \approx \uparrow \mathcal{D}_A \\ S(\mathfrak{g}) & \xleftarrow[\text{inc}]{} & S(\mathfrak{a}) \end{array}$$

The proof will now be developed. First we note that there is a natural identification between functions on A and functions on G which are (U, K)-invariant (U-invariant on the left and K-invariant on the right). In §2 we considered the ideal $\mathcal{J}(\mathfrak{k}) = \mathrm{IDO}(G)\tilde{\mathfrak{k}}$. Here we define similarly

$$(4)\qquad \mathcal{J}(\mathfrak{n}, \mathfrak{k}) = \tilde{\mathfrak{n}}\,\mathrm{IDO}(G) + \mathrm{IDO}(G)\tilde{\mathfrak{k}}.$$

Lemma 3.2. *We have* $\mathcal{J}(\mathfrak{n}, \mathfrak{k}) \subset \mathrm{Ker}(\mathrm{Iw}_A)_*$. *In other words, for* $f \in \mathrm{Fu}(A)$, *we have*

$$\mathcal{J}(\mathfrak{n}, \mathfrak{k})(f \circ \mathrm{Iw}_A) = 0 \quad \textit{restricted to } A.$$

Proof. Suppose a differential operator has the form $D \circ \tilde{w} = D\tilde{w}$, with $D \in \mathrm{IDO}(G)$ and $w \in \mathfrak{k}$. We have trivially $\tilde{w}(f \circ \mathrm{Iw}_A) = 0$ because $f \circ \mathrm{Iw}_A$ is constant on left cosets of K, and we just apply the definition of what $\tilde{w}$ means. Hence

$$D\tilde{w}(f \circ \mathrm{Iw}_A) = 0.$$

On the other hand, elements of $\mathrm{IDO}(G)$ are left-invariant operators, and consequently, if $f \in \mathrm{Fu}(A)$ (but here $f \in \mathrm{Fu}(U\backslash G)$ would suffice), for $D \in \mathrm{IDO}(G)$ the function $D(f \circ \mathrm{Iw}_A)$ is U-invariant, as was pointed out in Remark 2. Suppose a differential operator has the form $\tilde{v}D$ with $D \in \mathrm{IDO}(G)$ and $v \in \mathfrak{n}$. Let $F = D(f \circ \mathrm{Iw}_A)$. Since we have just seen that F is U-invariant, we get for $a \in A$:

$$\begin{aligned} \tilde{v}F(a) &= \frac{d}{dt}F(a \exp tv)\Big|_{t=0} \\ &= \frac{d}{dt}F(a(\exp tv)a^{-1}a)\Big|_{t=0} \\ &= 0 \end{aligned}$$

because A normalizes U, and the expression under the derivative is constant, so its derivative is 0. This proves the lemma.

Theorem 3.3. *Under the weak Iwasawa conditions, there is a direct sum decomposition*

$$\mathrm{IDO}(G) = \mathcal{D}_G(S(\mathfrak{a})) \oplus \mathcal{J}(\mathfrak{n}, \mathfrak{k}).$$

We have $\mathrm{Ker}(\mathrm{Iw}_A)_* = \mathcal{J}(\mathfrak{n}, \mathfrak{k})$, *and hence a linear isomorphism*

$$(\mathrm{Iw}_A)_* \colon \mathrm{IDO}(G)/\mathcal{J}(\mathfrak{n}, \mathfrak{k}) \to \mathrm{IDO}(A).$$

In other words, given $D \in \mathrm{IDO}(G)$ *there exists a unique polynomial* $P_\mathfrak{a} \in S(\mathfrak{a})$ *such that*

$$D \equiv \mathcal{D}_G(P_\mathfrak{a}) \mod \mathcal{J}(\mathfrak{n}, \mathfrak{k}).$$

For this polynomial, we have

$$(\mathrm{Iw}_A)_*(D) = \mathcal{D}_A(P_\mathfrak{a}).$$

Proof. Let $\{v_1, \dots, v_p\}$ be a basis of $\mathfrak{n}$; let $\{v_{p+1}, \dots, v_{p+r}\}$ be a basis of $\mathfrak{a}$; and let $\{v_{p+r+1}, \dots, v_N\}$ be a basis of $\mathfrak{k}$. By Corollary 1.3, every $D \in \mathrm{IDO}(G)$ has a unique (non-commutative) polynomial expression

$$D = \sum c_{(j)} \tilde{v}_1^{j_1} \cdots \tilde{v}_N^{j_N}.$$

From this expression, we see at once that $\mathrm{IDO}(G)$ is the sum of $\mathcal{D}_G(S(\mathfrak{a}))$ and $\mathcal{J}(\mathfrak{n}, \mathfrak{k})$, namely the $\mathcal{D}_G(S(\mathfrak{a}))$ component consists of the sum of all monomials in the polynomial expression which contain only powers of elements from the basis for $\mathfrak{a}$. The intersection of $\mathcal{D}_G(S(\mathfrak{a}))$ and $\mathcal{J}(\mathfrak{n}, \mathfrak{k})$ is $\{0\}$, as follows at once from (4) by applying an element of the intersection to all functions $f \circ \mathrm{Iw}_A$ with $f \in \mathrm{Fu}(A)$. Thus the sum is direct. There remains to see that $\mathrm{Ker}(\mathrm{Iw}_A)_* = \mathcal{J}(\mathfrak{n}, \mathfrak{k})$.

We let

$$\mathrm{pr}_{\mathfrak{a},G} = \mathrm{pr}_\mathfrak{a} \colon \mathrm{IDO}(G) \to \mathcal{D}_G(S(\mathfrak{a}))$$

be the projection. For $\mathcal{D} \in \mathrm{IDO}(G)$, we also denote its projection in $\mathcal{D}_G(S(\mathfrak{a}))$ by $D_\mathfrak{a}$, so $D_\mathfrak{a}$ is just the sum of the monomials containing only basis vectors from $\mathfrak{a}$ in the polynomial expression for D. Thus $D_\mathfrak{a}$ is the unique differential operator in $\mathcal{D}_G(S(\mathfrak{a}))$ such that

$$D \equiv D_\mathfrak{a} \mod \mathcal{J}(\mathfrak{n}, \mathfrak{k}).$$

We may call $D_\mathfrak{a}$ the **$\mathfrak{a}$–Iwasawa projection** of D. We let $D_A^{U,K} = \mathrm{pr}_{\mathfrak{a},A}(D)$ be the corresponding differential operator on A, and we now conclude that

$$\mathrm{Iw}_{A*}(D) = D_A^{U,K} = \mathrm{pr}_{\mathfrak{a},A}(D).$$

Thus Iw_{A*} satisfies the commutative diagram (3). Taking into account that the Harish-Chandra map

$$\mathcal{D}_A : S(\mathfrak{a}) \to \mathrm{IDO}(A)$$

is an isomorphism, we have therefore concluded the proof of Theorems 3.1 and 3.3.

We shall now study more closely the K-invariants, which will provide additional information from at least two points of view: the homogeneous space G/K and the multiplicative structure.

Remark. In Lemma 3.2, we did not assert that $\mathcal{J}(\mathfrak{n}, \mathfrak{k})(f \circ \mathrm{Iw}_A) = 0$. Only its restriction to A is 0. On the other hand, we do have

$$\mathcal{J}(\mathfrak{n}, \mathfrak{k})^K (f \circ \mathrm{Iw}_A) = 0.$$

Indeed, if $D \in \mathrm{IDO}(G)^K$ then D is (U, K)-invariant, so $D(f \circ \mathrm{Iw}_A)$ is also (U, K)-invariant. If $D \in \mathcal{J}(\mathfrak{n}, \mathfrak{k})^K$, then we can apply Lemma 3.2 to conclude that $D(f \circ \mathrm{Iw}_A) = 0$. Furthermore, the first part of the proof of Lemma 3.2 showed that $\mathcal{J}(\mathfrak{k})(f \circ \mathrm{Iw}_A) = 0$.

Lemma 3.4. *Let $w \in \mathfrak{k}$ and $D \in \mathrm{IDO}(G)^K$. Then $\tilde{w}$ and D commute.*

Proof. Let f be a function on G. We think of the variable $g \in G$ as lying in a chart. We write D_g for the action of D with respect to this variable. Then

$$\begin{aligned} D_g \frac{d}{dt}(f(g \exp tw) &= \frac{d}{dt} D_g(f(g \exp tw)) \\ &= \frac{d}{dt}(Df)(g \exp tw) \end{aligned}$$

because D is K-invariant. Evaluating at $t = 0$ concludes the proof.

Lemma 3.5. *The vector space $\tilde{\mathfrak{n}}\,\mathrm{IDO}(G)$ is stable under multiplication by $\mathcal{D}_G(S(\mathfrak{a}))$ on the left, that is*

$$\widetilde{S(\mathfrak{a})}\tilde{\mathfrak{n}}\,\mathrm{IDO}(G) = \tilde{\mathfrak{n}}\,\mathrm{IDO}(G).$$

Thus $\tilde{\mathfrak{n}}\,\mathrm{IDO}(G)$ is a module over $\mathcal{D}_G(S(\mathfrak{a}))$.

Proof. From **IW 2** it follows that $[\mathfrak{a}, \mathfrak{n}] \subset \mathfrak{n}$ so $[\tilde{\mathfrak{a}}, \tilde{\mathfrak{n}}] \subset \tilde{\mathfrak{n}}$. Hence for $v \in \mathfrak{n}$ and $D_\mathfrak{a} \in \mathcal{D}_G(S(\mathfrak{a}))$, we conclude inductively on the degree of $D_\mathfrak{a}$ that

$$D_\mathfrak{a}\tilde{v} \in \tilde{\mathfrak{n}}\,\mathrm{IDO}(G),$$

which proves the lemma.

Proposition 3.6. *Under the weak Iwasawa conditions, the vector space* $\mathcal{J}(\mathfrak{n}, \mathfrak{k})$ *is a right module of* $\mathrm{IDO}(G)^K$, *and* $\mathcal{J}(\mathfrak{n}, \mathfrak{k})^K$ *is a two-sided ideal of* $\mathrm{IDO}(G)^K$. *Hence the restriction of* $(\mathrm{Iw}_A)_*$ *to* $\mathrm{IDO}(G)^K$ *is a multiplicative homomorphism, so an algebra homomorphism into* $\mathrm{IDO}(A)$.

Proof. Let $D \in \mathrm{IDO}(G)^K$ and $E \in \mathcal{J}(\mathfrak{n}, \mathfrak{k})$. We can write

$$E = \tilde{v}D_1 + D_2\tilde{w} \qquad \text{with } v \in \mathfrak{n},\ w \in \mathfrak{k},\ D_1, D_2 \in \mathrm{IDO}(G).$$

By Lemma 3.4, it follows that $ED \in \mathcal{J}(\mathfrak{n}, \mathfrak{k})$. This proves that $\mathcal{J}(\mathfrak{n}, \mathfrak{k})$ is a right $\mathrm{IDO}(G)^K$-module, which is the first statement.

As to the other side, let $E \in \mathcal{J}(\mathfrak{n}, \mathfrak{k})^K$. Write $D = D_{\mathfrak{a}} + D'$ with $D' \in \mathcal{J}(\mathfrak{n}, \mathfrak{k})$ according to Theorem 3.3. Then by the first part of the proof,

$$DE = D_{\mathfrak{a}}E + D'E \equiv D_{\mathfrak{a}}E \mod \mathcal{J}(\mathfrak{n}, \mathfrak{k}).$$

But $D_{\mathfrak{a}}\tilde{v} \in \tilde{\mathfrak{n}}\,\mathrm{IDO}(G)$ by Lemma 3.5, so

$$DE \in \mathcal{J}(\mathfrak{n}, \mathfrak{k}).$$

Since both D, E are assumed K-invariant, it follows that $DE \in \mathcal{J}(\mathfrak{n}, \mathfrak{k})^K$, which concludes the proof of the proposition.

Example. Readers wanting to see a crucial and simple example of the computation of a direct image to A may now look at Chapter VII.

II, §4. USE OF THE CARTAN LIE DECOMPOSITION

This section is a direct continuation of §3, and we follow the same notation.

The next question to be addressed concerns the kernel from Theorem 3.3,

$$\mathrm{Ker}(\mathrm{Iw}_A)_*\,\mathrm{IDO}(G)^K = \mathcal{J}(\mathfrak{n}, \mathfrak{k})^K.$$

We shall exploit the Cartan Lie decomposition axiomatized in Chapter I, §3. Note that we don't use the full strength of **CA 3**. We shall use only the injectivity of

$$S(\mathfrak{p})^{c(K)} \to S(\mathfrak{a}).$$

The full strength will be used in Theorem 5.3 of Chapter III.

Also note that the Cartan Lie decomposition implies that G/K is reductive.

Theorem 4.1. *Suppose that G satisfies* **IW 1, IW 2** *and $\mathfrak{g}$ has a Cartan decomposition. Then*

$$\mathcal{J}(\mathfrak{n}, \mathfrak{k})^K = \operatorname{Ker}(\mathrm{Iw}_A)_* \cap \mathrm{IDO}(G)^K = \mathcal{J}(\mathfrak{k})^K.$$

Proof. The first equality comes from taking K-invariants in Proposition 3.3, so we are really concerned with the second equality.

Furthermore, we have by definition $\mathcal{J}(\mathfrak{k})^K \subset \mathcal{J}(\mathfrak{n}, \mathfrak{k})^K$, so the inclusion

$$\mathcal{J}(\mathfrak{k})^K \subset \operatorname{Ker}(\mathrm{Iw}_A)_* \cap \mathrm{IDO}(G)^K$$

already follows from the first equality (going back to Lemma 3.2).

Conversely, let $D \in \operatorname{Ker}(\mathrm{Iw}_A)_*$ and $D \in \mathrm{IDO}(G)^K$. By Proposition 2.4, we have

$$\begin{aligned} \mathrm{IDO}(G)^K &= \mathcal{D}_G(S(\mathfrak{p}))^{\mathbf{c}(K)} \oplus \mathcal{J}(\mathfrak{k})^K \\ &= \mathcal{D}_G(S(\mathfrak{p})^{\mathbf{c}(K)}) \oplus \mathcal{J}(\mathfrak{k})^K \end{aligned}$$

because $\mathcal{D}$ is a $\mathbf{c}(K)$-isomorphism. Write

$$D = \mathcal{D}_G(P) + D_{\mathfrak{k}} \qquad \text{with } P \in S(\mathfrak{p})^{\mathbf{c}(K)} \text{ and } D_{\mathfrak{k}} \in \mathcal{J}(\mathfrak{k})^K.$$

Suppose $P \neq 0$. We have $\deg P = \deg \mathcal{D}_G(P)$, and

$$P \equiv P_{\mathfrak{a}} \quad \operatorname{mod} \mathfrak{n}S(\mathfrak{p}) + S(\mathfrak{p})\mathfrak{k},$$

so

$$\mathcal{D}_G(P) = \mathcal{D}_G(P_{\mathfrak{a}}) + D_{\mathfrak{n},\mathfrak{k}} + D'$$

where $D_{\mathfrak{n},\mathfrak{k}} \in \mathcal{J}(\mathfrak{n}, \mathfrak{k}) \subset \operatorname{Ker}(\mathrm{Iw}_A)_*$ and $\deg D' < \deg \mathcal{D}_G(P)$ $(= \deg D)$. Hence

$$0 = (\mathrm{Iw}_A)_* \mathcal{D}_G(P_{\mathfrak{a}}) + (\mathrm{Iw}_A)_*(D') = \mathcal{D}_A(P_{\mathfrak{a}}) + (\mathrm{Iw}_A)_*(D'),$$

and $(\mathrm{Iw}_A)_*(D')$ has strictly smaller degree than $\mathcal{D}_A(P_{\mathfrak{a}})$ (by **CA 3**). This contradiction concludes the proof.

Let X be a homogeneous space for G, with isotropy group K from the Iwasawa and Cartan decomposition, and Theorem 2.5. The Iwasawa submersion $G \to A$ can be factored through X in a commutative diagram:

$$(5) \qquad \begin{array}{ccc} G & \longrightarrow & X \\ & \searrow \quad \swarrow & \\ & A & \end{array} \qquad \text{namely} \qquad \begin{array}{ccc} uak & \longrightarrow & [ua]I \\ & \searrow \quad \swarrow & \\ & a & \end{array}$$

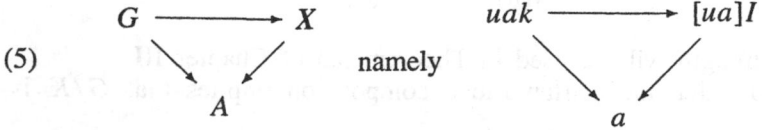

For each arrow, we have the direct image on the space of invariant differential operators. If $D \in \mathrm{IDO}(G)^K$, we let D_X and D_A be its direct image under the maps from G onto X and G onto A respectively, as in the following diagram which summarizes the situation:

(6)
$$\begin{array}{ccc} \mathrm{IDO}(G)^K/\mathcal{J}(\mathfrak{k})^K & \xrightarrow{\approx} & \mathrm{DO}(X)^G \\ & \searrow \quad \swarrow & \\ & \mathrm{IDO}(A) & \end{array}$$

such that
$$\begin{array}{ccc} D & \longmapsto & D_X \\ & \searrow \quad \swarrow & \\ & D_A & \end{array}$$

Theorem 4.2. *Suppose G has a weak Iwasawa and $\mathfrak{g}$ has a Cartan decomposition. In diagram (6) of direct images, all arrows are algebra homomorphisms. The top arrow is an isomorphism. The other two slanted arrows are injective. In particular,* $\mathrm{DO}(X)^G$ *is commutative.*

Proof. The isomorphism on top comes from Theorems 2.5 and 3.6. That all arrows are algebra homomorphisms comes from Theorem 3.6. By Theorems 2.5 and 4.1, the kernels of the two direct image maps

$$\mathrm{IDO}(G)^K \to \mathrm{DO}(X)^G \qquad \text{and} \qquad \mathrm{IDO}(G)^K \to \mathrm{IDO}(A)$$

are the same, so the injectivity follows, as does the commutativity since $\mathrm{IDO}(A)$ is commutative. This concludes the proof.

Remark. The commutativity of $\mathrm{DO}(X)^G$ can be derived from another point of view, due to Gelfand and Selberg. Cf. [Gel 50] and Chapter VII, Theorem 1.8.

In Chapter III, Theorem 5.3, we shall determine the image of $\mathrm{IDO}(G)^K$ in $\mathrm{IDO}(A)$ in the classical case of $\mathrm{GL}_n(\mathbf{R})$ or $\mathrm{SL}_n(\mathbf{R})$.

II, §5. THE HARISH-CHANDRA TRANSFORMS

We let G be a unimodular Lie group with a weak Iwasawa decomposition

$$G = UAK.$$

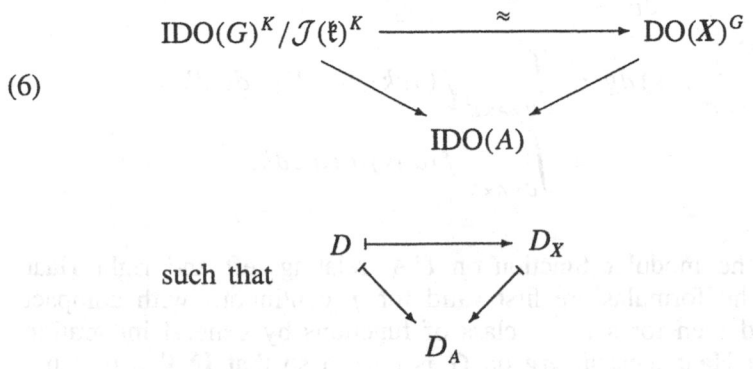

We let δ be the modular function on UA, and *we assume K compact*, with total Haar measure 1. Then we have the integral formulas (by definition):

INT 1.
$$\int_U f(ua)\,du = \delta(a)\int_U f(au)\,du.$$

INT 2.
$$\int_G f(g)\,dg = \int_{U\times A\times K} f(uak)\delta(a)^{-1}\,du\,da\,dk = \int_{U\times A\times K} f(auk)\,du\,da\,dk.$$

Thus δ is the modular function on UA, relating left and right Haar measure. The formulas are first valid for f continuous with compact support, and then for a larger class of functions by general integration theory. The Haar measure dg on G is chosen so that **INT 2** is valid. The function $\delta: A \to \mathbf{C}^*$ is a multiplicative continuous homomorphism on which we elaborate later. It was computed explicitly on $GL_n(\mathbf{R})$ in Chapter I, Propositions 1.1 and 1.2. We also call δ the **Iwasawa character**.

The main point of this section is to introduce the Harish-Chandra twist (or conjugation) on the direct image, where the conjugation is taken with $\delta^{1/2}$. Let:

$C_c(G, K) = C_c(G)^{c(K)}$ = space of continuous functions with compact support on G, invariant under conjugation by K, i.e. satisfying

$$f(k^{-1}gk) = f(g) \qquad \text{for all } k \in K \text{ and } g \in G.$$

For $f \in C_c(G, K)$ we define the **Harish transform** $\mathbf{H}f$ by the integral

$$\mathbf{H}f(a) = \delta(a)^{1/2}\int_U f(au)\,du.$$

Then $\mathbf{H}f$ is immediately seen to be in $C_c(A)$, so $\mathbf{H}$ is a linear map

$$\mathbf{H}: C_c(G, K) \to C_c(A).$$

Note. The Harish transform was originally defined by Harish-Chandra as an orbital integral. The equality with the above integral is given by Harish-Chandra [Har 58a], Lemma 17, p. 261. See Chapter III, Theorem 4.6 and the remarks following Theorem 4.8.

We let:

$C_c^\infty(K\backslash G/K) = C_c^\infty(G)^{K,K}$ = subspace of $C_c^\infty(G)$ consisting of those functions which are K-invariant on the left and on the right.

We shall consider the Harish transform on $C_c^\infty(K\backslash G/K)$, in which case we can differentiate under the integral sign, and $\mathbf{H}f$ is in $C_c^\infty(A)$.

We note that the conjugation map with the square root of the Iwasawa character

$$D \mapsto \delta^{-1/2} D \circ \delta^{1/2}$$

is an automorphism of $\mathbb{D}\mathrm{O}(A)$, *as* $\mathbf{R}$*-algebra.*

This would also be true for conjugation with any Lie group homomorphism (**character**)

$$\chi : A \to \mathbf{R}^+.$$

The invariance of $\chi^{-1} D \circ \chi$ is immediate from the multiplicativity $\chi(ba) = \chi(b)\chi(a)$. More structure will be given about δ as a character in the next chapter.

For $D \in \mathbb{D}\mathrm{O}(G)^K$ we define the **Harish-Chandra image** $\mathbf{h}_G(D)$ as the conjugation

$$\mathbf{h}(D) = \mathbf{h}_{\mathrm{Iw}}(D) = \mathbf{h}_G(D) = \delta^{-1/2} D_A \circ \delta^{1/2} = \delta^{-1/2}(\mathrm{Iw}_A)_*(D) \circ \delta^{1/2}.$$

Thus by Theorem 3.6,

$$\mathbf{h}_G : \mathbb{D}\mathrm{O}(G)^K \to \mathbb{D}\mathrm{O}(A)$$

is an algebra homomorphism with kernel $\mathcal{J}(\mathfrak{n}, \mathfrak{k})^K$. We call $\mathbf{h}_G$ the **Harish-Chandra homomorphism on the differential operators.**

Remark 1. For quite a while we shall deal only with $\mathbf{h}$ as defined above, but in the total picture, it is really the Harish-Chandra image associated with the Iwasawa decomposition. In Chapter VI and its follow up, we shall deal with a conjugation of the polar direct image, thus getting another Harish-Chandra image associated with the polar decomposition, but the conjugation will be taken with $\delta^{-1/2}$ on the right. We shall compare these two images in Chapter VI, Theorem 5.5, following Harish-Chandra. A major example will be given in Chapter VII.

Remark 2. *If the weak Iwasawa decomposition is also accompanied by a Cartan Lie decomposition, then the above maps can be defined on* *the homogeneous space $X = G/K$. We have an isomorphism given by restriction*

$$C_c^\infty(G)^{K,K} = C_c^\infty(K\backslash G/K) \xrightarrow{\approx} C_c^\infty(K\backslash X) = C_c^\infty(X)^K.$$

Note that it is essential here to view $C_c^\infty(K\backslash G/K)$ as K-bi-invariant functions on G as reflected by the notation on the left, because the quotient space

$$K\backslash G/K = K\backslash X$$

usually has singularities, and so the notion of differentiable functions on this space is not a priori defined. With this warning, the notation $C_c^\infty(K\backslash G/K)$ should not lead into trouble. Some serious questions do arise about the possible meaning of C^∞ functions on $K\backslash G/K$, and they will be considered for their own sake in Chapter VI, §2.

The Harish transform $\mathbf{H}$ may also be viewed as a linear map

$$\mathbf{H}_X : C_c^\infty(K\backslash X) \to C_c^\infty(A),$$

defined by the same formula as $\mathbf{H}_G$. Indeed, a function on X is "the same" as a function on G which is right K-invariant. Hence for such a function f, we may define

$$(\mathbf{H}_X f)(a) = \delta^{1/2}(a) \int_U f(au)\, du,$$

using the Iwasawa decomposition $A \times U \xrightarrow{\approx} G/K$ via $(a, u) \mapsto au$. If f_G denotes the function on G defined by $f_G(xk) = f(x)$, then

$$\mathbf{H}_G f_G = \mathbf{H}_X f.$$

Similarly, the Harish-Chandra mapping on invariant differential operators is also defined on X, namely we have

$$\mathbf{h}_X : \mathrm{DO}(X)^G \to \mathrm{IDO}(A)$$

defined by the same formula as $\mathbf{h}_G$, and $\mathbf{h}_X$ *is an injective algebra homomorphism.*

The next results will be stated on the group, under the weak Iwasawa decomposition assumptions, but if there is a Cartan decomposition then they hold on X as per the above isomorphisms.

Let $M = C_c^\infty(K\backslash G/K)$. Then M is a module over the algebra $\mathcal{D} = \mathrm{IDO}(G)^K$. Let $M' = C_c^\infty(A)$ and $\mathcal{D}' = \mathrm{IDO}(A)$. We shall now see that

$$(\mathbf{H}, \mathbf{h}) : (M, \mathcal{D}) \to (M', \mathcal{D}')$$

is a homomorphism of modules over the respective algebras. A prototype of the following result first occurred in Harish-Chandra [Har 57], Theorem 3.

Proposition 5.1. *Let G be a Lie group with a weak Iwasawa decomposition. For $D \in \mathbb{D}\mathrm{O}(G)^K$ on $C_c^\infty(K\backslash G/K)$, we have*

$$\mathbf{H} \circ D = \mathbf{h}(D) \circ \mathbf{H}.$$

Proof (Helgason [Hel 64], see also [Hel 84], Chapter II, Lemma 5.21, (39).) Let $f \in C_c^\infty(K\backslash G/K)$ be a K-bi-invariant function on G. Define the function F on G by

$$F(g) = \delta_G^{-1/2}(g) \int_U f(ug)\, du,$$

where δ_G is the Iwasawa lift of δ to G, so U-invariant on the left and K-invariant on the right. So $\delta_G = \delta \circ \mathrm{Iw}_A$, where $\mathrm{Iw}_A\colon G \to A$ is the Iwasawa projection on A. Then

$$(*) \qquad D(\delta_G^{1/2}F)(g) = D_g \int_U f(ug)\, du = \int_U (Df)(ug) du.$$

The function $\delta_G^{1/2}F$ is left U-invariant, right K-invariant, and we have

$$(**) \qquad F(a) = \mathbf{H}f(a) \qquad \text{for } a \in A.$$

(Note the order of the variables $f(ua)$ in the integral for $F(a)$, so that reversing this order introduces a factor $\delta(a)$, and thus yields the equality as stated.) Applying the definition of the direct image to $D(\delta_G^{1/2}F)$, we get

$$\begin{aligned}(\mathbf{H}Df)(a) = \delta(a)^{1/2} \int_U (Df)(au)\, du &= \delta_G^{-1/2}(a) \int_U (Df)(ua)\, du \\ &= \delta_G^{-1/2} D_G(\delta_G^{1/2}F)(a) \qquad \text{by } (*) \\ &= \delta_A^{-1/2} D_A(\delta_A^{1/2}\mathbf{H}f)(a) \quad \text{by } (**) \\ &= (\mathbf{h}(D)\mathbf{H}f)(a),\end{aligned}$$

which concludes the proof.

Remarks. As we shall see in the next chapter, there exists another expression for the Harish transform $\mathbf{H}$, in terms of an "orbital" integral on $A\backslash G$. Harish-Chandra originally proved the formula of Proposition 5.1 in terms of this other expression. Helgason gave the much shorter direct proof in terms of the U-integral. Of course, the orbital integral expression is defined only on regular elements, whereas the U-integral is everywhere defined, but certain identities on regular elements extend by continuity to all elements, as in the Harish-Chandra W-invariance proof reproduced in Theorem 4.8 of Chapter III. Cf. the Remarks on Terminology at the end of Chapter III, §4.

Both Harish-Chandra and Helgason work in the specific context of semisimple Lie groups. We have axiomatized a number of results which are thereby multivalent. We place ourselves in the general situation of a Lie group with a weak Iwasawa decomposition, no matter where it comes from.

Harish-Chandra originally proved his intertwining formula for differential operators in the center of IDO(G), so IDO$(G)^G$. Helgason worked with IDO$(G)^K$, which may be a different algebra in some exceptional cases, see for instance [Hel 64], Proposition 7.4 and Theorem 7.5, which shows however the equality in the classical cases. In any case, Harish-Chandra's formula served as a prototype, setting up the pervasive structure of the conjugation by the square root of the Iwasawa character. Chevalley's theorem constituted an important background.

Before giving more formulas about the Harish transform, we make some general remarks about K-invariant projections.

Let G be a locally compact unimodular group with compact subgroup K having Haar measure 1. Given a function f on G, we have its two K-**projections** f^K and ${}^K f$ on K-right invariant and K-left-invariant functions, by means of the corresponding integral over K. Thus

$$f^K(x) = \int_K f(xk)\,dk \qquad \text{and} \qquad {}^K f(x) = \int_K f(kx)\,dk.$$

We let ${}^K f^K$ be the corresponding K-bi-invariant function, obtained by composing the two integral operators.

If f is K-right-invariant and φ is any function, then

$$\int_G f(x)\varphi(x)\,dx = \int_G f(x)\varphi^K(x)\,dx,$$

because

$$\begin{aligned}\int_G f(x)\varphi(x)\,dx &= \int_K\int_G f(x)\varphi(x)\,dx\,dk \\ &= \int_K\int_G f(x)\varphi(xk)\,dx\,dk \\ &= \int_G f(x)\varphi^K(x)\,dx.\end{aligned}$$

A similar statement holds on the left. In particular:

If f is K-bi-invariant, then

$$\int_G f(x)\varphi(x)\,dx = \int_G f(x)\,{}^K\varphi^K(x)\,dx.$$

These relationships hold whenever the integrals are absolutely convergent, e.g. if f is continuous and φ has compact support, or if f is in L^1 and φ is bounded. In practice, such convergence will always be clearly satisfied. The theory to be developed is not delicate from this point of view.

We are now ready to describe the transpose of the Harish transform. We deal with the Haar measure on $G = UAK$ having a weak Iwasawa decomposition, and we shall use the integral formula **INT 2**. We have symmetric scalar products on G and on A given by the integral. For instance, for φ, ψ functions on A, we define

$$[\varphi, \psi]_A = \int_A \varphi(a)\psi(a)\,da$$

and for functions f, h on G,

$$[f, h]_G = \int_G f(x)h(x)\,dx.$$

The **transpose** of an operator

$$H\colon \text{functions on } G \to \text{functions on } A$$

is an operator

$${}^tH\colon \text{functions on } A \to \text{functions on } G,$$

and is defined by the formula

$$[\varphi, Hf]_A = [{}^tH\varphi, f]_G.$$

We shall of course specify which spaces are involved. As usual in integration theory, one starts with C_c, and then one extends to cases when the integral is absolutely convergent.

Proposition 5.2. *Let $G = UAK$ have a weak Iwasawa decomposition (K compact), and view the Harish transform as a linear map*

$$\mathbf{H}\colon C_c(K\backslash G/K) \to C_c(A) \subset C(A).$$

Then its transpose

$${}^t\mathbf{H}\colon C(A) \to C(K\backslash G/K)$$

on the spaces of continuous functions is the left K-projection of the twisted Iwasawa lifting, that is, for $\varphi \in C(A)$, $\varphi_G(auk) = \varphi_G(uak) = \varphi(a)$, we have

$$ {}^t\mathbf{H}\varphi = {}^K(\varphi\delta^{1/2})_G. $$

Proof. Routinely from the definitions, we have absolutely convergent integrals

$$
\begin{aligned}
&[{}^K(\varphi\delta^{1/2})_G, f]_G \\
&= \int_G\int_K \varphi_G(kx)\delta_G^{1/2}(kx)f(x)\,dk\,dx \qquad (\text{for } f \in C_c(K\backslash G/K)) \\
&= \int_K\int_G \varphi_G(x)\delta_G^{1/2}(x)f(k^{-1}x)\,dx\,dk \qquad (\text{by } x \mapsto k^{-1}x) \\
&= \int_K\int_G \varphi_G(x)\delta_G^{1/2}(x)f(x)\,dx\,dk \qquad (\text{by } K\text{-invariance of } f) \\
&= \int_A\int_U \varphi(a)\delta^{1/2}(a)f(au)\,da\,du \qquad (\text{by } dx = da\,du) \\
&= [\varphi, \mathbf{H}f]_A \qquad (\text{by definition})
\end{aligned}
$$

thereby concluding the proof.

Note that so far, we have said nothing about the injectivity or surjectivity of the Harish transform.

II, §6. THE TRANSPOSE AND INVOLUTION

Let M be a manifold with a volume form, and D a differential operator. As usual, we can deal with the hermitian integral scalar product, or the bilinear symmetric integral scalar product, given by the integral without the extra complex conjugate, with respect to the volume $d\mu$. We let D^* be the **adjoint** of D with respect to the hermitian product, and ${}^tD$ the **transpose** of D with respect to the symmetric scalar product. Thus for C^∞ functions ψ_1, ψ_2 for which the following integrals are absolutely convergent, we have by definition

$$ \int_M (D\psi_1)\psi_2\,d\mu = \int_M \psi_1({}^tD\psi_2)\,d\mu. $$

We shall denote the symmetric scalar product by $[\psi_1, \psi_2]$, to distinguish it from the hermitian one $\langle\psi_1, \psi_2\rangle$. Then the transpose formula reads

$$ [D\psi_1, \psi_2] = [\psi_1, {}^tD\psi_2]. $$

The existence of the transpose in general is a simple routine matter. Let Ω be a volume form on a Riemannian manifold. In local coordinates $x_1, \ldots, x_N$ on a chart which in Euclidean space is a rectangle, say, we can write

$$\Omega(x) = \beta(x_1, \ldots, x_N)dx_1 \wedge \cdots \wedge dx_N.$$

We suppose the coordinates oriented so that the function β is positive. Let D be a monomial differential operator, so in terms of the coordinates

$$D = \gamma \partial_{j_1} \cdots \partial_{j_m},$$

where γ is a function, and ∂_j is the partial derivative with respect to the j-th variable. Then integrating over the manifold, if γ or φ or ψ has compact support in the chart, we can integrate by parts and the boundary terms will vanish, so we get

$$\begin{aligned}\int (D\varphi)\psi\Omega &= \int \gamma\psi\beta(\partial_{j_1} \cdots \partial_{j_m}\varphi)\, dx_1 \wedge \cdots \wedge dx_N \\ &= (-1)^m \int \partial_{j_1} \cdots \partial_{j_m}(\gamma\psi\beta)\varphi\, dx_1 \wedge \cdots \wedge dx_N \\ &= (-1)^m \int \frac{1}{\beta}\partial_{j_1} \cdots \partial_{j_m}(\gamma\beta\psi)\varphi\Omega.\end{aligned}$$

Thus we find

Proposition 6.1. *In local coordinates, suppose*

$$\Omega(x) = \beta(x_1, \ldots, x_N)\, dx_1 \wedge \cdots \wedge dx_N \quad \textit{and} \quad D = \gamma\partial_{j_1} \cdots \partial_{j_m}.$$

If γ, or φ, or ψ has compact support in the chart, then

$$^tD\psi = (-1)^m \frac{1}{\beta}\partial_{j_1} \cdots \partial_{j_m}(\gamma\beta\psi).$$

The formula holds under more general conditions, needed to insure both the absolute convergence of the integrals appearing in the above proof, and the vanishing of the boundary term in the integration by parts. Conditions will be given precisely in each concrete case that we consider.

An arbitrary differential operator is a sum of operators of the above type, so the transpose can be computed by means of the above formula.

Proposition 6.2. *Let G be a group acting on the manifold and leaving a measure $d\mu$ invariant. If D is a differential operator invariant under the action of G then its transpose ^tD is also invariant.*

Proof. Abstract nonsense about invariant scalar products. We consider the symmetric scalar product $[\varphi, \psi]$ given by the integral of the product of two functions φ, ψ. Using exponential notation for the action of the group, we note that $[\varphi^g, \psi^g] = [\varphi, \psi]$, i.e. the scalar product is invariant under the action of the group. Let D be a linear map on the vector space of functions, and $D^g = D$ for all $g \in G$. Then

$$[D\varphi, \psi] = [D^g\varphi, \psi] = [D\varphi^{g^{-1}}, \psi^{g^{-1}}] = [\varphi^{g^{-1}}, {}^tD\psi^{g^{-1}}] = [\varphi, ({}^tD)^g\psi].$$

This equality being true for all φ, ψ proves the proposition.

The formula for the transpose of the direct image will be given at the beginning of the next chapter, because as far as we know, it requires using an additional techniquc (characters). See Proposition 1.6 of Chapter III.

Next we relate the transpose to the Harish-Chandra linear isomorphism

$$\mathcal{D}\colon S(\mathfrak{g}) \to \mathrm{IDO}(G) \qquad \text{also denoted} \qquad \mathcal{D}(P) = \tilde{P},$$

for an arbitrary Lie group G. Harish called the following identity "clear" in [Har 58a], p. 267, without giving any argument.

Proposition 6.3. *Let G be a unimodular Lie group. Let $P \in S(\mathfrak{g})$ be a polynomial homogeneous of degree d. Then taking transpose with respect to Haar measure, we have on $C_c^\infty(G)$:*

$${}^t(\mathcal{D}(P)) = (-1)^d\mathcal{D}(P).$$

Or alternatively, if P is not necessarily homogeneous and P^- is defined by $P^-(X) = P(-X)$, then

$${}^t(\mathcal{D}(P)) = \mathcal{D}(P^-).$$

Proof. It suffices to prove the proposition when P is a monomial $v_1 \cdots v_d$ with $v_j \in \mathfrak{g}$. The Harish-Chandra map is then the symmetrization, and is a sum of terms $\mathcal{D}(v_{i_1}) \cdots \mathcal{D}(v_{i_d})$, taken over all permutations of the factors. Thus it suffices to prove the lemma when $P = v \in \mathfrak{g}$. In this case, we go back to the definition, whereby for any function f we have

$$(\mathcal{D}(v)f)(x) = \frac{d}{dt} f(x \cdot \exp(tv))\bigg|_{t=0}.$$

For any two functions φ, ψ we have $\mathcal{D}(v)(\varphi\psi) = \varphi\mathcal{D}(v)\psi + \psi\mathcal{D}(v)\varphi$, so from the definition of the transpose, one needs to prove only that for any function $f \in C_c^\infty(G)$ one has

$$\int_G \frac{d}{dt} f(x \cdot \exp(tv))\Big|_{t=0} \, d\mu(x) = 0.$$

We are indebted to Shah for the following direct argument avoiding Stokes' theorem. We take the derivative d/dt outside the integral sign, and then use the right invariance of Haar measure $d\mu(x)$ to cancel the factor $\exp(tv)$ in the original integral expression, so that this expression is constant with respect to t. Hence one gets the value 0 after differentiating with respect to t, as desired.

Remark. The arguments given in the proof of Theorem 6.3 are valid under weaker conditions than C_c^∞, since the only thing needed is absolute convergence of all the integrals involved. We deal with the matter systematically in Chapter X, Lemma 6.1.

The rest of this section is also derived from [Har 58a], p. 267.

Let G be a unimodular Lie group with an Iwasawa and Cartan Lie decomposition

$$G = UAK \qquad \text{and} \qquad \mathfrak{g} = \mathfrak{p} + \mathfrak{k}.$$

Let θ be an **involution** of G, i.e. a Lie group automorphism such that $\theta^2 = \mathrm{id}$ but $\theta \neq \mathrm{id}$. We say that θ is **Cartan Lie adapted** if

$$\theta_{\mathfrak{g}} = T_e\theta = \mathrm{id} \text{ on } \mathfrak{k} \qquad \text{and} \qquad T_e\theta = -\mathrm{id} \text{ on } \mathfrak{p}.$$

If we let $\mathbf{t}$ be defined by $\mathbf{t}x = \theta x^{-1}$, then $\mathfrak{p}$ (resp. $\mathfrak{k}$) consist of the symmetric (resp. skew symmetric) elements of $\mathfrak{g}$ with respect to $\mathbf{t}$.

Example. $G = \mathrm{SL}_n(\mathbf{R})$ and $\theta x = {}^t x^{-1}$. Then K = special real unitary group.

Let θ be Cartan Lie adapted. By the chain rule, we have

(1) $$\theta_{\mathfrak{g}} v = -\mathbf{t}v \qquad \text{for } v \in \mathfrak{g}.$$

We now want to see the effect θ_* on the corresponding differential operators.

Proposition 6.4. *Let G be a unimodular Lie group with an Iwasawa and Cartan Lie decomposition, and a Cartan Lie adapted involution. Let θ_* be the induced mapping (direct image) on invariant differential operators. Then:*

(2) $\theta_* \tilde{v} = (\theta_{\mathfrak{g}} v)^{\sim}$ *for all* $v \in \mathfrak{g}$.

(3) $\theta_*\tilde{v} = -\tilde{v}$ *for* $v \in \mathfrak{p}$.

(4) $\theta_*\tilde{w} = \tilde{w}$ *for* $w \in \mathfrak{k}$.

(5) $\theta_*\tilde{P} = (-1)^d\tilde{P}$ *for* $P \in S(\mathfrak{p})$ *homogeneous of degree* d.

Thus θ_* *preserves the direct sum decomposition,*

$$S(\mathfrak{g})^{\sim} = S(\mathfrak{p})^{\sim} \oplus \mathrm{IDO}(G)\tilde{\mathfrak{k}},$$

and for $P \in S(\mathfrak{p})$ *from* (5) *we get*

$$\theta_*\mathcal{D}_G(P) = \mathcal{D}_G(\theta_{S(\mathfrak{g})}P).$$

Proof. Note that (2) is a repetition of Proposition 1.5, which is being made more explicit in terms of ± 1-eigenspaces. Now we prove (3).

By definition (functoriality), we know that $(\theta_*\tilde{v})f(\theta x) = (\tilde{v}(f \circ \theta))(x)$. Then

$$\begin{aligned}
(\tilde{v}(f \circ \theta))(x) &= \frac{d}{dt}(f \circ \theta)(x \cdot \exp(tv))\Big|_{t=0} \\
&= \frac{d}{dt}f(\theta x \cdot \theta(\exp tv))\Big|_{t=0} \\
&= \frac{d}{dt}f(\theta x \cdot \exp(t\theta_{\mathfrak{g}}v))\Big|_{t=0} \\
&= \frac{d}{dt}f(\theta x \cdot (-tv))\Big|_{t=0} \qquad \text{[by the Lie adaptation]} \\
&= -(\tilde{v}f)(\theta x),
\end{aligned}$$

which proves (3). Then (4) is proved similarly, using $\theta_{\mathfrak{g}}w = w$ for $w \in \mathfrak{k}$. As to (5), the automorphism θ induces an algebra automorphism on the algebra of invariant differential operators. If P is a homogeneous polynomial, then $\mathcal{D}(P) = \tilde{P}$ as a symmetrization is a sum of monomials of the given degree d, and hence (5) follows from (3). This concludes the proof.

Proposition 6.5. *Assumptions being as in Proposition* 6.4, *we have*

$$\theta_*D = {}^tD \qquad \text{for } D \in \mathcal{D}(S(\mathfrak{p})) = S(\mathfrak{p})^{\sim}.$$

Proof. Apply Proposition 6.3 to Proposition 6.4(5).

We defined the notion of being Lie adapted to make some logical dependencies clear at the beginning. Using the exponential map, and the chain rule, one sees that on the connected component K^0 of K, we have the global condition that θ is the identity. It is customary then to make the global assumption that θ is the identity on all of K. Since A is connected, the Lie condition that $T_e\theta = -\text{id}$ on $\mathfrak{a}$ is equivalent with $\theta(a) = a^{-1}$ for all $a \in A$, so we don't have to assume anything more to get the desired effect on A from the effect on the Lie algebra. We shall now deduce consequences from these global properties.

Assume that θ leaves K elementwise fixed. Then θ induces an involution on the homogeneous space $X = G/K$ by the formula

$$\theta_{G/K}(xK) = \theta(x)K.$$

Sometimes, we may abbreviate the notation and write simply θ instead of $\theta_{G/K}$ or θ_X. By the commutative diagrams of §2, and Theorem 2.3, we get:

Corollary 6.6. *Assume globally that θ leaves K elementwise fixed. Notation and assumptions being as in Proposition* 6.4, *for*

$$D \in \mathrm{DO}(G/K)^G$$

and K compact, we have

$$(\theta_{G/K})_*(D) = {}^tD,$$

where the transpose is taken with respect to Haar measure on G/K.

As in §5, we have the Harish-Chandra homomorphism

$$\mathbf{h}\colon \mathbb{D}\mathrm{O}(G)^K \to \mathrm{DO}(G/K)^G \hookrightarrow \mathbb{D}\mathrm{O}(A).$$

In Chapter III, Theorem 1.6, we shall prove that for $D \in \mathbb{D}\mathrm{O}(G)^K$, we have

$$\mathbf{h}({}^tD) = {}^t\mathbf{h}(D). \tag{6}$$

Here we use this result to prove a more complete tabulation of commutation rules for θ. The tabulation won't be used until Chapter VI.

Proposition 6.7. *Let $G = UAK$ be an Iwasawa decomposition, together with a Cartan Lie decomposition, with K compact, and a Cartan Lie adapted involution such that $\theta = \text{id}$ on K. Then for $D \in \mathbb{D}\mathrm{O}(G)^K$, we have*

$$\mathbf{h}(\theta_*D) = (\theta_A)_*\mathbf{h}(D), \tag{7}$$

$$(\theta_A)_*(\mathrm{Iw}_A)_*(D) = (\mathrm{Iw}_A)_*(\theta_*D), \tag{8}$$

$$(\theta_A)_*(\mathrm{Iw}_A)_*(D) = \delta^{-1/2}\mathbf{h}(D)\circ\delta^{1/2} = \delta^{-1}(\mathrm{Iw}_A)_*(D)\circ\delta. \tag{9}$$

Proof. Recall that $\mathcal{J}(\mathfrak{k}) = \mathrm{IDO}(G)\tilde{\mathfrak{k}}$ is in the kernel of $\mathbf{h}$ and $(\mathrm{Iw}_A)_*$, so both formulas need to be proved only when $D \in \mathcal{D}(S(\mathfrak{p}))$. Then by Proposition 6.5 and (6),

$$\mathbf{h}(\theta_* D) = \mathbf{h}({}^t D) = {}^t\mathbf{h}(D) = (\theta_A)_*\mathbf{h}(D),$$

which proves (7). For (8), by linearity and Theorem 3.3 we can assume $D = \mathcal{D}(P)$ with P homogeneous in $S(\mathfrak{a})$, so (8) comes from Proposition 6.4(5). Finally for (9), we take the transpose in two ways. First by reversing the order in an ordinary way, and then using Proposition 6.5 on A, we get

$${}^t(\mathbf{h}(\theta_* D)) = \delta^{1/2}\,{}^t((\mathrm{Iw}_A)_* D) \circ \delta^{-1/2} = \delta^{1/2}((\theta_A)_*(\mathrm{Iw}_A)_* D) \circ \delta^{-1/2}.$$

Second, applying Proposition 6.5 as such,

$${}^t(\mathbf{h}(\theta_* D)) = \mathbf{h}({}^t(\theta_* D)) = \mathbf{h}(D).$$

This proves (9) and concludes the proof of Proposition 6.7.

Proposition 6.8. *Let $D \in \mathrm{IDO}(G)$. There is a unique element $Q \in S(\mathfrak{a})$ such that*

$$D \equiv \mathcal{D}_G(Q) \quad \bmod \theta_*(\tilde{\mathfrak{n}})\,\mathrm{IDO}(G) + \mathrm{IDO}(G)\tilde{\mathfrak{k}}.$$

Furthermore,

$$\mathcal{D}_A(Q) = \delta^{-1/2}\mathbf{h}(D) \circ \delta^{1/2}.$$

Proof. The existence and uniqueness are immediate since θ_* is an automorphism of $\mathrm{IDO}(G)$ of order 2, and we proved the corresponding statement in Theorem 3.3 with $\tilde{\mathfrak{n}}$ instead of $\theta_*(\tilde{\mathfrak{n}})$. It suffices to prove the second formula when $D \in S(\mathfrak{p})^{\sim}$, because of the direct sum decomposition

$$\mathcal{D}(S(\mathfrak{g})) = \mathcal{D}(S(\mathfrak{p})) \dotplus \mathrm{IDO}(G)\tilde{\mathfrak{k}},$$

and the fact that $\mathrm{IDO}(G)\tilde{\mathfrak{k}}$ is in the kernel of $\mathbf{h}$. Now applying θ_* to the first relation, we see that

$$\theta_*\mathcal{D}_A(Q) = (\mathrm{Iw}_A)_*(\theta_* D).$$

Therefore

$$\delta^{-1/2}\theta_*\mathcal{D}_A(Q) \circ \delta^{1/2} = \mathbf{h}(\theta_* D).$$

The transpose then gives

$$\begin{aligned}\delta^{1/2}\,{}^t(\theta_*\mathcal{D}_A(Q))\circ\delta^{-1/2} &= {}^t\mathbf{h}(\theta_*D)\\ &= (\theta_A)_*\mathbf{h}(\theta_*D) \quad \text{by Proposition 6.5}\\ &= \mathbf{h}(D) \qquad\qquad \text{by Proposition 6.7(7).}\end{aligned}$$

Proposition 6.5 also tells us that ${}^t(\theta_*\mathcal{D}_A(Q)) = \mathcal{D}_A(Q)$. This concludes the proof.

It may at some point be convenient to let

$$(\mathrm{Iw}_A^\theta)_*(D) = \mathcal{D}_A(Q),$$

where Q is the polynomial in the proposition. The second formula in Proposition 6.8 then reads

$$(\mathrm{Iw}_A^\theta)_*(D) = \delta^{-1/2}\mathbf{h}(D)\circ\delta^{1/2}.$$

In applications, although UAK is usually the ordinary Iwasawa decomposition, with U being the upper triangular unipotents, the above results apply also to the case when U is the subgroup of lower triangular unipotents. This symmetry does not play a role in many instances, but when it does, it is significant and does not just provide a mirror image of what has already been done. It goes beyond. For an application, see Chapter VI, Proposition 5.5.

CHAPTER III

Characters, Eigenfunctions, Spherical Kernel and W-Invariance

To go further in describing the invariant differential operators, we have to introduce characters and eigenfunctions. We deal with three types of characters:

- characters on A, that is continuous homomorphism $A \to \mathbf{C}^*$;
- characters on $\mathfrak{a}$, so functionals $\mathfrak{a} \to \mathbf{R}$ or $\mathfrak{a}_{\mathbf{C}} \to \mathbf{C}$;
- algebra characters, namely $\mathbf{C}$-characters of the algebra of invariant differential operators, that is algebra homomorphisms into $\mathbf{C}$.

The latter arise from eigenfunctions for such algebras. Our first concern is to complete Theorem 3.8 of Chapter II by determining the image of the direct image of invariant differential operators to A. The answer in Theorem 4.3 is most easily formulated by using a twist, or more precisely a conjugation by a certain character, due to Harish-Chandra. After this conjugation, the image is precisely the algebra $\mathrm{IDO}(A)^W$ of invariant differential operators on A which are also invariant under the so-called Weyl group W. In our case, W is the group of permutations of the variables. Except for §8, this chapter is again essentially due to Harish-Chandra [Har 58].

III, §1. CHARACTERS

There are two conventions about multiplicative characters, which in the most general setting are continuous homomorphisms of an abelian group into either the group of complex numbers of absolute value 1, or the multiplicative group of complex numbers. In one terminology, the former are called **characters** and the latter are called **quasi characters**. In this book, we adopt another convention whereby the former are called **unitary characters**, and the latter are called simply **characters**. This is done because we are concerned almost entirely with the latter type, so it shortens language to omit the word "quasi."

Characters of $\mathbf{R}^+$ into $\mathbf{C}^*$ are all of the form

$$a \mapsto a^s$$

with some complex number s. Given an isomorphism $A \xrightarrow{\approx} \mathbf{R}^+ \times \cdots \times \mathbf{R}^+$, characters are of the form

$$a = (a_1, \ldots, a_r) \mapsto a_1^{s_1} \cdots a_r^{s_r},$$

with some r-tuple $(s_1, \ldots, s_r)$ of complex numbers. We write such a character as χ_s, and we use vector notation, so by definition

$$\chi_s(a) = \chi_{s_1,\ldots,s_r}(a_1, \ldots, a_r) = a_1^{s_1} \cdots a_r^{s_r}.$$

By Proposition 4.1 of Chapter IV, the algebra of invariant differential operators on A is generated by the operators

$$\mathcal{D}_i = a_i \frac{\partial}{\partial a_i} = \mathcal{D}(E_{ii}).$$

We note that χ_s is an eigenfunction of $\mathcal{D}_i$, namely

$$\mathcal{D}_i \chi_s = s_i \chi_s,$$

so if $D = P(\mathcal{D}_1, \ldots, \mathcal{D}_r)$, then χ_s is an eigenfunction of D with eigenvalue

$$\mathrm{ev}(D, \chi_s) = \mathrm{ev}_s(D) = P(s_1, \ldots, s_r) = P(s).$$

In particular, we have the easy result on A: *If $D \in \mathrm{IDO}(A)$ and $D\chi = 0$ for all characters χ, in other words $\mathrm{ev}(D, \chi) = 0$ for all χ, then $D = 0$.* However, we shall need a stronger version as follows:

Theorem 1.1. *Let $D \in \mathrm{IDO}(A)$. Let Z be a subset of $\mathbf{C}^r$ such that if P is a polynomial and $P(s) = 0$ for all $s \in Z$, then P is identically zero. If $D\chi_s = 0$ for all $s \in Z$, then $D = 0$.*

A set Z as in Theorem 1.1 is said to be **Zariski dense**. We shall see the need for the stronger version of Theorem 1.1 in Chapter IV, where Z can be taken to be a half space. This stronger version will propagate via Theorem 1.3.

The parametrization of characters by the variables $s_1, \ldots, s_r$ depends on a choice of basis, and there are many such choices. Different bases will play different roles in the sequel. See for instance Chapter V, §7. In more invariant terms, let

$$\zeta : \mathfrak{a} \to \mathbf{C} \qquad \text{so } \zeta \in \mathfrak{a}_{\mathbf{C}}^{\vee}$$

be a linear functional, complex valued. Then ζ gives rise to the character defined by

$$\chi_{\zeta}(a) = a^{\zeta}.$$

If the functional ζ has coordinates $(s_1, \ldots, s_r)$ under the isomorphism

$$A \xrightarrow{\approx} \mathbf{R}^+ \times \cdots \times \mathbf{R}^+,$$

then we write also

$$\chi_{\zeta}(a) = \chi_s(a) = a_1^{s_1} \cdots a_r^{s_r}.$$

In later work, we pick ζ pure imaginary, that is, $\zeta = \mathbf{i}\lambda$ with $\lambda \in \mathfrak{a}^{\vee}$.
We now continue the study of direct images.

We let G and $\mathfrak{g}$ have an Iwasawa and Cartan decomposition as defined in Chapter I, §1 and §3. We suppose X is a homogeneous space for G with isotropy group K from these decompositions. We abbreviate

$$(\mathrm{Iw}_A)_*(D) = D_A.$$

By a **character** on X we shall simply mean a function which is the pull-back of a character χ on A under the Iwasawa projection. We denote this pull-back to X by χ_X, and similarly we let χ_G be the pull-back of χ to G. Thus

$$\chi_G(px) = \chi(p)\chi_G(x) \qquad \text{for } p \in P.$$

Such characters were introduced by Selberg [Sel 56], p. 55. We have the formulas

$$\text{(1a)} \qquad (D\chi_G)(a) = (D_A\chi)(a) \qquad \text{for } D \in \mathrm{IDO}(G)^K \text{ and } a \in A;$$

and

$$\text{(1b)} \qquad (D\chi_X)([a]I) = (D_A\chi)(a).$$

These just express the definition of the direct image as in formula (1) of Chapter II, §3.

We recall here a remark made when we introduced the Iwasawa decomposition. The remark was made for (U, K)-invariant functions, and applies to the extensions χ_G or χ_X of a character. Thus for $D \in \mathrm{IDO}(G)^K$, $D\chi_G$ is also (U, K)-invariant, and $D\chi_X$ is U-invariant. Thus the values given in (1a), (1b) above apply to the functions $D\chi_G$ and $D\chi_X$ viewed as functions on G and X respectively.

Let $\mathcal{A}$ be an algebra of operators on some function space. A function f is called an **eigenfunction** of the algebra if for each operator D there is a number $\mathrm{ev}(D, f)$ such that $Df = \mathrm{ev}(D, f)f$. Then $\mathrm{ev}(D, f)$ is called the **eigenvalue** of D on f. The map $D \mapsto \mathrm{ev}(D, f)$ is called the **eigencharacter** of f, and is an algebra homomorphism which we also denote

$$\mathrm{ec}_f = \mathrm{ev}_f : \mathcal{A} \to \mathbf{C}.$$

In this chapter, $\mathcal{A}$ is mostly $\mathrm{DO}(X)^G$ and $\mathrm{IDO}(G)^K$.

Theorem 1.2. *Characters on X are eigenfunctions of* $\mathrm{DO}(X)^G$, *with the same eigenvalues as their direct images on A. That is, with the above notation:*
For $D \in \mathrm{DO}(X)^G$, $D\chi_X = \mathrm{ev}(D_A, \chi)\chi_X$.
For $D \in \mathrm{IDO}(G)^K$, $D\chi_G = \mathrm{ev}(D_A, \chi)\chi_G$.

Proof. Immediate from Theorem 1.1 and (1a), (1b).

The eigenfunction property of characters can be seen from a very different viewpoint, cf. Proposition 2.1 of Chapter IV.

Next we see that invariant differential operators on X are characterized by their effect on (multiplicative) characters.

Theorem 1.3. *Assume G has an Iwasawa and Cartan decomposition. Let* $D \in \mathrm{DO}(X)^G$. *If* $D\chi_X = 0$ *for all characters* χ_X *on* X, *then* $D = 0$. *In fact, if there exists a Zariski dense set Z in* $\mathbf{C}^r$ *such that* $D\chi_{s,X} = 0$ *for all* $s \in Z$, *then* $D = 0$.

Proof. For $\chi = \chi_s$, and all $a \in A$, we have

$$0 = (D\chi_X)(a) = (D_A\chi)(a),$$

so $D_A = 0$ by Theorem 1.1. The proof is concluded by Theorem 4.2 of Chapter II.

Remark 1. Theorem 1.3, in various degrees of generality, is due to Selberg and Harish-Chandra, in various contexts. Maass [Maa 71] gives a direct proof on Pos_n, pp. 71–75, which actually does something somewhat more generally, but never mind this for the moment. As Maass says, he develops "a constructive method for the computation of the eigenvalues" of the Selberg–Maass generators for the algebra of invariant differential operators on $X = \text{Pos}$. The proof is instructive and short, but it is very ad hoc, and requires an induction and some explicit formulas, which prompted us to use the Lie theoretic generators. The determination of the kernels of the direct images going to X and A is a more systematic and transparent procedure.

Remark 2. Let h be a function on A, eigenfunction of $\text{IDO}(A)$. Then h is a character. Indeed, using the standard generators $\mathcal{D}_i = a_i\partial/\partial a_i$ and transferring to the additive groups by the exponential maps, it follows directly from calculus that h is a product of exponentials in the coordinates. But on X, the lack of surjectivity of the imbedding

$$\text{DO}(X)^G \hookrightarrow \text{IDO}(A)$$

does not allow for the same conclusion about eigenfunctions of the algebra $\text{DO}(X)^G$. Information on such eigenfunctions depends on knowing the image of $\text{DO}(X)^G$ in $\text{IDO}(A)$, to be determined in Theorem 5.3. The description of the eigenfunctions will then be given in §7. "Generically" (in a sense to be explained), the eigenfunctions are characters, but there are exceptional cases having to do with "ramification" when polynomial linear combinations of characters occur as eigenfunctions. See Theorem 7.8.

In the above comments, we wanted to emphasize the homogeneous space. However, a number of results do not need the Cartan decomposition, just an Iwasawa decomposition, to which we now revert for the next items. Readers by now should be able to formulate these results at once on G/K whenever the Cartan decomposition also exists.

Let h be a continuous function on A and let h_G be its Iwasawa lift to G, that is

$$h_G = h \circ \text{Iw}_A \qquad \text{so} \qquad h_G(uak) = h(a).$$

In the applications, h is a character χ. Then h_G is a function on G, U-invariant on the left and K-invariant on the right. *Assume that K is compact.* As in Chapter II, §5, we let the K-bi-invariant projection of h_G be the integral

$$(2) \qquad {}^K h_G(g) = \int_K h_G(kg)\,dk,$$

where dk is a Haar measure on K, which we normalize to have total measure 1 unless otherwise specified. Such a K-projection occurs in [Gel 50], [God 52], and [Sel 56]. Note that ${}^K h_G(e) = h(e)$. The following result essentially comes from these references:

Proposition 1.4. *Assume K compact. Let $h \in C^\infty(A)$ be an eigenfunction of $\mathrm{IDO}(A)$, for instance, a character on A. Then ${}^K h_G$ is an eigenfunction of $\mathrm{IDO}(G)^K$ with the same eigenvalues as h. In other words, for $D \in \mathrm{IDO}(G)^K$,*

$$D({}^K h_G) = \mathrm{ev}(D_A, h)\,{}^K h_G.$$

Proof. We differentiate the integral (2) by a differential operator D_g (with respect to the variable g), which we may bring under the integral sign since the domain of integration is compact and there is no convergence problem. By the invariance of D_g, for any function f on G, U-invariant on the left and K-invariant on the right, we have

$$D_g f(kg) = D(f \circ L_k)(g) = (Df)(kg).$$

Let $f = h_G$. Then $(Df)(kg) = \lambda f(kg)$, where $\lambda = \mathrm{ev}(D, h_G)$ is the eigenvalue of h_G for D, or h for D_A, same thing by formula (1) and Remark 2 of Chapter II, §3. We then integrate over K to conclude the proof.

We shall now use the Iwasawa character δ as in Chapter II, §5. We don't need to know more about it for now, but more information will be given in §2.

Let χ be a character on A, and χ_G its Iwasawa extension to G. Following Harish-Chandra we define the **Harish-Chandra spherical function** φ_χ to be the left K-projection of $(\chi\delta^{1/2})_G$, that is

$$\varphi(\chi, g) = \varphi_\chi(g) = \int_K (\chi\delta^{1/2})_G(kg)\,dk = \int_K (\chi\delta^{1/2})(kg)_A\,dk.$$

We call the function $(\chi, g) \mapsto \varphi(\chi, g)$ of two variables the **spherical kernel**. Like any kernel function, it gives rise to an integral operator called the **spherical transform**

$$\mathbf{S}\colon C_c(K\backslash G/K) \to \text{functions on the character group}$$

by the formula, for $f \in C_c(K\backslash G/K)$:

$$\mathbf{S}f(\chi) = \int_G \varphi(\chi, g) f(g)\,dg.$$

Remark. If we parametrize the characters by r complex variables as in §1, then $\chi = \chi_s$, and φ becomes a function $\varphi(s, g)$, so $\mathbf{S}f$ is viewed as a function of r complex variables, whose values we write $(\mathbf{S}f)(s)$. A lot of things are now happening simultaneously, and unfortunately they have to be projected in a totally ordered way on the page axis.

Cf. Chapter IV, §6 and §7, for the pursuit of the point of view of the present remark.

We note the trivial fact that the spherical kernel itself is K-bi-invariant. Indeed, the Iwasawa projection

$$kg \mapsto (kg)_A$$

is K-invariant on the right, and the integral over K causes K-invariance on the left. Thus if G is unimodular and if we convolve the spherical kernel with any function in $C_c(G)$, then we obtain a K-bi-invariant function. Furthermore, if $G = KAK$ as will be the case later, then a K-bi-invariant function can be viewed by restriction as a function on A. If $k \in K$ and conjugation by k leaves A stable, then the K-bi-invariance shows that

$$\varphi_\chi(a^k) = \varphi_\chi(a),$$

so $\varphi_\chi(a)$ is fixed under the action of such k. The group of such k will be discussed more extensively as the Weyl group in §3.

In Chapter II, §5, we defined the Harish-Chandra image $\mathbf{h}(D)$ for $D \in \mathrm{IDO}(G)^K$. We are now in a position to determine the eigenvalues of $\mathbf{h}(D)$ on characters.

Proposition 1.5 (Harish-Chandra [Har 58a]). *Let $G = UAK$ be an Iwasawa decomposition with K compact. For $D \in \mathrm{IDO}(G)^K$ and characters χ of A, we have*

$$\mathrm{ev}(D, \varphi_\chi) = \mathrm{ev}(D_A, \chi\delta^{1/2}) = \mathrm{ev}(\mathbf{h}(D), \chi).$$

Proof. By Proposition 1.4 applied to $\chi\delta^{1/2}$ we have

$$D\varphi_\chi = \mathrm{ev}(D_A, \chi\delta^{1/2})\varphi_\chi,$$

which is the first eigenvalue equality. Next, we have

$$\mathbf{h}(D)(\chi) = \delta^{-1/2} D_A \circ \delta^{1/2}(\chi) = \mathrm{ev}(D_A, \chi\delta^{1/2})\chi,$$

which gives the second equality of eigenvalues and concludes the proof.

Proposition 1.6. *Under the same conditions, let $D \in \mathrm{IDO}(G)^K$. Then*

$$\mathbf{h}({}^tD) = {}^t\mathbf{h}(D).$$

Proof. By Theorem 1.1, it suffices to show that $\mathbf{h}({}^tD)$ and ${}^t\mathbf{h}(D)$ have the same eigenvalues on characters. For every spherical function

φ_χ and all $f \in C_c^\infty(K\backslash G/K)$, we have

$$\begin{aligned}
[{}^tD\varphi_\chi, f]_G &= [\varphi_\chi, Df]_G \\
&= [\chi, \mathbf{H}Df]_A \quad \text{by Proposition 5.2 of Chapter II} \\
&= [\chi, \mathbf{h}(D)\mathbf{H}f]_A \quad \text{by Proposition 5.1 of Chapter II} \\
&= [{}^t\mathbf{h}(D)\chi, \mathbf{H}f]_A \quad \text{by definition of transpose} \\
&= \mathrm{ev}\,{}^t\mathbf{h}(D), \chi)[\chi, \mathbf{H}f]_A.
\end{aligned}$$

Since φ_χ is an eigenfunction of the invariant differential operators by Propositions 1.4 and 1.5, we also get by Proposition 5.2 of Chapter II,

$$[{}^tD\varphi_\chi, f]_G = \mathrm{ev}({}^tD, \varphi_\chi)[\varphi_\chi, f]_G = \mathrm{ev}({}^tD, \varphi_\chi)[\chi, \mathbf{H}f]_A.$$

Trivially, given χ there exists f such that $[\varphi_\chi, f]_G \neq 0$ so $[\chi, \mathbf{H}f]_A \neq 0$. Hence

$$\mathrm{ev}({}^tD, \varphi_\chi) = \mathrm{ev}({}^t\mathbf{h}(D), \chi).$$

By Proposition 1.5, the left side is equal to $\mathrm{ev}(\mathbf{h}({}^tD), \chi)$. Thus

$$\mathrm{ev}(\mathbf{h}({}^tD), \chi) = \mathrm{ev}({}^t\mathbf{h}(D), \chi) \qquad \text{for all } \chi.$$

Hence $\mathbf{h}({}^tD) = {}^t\mathbf{h}(D)$, thus concluding the proof of the proposition.

Finally we give an eigenvalue formula for the spherical transform. This gives us the opportunity to make a general comment about the transpose on G and on $X \approx G/K$. Let $[f, \varphi]_G$ denote the symmetric scalar product of functions on G with respect to Haar measure dg. There is a constant c such that $dg = c \cdot dk\, d\mu(x)$. If $f \in C_c(K\backslash G/K)$ and $\varphi \in C(K\backslash G/K)$, then we have

$$[f, \varphi]_G = c[f_X, \varphi_X]_X,$$

where f_X is the corresponding K-invariant function on X, so in $C_c(K\backslash X)$. If $D \in \mathrm{IDO}(G)^K \approx \mathrm{DO}(X)^G$, and $f \in C_c^\infty(K\backslash G/K) \approx C_c^\infty(K\backslash X)$, $\varphi \in C^\infty(K\backslash G/K)$, then

$$(3) \qquad [Df, \varphi]_G = c[D_X f_X, \varphi_X]_X = c[f_X, {}^tD_X\varphi_X]_X.$$

In practice, one often omits the subscript, and these formulas may be written

$$\int_G Df(g)\varphi(g)\,dg = c\int_X (Df)(x)\varphi(x)\,d\mu(x) = c\int_X f(x)\,{}^tD\varphi(x)\,d\mu(x).$$

Proposition 1.7. *Let $D \in \mathrm{IDO}(G)^K$. Then for $f \in C_c^\infty(K\backslash G/K)$, we have*

$$\mathbf{S}(Df)(\chi) = \mathrm{ev}({}^tD_X, \varphi_\chi)\mathbf{S}f(\chi).$$

Proof. We just use $\varphi = \varphi_\chi$ in (3), and the fact that φ_χ is an eigenfunction of ${}^tD_X$ (viewing φ_χ as a function on X). The stated formula drops out.

Remark. If one parametrizes the characters as usual by r complex variables $s = (s_1, \ldots, s_r)$, then $\mathrm{ev}({}^tD, \varphi_{\chi_s})$ is a polynomial in these variables. In fact, if $\chi = \chi_s$ and P is the polynomial such that $\mathbf{h}({}^tD) = \mathcal{D}(P)$,

$$\mathrm{ev}({}^tD, \varphi_{\chi_s}) = P(s). \tag{4}$$

III, §2. THE ($\mathfrak{a}, \mathfrak{n}$)-CHARACTERS AND THE IWASAWA CHARACTER

We shall tabulate some formalism on A independently of subsequent applications. This section is not needed for the rest of this chapter, but it throws light on the Iwasawa character.

We let $G = \mathrm{GL}_n(\mathbf{R})$ or $\mathrm{SL}_n(\mathbf{R})$, with its Iwasawa decomposition

$$G = UAK.$$

We have the corresponding decomposition of the Lie algebra $\mathfrak{g} = \mathrm{Mat}_n(\mathbf{R})$ or $\mathfrak{g} = \mathrm{Mat}_n(\mathbf{R})_0$ (the subscript zero indicating matrices with trace 0),

$$\mathfrak{g} = \mathfrak{n} + \mathfrak{a} + \mathfrak{k},$$

where $\mathfrak{n}$ is the space of strictly upper triangular (nilpotent) matrices, $\mathfrak{a}$ is the space of diagonal matrices, and $\mathfrak{k} = \mathrm{Sk}$ is the space of skew-symmetric matrices. Let $\{E_{ij}\}$ $(i < j)$ be the basis of $\mathfrak{n}$ such that E_{ij} is the matrix with ij-component 1, and every other component equal to 0. For $v \in \mathfrak{a}$, $v = \mathrm{diag}(x_1, \ldots, x_n)$, the regular representation of $\mathfrak{a}$ on $\mathfrak{g}$ yields

$$[v, E_{ij}] = (x_i - x_j)E_{ij}. \tag{1}$$

Thus the single element E_{ij} is a basis for the α_{ij}-eigenspace, where α_{ij} is the character on $\mathfrak{a}$ given by

$$\alpha_{ij}(v) = x_i - x_j.$$

These characters are all distinct for $i < j$, and so each element E_{ij} $(i < j)$ is the basis of the α_{ij}-eigenspace, which is one-dimensional. We then get a decomposition of $\mathfrak{n}$ as a direct sum of these eigenspaces for the regular representation of $\mathfrak{a}$. The characters α_{ij} $(i < j)$ will be called the **relevant characters**, or **($\mathfrak{a}, \mathfrak{n}$)-characters**. We usually omit the indexing by ij, and write simply α. The set of these characters will be denoted by $\mathcal{R}(\mathfrak{n})$. Note that $\mathcal{R}({}^t\mathfrak{n})$ consists of all the characters $\alpha_{ji} = -\alpha_{ij}$, as it were. In any case, omitting the double index, we write the direct sum decomposition as

$$\mathfrak{n} = \bigoplus_{\alpha \in \mathcal{R}(\mathfrak{n})} \mathfrak{n}_\alpha,$$

where $\mathfrak{n}_\alpha$ is the eigenspace with basis element E_α.

We then get the **trace** of the representation of $\mathfrak{a}$ on $\mathfrak{n}$, namely

$$\tau = \sum_{\alpha \in \mathcal{R}(\mathfrak{n})} \alpha = \text{sum of the relevant characters.}$$

We could define the **Iwasawa character** $\delta = \delta_{\mathrm{Iw}}$ on A by the formula

(2) $\quad \delta(a) = e^{\tau(\log a)} \quad$ or equivalently $\quad \tau(\log a) = \log \delta(a).$

Then explicitly,

$$\delta_{\mathrm{Iw}}(a) = \delta(a) = \prod_{i<j} a_i/a_j = \prod_{\alpha \in \mathcal{R}(\mathfrak{n})} a^\alpha = a^\tau.$$

Thus δ is the multiplicative version of the additive trace character in the dual space $\mathfrak{a}^\vee$. From Chapter I, §1, (3), we see that δ is the modular character. The eigenspace decomposition gives the Lie algebra way of deriving the measure computation of Chapter I, §2. This way will be used again in the next section for a further integral formula, so what we just did puts the reader in the right frame of mind for what follows, computing a Jacobian by the effect of the tangent map on the Lie algebra via its eigenspace decomposition.

The half trace will be especially important. Thus we let

$$\rho = \tfrac{1}{2}\tau = \tfrac{1}{2}\log\delta = \tfrac{1}{2}\sum_{\alpha \in \mathcal{R}(\mathfrak{n})} \alpha \qquad \text{so multiplicatively} \qquad a^\rho = \delta^{1/2}(a).$$

For $H = \mathrm{diag}(h_1, \ldots, h_n)$ we define

$$\lambda_i(H) = h_1 + \cdots + h_i.$$

Directly from the definitions, and from $\tau = \sum \alpha_{ij}$ $(i < j)$ (cf. Chapter I, §2, (3), (7)) we get

$$\boxed{\rho = \lambda_1 + \cdots + \lambda_r.}$$

In Chapter I, §4, we saw that $\{\lambda_1, \dots, \lambda_r\}$ is the dual basis of $\{\alpha_1, \dots, \alpha_r\}$, where $\alpha_i = \alpha_{i,i+1}$ $(i = 1, \dots, r)$.

Remark. The Iwasawa character can be defined in a wider context. Suppose a group has an Iwasawa decomposition as given in §2. We consider the regular representation of $\mathfrak{a}$ on $\mathfrak{n}$. Given $u \in \mathfrak{a}$, this representation is defined by

$$v \mapsto [u, v],$$

and in a linear representation as we have in practice, $[u, v] = uv - vu$. Let α be an additive character on $\mathfrak{a}$, that is, a functional into $\mathbf{R}$. We let $\mathfrak{n}_\alpha$ be the α-eigenspace, that is, the subspace of $\mathfrak{n}$ consisting of all v such that $[u, v] = \alpha(u)v$ for all $u \in \mathfrak{a}$. We say that the Iwasawa decomposition is **Lie semisimple** if it satisfies the condition:

LSS. The Lie algebra $\mathfrak{n} = \mathrm{Lie}(U)$ is a direct sum

$$\mathfrak{n} = \bigoplus_{\alpha \in \mathcal{R}(\mathfrak{n})} \mathfrak{n}_\alpha,$$

where the sum is taken over a finite set $\mathcal{R}(\mathfrak{n})$ of non-zero characters of $\mathfrak{a}$, and $\mathfrak{n}_\alpha \neq 0$.

The set $\mathcal{R}(\mathfrak{n})$ is called the set of **relevant** or **regular characters** (for the action of $\mathfrak{a}$ on $\mathfrak{n}$) or also the **$(\mathfrak{a}, \mathfrak{n})$-characters.**

Let $m(\alpha) = \dim \mathfrak{n}_\alpha$. Then we define the **$(\mathfrak{a}, \mathfrak{n})$-trace** to be

$$\tau = \sum_{\alpha \in \mathcal{R}(\mathfrak{n})} m(\alpha)\alpha \qquad \text{and let} \qquad \rho = \tau/2.$$

With this notation, we may then define the **Iwasawa character** $\delta = \delta_{\mathrm{Iw}}$ on A in the more general context to be

$$\delta(a) = e^{\tau(\log a)} = \prod_{\alpha \in \mathcal{R}(\mathfrak{n})} a^{m(\alpha)\alpha} \qquad \text{so} \qquad a^\rho = \delta^{1/2}(a).$$

In the case of $\mathrm{GL}_n(\mathbf{R})$ or $\mathrm{SL}_n(\mathbf{R})$, the multiplicities $m(\alpha)$ are all equal to 1, so they don't appear explicitly in the formula. The general context here introduces a few more symbols, and at this point we wish to continue emphasizing the standard situation of $\mathrm{GL}_n(\mathbf{R})$ or $\mathrm{SL}_n(\mathbf{R})$, so we merely indicate to the reader how to carry along the more general situation.

Further aspects of the semisimple decomposition will be met in Chapter VI, §1, when the eigenspace decomposition will be used to compute further Jacobian factors for Haar measures.

We shall deal with the Harish-Chandra isomorphism of Chapter II, §1,

$$\mathcal{D} = \mathcal{D}_A : S(\mathfrak{a}) \to \mathrm{IDO}(A) \quad \text{induced by} \quad v \mapsto \tilde{v} = \mathcal{D}(v) \quad \text{for } v \in \mathfrak{a}.$$

We shall encounter a "logarithmic derivative". It's just as well to tabulate some general formulas for an arbitrary Lie group G. Let $v \in \mathrm{Lie}(G)$ and let $\tilde{v}$ be the differential operator $\mathcal{D}(v)$ defined in Chapter II, §1. For any positive function φ on an open subset of G, we define its v-**logarithmic derivative** by the next formula:

$$\text{(3)} \qquad \tilde{v}(\log \varphi) = \varphi^{-1}\tilde{v}(\varphi),$$

$$\text{(4)} \quad \varphi^{-1}\tilde{v} \circ \varphi = \tilde{v} + \tilde{v}(\log \varphi) \quad \text{and} \quad \varphi\tilde{v} \circ \varphi^{-1} = \tilde{v} - \tilde{v}(\log \varphi),$$

$$\text{(5)} \qquad \tilde{v} \circ q = q\tilde{v} + \tilde{v}(q) \quad \text{for any function } q.$$

The proof of equality in (3) is immediate from the definitions, namely

$$\frac{d}{dt} \log \varphi(x \exp tv)\Big|_{t=0} = \varphi^{-1}(x) \frac{d}{dt}\varphi(x \exp tv)\Big|_{t=0}.$$

Formulas (4), (5) are then immediate from the fact that $\tilde{v}$ is a derivation. As in calculus, the map

$$\varphi \mapsto \tilde{v}(\log \varphi)$$

is a homomorphism from the multiplicative group of positive functions to the additive group of functions.

As in §1, let χ_ζ be the character defined by $\chi_\zeta(a) = a^\zeta$. Then we have

$$\text{(6)} \qquad \tilde{v}(\log \chi_\zeta) = \zeta(v).$$

This is freshman calculus, because $\log \chi_\zeta = \zeta \circ \log$ and

$$\tilde{v}(\zeta \circ \log) = \frac{d}{dt} \exp(\zeta(tv))\Big|_{t=0} = \zeta(v).$$

Note that $\chi_\tau = \delta$ and $\chi_{\tau/2} = \delta^{1/2} = \chi_\rho$. The conjugation

$$D \mapsto \delta^{-1/2} D \circ \delta^{1/2}$$

is an automorphism of the $\mathbf{R}$-algebra $\mathrm{IDO}(A)$, and it will play a major role in what follows. Although the next result will not be needed until Chapter VII, include it here because it answers an obvious, elementary, and basic question: How is such conjugation represented on the polynomial algebra under the map $\mathcal{D}$?

Let λ be a real functional on $\mathfrak{a}$, i.e. a linear map of $\mathfrak{a}$ into $\mathbf{R}$. Then λ induces first a linear map of $\mathfrak{a}$ into $S(\mathfrak{a})$, given by

$$v \mapsto v + \lambda(v) \qquad \text{for } v \in \mathfrak{a}.$$

By the general functorial properties of the commutative algebra $S(\mathfrak{a})$, this implies that there is a unique $\mathbf{R}$-algebra homomorphism

$$S(\lambda)\colon S(\mathfrak{a}) \to S(\mathfrak{a})$$

(immediately verified to be an automorphism) such that for all $v \in \mathfrak{a}$ we have

$$S(\lambda)v = v + \lambda(v).$$

Let $P \in S(\mathfrak{a})$. We often write $P^{S(\lambda)}$ instead of $S(\lambda)P$.

Proposition 2.1. *Let $\chi_\lambda(a) = a^\lambda$. Under the Harish-Chandra isomorphism*

$$\mathcal{D}\colon S(\mathfrak{a}) \to \mathrm{IDO}(A),$$

the conjugation $D \mapsto \chi_\lambda^{-1} D \circ \chi_\lambda$ corresponds to the automorphism $S(\lambda)$. In other words, for $P \in S(\mathfrak{a})$, we have

$$\mathcal{D}(P^{S(\lambda)}) = \chi_\lambda^{-1}\mathcal{D}(P) \circ \chi_\lambda.$$

In particular, for $\lambda = \tau/2 = \rho$,

$$\mathcal{D}(P^{S(\rho)}) = \delta^{-1/2}\mathcal{D}(P) \circ \delta^{1/2}.$$

Proof. Since D is an algebra isomorphism, it suffices to prove the relation when $P = v \in \mathfrak{a}$. Then we apply (4) and use (6) to conclude the proof.

III, §3. THE WEYL GROUP

On GL_n the Weyl group can simply be defined as the group of permutation of the coordinates of A. However, already on SL_n, this naive definition requires some comments, as will become apparent below.

We let $G = \mathrm{GL}_n(\mathbf{R})$ or $\mathrm{SL}_n(\mathbf{R})$. We have the Iwasawa decomposition

$$G = UAK.$$

Thus A is the group of diagonal matrices with positive diagonal elements.

We let

M' = normalizer of A in K;

M = centralizer of A in K;

$W = M'/M$ = **Weyl group**.

Among other things, we shall prove that W is isomorphic in a natural way to the group of permutations of the coordinates of A.

We let

$$\mathfrak{g} = \mathfrak{a} + \mathfrak{n} + \mathfrak{k}$$

be the direct sum decomposition of the Lie algebra. Thus:

$\mathfrak{g} = \mathrm{Mat}_n(\mathbf{R})$ for $GL_n(\mathbf{R})$, and $\mathrm{Mat}_n(\mathbf{R})_0$ (matrices with trace 0) for $SL_n(\mathbf{R})$;

$\mathfrak{a}$ = subspace of $\mathfrak{g}$ consisting of all diagonal matrices;

$\mathfrak{n}$ = subspace of $\mathfrak{g}$ consisting of all strictly upper triangular matrices;

$\bar{\mathfrak{n}} = {}^t\mathfrak{n}$ = subspace of strictly lower triangular matrices;

$\mathfrak{k}$ = subspace of skew-symmetric matrices i.e. Z such that ${}^tZ = -Z$.

We define an element H of $\mathfrak{a}$ to be **regular** if $\alpha(H) \neq 0$ for all $(\mathfrak{a}, \mathfrak{n})$ characters α; or equivalently, H as a diagonal matrix has distinct diagonal components. We let:

$\mathfrak{a}'$ = set of regular elements in $\mathfrak{a}$, so $A' = \exp \mathfrak{a}'$;

A' = set of regular elements in A, that is, $a \in A$ such that $a^\alpha \neq I$ for all α, or equivalently, $a = \mathrm{diag}(a_1, \ldots, a_n)$ having all components distinct.

We note that

$$A' = \exp \mathfrak{a}'.$$

Furthermore exponentiation transforms the additive character α to the character $a \mapsto a^\alpha$ on the multiplicative group A.

The next results show that in order to verify commutation will all elements of $\mathfrak{a}$, it suffices to do so for a single regular element, in various contexts.

Proposition 3.1. *Let H be regular in $\mathfrak{a}$. Then:*

(i) *The centralizer of H in $\mathfrak{g}$ is $\mathfrak{a}$; that is, if $X \in \mathfrak{g}$ and $XH = HX$, then $X \in \mathfrak{a}$.*

(ii) *Let* $H_{\mathfrak{g}}$ *be the regular representation of* H *on* $\mathfrak{g}$, *defined by*

$$H_{\mathfrak{g}}(Z) = [H, Z].$$

Then $H_{\mathfrak{g}}$ *induces a linear automorphism of* $\mathfrak{n}$, *also written* $H_{\mathfrak{n}}$.

Proof. We can write $X = D + Z_1 + Z_2^-$ where D is diagonal, $Z_1 \in \mathfrak{n}$, and $Z_2^- \in \bar{\mathfrak{n}}$ is strictly lower diagonal. Then $DH = HD$, so $XH = HX$ implies that

$$0 = Z_1 H - HZ_1 + Z_2^- H - HZ_2^-.$$

But $[Z_1, H] \in \mathfrak{n}$ and $[Z_2^-, H] \in \bar{\mathfrak{n}}$ so each one is 0. So we are reduced to showing the centralizer of H in $\mathfrak{n}$ and $\bar{\mathfrak{n}}$ is 0. Note that $\mathfrak{n}$ is the direct sum of the α-eigenspaces for the relevant characters. By the regularity of H, we have $\alpha(H) \neq 0$ for all α, so $H_{\mathfrak{g}}$ induces a linear automorphism of each α-eigenspace, whence a linear automorphism of $\mathfrak{n}$. This concludes the proof.

Remark. Although Proposition 3.1 was stated for the real Lie algebra, the same proof is valid algebraically over **C**, or for that matter over any field.

Corollary 3.2. *The group* M *consists of the diagonal matrices with* ± 1 *on the diagonal.*

Proof. By Proposition 3.1 a matrix which centralizes A has to be diagonal, and if it lies in K then all its components must be ± 1 since for such an element k we have ${}^t k = k^{-1}$ so $a = a^{-1}$, as desired.

Corollary 3.3. *Let* $k \in K$ *and* $H \in \mathfrak{a}'$ *regular in* $\mathfrak{a}$. *Let* H^k *be the conjugation of* H *by* k. *If* $H^k \in \mathfrak{a}$ *then* $k \in M'$.

Proof. By Proposition 3.1 and the fact that k-conjugation is an automorphism, $\mathfrak{a}^k$ is the centralizer of H^k. But $H^k \in \mathfrak{a}$ and H^k has distinct eigenvalues, so H^k is regular, whence $\mathfrak{a}$ is the centralizer of H^k, and thus finally $\mathfrak{a} = \mathfrak{a}^k$, whence $k \in M'$, thus proving the corollary.

Of course, we have similar results applying multiplicatively. In the concrete case of $\mathrm{GL}_n(\mathbf{R})$, say, Proposition 3.1 and Corollary 3.2 also apply multiplicatively, for instance:

Let H *be a regular diagonal matrix. If* $x \in \mathrm{GL}_n(\mathbf{R})$ *and* $xHx^{-1} = H$ *then* x *is diagonal.*

Conjugation by an element of K preserves eigenvalues, i.e. it preserves the characteristic polynomial of a matrix. Hence if such a conjugation leaves A stable, i.e. lies in M', then it induces a permutation of the diagonal elements. Thus we obtain a homomorphism

$$M' \to \text{permutation group of the coordinates}$$

whose kernel is M.

Theorem 3.4. *The map $M'/M \to$ permutation of the coordinates is an isomorphism for* $\mathrm{SL}_n(\mathbf{R})$ *as well as* $\mathrm{GL}_n(\mathbf{R})$.

Proof. The group $\mathrm{GL}_n(\mathbf{R})$ contains the group of matrices permuting the coordinates of the diagonal matrices. For instance matrices having one component equal to 1 in each row represent such permutations, but they may not be in $\mathrm{SL}_n(\mathbf{R})$. One may change the sign of some components, or better, one may use generators

$$\tau_i = \begin{pmatrix} \underbrace{\begin{matrix} 1 & & \\ & \ddots & \\ & & 1 \end{matrix}}_{i-1} & & & & 0 \\ & \begin{matrix} 0 & 1 \\ -1 & 0 \end{matrix} & & & \\ & & 1 & & \\ & & & \ddots & \\ 0 & & & & 1 \end{pmatrix}.$$

which represent the permutations $(i, i+1)$ and lie in $\mathrm{SL}_n(\mathbf{R})$. However, the square of such a matrix has $-I_2$ in the $(i, i+1)$ place, so it is in M but is not the identity matrix. Hence the set of these generators is not a group.

For $\mathrm{SL}_n(\mathbf{R})$, we conclude however that M'/M is indeed the permutation group, thus proving Theorem 3.4.

Unlike the case of $\mathrm{GL}_n(\mathbf{R})$, this permutation group does not always split, in other words it is not necessarily represented by a subgroup of M', as one sees directly already on $\mathrm{SL}_2(\mathbf{R})$. Varshavsky pointed out to us that for n odd, there is a splitting, namely let σ be a permutation matrix in $\mathrm{GL}_n(\mathbf{R})$. Then $\sigma(-1)^{\mathrm{sign}\,\sigma}$ is in $\mathrm{SL}_n(\mathbf{R})$ for n odd, so these adjusted permutation matrices split the Weyl group for n odd, that is, they split the exact sequence

$$O \to M \to M' \to W \to 0.$$

More on W. For a natural fundamental domain of the action of W on A', cf. Chapter I, Proposition 4.3, Chapter VI, Propositions 1.1 and 1.2. For a particularly significant element of the Weyl group, see Chapter V, Proposition 2.1.

III, §4. ORBITAL INTEGRAL FOR THE HARISH TRANSFORM

This section gives some integral formulas due to Harish-Chandra [Har 58] except as otherwise noted. We suppose $G = \mathrm{GL}_n(\mathbf{R})$ or $\mathrm{SL}_n(\mathbf{R})$. The main property to be used which has not yet been listed is the following property of nilpotence:

NIL. The Lie algebra $\mathfrak{n}$ is nilpotent, and $U = \exp \mathfrak{n}$ is therefore unipotent.

This nilpotency property will now influence the situation in various ways, starting with integral formulas which we shall recall immediately, and going on with proofs such as the one for Theorem 4.1. Thus things would become heavier to carry along axiomatically, and our priority was to keep things no more complicated than they are on $\mathrm{GL}_n(\mathbf{R})$ or $\mathrm{SL}_n(\mathbf{R})$.

In practice, A is given coordinatized as a product of positive real half lines. In this case, A has a natural Haar measure as a product of the natural measures on each factor in terms of the coordinates. To avoid confusion, we then denote the Haar measure by d^*a. Writing a in terms of coordinates $a = \mathrm{diag}(a_1, \ldots, a_n)$, we have for instance

$$d^*a = \prod_{i=1}^{n-1} da_i/a_i.$$

Integration over A with respect to d^*a then amounts to integration with respect to $\prod da_i/a_i$ from 0 to ∞.

For $a \in A$ we let $\log a$ be the element of $\mathfrak{a}$ such that $\exp(\log a) = a$. For a character λ on $\mathfrak{a}$, we shall use the notation

$$a^\lambda = e^{\lambda(\log a)}.$$

By definition, an element $a \in A$ is **regular** if and only if $a^\alpha \neq 1$ for all $\alpha \in \mathcal{R}(\mathfrak{n})$.

Theorem 4.1. *Let*

$$J(a) = \prod_{\alpha \in \mathcal{R}(\mathfrak{n})} |1 - a^{-\alpha}| = \prod_{i<j} |1 - a_j/a_i|.$$

Then for $f \in C_c(G)$ *and a regular element* $a \in A$, *we have*

$$\int_U f(u)\,du = J(a) \int_U f(a^{-1}uau^{-1})\,du.$$

The complete proof will take up to Theorem 4.5. Note the expression $a^{-1}uau^{-1}$ in the above expression. We shall be led to study the map

$$u \mapsto a^{-1}uau^{-1} \quad \text{denoted by } u^{(a)}u^{-1},$$

using $u^{(a)} = a^{-1}ua$ for the conjugation operation. The expression for $J(a)$ will come as the Jacobian determinant of the above map, which we determine in the next proposition. The integral formula is an immediate consequence of the computation of this Jacobian, and the fact that the above map is a differential isomorphism for a regular element a. The change of variables formula from calculus then applies, namely

$$\int f(y)\,dy = \int f(y(x))\frac{dy}{dx}\,dx.$$

The general version of the next theorem is due to Harish-Chandra, cf. the historical note in [Wal 73], p. 213.

Proposition 4.2. *Let* $a \in A$, *and let* $F_a : U \to U$ *be the map such that*

$$F_a(u) = a^{-1}uau^{-1}.$$

(i) *The absolute value of the Jacobian determinant of* F_a *is*

$$J(a) = JF_a = \prod_{\alpha \in \mathcal{R}(\mathfrak{n})} |1 - a^{-\alpha}| = \prod_{i<j} |1 - a_j/a_i|.$$

(ii) *Let* a *be a regular element (i.e. all its coordinates are distinct). Then* F_a *is a differential automorphism (actually a polynomial automorphism).*

Proof. We first compute the tangent map $TF_a(u)X$ for $u \in U$ and $X \in \mathfrak{n}$. We have

$$\begin{aligned} F_a(u \exp tX) &\equiv a^{-1}u(I + tX)a(I - tX)u^{-1} \mod t^2 \\ &\equiv F_a(u) + t(a^{-1}uXau^{-1} - a^{-1}uaXu^{-1}) \mod t^2. \end{aligned}$$

Hence

$$TF_a(u)X = a^{-1}u(Xa - aX)u^{-1}.$$

Now let $\{E_{ij}\}$ $(i < j)$ be the usual basis of unit vectors in $\mathfrak{n}$. For $X = E_{ij}$ in the above expression, we find

$$\begin{aligned} TF_a(u)E_{ij} &= a^{-1}u(a_j - a_i)E_{ij}u^{-1} \\ &= a^{-1}ua(a_j/a_i - 1)E_{ij}u^{-1}. \end{aligned}$$

Both elements $a^{-1}ua$ and u^{-1} are unipotent, in U, so of the form $I + Z$ with Z nilpotent. Let $v \in U$. Let L_v and R_v be multiplication by v on the left and right of $\mathfrak{n}$. Then

$$\det L_v = \det(I + Z) = 1 = \det R_v.$$

Hence

$$\det TF_a(u) = \prod_{i<j}(a_j/a_i - 1),$$

as desired. In particular, if a is regular, then this determinant is $\neq 0$, and hence F_a is a local isomorphism. Continuing to assume a regular, we note that F_a is injective. Indeed, suppose $F_a(u) = F_a(v)$ with $u, v \in U$. Then

$$a^{-1}uau^{-1} = a^{-1}vav^{-1},$$

from which it follows that $v^{-1}u$ commutes with a. But all the diagonal elements of a are distinct, so this is impossible unless $u = v$. (This comes from a direct computation. More structurally, see Lemma 4.3 below.) Hence F_a is injective, so F_a is a differential isomorphism of U with its image.

For the surjectivity, we find it more instructive to view the map $F_a(u)$ in a larger context, and more algebraically, in a way similar to that of [Lan 56]. We fix a regular in A. Given $u \in U$ we define the map

(1) $$F_{a,u}\colon U \to U \quad \text{by} \quad F_{a,u}(v) = u^{(a)}vu^{-1},$$

where $u^{(a)} = a^{-1}ua$. We write simply

$$F_{a,u}(v) = u \cdot v \quad \text{(abbreviating } u \cdot_a v).$$

The notation is justified because an immediate direct verification shows that the map $v \mapsto u \cdot v$ is an action of U on itself, that is, $u_1 \cdot (u \cdot v) = (u_1u) \cdot v$ and $I \cdot v = v$ for all $u_1, u, v \in U$. The action is in the category of affine algebraic groups. The map $F_{a,u}$ is in fact a polynomial map on the affine coordinate ring of U. Furthermore, we need not restrict ourselves to the real points of the algebraic group. Write now $U(\mathbf{R})$ for what we have denoted by U, and let $U(\mathbf{C})$ be the group of complex upper triangular unipotent matrices. Then the above action is valid for points in $U(\mathbf{C})$. Furthermore, the injectivity mentioned before for $U(\mathbf{R})$ is valid for $U(\mathbf{C})$, with the same proof. The relevant structure here is completely algebraic. Actually, we shall need a stronger version of this injectivity.

Lemma 4.3. *Let $v \in \mathrm{Tri}^{+}(\mathbf{C})$ be a regular upper triangular matrix, that is with distinct non-zero diagonal elements. Then the centralizer of v in the group of unipotent matrices $U(\mathbf{C})$ is the identity.*

Proof. Suppose $u \in U(\mathbf{C})$ and $uv = vu$. Write $u = I + Z$ and $v = D + X$, where Z, X are strictly upper triangular. Then $uv = vu$ is equivalent with

$$DZ - ZD = XZ - ZX \qquad \text{or also} \qquad [D, Z] = [X, Z].$$

But $Z \mapsto [D, Z]$ does not annihilate any component of the first "diagonal" where Z is non-zero, whereas $[X, Z]$ does. Hence the only element Z for which $[D, Z] = [X, Z]$ is $Z = 0$, which proves the lemma.

Proposition 4.4. *Fix $v_0 \in U(\mathbf{C})$. Then the isotropy group of elements $u \in U(\mathbf{C})$ such that $u \cdot v_0 = v_0$ is trivial, i.e. consists only of the identity.*

Proof. By definition, such an element u satisfies $u^{(a)}vu^{-1} = v$, which is equivalent with

$$u(av_0) = (av_0)u.$$

We apply the lemma with $v = av_0$ to conclude the proof.

Remark. Suppose we want to solve $u \cdot I = v$ with given $v \in U(\mathbf{R})$. By the injectivity, it suffices to solve this equation with $u \in U(\mathbf{C})$, because if a is real, then u must be real. If it were not, we would take the complex conjugate of the equation to contradict injectivity. Of course, we could work completely algebraically, taking u complex from the start. All arguments are algebraic.

We are now ready for the main algebraic theorem concerning our operation.

Theorem 4.5. *Fix a regular element $a \in A$. Then the map*

$$u \mapsto u \cdot I = u^{(a)}u^{-1}$$

is an algebraic automorphism of $U(\mathbf{C})$ as affine variety with itself. Thus $U(\mathbf{C})$ is a principal homogeneous space over itself, under the operation

$$F_{a,u}\colon v \mapsto u \cdot v.$$

Proof. All that remains to be proved is the surjectivity, i.e. every element $v_0 \in U(\mathbf{C})$ is of the form $v_0 = u^{(a)}u^{-1}$ for some $u \in U(\mathbf{C})$. By the local result concerning the map $u \mapsto u \cdot I$, we know that this map is a differential isomorphism of a neighborhood of the identity in $U(\mathbf{R})$. But such a neighborhood contains (non countably) many generic elements. In other words, let k be a finitely generated subfield of $\mathbf{C}$ containing all coordinates of v_0 and a. Then there exists (non countably) many elements $v_1 \in U(\mathbf{R})$ such that $k(v_1)/k$ has transcendence degree equal to $\dim U = n(n-1)/2$. Therefore there exists $u_1 \in U(R)$ such that $v_1 = u_1 \cdot I$. Then we conclude that for any $v \in U(\mathbf{C})$, v generic over k (i.e. $k(v)/k$ having transcendence degree equal to $\dim U$), there exists $u \in U(\mathbf{C})$ such that $u \cdot I = v$. Now

$$k(v_1 \cdot v_0) = k(v_1),$$

because each element v_1 and $v_1 \cdot v_0$ is a polynomial function of the other with coefficients in k, for instance, $v_1 = v_1^{-1} \cdot v_1 \cdot v_0$. Let $v = v_1 \cdot v_0 = u \cdot I$. Then

$$v_0 = v_1^{-1} \cdot (v_1 \cdot v_0) = (v_1^{-1} u) \cdot I,$$

which concludes the proof.

Note that Theorem 4.5 also concludes the proof of surjectivity which was the only part of Proposition 4.2 left open. By the same token, Theorem 4.1 is also proved.

Remark. The above considerations about the principal homogeneous space with the map $F_{a,u}$ apply quite generally to the situation of a semisimple Lie group and the Iwasawa decomposition of its Lie algebra. Readers acquainted with this general framework will immediately see that the proofs we have given go through mutatis mutandis.

We have now done on $\mathrm{GL}_n(\mathbf{R})$ or $\mathrm{SL}_n(\mathbf{R})$ the first formula of [Lan 75/85], Chapter V, by a method which also works for arbitrary semisimple Iwasawa decompositions. We may then proceed as in that reference, because the next formulas just come from Theorem 4.1 and abstract nonsense as follows.

We let $G = \mathrm{GL}_n(\mathbf{R})$ or $\mathrm{SL}_n(\mathbf{R})$ with its Iwasawa decomposition, and the corresponding decomposition of Haar measure from **INT 2** (Chapter V, §5). The measure on A comes from the natural measure on the multiplicative group as explained at the beginning of the section. The measure du is the euclidean measure with respect to the euclidean coordinates of an element $u = I + X$ with $X \in \mathbf{R}^{\dim U}$. The measure on K has total measure 1.

We let $d\dot{x}$ be the the measure on the coset space $A\backslash G$ such that Fubini's theorem is valid, that is, $dx = da\, d\dot{x}$ (integrating successively on A and $A\backslash G$ yields the integral on G).

If Φ is a function on G, we let ${}^A\Phi$ be its (left) A-projection, that is

$$ {}^A\Phi(x) = \int_A \Phi(ax)\,da. $$

Theorem 4.6. *Let $f \in C_c(G)$ and let $a \in A$ be a regular element. Then the function*

$$ x \mapsto f(x^{-1}ax) $$

has compact support on $A\backslash G$, and we have

$$ J(a) \int_{A\backslash G} f(x^{-1}ax)\,d\dot{x} = \int_{K\times U} f(kauk^{-1})\,du\,dk. $$

Proof. Let $\varphi(x) = f(x^{-1}ax)$. Using the fact that F_a is a topological isomorphism (differentiability is not needed here) from Proposition 4.2(ii), one sees directly from the Iwasawa decomposition that φ has compact support in $A\backslash G$. Indeed, let S be the compact support of f. Write an element g in the form $g = buk$ with $u \in U$, $b \in A$, $k \in K$. Then

$$ k^{-1}u^{-1}b^{-1}abuk \in S \qquad \text{whence} \qquad u^{-1}au \in KSK \quad \text{(compact).} $$

By Proposition 3.1, it follows that u lies in a compact set, whence φ has compact support in $A\backslash G$ as asserted. Let $\Phi \in C_c(G)$ be such that ${}^A\Phi = \varphi$. Then using $dg = da\,du\,dk$ we get:

$$ \begin{aligned} \int_{A\backslash G} \varphi(x)d\dot{x} &= \int_G \Phi(x)\,dx = \int_U \int_K \varphi(uk)\,du\,dk \\ &= \int_U \int_K f(k^{-1}u^{-1}auk)\,du\,dk \\ &= \int_U \int_K f(k^{-1}aa^{-1}u^{-1}auk)\,du\,dk \\ &= \frac{1}{J(a)} \int_K \int_U f(k^{-1}auk)\,du\,dk \quad \text{by Theorem 4.1.} \end{aligned} $$

This proves Theorem 4.6.

It is convenient to transform the Jacobian factor in Theorems 4.1 and 4.6 as follows. For a regular, we define

$$ D(a) = \prod_{\alpha\in\mathcal{R}(\mathfrak{n})} (a^{\alpha/2} - a^{-\alpha/2}). $$

Then

$$ |D(a)| = |D(a^{-1})|. $$

Also

$$(2)\qquad \begin{aligned} J(a) &= \prod_{\alpha\in\mathcal{R}(\mathfrak{n})} |a^{-\alpha}-1| \\ &= \prod_{\alpha\in\mathcal{R}(\mathfrak{n})} a^{-\alpha/2}|a^{\alpha/2}-a^{-\alpha/2}| \\ &= a^{-\tau/2}|D(a)| = \delta(a)^{-1/2}|D(a)|. \end{aligned}$$

Therefore Theorem 4.6 becomes:

Corollary 4.7. *For a regular,*

$$|D(a)|\int_{A\backslash G} f(x^{-1}ax)\,d\dot{x} = \delta(a)^{1/2}\int_K\int_U f(kauk^{-1})\,du\,dk.$$

Let $C_c(G,K)$ be the space of continuous functions with compact support on G, invariant under conjugation by K, i.e. satisfying

$$f(k^{-1}xk) = f(x) \qquad \text{for all } k\in K \text{ and } x\in G.$$

For $f\in C_c(G,K)$ the **Harish transform** $\mathbf{H}f$ is now seen to have another integral expression coming from Theorem 4.6, namely

$$\mathbf{H}f(a) = \delta(a)^{1/2}\int_U f(au)\,du = |D(a)|\int_{A\backslash G} f(x^{-1}ax)\,d\dot{x}.$$

The first expression is valid for all $a\in A$, the second expression only for a regular. Corollary 4.7 is the source of the appearance of the factor $\delta^{1/2}$ for the rest of this chapter.

The next theorem in the general situation of semisimple Lie groups is due to Harish-Chandra [Har 58a], Lemma 17, p. 261.

Theorem 4.8. *For $f\in C_c(G,K)$ the Harish transform* $\mathbf{H}f$ *is invariant under the Weyl group, that is*

$$\mathbf{H}f([w]a) = \mathbf{H}f(a) \qquad \textit{for } w\in W \textit{ and } a\in A.$$

Proof. First we note that trivially, $|D(a)|$ is invariant under the Weyl group. Second, by the measure preserving property, we can change variables in the second integral expressing the Harish transform when a is regular, so for such elements the invariance of the Harish transform under the Weyl group is clear. For arbitrary a, the invariance follows by continuity. This concludes the proof.

The Harish transform is therefore a linear map

$$\mathbf{H}: C_c(G, K) \to C_c(A)^W,$$

where the upper index W means invariance under the Weyl group.

In this chapter, we are principally interested in using some basic properties of the Harish transform to get properties of invariant differential operators, so we do not go fully into the Harish transform at this time. In Chapter IV, we shall list some of the properties which can be proved more easily and directly in the framework of convolutions. The same remarks apply to the spherical transform of the next section. Our main goal here is Theorem 5.3.

Remarks on terminology. As in [Lan 75/85] the notation $\mathbf{H}f$ is used for what Harish-Chandra writes F_f. Calling $\mathbf{H}$ the **Harish transform** dates back to this reference, and Wallach [Wal 88] calls it the **Harish-Chandra transform.** On the other hand, $\mathbf{H}$ is called the **Abel transform** in [GaV 88]. The **spherical transform** (see the next section) is called by this name in [Lan 75/85] and in Helgason [Hel 84], but is called both the **Harish-Chandra transform** and **spherical transform** in [GaV 88].

Harish-Chandra uses the notation F_f for what we have called the Harish transform. He uses it indiscriminately either for the integral over U or for the orbital integral of Theorem 4.6. Cf. for instance, [Har 58a], Lemma 17, p. 261; or for a variation, [Har 58b], Corollary 4, p. 595, as well as a few lines below, and also p. 596, etc. Helgason in both his books follows the Harish-Chandra notation using F_f for both integrals, and reproduces Harish's proof for the equality of the two integrals on regular elements. See [Hel 62], Chapter X, Theorem 1.15; and [Hel 54], Chapter I, Theorem 5.7, formula (7), as well as Chapter II, p. 304, and Chapter IV, §7, (6), p. 450.

III, §5. *W*-INVARIANCE OF THE HARISH AND SPHERICAL TRANSFORMS

To start, let G be a Lie group having just an Iwasawa decomposition $G = UAK$, with K compact. Let $r = \dim_{\mathbf{R}}(A)$.

Let $F \in C_c(A)$. Suppose we have chosen coordinates for A, so an isomorphism

$$A \xrightarrow{\approx} \mathbf{R}^+ \times \cdots \times \mathbf{R}^+.$$

Then characters on A are all of type $a \mapsto a^s$ with an r-tuple of complex variables $s = (s_1, \ldots, s_r)$. In terms of these coordinates, we define the **Mellin transform** $\mathbf{M}F$ by

$$\mathbf{M}F(s) = \int_A F(a)a^s\, d^*a = \int_0^\infty \cdots \int_0^\infty F(a) \prod a_i^{s_i} \prod da_i/a_i.$$

Thus the Mellin transform is a linear map

$$\mathbf{M}\colon C_c(A) \to \mathrm{Hol}(\mathbf{C}^r).$$

We do not discuss here the general deeper theorem about Mellin inversion, but we just want a commutative diagram with the Harish transform. The expression of the Mellin transform in terms of several complex variables has the obvious advantage to fit the standard notion from one variable. However, for this chapter and elsewhere, it is also valuable to express it in terms of characters without the complex variables. Thus we define the **Mellin transform** $\mathbf{M}F$ on characters of A by the integral scalar product

$$\mathbf{M}F(\chi) = \int_A F(a)\chi(a)\, da = [F, \chi]_A,$$

where da is our standard Haar measure on A, which we write d^*a only if we are forced to use coordinates a_i in $(0, \infty)$. Both interpretations, as an integral transform and as a scalar product, will be used.

Let $A^\vee_{\mathbf{C}}$ be the space of complex valued characters on A. We define the **spherical transform**

$$\mathbf{S}\colon C_c(K\backslash G/K) \to \mathrm{Hol}(A^\vee_{\mathbf{C}}) \approx \mathrm{Hol}(\mathbf{C}^r)$$

to be the convolution operator with the spherical kernel, or the **symmetric scalar product** of the spherical function φ_χ with f, that is

$$(\mathbf{S}f)(\chi) = \int_G \varphi(\chi, x) f(x)\, dx = (\varphi * f)(\chi) = [\varphi_\chi, f]_G.$$

If $\chi(a)$ is seen as $\chi_s(a) = a^s$, then $\mathbf{S}f$ is a holomorphic function of r complex variables. We shall compose the Mellin transform and the Harish transform. In this section, the integrals will converge absolutely in a trivial manner, and we shall express them in terms of the variables g and χ, rather than g and s. The next result is from [Har 58b], Lemma 43, p. 596.

Proposition 5.1. *We have* $\mathbf{S} = \mathbf{MH}$ *on* $C_c(K\backslash G/K)$, *in other words we have a commutative diagram:*

$$\begin{array}{ccc} C_c(K\backslash G/K) & \xrightarrow{\mathbf{H}} & C_c(A) \\ & {}_{\mathbf{S}}\searrow \quad \swarrow_{\mathbf{M}} & \\ & \mathrm{Hol}(A_{\mathbf{C}}) & \end{array}$$

Alternatively, for $f \in C_c(K\backslash G/K)$, *we have the transpose relation*

$$[\varphi_\chi, f]_G = [\chi, \mathbf{H}f]_A.$$

Proof. This is merely the transpose relation of Chapter II, Proposition 5.2, together with the current definitions with characters. Indeed, by that proposition,

$$(\mathbf{S}f)(\chi) = [\varphi_\chi, f]_G = [{}^t\mathbf{H}\chi, f]_G = [\chi, \mathbf{H}f]_A = \mathbf{MH}f(\chi),$$

as asserted.

The above is just abstract nonsense. It is a more elaborate and deeper matter to consider the above transforms on C^∞ functions with G more special, and to prove Harish-Chandra's theorems about their images and their inversions. We postpone this, and here limit ourselves to simpler applications. We shall need Theorem 4.5, so *we assume* $G = \mathrm{GL}_n(\mathbf{R})$ or $\mathrm{SL}_n(\mathbf{R})$.

The first one has to do with the Weyl group W. For $w \in W$, we have an action $[w]$ on characters, given

$$([w]\chi)(a) = \chi([w^{-1}]a) = \chi(w^{-1}aw).$$

Theorem 5.2 (Harish-Chandra). *The spherical kernel is invariant under the Weyl group as a function of characters. In other words, for* $w \in W$ *and for a character* χ *on* A, *we have*

$$\varphi([w]\chi, x) = \varphi(\chi, x) \qquad \textit{for all } x \in G.$$

Furthermore, the image of the spherical transform is invariant under W, *that is,* $\mathbf{S}f$ *is invariant under* W *for all* $f \in C_c(K\backslash G/K)$. *Hence we have the more precise commutative diagram*

$$\begin{array}{ccc} C_c(K\backslash G/K) & \xrightarrow{\mathbf{H}} & C_c(A)^W \\ & {}_{\mathbf{S}}\searrow \quad \swarrow_{\mathbf{M}} & \\ & \mathrm{Hol}(A^\vee_{\mathbf{C}})^W & \end{array}$$

Proof. We first prove that if F is a function on A invariant under the Weyl group, then $\mathbf{M}F$ is invariant under the Weyl group. Indeed, for $w \in W$,

$$\begin{aligned}\mathbf{M}F([w]\chi) &= \int_A F(a)([w]\chi)(a)\,da \\ &= \int_A F(a)\chi([w^{-1}]a)\,da.\end{aligned}$$

We make the transformation $a \mapsto [w]a$, which preserves the Haar measure on A, and use the assumption on F to prove our first statement. To prove that

$$\varphi([w]\chi, x) = \varphi(\chi, x) \qquad \text{for all } x \in G,$$

It suffices to prove that the scalar product of both functions of x with an arbitrary function $f \in C_c(K\backslash G/K)$ are equal. But these scalar products are just the convolutions with φ, in other words the spherical transform. By Proposition 5.1, we know that

$$\mathbf{S}f = \mathbf{MH}f \qquad \text{for } f \in C_c(K\backslash G/K),$$

and by Theorem 4.8 we know further that $\mathbf{H}f$ is invariant under the Weyl group. We then put $F = \mathbf{H}f$ and apply the first step to conclude the proof of the theorem.

For the next theorem, we need to use Theorem 4.5, so we assume G is special. From Chapter II, §2, (14a), we recall that $\mathcal{J}(\mathfrak{k}) = \mathrm{IDO}(G)\tilde{\mathfrak{k}}$.

Theorem 5.3 (Harish-Chandra [Har 58a]). *Suppose* $G = \mathrm{GL}_n(\mathbf{R})$ *or* $\mathrm{SL}_n(\mathbf{R})$. *The maps* $\mathbf{h}_G$ *and* $\mathbf{h}_X$ *given by*

$$\mathbf{h}\colon D \mapsto \delta^{-1/2} D_A \circ \delta^{1/2}$$

have images invariant under the Weyl group, and give **R**-*algebra isomorphisms*

$$\mathrm{IDO}(G)^K/\mathcal{J}(\mathfrak{k})^K \approx \mathrm{DO}(X)^G \xrightarrow{\mathbf{h}} \mathrm{IDO}(A)^W.$$

Proof. For $w \in W$ and $D \in \mathrm{IDO}(G)^K$, abbreviate

$$\mathbf{h}(D)^w = [w^{-1}] \circ \mathbf{h}(D) \circ [w].$$

By Theorem 1.1, to prove that $\mathbf{h}(D)^w = \mathbf{h}(D)$ it suffices to prove that $\mathbf{h}(D)^w$ and $\mathbf{h}(D)$ have the same eigenvalues when applied to characters. Replace χ by $[w]\chi$ in the first equality of Proposition 1.5. By Theorem 5.2, $\varphi_\chi = \varphi_{[w]\chi}$, and hence

$$\mathrm{ev}(D_A, [w]\chi\delta^{1/2}) = \mathrm{ev}(D_A, \chi\delta^{1/2}),$$

so for all χ we obtain

$$\delta^{-1/2} D_A \circ \delta^{1/2}([w]\chi) = \mathrm{ev}(D_A, \chi\delta^{1/2})[w]\chi$$

so

$$\mathbf{h}(D)^w \chi = \mathrm{ev}(D_A, \chi\delta^{1/2})\chi.$$

These relations mean precisely that $\mathbf{h}(D)^w$ and $\mathbf{h}(D)$ have the same eigenvalues on characters, so they are equal, and $\mathbf{h}(D)$ is therefore W-invariant.

There remains only to prove the surjectivity of $\mathbf{h}\colon D \mapsto \delta^{-1/2} D_A \circ \delta^{1/2}$ onto $\mathrm{IDO}(A)^W$. We consider the situation on the graded algebras. We recall the Chevalley linear isomorphism

$$\mathrm{Ch}\colon S(\mathfrak{p})^{c(K)} \to S(\mathfrak{a})^W$$

given by the projection $P \mapsto P_{\mathfrak{a}}$ of a polynomial on its $\mathfrak{a}$-part, as in Chapter II, **CA 3**.

Lemma 5.4. *The following diagram is commutative, on the graded algebras*:

$$\begin{array}{ccc}
\mathrm{gr}(\mathrm{DO}(X)^G) & \xrightarrow{\mathrm{gr}(\mathbf{h}_X)} & \mathrm{gr}(\mathrm{IDO}(A)^W) \\
\Big\downarrow {\scriptstyle \mathrm{gr}(\mathcal{D}_{\mathfrak{p},X})} & & \Big\uparrow {\scriptstyle \mathrm{gr}(\mathcal{D}_A)} \\
\mathrm{gr}(S(\mathfrak{p})^{c(K)}) & \xrightarrow[\mathrm{gr}(\mathrm{Ch})]{} & \mathrm{gr}(S(\mathfrak{a})^W)
\end{array}$$

All the arrows are isomorphisms.

Proofs. We know from Chapter II, Theorem 2.3, that $\mathcal{D}_{\mathfrak{p},X}$ is a linear isomorphism, and it preserves the degree filtration, so $\mathrm{gr}(\mathcal{D}_{\mathfrak{p},X})$ is an isomorphism. The bottom arrow comes from the Chevalley isomorphism, so $\mathrm{gr}(\mathrm{Ch})$ is an isomorphism. The vertical arrow $\mathrm{gr}(\mathcal{D}_A)$ is the same as $\mathrm{gr}(\delta^{-1/2}\mathcal{D}_A \circ \delta^{1/2})$ because on the graded algebra, conjugation by $\delta^{-1/2}$ is the identity. Using $\mathrm{gr}(\delta^{-1/2}\mathcal{D}_A \circ \delta^{1/2})$ from $\mathrm{gr}(S(\mathfrak{a}))$ to $\mathrm{gr}(\mathrm{IDO}(A))$ (without the W) shows that the diagram commutes. Leaving out the conjugation by $\delta^{1/2}$ and using the first part of the proof which shows that the image of $\mathbf{h}_X$ is contained in $\mathrm{IDO}(A)^W$, we conclude that the diagram of the lemma commutes on the graded rings, as asserted. Since all arrows except possibly $\mathrm{gr}(\mathbf{h}_X)$ have been seen to be isomorphism, it follows that $\mathrm{gr}(\mathbf{h}_X)$ is also an isomorphism, thus concluding the proof of the lemma.

We may now conclude the proof of Theorem 5.3. The map $\mathbf{h}$ is an algebra homomorphism, preserving the degree filtration, and $\mathrm{gr}(\mathbf{h})$ is an isomorphism. By the general fact recalled as Lemma 1.4 of Chapter II it follows that $\mathbf{h}$ is an algebra isomorphism, thus concluding the proof.

Remark. Harish-Chandra went directly from $\mathrm{IDO}(G)^K$ to $\mathrm{IDO}(A)^W$. That one can interpolate $\mathrm{DO}(X)^G$ in between stems from Helgason [Hel 59], as given here in Theorem 2.5 of Chapter II. See [Hel 84], Chapter II, Theorem 5.18. However, all the topics were discovered in an intertwined way, with Chevalley's theorem as part of the essential backdrop, as well as Harish-Chandra's proof of this theorem, reproduced in [Hel 62], Chapter X, Theorem 6.10. Concerning the situation, Helgason comments ([Hel 62], *Notes*, p. 455): "The integral formula for the spherical function (Theorem 6.16) was found by Harish-Chandra. ... The proof in the text which also leads to the isomorphism between $\mathrm{DO}(G/K)$ and $I(\mathfrak{h}_{\mathfrak{p}_0})$ is from Harish-Chandra ..."

As remarked after Theorem 1.3, the only eigenfunctions of $\mathrm{IDO}(A)$ are the characters. We want to analyze the analogous situation on $X \approx G/K$. A U-invariant function f on X corresponds to a function f_A on A by restriction. By the general formalism of the direct image, such a function f is an eigenfunction of $\mathrm{DO}(X)^G$ if and only if f_A is an eigenfunction of $(\mathrm{Iw}_A)_* \mathrm{DO}(X)^G$, and the eigenvalues are the same. More precisely, for $D \in \mathrm{DO}(X)^G$, we have $\mathrm{ev}(D, f) = \mathrm{ev}(D_A, f_A)$, that is

(1) $Df = \mathrm{ev}(D, f)f$ if and only if $D_A f_A = \mathrm{ev}(D, f) f_A$.

Let us fix f (or, equivalently, f_A), What happens under the Harish twist? Let

$$\mathrm{ev}_{X,f} : \mathrm{DO}(X)^G \to \mathbf{C}$$

be given by $\mathrm{ev}_{X,f}(D) = \mathrm{ev}(D, f)$. Then $\mathrm{ev}_{X,f}$ is an algebra homomorphism of $\mathrm{DO}(X)^G$ into $\mathbf{C}$, also called a **C-character**. Directly from the definition of $\mathbf{h}_X$, we have

(2) $$\mathbf{h}_X(D)(\delta^{-1/2} f_A) = \mathrm{ev}(D_A, f_A)\delta^{-1/2} f_A$$

Lemma 5.5. *A U-invariant function $f \in \mathrm{Fu}(X)$ is an eigenfunction of $\mathrm{DO}(X)^G$ if and only if $\delta^{-1/2} f_A$ is an eigenfunction of $\mathrm{IDO}(A)^W$, with the same eigenvalues.*

As an application of Theorem 5.3, we get a converse for Theorem 5.2, also due to Harish-Chandra.

Theorem 5.6. *Let χ, χ' be characters of A. The following* **W-equivalence** *conditions are equivalent:*

WE 1. *There is some $w \in W$ such that $\chi' = [w]\chi$.*

WE 2. *The spherical functions are equal, that is, $\varphi_\chi = \varphi_{\chi'}$.*

WE 3. $\mathrm{ev}(D, \chi\delta^{1/2})_X) = \mathrm{ev}(D, \chi'\delta^{1/2})_X)$ *for all $D \in \mathrm{DO}(X)^G$.*

WE 4. $\mathrm{ev}(D, \chi) = \mathrm{ev}(D, \chi')$ *for all $D \in \mathrm{IDO}(A)^W$.*

Proof. That **WE 1** implies **WE 2** is Theorem 5.2. Assume **WE 2**, which by definition means ${}^K(\chi\delta^{1/2})_G = {}^K(\chi'\delta^{1/2})_G$. Then for all $D \in \mathrm{DO}(X)^G$ we get the equality of eigenvalues

$$\mathrm{ev}(D_A, \chi\delta^{1/2}) = \mathrm{ev}(D_A, \chi'\delta^{1/2}),$$

which gives **WE 3** by Theorem 5.3 and then **WE 4** by Lemma 5.5. Finally assume **WE 4**. Each $D \in \mathrm{IDO}(A)^W$ corresponds to a unique symmetric polynomial P such that $D\chi_s = P(s)\chi_s$ for all $s \in \mathbf{C}^n$. Write $\chi = \chi_s$ and $\chi' = \chi_{s'}$. By **WE 4**, $P(s) = P(s')$. If s, s' do not differ by a permutation, then the polynomials

$$\prod (T - s_i) \quad \text{and} \quad \prod (T - s_i')$$

are distinct, so there is some elementary symmetric polynomial I such that $I(s) \neq I(s')$, which proves that **WE 4** implies **WE 1**, and concludes the proof of the theorem.

INTRODUCTION TO THE NEXT THREE SECTIONS

At this point several things are happening in parallel, but as usual, and unfortunately, they have to be projected in a totally ordered way on the page axis. The next three sections are independent of each other, and each contributes to one aspect of the topics we have just treated.

In §6 we give the Selberg–Harish-Chandra characterization of spherical functions.

In §7 we give a Jacobian formula from Harish, and an application from Helgason to an integral formula for the spherical functions.

In §8 we return to Lemma 5.5, which raised the question of describing all eigenfunctions of the algebra $\mathrm{IDO}(A)^W$, consisting of the invariant differential operators on A fixed under the Weyl group (permutation of the variables), because the characters may only constitute some of the eigenfunctions. However, the problem is down to earth, because A is a product of positive real lines.

At this point, dealing only with A, we can transfer the question to the tangent space at the origin, which is just a euclidean space. The algebra of invariant differential operators on A corresponds to the algebra of additive translation invariant differential operators on euclidean space, which is simply the polynomial algebra $\mathbf{C}[\partial_1, \dots, \partial_n]$ where ∂_i are the partial derivatives in the ordinary variables, corresponding to a choice of basis. Thus the problem is now to determine the space of functions on $\mathbf{R}^n$, which are eigenfunctions of $\mathbf{C}[\partial_1, \dots, \partial_n]^W$, namely the algebra of polynomial differential operators invariant under the group W of permutation of the variables.

III, §6. *K*-BI-INVARIANT FUNCTIONS AND UNIQUENESS OF SPHERICAL FUNCTIONS

We begin with a result which will depend only on Theorem 2.5 of Chapter II, but which we postponed till now to join it with the present context of characters and eigenfunctions. The following theorem is essentially due to Selberg [Sel 56].

We are interested in the eigenspaces for the action of $\mathrm{IDO}(G)^K$. Let λ be a $\mathbf{C}$-character of $C^\infty(K\backslash G/K)$. We denote by

$$C^{\mathrm{an}}(K\backslash G/K, \lambda)$$

the λ-eigenspace of (real) analytic K-bi-invariant functions f having eigencharacter λ, that is, $Df = \lambda(D)f$ for all $D \in \mathrm{IDO}(G)^K$.

Theorem 6.1. *Let G be a connected Lie group and K a compact subgroup. (Lie groups are here assumed real analytic.) Then for each λ, the eigenspace $C^{\mathrm{an}}(K\backslash G/K, \lambda)$ is one-dimensional or zero. Let f be in this space. If $f(e) = 0$ then f is the zero function. If $f(e) \neq 0$ then f is a basis for the space. In particular, if the eigenspace is not zero, then there is a unique basis element f such that $f(e) = 1$. If f_1, f_2 are elements of the eigenspace and $f_1(e) = f_2(e)$ then $f_1 = f_2$.*

Proof. The last statement is obtained from the first by considering the function $f = f_1 - f_2$. Then the one-dimensionality of the eigenspace is also seen to follow from the first statement, which we now prove. It will suffice to prove that the Taylor series for f at the origin is 0. By Chapter II, §1, (1_m), we know the Taylor series for f, namely for sufficiently small vectors v, we have

$$f(\exp v) = \sum \frac{1}{m!}(\tilde{v}^m f)(e).$$

Let $\mathbf{c}(k)$ be conjugation by an element $k \in K$ as usual. Then by functoriality, for any $D \in \mathrm{IDO}(G)$, denoting the action exponentially, we have

$$(*) \qquad (Df)(e) = (D^{\mathbf{c}(k)} f^{\mathbf{c}(k)})(e^{\mathbf{c}(k)}) = (D^{\mathbf{c}(k)} f)(e)$$

by the K-invariance of f. Using Remark 2 below, write

$$D^K = \int_K D^{\mathbf{c}(k)}\, dk \qquad \text{so that} \qquad D^K \in \mathrm{IDO}(G)^K.$$

Then $(D^K f)(e) = \lambda(D^K) f(e) = 0$ by the eigenfunction assumption, and the assumption $f(e) = 0$. Integrating formula $(*)$ over $k \in K$ and using the normalization that the total measure of K is 1, we get

$$(Df)(e) = (D^K f)(e) = 0.$$

Now put $D = \tilde{v}^m$. Then for sufficiently small v we get $f(\exp v) = 0$, whence $f = 0$, as was to be shown.

Remark 1. In practice, one may assume only that the function f is C^∞, because there is an invariant real analytic Riemannian metric on G/K, the Laplacian is among the invariant differential operators, and thus f being an eigenfunction satisfies an elliptic differential equation. We then invoke the standard black box of the regularity theorem, of Morrey–Nirenberg, that f is real analytic. The proof is reproduced elegantly and briefly in [Bers 64].

Remark 2. Taking the integral is legitimate for the simple reason that the differential operators $D^{\mathbf{c}(k)}$ all lie in a finite dimensional vector space. Indeed, an invariant differential operator on G is determined by its polynomial differential operator at the origin, and under conjugation the degree is preserved, so these polynomials lie in a finite dimensional space.

In light of Theorem 6.1, we define a **spherical function** on G to be a real analytic function f satisfying:

SPH$_G$ 1. The function of f is K-bi-invariant.

SPH$_G$ 2. The function f is an eigenfunction of the algebra $\mathrm{IDO}(G)^K$.

SPH$_G$ 3. $f(e_G) = 1$.

We repeat Theorem 6.1 in the context of homogeneous spaces. The statement comes from the isomorphism of Chapter II, Theorem 2.5.

Theorem 6.2. *Let G be a connected unimodular Lie group and K a compact subgroup. Let X be a homogeneous space isomorphic to G/K, and reductive. Let f be a K-invariant real analytic function on X, eigenfunction of* $\mathrm{DO}(X)^G$. *Let I be the origin in X (corresponding to the identity coset K). If $f(I) = 0$ then f is the zero function. Hence if f_1, f_2 are real analytic K-invariant eigenfunctions with the same eigencharacter, and $f_1(I) = f_2(I)$, then $f_1 = f_2$.*

Following Harish-Chandra and Selberg, we define a function f on X to be **spherical** if it is real analytic and satisfies:

SPH 1. The function f is K-invariant (for the action of K on the left).

SPH 2. The function f is an eigenfunction of the algebra $\mathrm{DO}(X)^G$.

SPH 3. $f(I) = 1$.

Example. If G has an Iwasawa and Cartan decomposition, and if ρ is a character on X, then ${}^K\rho$ is a spherical function. If χ is a character on A, then ${}^K\chi_G$ viewed as a function on X is a spherical function.

The above example uses implicitly that the functions φ_χ are real analytic. A formal proof follows from the next lemma.

Lemma 6.3. *Let G be a Lie group and K a compact subgroup. Let f be real analytic on G. Then f^K is real analytic.*

Proof. One writes down the power series expansion locally near a point p of G and near a point q of K,

$$f(xk) = F(x, k) = \sum \varphi_{(i,j)}(x, k) M_{(i)}(k) M_{(j)}(x)$$

in terms of coordinates $(x) = (x_1, \ldots, x_n)$ and $(k) = (k_1, \ldots, k_m)$ near p and q, respectively. The $\varphi_{(i,j)}$ are functions, and $M_{(i)}, M_{(j)}$ are monomials. We then use a continuous partition of unity on K, so without loss of generality we can assume that the coefficients of F as power series in (x), with coefficients which are functions of (k), have support in a small neighborhood of q. Integrating over (k) yields a power series in (x) with constant coefficients, thus proving the lemma.

Of course, the lemma also applies to the left K-projection ${}^K f$ and the bi-K-projection ${}^K f^K$. Characters are obviously analytic, so the functions φ_χ are analytic just from this point of view. No need for the black box of the regularity theorem.

Suppose G has an Iwasawa and Cartan decomposition. From the isomorphism $G/K \approx X$, and the general theory (Chapter II, Theorem 2.5) we have a natural isomorphism, even a safe identification, between spherical functions on X and spherical functions on G. Indeed, the ideal $\mathcal{J}(\mathfrak{k})^K$ annihilates K-bi-invariant functions, so a K-bi-invariant eigenfunction of $\mathrm{IDO}(G)^K$ is really an eigenfunction of $\mathrm{IDO}(G)^K/\mathcal{J}(\mathfrak{k})^K \approx \mathrm{DO}(X)^G$. The functions φ_χ are spherical functions in the above generalized sense by Proposition 1.4.

We say that a pair (G, K) is **analytically elliptic** if G, K are analytic Lie groups, and $\mathrm{DO}(X)^G$ or $\mathrm{IDO}(G)^K$ contains an elliptic analytic differential operator. Then Theorems 6.1 and 5.2 immediately yield an axiomatized statement essentially due to Selberg [Sel 56], p. 53, and Harish-Chandra [Har 58a] in the classical cases of symmetric spaces and semisimple Lie groups.

Theorem 6.4. *Let (G, K) be analytically elliptic, G unimodular connected, and $X \approx G/K$ reductive. Let λ be a $\mathbf{C}$-character of $\mathrm{IDO}(G)^K = \mathrm{DO}(X)^G$. Then the eigenspace in*

$$C^\infty(K\backslash G/K) = C^\infty(X)^K$$

of these algebras having the given eigencharacter λ consists of real analytic functions and has dimension 0 or 1. If the dimension is 1, there exists a unique basis f such that $f(e)$ (or $f(I)) = 1$.

For $G = \mathrm{GL}_n(\mathbf{R})$ or $\mathrm{SL}_n(\mathbf{R})$, the only spherical functions are the Harish-Chandra functions φ_χ.

Proof. All but the last statement are contained in the previous results of this section. For the last statement, we need something depending on more structure, namely we need to know that given f there exists a character χ on A such that f and φ_χ have the same eigencharacter, so we can apply Theorem 6.1. We invoke Theorem 5.3 (which depended on Theorem 5.2, and back to Theorem 4.8) to conclude that the $\mathbf{C}$-characters of $\mathrm{DO}(X)^G$ are in natural bijection with the $\mathbf{C}$-characters of $\mathrm{IDO}(A)^W$. By Proposition 1.4, such a character is the eigencharacter of some spherical function φ_χ for some χ. This concludes the proof.

III, §7. INTEGRATION FORMULAS AND THE MAP $x \mapsto x^{-1}$

We want to extend the formalism of spherical functions, and to do so we need more integral formulas. This follows the pattern we have encountered systematically. Readers may skip this section until its results are used later.

By a **pre-Iwasawa decomposition** of a Lie group G we mean an expression $G = PK$ as a differential product of two closed subgroups P, K, such that G and K are unimodular. We suppose the measures normalized so that $dx = dp \otimes dk$ (no fudge factor for $x = pk$). We let δ_P be the modular function on P, so that for $p_1 \in P$ we have

$$\int_P f(pp_1)\,dp = \delta_P(p_1) \int_P f(p)\,dp.$$

We let

$$\mathrm{Iw}_K: G \to K, \quad \mathrm{Iw}_K(pk) = k \quad \text{and} \quad \mathrm{Iw}_P: G \to P, \quad \mathrm{Iw}_P(pk) = p$$

be the projections on each factor. Of course, if $P = UA$ as in the Iwasawa case, we have the additional projections Iw_U and Iw_A. Note that Iw_A is independent of the order in which we write an element $ua = au'$ with $a \in A$ and $u, u' \in U$.

Lemma 7.1. *Let $\varphi \in C_c(P)$ be such that $\int_P \varphi(p)\,dp = 1$. Then for $z \in G$,*

$$\int_P \varphi(\mathrm{Iw}_P(pz))\,dp = \delta_P(\mathrm{Iw}_P(z)).$$

Proof. This is immediate from the definition of Iw_P and the fact that

(1) $$\mathrm{Iw}_P(pz) = p\,\mathrm{Iw}_P(z).$$

We look closer at the projection Iw_K. Since $P \times K \to PK = G$ is a differential isomorphism, it follows that for each $x \in G$, the map

$$F_x: K \to K \qquad \text{given by} \qquad F_x(k) = \mathrm{Iw}_K(kx^{-1})$$

is a differential automorphism. The map $x \mapsto F_x$ is a representation of G as a group of permutations of K, that is:

(2) $$F_x \circ F_z = F_{xz} \qquad \text{and so} \qquad F_{x^{-1}} = (F_x)^{-1}.$$

Thus we obtain the corresponding formulas for the **absolute value of the Jacobian determinant**, denoted JF_x. Especially we get the formula

(3) $$(JF_x)^{-1} = J(F_{x^{-1}}).$$

Calculus tells us that

$$\int f(u(t))\frac{du}{dt}\,dt = \int f(u)\,du,$$

and this formula is equivalent to

$$\int f(u(t))\,dt = \int f(u)\frac{dt}{du}\,du \quad \text{with} \quad dt/du = \frac{1}{du/dt}.$$

We shall now give the explicit form for these formulas, using F_x instead of u.

Lemma 7.2 (Harish-Chandra). *We have*

$$JF_x = \delta_P(\mathrm{Iw}_P(kx^{-1})).$$

In other words, if we put $k' = F_x(k)$ or $kx^{-1} = pk'$ with $p \in P$, then

$$\frac{dk'}{dk} = \delta(p).$$

Or still put another way, for $f \in C_c(K)$,

$$\int_K f(\mathrm{Iw}_K(kx^{-1}))\delta_P(\mathrm{Iw}_P(kx^{-1}))\,dk = \int_K f(k')\,dk'.$$

Proof. Let $\varphi \in C_c(P)$ be such that $\int_P \varphi(p)\,dp = 1$. Then

$$\begin{aligned}
\int_K f(k)\,dk &= \int_P\int_K f(k)\varphi(p)\,dp\,dk = \int_G (f(\mathrm{Iw}_K(z))\varphi(\mathrm{Iw}_P(z))\,dz \\
&= \int_G f(\mathrm{Iw}_K(zx^{-1}))\varphi(\mathrm{Iw}_P(zx^{-1}))\,dz \quad \text{by } z \mapsto zx^{-1} \\
&= \int_P\int_K f(\mathrm{Iw}_K(pkx^{-1}))\varphi(\mathrm{Iw}_P(pkx^{-1}))\,dk\,dp \quad \text{by } z = pk \\
&= \int_K f(\mathrm{Iw}_K(kx^{-1}))\left[\int_P \varphi(\mathrm{Iw}_P(pkx^{-1}))\,dp\right]dk \\
&= \int_K f(\mathrm{Iw}_K(kx^{-1}))\delta_P(\mathrm{Iw}_P(kx^{-1}))\,dk \quad \text{by Lemma 7.1,}
\end{aligned}$$

thus proving the lemma.

From (3) and Lemma 7.2 we may rewrite the integral formula in the form

Lemma 7.2′.

$$\int_K f(\mathrm{Iw}_K(kx^{-1}))\,dk = \int_K f(k)\delta_P(\mathrm{Iw}_P(kx))\,dk.$$

Next we suppose that $P = UA$ with A abelian normalizing U, so we have the projection Iw_A. Directly from the definitions, translating old remarks into the current notation yields

$$\delta_P(\mathrm{Iw}_P(z)) = \delta(\mathrm{Iw}_A(z)) \quad \text{for all } z \in G,$$

where δ is a character on A. With more groups coming in, it may be useful to tabulate a cocycle formalism of the Iwasawa projection with respect to various multiplicative translations. For $k \in K$ and $x, y \in G$ we have

$$\text{(4)} \qquad \mathrm{Iw}_A(kxy) = \mathrm{Iw}_A(kx)\,\mathrm{Iw}_A(\mathrm{Iw}_K(kx)y).$$
$$\text{(5)} \qquad \mathrm{Iw}_A(kx^{-1}) = (\mathrm{Iw}_A(\mathrm{Iw}_K(kx^{-1})x))^{-1}.$$

For (4), we write

$$kx = u_1a_1k_1 \quad \text{and}$$
$$kxy = u_1a_1(k_1y) = u_1a_1u_2a_2k_2 = u_1\mathbf{c}(a_1)u_2a_1a_2k_2.$$

Thus both sides of (4) are equal to a_1a_2 directly from the definitions. Then (5) follows by putting $y = e$, replacing x by x^{-1}, and noting that $\mathrm{Iw}_A(k) = 1$.

Next we come to spherical functions and the spherical kernel $\varphi(\chi, x)$. We prove a basic property of the spherical transform $\mathbf{S}$, which is the integral transform associated with the spherical kernel, so by definition,

$$(\mathbf{S}f)(\chi) = (\varphi * f)(\chi) = \int_G \varphi(\chi, x) f(x)\, dx.$$

Proposition 7.3. *Let G be a Lie group with an Iwasawa decomposition $G = UAK$ and K compact. Let φ be the spherical kernel. Then:*

(i) (Harish-Chandra [Har 58a]). *For all $x \in G$,*

$$\varphi(\chi, x^{-1}) = \varphi(\chi^{-1}, x).$$

(ii) (Helgason [Hel 70]). *For all $x, y \in G$,*

$$\varphi_\chi(x^{-1}y) = \int_K (\chi^{-1}\delta^{1/2})(\mathrm{Iw}_A(kx)) \cdot (\chi\delta^{1/2})(\mathrm{Iw}_A(ky))\, dk.$$

Proof. Note that (i) is a consequence of (ii), obtained by putting $y = e$. For (ii), it will be convenient to use the notation χ_G for the Iwasawa extension of a character χ on A. In other words, if $x = uak$, we define

$$\chi_G(x) = \chi(a) = \chi(ua).$$

For a given $x \in G$ and k variable in K, we put

$$kx^{-1} = p_k k' \quad \text{so that} \quad k = p_k k' x.$$

Then for $y \in G$, $\chi_G(p_k k' y) = \chi(p_k)\chi_G(k'y)$, and so using Lemma 7.2,

$$\varphi_\chi(x^{-1}y) = \int_K (\chi\delta^{1/2})_G(kx^{-1}y)\,dk$$
$$= \int_K (\chi_G\delta_G^{1/2})(kx^{-1}y)(\chi_G\delta_G^{1/2})(k)\,dk$$
$$= \int_K \chi(p_k)\delta^{1/2}(p_k)(\chi_G\delta_G^{1/2})(k'y)\chi(p_k)\delta^{1/2}(p_k)(\chi\delta^{1/2})_G(k'x)\delta^{-1}(p_k)\,dk'$$
$$[\text{because } dk/dk' = \delta^{-1}(p_k)]$$
$$= \int_K \chi^2(p_k)(\chi\delta^{1/2})_G(k'y)(\chi\delta^{1/2})_G(k'x)\,dk'.$$

But $k'x = p_k^{-1}k$ so $\chi_G(k'x) = \chi(p_k)^{-1}$, and $\chi^2(p_k) = \chi_G^{-2}(k'x)$. Hence the desired formula drops out with k' instead of k throughout, thus proving the proposition.

Corollary 7.4. *Let $f \in C_c^\infty(K\backslash G/K)$. Define $f^*(x) = \overline{f(x^{-1})}$. Then*

$$\mathbf{S}(f^*)(\chi) = (\mathbf{S}\bar{f})(\chi^{-1}).$$

In particular, if χ is unitary, then

$$\mathbf{S}(f^*)(\chi) = \overline{\mathbf{S}f(\chi)}.$$

Proof. By definition,

$$\mathbf{S}(f^*)(\chi) = \int_G \varphi(\chi, x)\overline{f(x^{-1})}\,dx$$
$$= \int_G \varphi(\chi^{-1}, x)\overline{f(x)}\,dx \quad \text{by } x \mapsto x^{-1} \text{ and Proposition 7.3(i)}$$
$$= (\mathbf{S}\bar{f})(\chi^{-1}) \quad \text{by definition.}$$

This proves the first formula, and the second follows since for unitary χ, we have $\chi^{-1} = \bar{\chi}$.

The anti-Iwasawa decomposition

Through the book, we mostly phrase results in terms of the Iwasawa decomposition $G = UAK$ with K compact. However, in Lemma 7.2 we already consider one of the most basic properties having to do with the **anti-Iwasawa decomposition** $G = KAU$. It amounts to investigating the map $x \mapsto x^{-1}$ in some concrete cases to see how it affects past formulas. We continue to deal with left invariant differential operators. In Chapter II we considered those which additionally are K-fixed. Here we do not impose this restriction and give some formulas valid in general. Harish-Chandra and others following him use principally the anti-Iwasawa decomposition. Let

$$x = k'a'u' \qquad \text{with } k' \in K,\ a' \in A,\ u' \in U,$$

be the anti-Iwasawa decomposition. We write the anti-Iwasawa projection on A as

$$\mathrm{Iw}'_A(x) = a' \qquad \text{as distinguished from} \qquad \mathrm{Iw}_A(x) = x_A = a.$$

Let $a' = \exp(H)$ with $H \in \mathfrak{a}$. Then we write

$$H = H(x) = \log \mathrm{Iw}'_A(x).$$

The two projections are related by the simple formula

$$\text{(6)} \qquad \mathrm{Iw}'_A(x^{-1}) = \mathrm{Iw}_A(x)^{-1} \qquad \text{or also} \qquad H(x^{-1}) = -\log a.$$

As in §2, let $\rho = \frac{1}{2}\log\delta$. Writing $\chi = \chi_\zeta$ with a character $\zeta \in \mathfrak{a}^\vee_{\mathbf{C}}$, we write $\varphi(\zeta, x)$ for the spherical function $\varphi(\chi_\zeta, x)$, defined by the Harish-Chandra integral

$$\varphi_\zeta(x) = \varphi(\zeta, x) = \int_K \mathrm{Iw}_A(kx)^{\zeta+\rho}\, dk.$$

Proposition 7.5. *With respect to the anti-Iwasawa decomposition,*

$$\varphi(\zeta, x) = \int_K e^{(\zeta-\rho)H(xk)}\, dk = \int_K \mathrm{Iw}'_A(xk)^{\zeta-\rho}\, dk.$$

Proof. This is immediate from the Harish-Chandra formula of Theorem 7.3(i) and (6) above, namely

$$\varphi(\zeta, x) = \varphi(-\zeta, x^{-1}) = \int_K \mathrm{Iw}_A(kx^{-1})^{-\zeta+\rho}\, dk = \int_K \mathrm{Iw}'_A(xk^{-1})^{\zeta-\rho}\, dk,$$

and finally using the invariance of dk under $k \mapsto k^{-1}$. QED.

The formula of Proposition 7.5 is what readers will mostly find in the literature, including Anker's paper [Ank 91] which will play the major role in Chapter X. Note how minus signs are shuffled via both Theorem 7.3(i) and the map $x \mapsto x^{-1}$ on G, changing Iwasawa into anti-Iwasawa.

We define the function g_ζ on G by

$$g_\zeta(x) = \mathrm{Iw}'_A(x)^{\zeta-\rho},$$

so that

$$\varphi_\zeta(x) = \int_K g_\zeta(xk)\,dk = \int_K g_\zeta(xk^{-1})\,dk.$$

We add one more eigenfunction theorem to our list in connection with the anti-Iwasawa decomposition.

Proposition 7.6. *The function g_ζ is an eigenfunction of $\mathcal{D}_G(S(\mathfrak{a}))$, and in fact, if $E = \mathcal{D}_G(P)$ with $P \in S(\mathfrak{a})$, then*

$$Eg_\zeta = P(\zeta - \rho)g_\zeta.$$

Proof. It suffices to prove the relation for an operator $\tilde{H}$ with $H \in \mathfrak{a}$. By definition,

$$\begin{aligned}(\tilde{H}g_\zeta)(x) &= \frac{d}{dt}\mathrm{Iw}'_A(x\exp(tH))^{\zeta-\rho}\Big|_{t=0} = \frac{d}{dt}\mathrm{Iw}'_A(au\exp(tH))^{\zeta-\rho}\Big|_{t=0}\\ &= \frac{d}{dt}(a\exp(tH))^{\zeta-\rho}\Big|_{t=0} \qquad \text{because } A \text{ normalizes } U\end{aligned}$$

so the identity is reduced to the identity on A, where it is immediate.

Remark. We are using systematically left invariant differential operators, so to apply the above argument, we must have the U-component on the right side, whence the need for the anti-Iwasawa decomposition. The proposition will be used in Chapter X, Lemma 4.1, which may be seen as a continuation of the present material on anti-Iwasawa. It is not true that φ_ζ is an eigenfunction of $\mathcal{D}_G(S(\mathfrak{a}))$, let alone of $\mathrm{IDO}(G)$. Of course, Proposition 1.5 showed how it is an eigenfunction of $\mathrm{IDO}(G)^K$. Proposition 7.6 and its continuation in Chapter X give information on the way $D\varphi_\zeta$ depends on φ_ζ and D, using the anti-Iwasawa decomposition.

III, §8. W-HARMONIC POLYNOMIALS AND EIGENFUNCTIONS OF W-INVARIANT DIFFERENTIAL OPERATORS ON A

This section is an entirely self-contained piece of elementary algebra. For our immediate purposes, we want to determine invariants associated with the polynomial algebra over the reals, the group of permutation of variables and certain subgroups, and the differential operators obtained by substituting the partial derivatives for the variables. In line with our overall point of view, we do not want generality to add complications to results which are as transparent as they can be in the concrete case of SL_n, which is our main concern. So the exposition of the chapter starts from the standard situation of n real variables, and W is the group of permutation of the variables. The existence of the elementary symmetric polynomials will then be used automatically. Readers may note that much less is used, and the appropriate generalizations will be discussed at the end of the section.

Let k be a field of characteristic 0. Let $\mathrm{Pol}(n, d)$ denote the vector space of homogeneous polynomials of degree d in n variables $x_1, \ldots, x_n$. For an n-tuple of integers $(j_1, \ldots, j_n)$ with $j_i \geqq 0$ all i, we denote by $M_{(j)}$ the monomial

$$M_{(j)}(x) = x_1^{j_1} \cdots x_n^{j_n}.$$

The number of such monomials of degree d is the binomial coefficient

$$\binom{n-1+d}{n-1},$$

which is also the dimension of $\mathrm{Pol}(n, d)$.

Let $(\partial) = (\partial_1, \ldots, \partial_n)$ be the n-tuple of partial derivatives with respect to the n variables. We can define $P(\partial) = P(\partial_1, \ldots, \partial_n)$ as usual. For $P, Q \in \mathrm{Pol}(n, d)$ define the scalar product

$$\langle P, Q \rangle = P(\partial) Q(0).$$

Proposition 8.1.

(i) *This product is symmetric and non-degenerate.*

(ii) *If the ground field is* **R**, *then the product is positive definite.*

(iii) *The monomials of given degree d form an orthogonal basis for* $\mathrm{Pol}(n, d)$. *Monomials of different degrees are orthogonal.*

(iv) *The map* $P \mapsto P(\partial)$ *is a linear isomorphism of* $\mathrm{Pol}(n, d)$ *with its dual space.*

(v) *For* $R \in \mathrm{Pol}(n, m)$ *and* $Q \in \mathrm{Pol}(n, n-m)$ *we have*

$$\langle R(\partial) P, Q \rangle = \langle P, RQ \rangle,$$

which determines the adjoint of multiplication by a polynomial.

Proof. Left as an exercise in elementary algebra. Just as an example that nothing deep is going on, we show how to do (v). We have

$$\langle P, RQ \rangle = (RQ)(\partial) P(0) = Q(\partial) R(\partial) P(0) = \langle R(\partial) P, Q \rangle,$$

because partials commute. QED.

Proofs of subsequent properties will be given as usual.

We abbreviate

$$S = \mathrm{Pol}_n = \bigoplus_{d=1}^{\infty} \mathrm{Pol}(n, d)$$

for the algebra of polynomials in n variables. We also abbreviate

$$S^{(d)} = \mathrm{Pol}(n, d)$$

for the subspace of homogeneous elements of degree d (other than 0). We let:

S^W = subalgebra of W-invariant polynomials, that is, symmetric polynomials.

S_+^W = subspace of W-invariant polynomials with zero constant term.

Of course, S^W is generated by the elementary symmetric polynomials. As far as we are told, it goes back to Kronecker that S is free over S^W. It is a simple exercise to show that the monomials

$$x_1^{r_1} \cdots x_n^{r_n} \qquad \text{with } 0 \leqq r_i \leqq n - i$$

form a basis. (Yes, there are $n!$ of them.) However, we want more, in two directions. So we define a polynomial H to be W-**harmonic** if and only if

$$Q(\partial)H = 0 \qquad \text{for all } Q \in S_+^W.$$

We let:

Har_W = vector space of W-harmonic polynomials;

$\mathrm{Har}_W^{(d)}$ = vector subspace of homogeneous elements of degree d.

Then

$$\mathrm{Har}_W = \bigoplus_{d=0}^{\infty} \mathrm{Har}_W^{(d)}.$$

In other words, a polynomial is W-harmonic if and only if all its homogeneous components are W-harmonic. This is immediate from the fact that homogeneous polynomials of different degrees are orthogonal.

Theorem 8.2 (Chevalley [Che 55]). *Over the reals, we have*

$$S = S^W \mathrm{Har}_W.$$

Furthermore, $\dim \mathrm{Har}_W = |W|$, *and* S *is free of dimension* $|W|$ *over* S^W, *so the above product gives a linear isomorphism*

$$S \approx S^W \otimes \mathrm{Har}_W .$$

Proof. First we show that

$$S = S_+^W S \oplus \mathrm{Har}_W \quad \text{(orthogonal direct sum).} \tag{1}$$

Let P be a polynomial orthogonal to $S_+^W S$. For $Q \in S_+^W$ we get

$$0 = \langle QS, P \rangle = \langle S, Q(\partial)P \rangle,$$

from which it follows that $Q(\partial)P = 0$, whence $P \in \mathrm{Har}_W$. The reverse inclusion follows by reversing the steps in this argument, so (1) is proved. Since homogeneous polynomials of different degrees are orthogonal, we get the homogeneous version for each positive integer d, namely

$$S^{(d)} = \sum_{r=1}^{d} (S^W)^{(r)} S^{(d-r)} \oplus \mathrm{Har}_W^{(d)} . \tag{1_d}$$

We may now repeat inductively to get the first assertion of the theorem, namely

$$S = S^W \mathrm{Har}_W .$$

Next we need two lemmas about the ideal $S_+^W S$ of S, following Chevalley.

Lemma 8.3. *Let* $Q_1, \ldots, Q_m \in S^W$ *be such that* Q_1 *does not belong to the ideal* $(Q_2, \ldots, Q_m)$ *of* S^W. *Let* $P_1, \ldots, P_m$ *be homogeneous elements of* S *such that* $\sum_{\nu=1}^m P_\nu Q_\nu = 0$. *Then* $P_1 \in S_+^W S$.

Proof. Let A be the projection operator on W-invariants, namely

$$A(P) = \frac{1}{|W|} \sum_{w \in W} w(P).$$

We apply this operator to the linear relation and get

$$A(P_1)Q_1 + \cdots + A(P_m)Q_m = 0.$$

If $\deg P_1 = 0$, so P_1 is constant, then $P_1 = A(P_1)$, contradicting the hypothesis. Hence $\deg P_1 > 0$. Assume the lemma by induction for all relations

$$\sum_{\nu=1}^{m} R_\nu Q_\nu = 0$$

with homogeneous R_ν and $\deg R_1 < \deg P_1$. Let τ be a transposition of two variables, say x_j, x_k. Then $\tau(P_\nu) - P_\nu$ is divisible by $x_j - x_k$, that is

$$\tau(P_\nu) - P_\nu = (x_j - x_k)R_\nu$$

for some homogeneous polynomial R_ν of degree smaller than $\deg P_\nu$, and we have the linear relation

$$R_1 Q_1 + \cdots + R_m Q_m = 0.$$

By induction, $R_1 \in S_+^W S$, or in other words, $\tau(P_1) \equiv P_1 \bmod S_+^W S$. Since W is generated by transpositions, it follows that

$$w(P_1) \equiv P_1 \quad \bmod S_+^W S \qquad \text{for all } w \in W.$$

Hence $A(P_1) \equiv P_1 \bmod S_+^W S$. Since P_1 is homogeneous of strictly positive degree, the same is true for $A(P_1)$. Hence $A(P_1) \in S_+^W S$, whence finally $P_1 \in S_+^W S$, which proves the lemma.

Lemma 8.4. *Let $H_1, \ldots, H_m$ be homogeneous elements of S whose residue classes* mod $S_+^W S$ *are linearly independent over the constants. Then $H_1, \ldots, H_m$ are linearly independent over S^W.*

Proof. Let $R_1 H_1 + \cdots + R_m H_m = 0$ be a relation of linear dependence with $R_i \in S^W$ not all 0. Without loss of generality, we may assume none of the coefficients $R_1, \ldots, R_m = 0$. Furthermore we may also assume that the polynomials R_i are homogeneous, and $\deg R_i + \deg H_i = d$ is constant, independent of i.

Let $I_1, \ldots, I_n$ be the elementary symmetric polynomials, which are homogeneous generators of the algebra S^W. The monomials $I_1^{j_1} \cdots I_n^{j_n}$ are elements of S, and $\deg I_1^{j_1} \cdots I_n^{j_n}$ refers to the degree in S. Let $Q_0, Q_1, \ldots$ be an ordering of these monomials by increasing degree, with $Q_0 = 1$. For each $i = 1, \ldots, m$ we have a linear expression

$$R_i = \sum_{\nu \geqq 0} c_{i\nu} Q_\nu$$

with constant coefficients $c_{i\nu}$, and $c_{i\nu} = 0$ if $\deg R_i \neq \deg Q_\nu$. Collecting terms yields

$$0 = \sum_{i=1}^m R_i H_i = \sum_{\nu \geqq 0} P_\nu Q_\nu \qquad \text{with} \quad P_\nu = \sum_{i=1}^m c_{i\nu} H_i,$$

and P_ν is homogeneous of degree $d - \deg Q_\nu$. We prove that $P_\nu = 0$ for all ν by induction. For $\nu = 0$, we note that $Q_0 \notin (Q_1, Q_2, \ldots)$, so by Lemma 8.3, $P_0 \in S_+^W S$, which contradicts the linear independence assumption on $H_1, \ldots, H_m$. Inductively, suppose that $P_0, \ldots, P_{r-1} = 0$. We can then argue in exactly the same way, because

$$Q_r \notin (Q_{r+1}, Q_{r+2}, \ldots).$$

This concludes the proof of Lemma 8.4.

Theorem 8.2 is now an immediate consequence of (1) and Lemma 8.4. Indeed, (1) shows that homogeneous linearly independent W-harmonic polynomials $H_1, \ldots, H_m$ satisfy the hypothesis of Lemma 8.4, and hence these polynomials are also linearly independent over S^W. The number of such polynomials is at most $|W|$ by Galois theory. The first identity $S = S^W \operatorname{Har}_W$ shows that $\dim \operatorname{Har}_W$ must be $|W|$, and that a basis of Har_W over $\mathbf{R}$ is a basis of S over S^W, thus concluding the proof.

Remark 1. The theorem actually applies over the rationals, and then follows at once over any field of characteristic 0.

Remark 2. Chevalley's paper obtains the result in general for reflection groups, i.e. groups generated by reflections. See the end of the section. This hypothesis comes into play at the step where he uses the divisibility argument in Lemma 8.3. There is a further general context for Lie groups, with a systematic account in [Hel 84], Chapter III, containing the Chevalley results, and others by Harish-Chandra, Helgason, Kostant, Maass, Steinberg, etc. Cf. Helgason's historical notes for his Chapter III. The general context includes the classical case of the orthogonal group, and its invariants which are polynomials in the Laplacian. Harmonicity in this case has its classical meaning: annihilated by the Laplacian.

We now give applications of Theorem 8.2 to the invariant differential operators. We start with $\mathbf{R}^n$ and its polynomial algebra $S = \mathbf{R}[x_1, \ldots, x_n]$ with variables $x_1, \ldots, x_n$. The algebra of invariant differential operators on $\mathbf{R}^n$ (invariant under translations) is the algebra of polynomial differential operators with constant coefficients,

$$\operatorname{IDO}(\mathbf{R}^n) = S(\partial).$$

An element of $S(\partial)$ is just a polynomial differential operator

$$P(\partial) = P(\partial_1, \ldots, \partial_n), \qquad \text{with } \partial_i = \partial/\partial x_i.$$

We define a W-**harmonic function** f on $\mathbf{R}^n$ just as we did harmonic polynomials; it is a C^∞ function such that $P(\partial)f = 0$ for all $P \in S_+^W$.

Theorem 8.5. *The W-harmonic functions on $\mathbf{R}^n$ are just the polynomial harmonic functions, so just the elements of* Har_W.

Proof. Let $I_1, \ldots, I_n$ be the elementary symmetric functions of the variables $x_1, \ldots, x_n$. Then for each i, x_i is a root of the symmetric polynomial,

$$0 = x_i^n - I_1 x_i^{n-1} + \cdots + (-1)^n I_n.$$

Hence $x_i^n \in S_+^W S$ for $i = 1, \ldots, n$, and therefore

$$\partial_i^n \in S(\partial) S_+^W(\partial),$$

so ∂_i^n annihilates every W-harmonic function. Thus finally, if f is W-harmonic, then $\partial_i^n f = 0$ for all i, whence f is a polynomial.

All that remains to be done is to put together the above results with some standard algebra, as in Lang's *Algebra*. We let $S = \mathbf{R}[x_1, \ldots, x_n]$ be the polynomial ring as above, and W the group of automorphisms permuting the variables. By a **C-character** of S we mean an algebra homomorphism of S into $\mathbf{C}$. The $\mathbf{C}$-characters λ are in bijection with n-tuples of complex numbers $(s_1, \ldots, s_n)$, namely for a polynomial $P \in S$,

$$\lambda_s(P) = P(s_1, \ldots, s_n) = P(s) \in \mathbf{C}.$$

The ring S is integral over the ring of invariants S^W. By *Algebra*, Chapter VII, Proposition 3.1, a homomorphism $\lambda: S^W \to \mathbf{C}$ extends to S, and all such extensions are conjugate under the Galois group W (Chapter VII, Proposition 2.1 and Corollary 2.6). Thus two $\mathbf{C}$-characters of S induce the same $\mathbf{C}$-character of S^W if and only if they are conjugate under the Galois group W. These results will be used without further reference.

Let $\mathbf{S}^n$ be affine n-space viewed as an algebraic variety. Its set of points in $\mathbf{C}$ is denoted by $\mathbf{S}^n(\mathbf{C})$. By definition, these points are the $\mathbf{C}$-characters of the coordinate ring $\mathbf{R}[x_1, \ldots, x_n]$. The ring S^W is the coordinate ring of $\mathbf{S}^n(\mathbf{C})/W$, and a homomorphism of S^W into $\mathbf{C}$ is then a point of this quotient space. We have $S^W = \mathbf{C}[I_1, \ldots, I_n]$. Let $\mu: S^W \to \mathbf{C}$ be a homomorphism, so a point on the quotient variety $\mathbf{S}^n(\mathbf{C})/W$, which is actually also affine n-space, with affine coordinates $I_1, \ldots, I_n$. Then μ has at most $|W|$ extensions to homomorphisms $\lambda: S \to \mathbf{C}$, that is, to points of $\operatorname{spec} \mathbf{R}[x_1, \ldots, x_n]$ in $\mathbf{C}$. If λ is an extension of μ, then the subgroup W_λ of W leaving λ fixed is called the **inertia group**. If this inertia group is non-trivial, then one says that μ is **ramified**, with **ramification index** equal to the order of this inertia group. This is the terminology which arose from a similar situation in algebraic number theory. We don't really need these remarks, but they illuminate the situation by giving it an algebraic geometric–topological context.

We shall say that λ_s or $s = (s_1, \ldots, s_n)$ is **regular** if the coordinates $s_1, \ldots, s_n$ are distinct. Equivalently, λ is regular if and only if the only element $w \in W$ such that $[w]\lambda = \lambda$ is the identity. Note that the action of W on λ is given by

$$([w]\lambda)(x) = \lambda([w^{-1}]x).$$

The terminology "regular" is the terminology used by the Lie industry to express the property that λ is unramified over its restriction to S^W, i.e. the inertia group W_λ is trivial.

We shall also denote a **C**-character of S^W by λ_W, so it is the restriction of some character λ of S.

Let $\mathfrak{a}$ be a vector space over $\mathbf{R}$ of dimension n, with basis $\{e_1, \ldots, e_n\}$. Then $\mathfrak{a} \approx \mathbf{R}^n$ via the coordinate vectors with respect to the basis, namely the expression

$$x = x_1 e_1 + \cdots + x_n e_n \in \mathfrak{a}, \qquad x_i \in \mathbf{R}.$$

By the **λ-eigenspace** or **λ_W-eigenspace** of $S^W(\partial)$ in $\mathrm{Fu}(\mathfrak{a})$ we mean the subspace of functions $f \in \mathrm{Fu}(\mathfrak{a})$ such that

$$P(\partial)f = \lambda(P)f \qquad \text{for all } P \in S^W.$$

Elements of this eigenspace will be called λ or **λ_W-eigenfunctions** of $S^W(\partial)$, that is, eigenfunctions of the W-invariant polynomial differential operators, with corresponding eigencharacter λ_W. We denote this eigenspace by $\mathrm{Fu}(\mathfrak{a}, S^W(\partial), \lambda_W)$, or $\mathrm{Fu}(\mathfrak{a}, S^W(\partial), \lambda)$.

Example. Let $s = (s_1, \ldots, s_n)$ be an n-tuple of complex numbers. Let

$$\chi_s \colon \mathfrak{a} \to \mathbf{C}^*$$

be the exponential function defined by

$$\chi_s(x) = e^{s \cdot x} = e^{s_1 x_1} \cdots e^{s_n x_n} = a_1^{s_1} \cdots a_n^{s_n}.$$

Then χ_s is a λ_s-eigenfunction of $S^W(\partial)$. Instead of writing χ_s we also write χ_{λ_s} or χ_λ, so we have

$$\chi_s \in \mathrm{Fu}(\mathfrak{a}, S^W(\partial), \lambda_s) \qquad \text{also written} \qquad \chi_\lambda \in \mathrm{Fu}(\mathfrak{a}, S^W(\partial), \lambda_W).$$

In other words, the exponential functions are eigenfunctions for the action of $S^W(\partial)$. It could be proved a priori that any eigenfunction for this action is what we may call an **exponential polynomial** with **polynomial coefficients**, that is, a linear combination

$$f = \sum p_s \chi_s \qquad \text{or} \qquad f(x) = \sum p_s(x) e^{s \cdot x}$$

with $p_s(x) \in \mathbf{C}[x_1, \ldots, x_n]$ and s ranging over a finite number of elements of $\mathbf{C}^n$. Cf. the proof of Theorem 8.5. On the other hand, from the general regularity theorem for elliptic operators, one also sees the weaker fact that an eigenfunction is necessarily real analytic. This is the only property we shall use in subsequent arguments, which will then describe eigenfunctions as exponential polynomials in a much more precise fashion. We start with the regular (unramified) case.

Theorem 8.6 (Steinberg [Ste 64]). *For every* **C**-*character* λ *of* S *we have*

$$\dim \mathrm{Fu}(\mathfrak{a}, S^W(\partial), \lambda_W) = |W|.$$

If λ *is regular* (*unramified*), *then the elements* $[w]\chi_\lambda$ $(w \in W)$, *or equivalently* $\chi_{[w]\lambda}$ $(w \in W)$, *constitute a basis of* $\mathrm{Fu}(\mathfrak{a}, S^W(\partial), \lambda_W)$.

Proof. We first prove that the eigenspace has dimension $\leqq |W|$. Let $\{H_1, \ldots, H_N\}$ be a basis for Har_W. Map the eigenspace into $\mathbf{C}^n$ by

$$f \mapsto ((H_1(\partial)f)(0), \ldots, (H_N(\partial)f)(0)).$$

The map is linear, and it suffices to prove that it is injective. If the image of f is $(0, \ldots, 0)$, then $(H(\partial)f)(0) = 0$ for all $H \in \mathrm{Har}_W$. By Theorem 7.2, $S = S^W \mathrm{Har}_W$. Let $P \in S^W$. Then for some scalar λ we get

$$H(\partial)P(\partial)f = \lambda H(\partial)f,$$

so for all $Q \in S$ we get $(Q(\partial)f)(0) = 0$. Hence the Taylor series of f at 0 is identically zero. Since f is real analytic, it follows that $f = 0$, whence the injectivity and the bound $|W|$ on the dimension of the eigenspace. In the unramified case, the characters $[w]\chi_\lambda$ are distinct, so are linearly independent by elementary algebra (Artin's theorem), whence they must form a basis for the eigenspace, thus concluding the proof in this case.

In the ramified case, one has to use a more refined analysis of the inertia group. Polynomial linear combinations of characters then occur as eigenfunctions as we shall now see.

We define a W_λ-**harmonic function** h in the same way that we defined a W-harmonic function, namely a C^∞ function such that $P(\partial)h = 0$ for all polynomials $P \in S_+^{W_\lambda}$. We let

$$\mathrm{Har}_{W_\lambda} = \text{vector space of } W_\lambda\text{-harmonic polynomials.}$$

Since a W_λ-harmonic function is W-harmonic, it follows from Theorem 8.5 that the W_λ-harmonic polynomials coincide with the W_λ-harmonic functions.

If λ is regular, then $W_\lambda = \{\mathrm{id}\}$, so the W_λ-harmonic polynomials are just the constants. More generally:

Lemma 8.7. *Let $H \in \mathrm{Har}_{W_\lambda}$ be W_λ-harmonic. Then $H\chi_\lambda$ is a λ-eigenfunction, in other words*

$$\mathrm{Har}_{W_\lambda}\chi_\lambda \subset \mathrm{Fu}(\mathfrak{a}, S^W(\partial), \lambda_W).$$

For every $w \in W/W_\lambda$ we also have the inclusion

$$\mathrm{Har}_{W_{[w]\lambda}}\chi_{[w]\lambda} \subset \mathrm{Fu}(\mathfrak{a}, S^W(\partial), \lambda_W).$$

Proof. Let $\lambda = \lambda_s$. Let $P \in S^{W_\lambda}$. Let Q be the polynomial such that

$$Q(T) = P(T+s) - P(s),$$

where $T + s = (T_1 + s_1, \ldots, T_n + s_n)$. Then Q has 0 constant term. If w is a permutation of the variables leaving $(s_1, \ldots, s_n)$ fixed, then w also leaves P fixed, so w leaves Q fixed. Now consider $H \in \mathrm{Har}_{W_\lambda}$. Then $Q(\partial)H = 0$, and

$$\begin{aligned} P(\partial)(H(x)e^{s\cdot x}) &= e^{s\cdot x}e^{-s\cdot x}P(\partial)(e^{s\cdot x}H(x)) \\ &= e^{s\cdot x}P(\partial + s)H(x) \\ &= e^{s\cdot x}(P(\partial+s) - P(s))H(x) + e^{s\cdot x}P(s)H(x) \\ &= P(s)H(x)e^{s\cdot x}. \end{aligned}$$

This shows that $H\chi_\lambda$ is an eigenfunction, and proves the first inclusion of the lemma. The second one with an arbitrary w follows from the general formalism,

$$[w]P(\partial)([w]H\chi_{[w]\lambda}) = [w](P(\partial)(H\chi_\lambda)).$$

This concludes the proof.

Theorem 8.8. *For every* **C**-*character λ of S, we have*

$$\dim \mathrm{Har}_{W_\lambda} = |W_\lambda|.$$

The λ_W-eigenspace is a direct sum

$$\mathrm{Fu}(\mathfrak{a}, S^W, \lambda_W) = \bigoplus_{w\in W/W_\lambda} \mathrm{Har}_{W_{[w]\lambda}} \cdot \chi_{[w]\lambda}.$$

Proof. By Lemma 8.7, the sum of the terms

(*) $$\sum_{w\in W/W_\lambda} \mathrm{Har}_{W_{[w]\lambda}} \cdot \chi_{[w]\lambda}$$

is constrained in the λ-eigenspace. Distinct exponentials $\chi_1, \ldots, \chi_N$ are linearly independent over the polynomial functions. (Cf. Lemma 8.9.) We now need an extension of Theorem 8.2 to subgroups generated by transposition. Actually, Chevalley proved his theorem for an arbitrary finite group generated by reflections. The proof, which we reproduced, is valid in this more general case, but requires one more argument to provide the fact corresponding to the existence of the algebraically independent elementary symmetric polynomials $I_1, \ldots, I_n$. See below. Once this is available, the rest of the proof shows that

$$\dim \mathrm{Har}_{W_\lambda} = |W_\lambda|.$$

Then we can conclude from the sum (*) and the linear independence of the exponentials $\chi_{[w]\lambda}$ (with $w \in W_\lambda$) that the sum is direct, and that the dimension of this sum of subspaces is

$$(W : W_\lambda) \dim \mathrm{Har}_W = (W : W_\lambda)|W_\lambda| = |W|.$$

Since we showed at the beginning of the proof of Theorem 8.6 that the eigenspace has dimension $\leqq |W|$, it follows that the sum (*) is equal to the whole eigenspace, thus concluding the proof of the theorem.

Remark. The combination of Theorems 8.6 and 8.8 in the more general case of reflection groups is due to Steinberg [Ste 64]. Helgason [Hel 84], Chapter III, Theorem 3.13, gives an exposition of Steinberg's result which we found very useful, and we have followed its main line.

We shall now provide the missing arguments which allow the greater generality in Chevalley's and Steinberg's theorems.

First we take care of the lemma about the linear independence of characters over polynomials.

Lemma 8.9. *Let $s^{(1)}, \ldots, s^{(N)}$ be distinct n-tuples of complex numbers, and let*

$$\chi_i(x) = e^{s^{(i)} \cdot x} \qquad \text{for } x \in \mathbf{R}^n.$$

Then the functions $\chi_1, \ldots, \chi_N$ are linearly independent over the ring of polynomials.

Proof. We reduce the statement to one variable. For all $x \in \mathbf{R}^n$ outside a finite number of hyperplanes, the complex numbers $s^{(i)} \cdot x$ are distinct. Indeed, consider the set of differences $s^{(i)} - s^{(j)}$ for $i \neq j$. These differences are not 0. If $z^{(1)}, \ldots, z^{(M)}$ are non-zero n-tuples of complex numbers, then the set of $x \in \mathbf{R}^n$ such that $z^{(k)} \cdot x = 0$ for some k is contained in a finite union of real hyperplanes, so there are plenty of elements $x \in \mathbf{R}^n$ such that $x \cdot z^{(k)} \neq 0$ for all k. We just use

$z^{(k)} = s^{(i)} - s^{(j)}$ to get x such that the numbers $s^{(i)} \cdot x$ $(i = 1, \dots, N)$ are distinct. Now for such an element x, we consider the relation

$$P_1(tx)\chi_1(tx) + \cdots + P_N(tx)\chi_N(tx) = 0 \qquad \text{for all } t \in \mathbf{R}.$$

This relation has the form

$$q_1(t)e^{\alpha_1 t} + \cdots + q_N(t)e^{\alpha_N t} = 0$$

with $q_i(t) = p_i(tx)$ and $\alpha_i = s^{(i)} \cdot x$, so $\alpha_1, \dots, \alpha_N$ are distinct complex numbers. By the result in one variable (which is at the level of elementary calculus) we conclude that $q_1, \dots, q_N$ are the 0 polynomials. Thus $p_i(x) = 0$ for all $i = 1, \dots, N$ and all x outside a finite union of hyperplanes. Hence $p_i = 0$ for $i = 1, \dots, N$, which concludes the proof.

Next we give the missing steps for the more general version of Chevalley's theorem.

Let G be a subgroup of W. Everything we say would be valid for a finite group acting on a vector space in characteristic 0, and generated by reflections in hyperplanes. In our special case, we suppose G is generated by transpositions in W (the group of permutations of the variables). For example, the subgroups W_λ are generated by reflection. In fact, if we batch together into disjoint subsets the components which are equal, then W_λ consists of the permutation groups of these subsets, and so are generated by the transpositions in these subgroups.

As before, we let S be the polynomial ring, and S^G is the ring of invariants under G. We then have the ideal S^G_+ of S^G generalizing the ideal S^W_+ of S^W, and we also have the S-ideal S^G_+S generated by S^G_+ in S. Of course S^G_+S plays a key role.

Chevalley's theorem. *We have*

$$S = S^W \operatorname{Har}_W.$$

Furthermore, Har_W *is finite dimensional over* $\mathbf{R}$, *and actually*

$$\dim_R \operatorname{Har}_W = |W| = n!$$

is the number of elements in W. *If* $\{H_1, \dots, H_{|W|}\}$ *is a basis of* Har_W *over* $\mathbf{R}$, *then it is also a basis of* S *over* S^W. *Thus*

$$S = \sum_{j=1}^{|W|} S^W H_j.$$

The above theorem gives us a very clear picture of how S is generated linearly over S^W, in a much more significant way than the naive monomials $M_{(r)}(x)$ that we indicated previously. The proof will require several steps.

The first lemma is just Lemma 8.3 with W replaced by G. We repeat the statement.

Lemma 8.3G. *Let $Q_1, \ldots, Q_m \in S^G$ be such that Q_1 does not belong to the ideal $(Q_2, \ldots, Q_m)$ of S^G. Let $P_1, \ldots, P_m$ be homogeneous elements of S such that*

$$\sum_{i=1}^{m} P_i Q_i = 0.$$

Then $P_1 \in S_+^G S$.

Proof. No change in the proof from the previous version.

The next lemma is the main new step, which we copy from Chevalley.

Lemma 8.10. *Let $I_1, \ldots, I_m$ be homogeneous elements of S_+^G, generating the ideal $S_+^G S$, but no fewer of these elements generate this ideal. Then $I_1, \ldots, I_m$ are algebraically independent.*

Proof. Let $F(I_1, \ldots, I_m) = 0$ be a non-trivial polynomial relation of minimal degree between $I_1, \ldots, I_m$, with some polynomial F in m variables. Let $d_i = \deg I_i$. Without loss of generality, we may assume that the polynomial relation is homogeneous, namely there is an integer d such that if a monomial $I_1^{k_1} \cdots I_m^{k_m}$ occurs in the relation, then

$$k_1 d_1 + \cdots + k_m d_m = d.$$

Write $F = F(y) = F(y_1, \ldots, y_m)$ in terms of variables $y_1, \ldots, y_m$. Then the partials $\partial_i F = \partial F / \partial y_i$ are not all equal to 0 for $i = 1, \ldots, m$. Let

$$J_i = (\partial_i F)(I_1, \ldots, I_m),$$

so $J_i \in S^G$. After renumbering, say $J_1, \ldots, J_r$ generate $(J_1, \ldots, J_m)S^G$ but no smaller subset of $\{J_1, \ldots, J_r\}$ generates this ideal. Let

$$J_{r+j} = \sum_{i=1}^{r} A_{ji} J_i \qquad \text{with } A_{ji} \in S^G.$$

Taking $\partial / \partial x_k$ of $F(I_1, \ldots, I_m) = 0$ yields

$$0 = \sum_{i=1}^{m} (\partial_i F)(I_1, \ldots, I_m)(\partial I_i / \partial x_k) = \sum_{i=1}^{m} J_i (\partial I_i / \partial x_k).$$

By Lemma 8.3G we obtain

$$\partial I_i/\partial x_k + \sum_{j=1}^{m-r} A_{ji}(\partial I_{r+j}/\partial x_k) \in S_+^G S$$

for $i = 1, \ldots, r$ and $k = 1, \ldots, n$. (Note that the left side is homogeneous in the variables $x_1, \ldots, x_n$.) We multiply this last relation by x_k and sum to get

$$d_i I_i + \sum_{j=1}^{m-r} A_{ji} d_{r+j} I_{r+j} = \sum_{p=1}^{m} B_{ip} I_p,$$

where B_{ip} is homogenous, and belongs to the ideal generated by the variables x_k, so $B_{ip} = 0$ or $\deg B_{ip} \geqq 1$. In fact, for reasons of homogeneity, we have $B_{ip} = 0$ if I_p is not of strictly lower degree than I_i. Then we see that $d_i I_i$ $(i = 1, \ldots, r)$, and so I_i itself, belongs to the ideal generated by the other invariants I_j $(j \neq i)$, contradicting the minimality assumption of $I_1, \ldots, I_m$, and concluding the proof that $I_1, \ldots, I_m$ are algebraically independent.

Theorem 8.11. *Let G be a subgroup of W generated by transpositions. Let $I_1, \ldots, I_m$ be a minimal system of homogeneous elements of S_+^G generating $S_+^G S$. Then $m = n$, and $I_1, \ldots, I_n$ are algebraically independent algebra generators of S^G. In other words, S^G is the polynomial algebra generated by $I_1, \ldots, I_n$.*

Proof. Lemma 8.10 shows that $m \leqq n$. Since G is a finite group of automorphisms of the quotient field of S, its fixed field has the same transcendence degree as the field of all rational functions in $x_1, \ldots, x_n$, namely n. Hence it will suffice to prove that $I_1, \ldots, I_m$ are algebra generators of S^G. Let A_G be the averaging operator over elements of G, namely

$$A_G = \frac{1}{|G|} \sum_{g \in G} g.$$

Given a homogeneous element $P \in S_+^G$, by hypothesis there are $Q_i \in S$ such that

$$P = Q_1 I_1 + \cdots + Q_m I_m.$$

Applying A_G, we find

$$P = A_G(Q_1) I_1 + \cdots + A_G(Q_m) I_m.$$

The elements $A_G(Q_1), \ldots, A_G(Q_m)$ are in S^G, and are homogeneous of degree strictly less than $\deg P$, so we can repeat the argument by induction to conclude that P is in the algebra generated by $I_1, \dots, I_m$. This proves Theorem 8.11.

Theorem 8.11 provides us with the only property needed for the invariants, for instance the property used in the proof of Lemma 8.4. From here on, the previous arguments apply quite generally without further essential change. Thus we obtain the other Chevalley lemma and Steinberg's theorems as follows:

Lemma 8.4*G*. *Let $H_1, \dots, H_m$ be homogeneous elements of S whose residue classes* mod $S_+^G S$ *are linearly independent over the constants. Then $H_1, \dots, H_m$ are linearly independent over S^G.*

Proof. Use Theorem 8.11 about $I_1, \dots, I_n$; otherwise, no other change.

We define *G*-**harmonic functions** to be functions h on $\mathbf{R}^n$ such that

$$S_+^G(\partial)h = 0.$$

The space of G-harmonic polynomials is denoted by Har_G.

Theorem 8.5*G*. *The G-harmonic functions on $\mathbf{R}^n$ are just the polynomial G-harmonic functions, so just the elements of* Har_G.

Proof. Essentially the same as that of Theorem 8.5. Instead of the standard polynomial for elementary symmetric functions, we now use a polynomial of degree $N = |G|$, namely

$$\prod_{g \in G}(T - gx_i) = T^N - J_1 T^{n-1} + \cdots + (-1)^N J_N,$$

where $J_1, \dots, J_N \in S_+^G$ are G-invariants, and are homogeneous. The bound of n has to be replaced by a different bound depending on the order of G, but this does not affect the conclusion of the proof.

We now skip some intermediate results, and state the general Steinberg theorem,

Theorem 8.8*G*. *Let G be a subgroup of W generated by transpositions. Let λ be a* **C**-*character of S, and λ_G its restriction to S^G. Then*

$$\dim \mathrm{Har}_G = |G|.$$

The λ-eigenspace is a direct sum

$$\mathrm{Fu}(\mathfrak{a}, S^G, \lambda_G) = \bigoplus_{w \in G/G_\lambda} \mathrm{Har}_{G_{[w]\lambda}}\, \chi_{[w]\lambda}.$$

Proof. Same as before, all the tools are in.

Final remarks. Instead of a subgroup of W generated by transpositions, one can merely assume, as in Chevalley and Steinberg, that G is a group of linear transformations of a finite dimensional vector space in characteristic 0, and the group is generated by reflections on hyperplanes. This property was used to make the divisibility argument go through in Lemma 8.3. Otherwise, nothing else was used. In Theorem 8.8G, one needs to prove the additional property that G_λ is also generated by reflections to give the intermediate result

$$\dim \mathrm{Har}_{G_\lambda} = |G_\lambda|$$

As pointed out just before Lemma 8.3G, it is immediate that W_λ is generated by reflections.

CHAPTER IV

Convolutions, Spherical Functions and the Mellin Transform

We shall study integral operators, but these are not independent of the differential operators. We begin by formulating a general result which applies to both. There are other relations showing that eigenvalues of certain differential operators are also eigenvalues of certain integral operators. The chapter concerns the work of Gelfand [Gel 50], Godement [God 52b], Maass [Maa 55], [Maa 56], [Maa 71], and Selberg [Sel 56], especially pp. 48–58.

In Chapter III, §6, we gave some properties of the Harish transform related to the differential operators. In the present chapter, we give more properties related to convolution. In addition, previously we worked with eigenfunctions of the algebra of invariant differential operators. We now put the emphasis on convolution operators, eigenfunctions and characters. In particular, spherical functions could have been defined in terms of being eigenfunctions for certain convolution operators, starting with the work of Gelfand [Gel 50]. Harish-Chandra developed the theory in the context of semisimple Lie groups, especially in [Har 58a]. Godement [God 52b] gave arguments depending less on the geometry and more on the Haar measure and convolutions. We follow Gelfand to some extent, partly via the presentation in [Lan 75/85]. We stop short of the main result about spherical inversion, which requires a much more elaborate treatment than on $SL_2(\mathbf{R})$. We shall deal with this inversion in a later chapter leading into the spectral decomposition.

In [Lan 75/85], Chapters III and IV, the basic connection of spherical functions and representations of the group is given in appropriate generality. We do not repeat it here.

On the other hand, in preparation for the spherical inversion, we deal systematically with one of its components which occurs on euclidean spaces, namely the Mellin transform, in §7 and §8. In particular, we give a refined support theorem for the Mellin transform, at the level of ordinary advanced calculus and complex analysis. It provides the basis for the use of the Anker convex sets in the spherical inversion of Chapters X and XI, but §7 and §8 are entirely self-contained, independently of the group setting on SL_n (or more complicated groups). The point is that the spherical transform is the composite of the Harish transform and the Mellin transform, which is a more elementary basic gadget on euclidean space.

IV, §1. WEAKLY SYMMETRIC SPACES

The results of this section are due to Selberg [Sel 56]. We found useful the exposition in Maass [Maa 71].

Let X be a differential manifold, and G a Lie group acting on X. We write the operation of G on X as $(g, x) \mapsto gx$. Let S be an automorphism of X. We say that X is a **weak** (G, S)**-symmetric space** if the following conditions are satisfied, and in particular X is a G-homogeneous space.

WSS 1. We have $SGS^{-1} = G$ and $S^2 \in G$.

WSS 2. Given $x, y \in X$ there exists $g \in G$ such that

$$gx = Sy \quad \text{and} \quad gy = Sx.$$

Note that from **WSS 2**, it also follows that there exists $h \in G$ such that

$$hx = S^{-1}y \quad \text{and} \quad hy = S^{-1}x.$$

WSS 3. The isotropy group of a point in X is compact.

Example. $\mathrm{Pos}_n(\mathbf{R})$. Let $\mathrm{Pos}_n(\mathbf{R})$ be the space of symmetric positive definite real $n \times n$ matrices. Let

$$S\colon \mathrm{Pos}_n \to \mathrm{Pos}_n \quad \text{be the map defined by } S(p) = p^{-1}.$$

Then S is an isometry for the trace metric, and satisfies several properties which give an example of the more general spaces to be considered in this section. We list some of these. As usual, we let $G = GL_n(\mathbf{R})$. For $g \in G$, we have the action of G on Pos_n defined by $[g]p = gp\,{}^tg$. We shall verify the conditions for a weak symmetric space. Note that **WSS 3** is immediate since the isotropy group is (conjugate to) the orthogonal group.

S 1. We have $S^2 = \text{id}$ (so $S^2 \in G$) and $S[g]S^{-1} = [{}^t g^{-1}]$ for all $g \in G$.

Proof. That $S^2 = \text{id}$ is clear. For the conjugation property, we have

$$\begin{aligned} S[g]S^{-1}(p) &= S[g]p^{-1} = S(gp^{-1}\,{}^t g) \\ &= {}^t g^{-1} p g^{-1}, \quad \text{as was to be shown.} \end{aligned}$$

S 2. Given $p, q \in \text{Pos}_n$ there exists a symmetric element $g \in G$ such that

$$[g]p = q^{-1} \qquad \text{and} \qquad [g]q = p^{-1}.$$

Proof. For symmetric g, the second condition $gqg = p^{-1}$ is equivalent to the first, so we are reduced to finding g symmetric satisfying the first condition. But this first condition is equivalent with

$$\begin{aligned} & p^{1/2}gp^{1/2}p^{1/2}gp^{1/2} = p^{1/2}q^{-1}p^{1/2} \\ \iff & (p^{1/2}gp^{1/2})^2 = p^{1/2}q^{-1}p^{1/2} \\ \iff & p^{1/2}gp^{1/2} = (p^{1/2}q^{-1}p^{1/2})^{1/2} \\ \iff & g = p^{-1/2}(p^{1/2}q^{-1}p^{1/2})^{1/2}p^{-1/2}, \end{aligned}$$

which concludes the proof that Pos_n is weakly symmetric.

After this example, we return to the general considerations.

We suppose that X is a weak (G, S)-symmetric space.

We note that **WSS 3** (compactness) will not be used until Lemma 1.6. Let Fu be the abelian group of functions (always assumed C^∞) on X. Then translations by elements of G, or S, act on Fu contravariantly, that is, we have **left translation** $L_g\varphi$ of a function φ by $g \in G$, from the formula

$$(L_g\varphi)(x) = \varphi(g^{-1}x) \qquad \text{and} \qquad (S\varphi)(x) = \varphi(S^{-1}x).$$

This is forced by the requirement that isometries of X are defined to act trivially on numbers, and we want the rule

$$\varphi(x) = (L_g\varphi)(L_g x) = (L_g\varphi)(gx)$$

to hold for all x, and similarly with S instead of L_g.

Let φ be a function on $X \times X$. We say that φ is a **point pair invariant** if $\varphi(gx, gy) = \varphi(x, y)$ for all elements $(x, y) \in X \times X$ and $g \in G$. Thus the action of G on $X \times X$ is defined diagonally, and a point pair invariant is just a function on the product which is fixed under the action of G on the product; or in other words, $L_g\varphi = \varphi$.

Lemma 1.1. *Let φ be a point pair invariant. Then for all (x, y),*

$$\varphi(Sx, Sy) = \varphi(y, x) = \varphi(S^{-1}x, S^{-1}y)$$

$$\text{and} \qquad \varphi(Sx, y) = \varphi(S^{-1}y, x).$$

Proof. There exists $g \in G$ such that $gx = Sy$ and $gy = Sx$, so

$$\varphi(Sx, Sy) = \varphi(gy, gx) = \varphi(y, x),$$

thereby proving the first equality. The others follow equally trivially.

Let D be an operator on the space of functions, or a G-invariant subspace. If A is an automorphism of Fu or of this subspace, we write $[A]$ for the corresponding action on D, so that

$$[A]D = A \circ D \circ A^{-1}.$$

We also operate on the right, with the definition

$$D[A] = A^{-1} \circ D \circ A.$$

Thus the group of automorphisms of Fu acts as a group of automorphisms of endomorphisms of Fu, on either side. For our purposes, we say that D is G-**invariant** if $[L_g]D = D$ for all $g \in G$, or equivalently,

$$L_g \circ D = D \circ L_g,$$

i.e. D and L_g commute as operators on the space of functions.

Lemma 1.2. *If D is G-invariant, so are $[S]D$ and $D[S]$.*

Proof. There exists $h \in G$ such that $S^{-1} \circ L_g = L_h \circ S^{-1}$, so $S \circ L_h = L_g \circ S$. Then

$$S \circ D \circ S^{-1} \circ L_g = S \circ D \circ L_h \circ S^{-1} = S \circ L_h \circ D \circ S^{-1} = L_g \circ S \circ D \circ S,$$

thus proving the lemma for $[S]D$. The lemma for $D[S]$ follows the same way.

Let φ be a function on $X \times X$. Let D be an operator on a vector space of function on X, stable under the action of G. We write D_1 for the operator acting on φ with respect to the first variable. In other words, if we define φ_y by $\varphi_y(x) = \varphi(x, y)$, then

$$(D_1\varphi)(x, y) = (D\varphi_y)(x).$$

We call D_1 the **first partial operator**. Readers may wish to read the present development first referring to differential operators, which do not necessitate any further discussion of convergence for whatever we say. Then readers can verify that all arguments hold for integral operators considered subsequently, under an absolute convergence condition on the integrals.

Lemma 1.3. *Let D be invariant and φ point pair invariant. Then so is $D_1\varphi$.*

Proof. We have:

$$\begin{aligned}(D_1\varphi)(gx, gy) = (D_1\varphi_{gy})(gx) &= (L_{g^{-1}}(D\varphi_{gy}))(x)\\ &= (D(L_{g^{-1}}\varphi_{gy}))(x)\\ &= (D_1\varphi)(x, y).\end{aligned}$$

Note that we have used the formula

$$(L_{g^{-1}}\varphi_{gy})(x) = \varphi(gx, gy) = \varphi(x, y).$$

This proves the lemma.

It will be convenient to write

$$\varphi^x(y) = \varphi(x, y).$$

Then we have the formula for a point pair invariant φ:

(1) $$(S\varphi^x)(u) = \varphi^x(S^{-1}u) = \varphi(x, S^{-1}u) = \varphi(u, Sx).$$

These are immediate, just being Lemma 1.1 in the new notation. Next comes the key lemma. See its application in Proposition 2.5, aside from Theorem 1.8.

Lemma 1.4. *Let φ be a point pair invariant, and D an invariant operator, acting as a first or second partial D_1 or D_2 on φ. Let*

$$\tilde{D} = [S^{-1}]D = S^{-1} \circ D \circ S$$

on functions. Then $D_1 = \tilde{D}_2$, which can also be written

$$D_1 = S_2^{-1} \circ D_2 \circ S_2,$$

where S_2 is S acting on the second variable.

Proof. Note that $D(S\varphi^x)(u) = (D_1\varphi)(u, Sx)$. Then

$$\begin{aligned}(([S_2^{-1}]D_2)\varphi)(x, y) &= (S_2^{-1}(D(S\varphi^x)))(y)\\ &= (D(S\varphi^x))(Sy)\end{aligned}$$

$$= (D_1\varphi)(Sy, Sx)$$
$$= (D_1\varphi)(x, y)$$

this last equality because $D_1\varphi$ is a point pair invariant by Lemma 1.3, and Lemma 1.1 applies. This concludes the proof.

The next lemma is preparatory to proving that the algebra of invariant operators is commutative. The lemma proves this property on point pair invariants. It will remain to show that such invariants are sufficiently general to give the main result.

Lemma 1.5. *Let φ be a point pair invariant, and D, E invariant operators. Then D_1 and E_1 commute on φ, that is, $D_1E_1 = E_1D_1$ on φ.*

Proof. By Lemma 1.4, we have

$$D_1E_1 = D_1\tilde{E}_2 = \tilde{E}_2D_1 = E_1D_1,$$

as was to be shown.

The next step is to show how to concoct a point pair invariant from an arbitrary function in a natural way. Given $y \in X$, we denote the isotropy group of y in G by G_y. It is the group of elements $g \in G$ such that $gy = y$. We let φ be an arbitrary function on X, so a function of one X-variable. We shall now use the hypothesis that G_y is compact, **WSS 3**, *and we assume that Haar measure on G_y has total measure* 1. We then define a function of two variables

$$\varphi_2(x, y) = \int_{G_y} \varphi(hx)\,dh.$$

Note that if we put $K(y) = G_y$, then $\varphi_2(x, y)$ is the $K(y)$-invariant projection of φ, namely

$$\varphi_2(x, y) = {}^{K(y)}\varphi(x).$$

Fix $x_0 \in X$. For each $y \in X$ there is some $g = g_{y,x_0}$, depending on y and x_0, such that $gy = x_0$. We write such an element g as x_0y^{-1}. Of course, it is defined only up to left translation by an element of G_0. We define

$$\varphi^{\#}(x, y) = \varphi_2(x_0y^{-1}x, x_0) = \varphi_2(gx, x_0). \tag{2}$$

Then the averaging integral in the definition of φ_2 shows that $\varphi^{\#}(x, y)$ is independent of the element g chosen such that $gy = x_0$. Of course,

$\varphi^{\#}(x, y)$ depends on the original choice of x_0. We note that

$$(3) \qquad \varphi^{\#}(x, x_0) = \varphi_2(x, x_0).$$

Lemma 1.6. *The function $\varphi^{\#}$ is a point pair invariant, depending on the choice of x_0.*

Proof. From the definitions, for all $g \in G$,

$$\varphi^{\#}(gx, gy) = \varphi_2(x_0 y^{-1} g^{-1} gx, x_0) = \varphi_2(x_0 y^{-1} x, x_0) = \varphi^{\#}(x, y).$$

Lemma 1.7. *Let D be an invariant operator. Then for all x,*

$$(D_1\varphi_2)(x, x_0) = (D\varphi)_2(x, x_0) \quad \text{and} \quad D_1\varphi^{\#}(x_0, x_0) = (D\varphi)(x_0).$$

Proof. Routinely,

$$(D_1\varphi^{\#})(x, x_0) = (D_1\varphi_2)(x, x_0).$$

But we may write

$$\varphi_2(x, x_0) = \int_{G_{x_0}} \varphi(hx)\, dh = \int_{G_{x_0}} (L_{h^{-1}}\varphi)(x)\, dh.$$

We then operate under the integral sign (see below), and find

$$((D \circ L_{h^{-1}})\varphi)(x) = (L_{h^{-1}} \circ D\varphi)(x) = (D\varphi)(hx).$$

We find:

$$\begin{aligned}(D_1\varphi_2)(x, x_0) &= \int_{G_{x_0}} (D(L_{h^{-1}}\varphi))(x)\, dh = \int_{G_{x_0}} (D\varphi)(hx)\, dh \\ &= (D\varphi)_2(x, x_0).\end{aligned}$$

Evaluating at $x = x_0$ and using the fact that $hx_0 = x_0$ for $h \in G_{x_0}$, as well as the assumption that Haar measure of G_{x_0} is normalized to be 1, we get the desired formula.

Remark 1. In the above lemma, we assume that the operator D has the two properties: it is defined on the function φ_2 as a function of the first variable, and commutes with the integral. These conditions are immediate for differential and integral operators, which are the only ones considered in the sequel.

Theorem 1.8 (Selberg [Sel 56]). *The algebra of invariant differential operators on a weak symmetric space X is commutative.*

Proof. Given an arbitrary function φ, we have seen how to construct a point pair invariant $\varphi^{\#}$, depending on a given point x_0. Lemma 1.7 shows that at the given point x_0, the derivative D amounts to the first partial D_1 of $\varphi^{\#}$ at (x_0, x_0). Then we can use the commutativity of Lemma 1.5 to conclude the proof.

Remark 2. The proof of Lemma 1.7 depended only on being able to operate under the integral sign, i.e. on the averaging operation $\varphi \mapsto \varphi_2$ commuting with the operator. Therefore it applies to certain integral operators, as we shall see below when we integrate under the integral sign, i.e. commute two integrals. See Theorem 2.3.

Remark 3. The commutativity of the algebra of invariant differential operators was proved from the point of view of an Iwasawa decomposition in Chapter III.

IV, §2. CHARACTERS AND CONVOLUTION OPERATORS

We start with some general remarks on characters, which will occur in a new context, as well as the old one. Selberg recognized the role of characters as eigenfunctions [Sel 56], pp. 53–58.

Let G be a group and X a homogeneous space of G. Let $x_0 \in X$ be a chosen point in X. Depending on x_0, if f is a function on X, we can **lift** f to a function on G by defining $f(g) = f(gx_0)$. If this lifted function is a character on G (homomorphism into the non-zero complex numbers) then we call f a **character** on X **relative to** x_0. Note that if f is a function on X such that $f(x_0) \neq 0$ and there is a character χ of G such that

$$f(gx) = \chi(g)f(x) \qquad \text{for all } g \in G, x \in X,$$

then dividing f by the constant $f(x_0)$ gives the character χ on G.

Proposition 2.1. *Let G be a group and X a homogeneous space. Let D be a linear operator on a vector space of functions on X, stable under the action of G. We suppose that D is G-invariant. Let f be a character on the space relative to x_0. Then f is an eigenfunction of D, with eigenvalue $Df(x_0)$.*

Proof. The left invariance gives us

$$(D(L_g f))(x) = (Df)(g^{-1}x) \qquad \text{for all } g, x.$$

But $(L_g f)(x) = f(g^{-1}x) = \chi(g)^{-1}f(x)$, that is $L_g f = \chi(g)^{-1}f$.

Hence

$$(D(L_g f))(x) = \chi(g)^{-1}(Df)(x) = (Df)(g^{-1}x) \qquad \text{for all } g, x.$$

In particular, for $x = gx_0$ we get

$$(Df)(x) = (Df)(x_0)\chi(g) = (Df)(x_0)f(x),$$

which proves the proposition.

In the applications, G is a topological group and **characters** are defined to be continuous, at the very least. Actually G is mostly a Lie group, even real analytic, and characters are as smooth as can be.

Standard example. We take $X = \mathrm{Pos}_n$ and $G = \mathrm{Tri}^+$ to be the group of upper triangular matrices with positive diagonal elements. Then X is a principal homogeneous space for Tri^+, and the characters we have considered previously are those which are trivial on the unipotent subgroup.

Let μ be a Haar measure on the homogeneous space X. Let φ be a measurable function on $X \times X$, and let f be a measurable function on X. We define the **convolution**

$$\varphi * f(x) = C_\varphi(f)(x) = \int_X \varphi(x, y) f(y)\, d\mu(y),$$

whenever the integral is absolutely convergent. Thus we view φ as a "kernel" defining an integral operator. In all our applications, φ is at least continuous. The following properties are proved formally and are valid under conditions of absolute convergence:

We suppose throughout that φ is a point pair invariant.

CONV 1. The operator C_φ commutes with the action of G on functions.

Proof. Let $a \in G$. Then

$$\begin{aligned} C_\varphi(L_a f)(x) &= \int \varphi(x, y) f(a^{-1}y)\, d\mu(y) \\ &= \int \varphi(x, ay) f(y)\, d\mu(y) && \text{by invariance of Haar measure} \\ &= \int \varphi(a^{-1}x, y) f(y)\, d\mu(y) && \text{by point pair invariance,} \end{aligned}$$

thus proving the property.

Note that the converse is also true, and the situation is simply such that an operator defined by a kernel function commutes with the group action if and only if the kernel function is invariant under the group, i.e. is a point pair invariant.

Proposition 2.2. *Let f be a character on X, relative to a point x_0. Suppose that the convolution $\varphi * f$ is absolutely convergent. Then f is eigenfunction of C_φ, with eigenvalue $C_\varphi(f)(x_0)$.*

Proof. Apply Proposition 2.1.

CONV 2. If ψ is another point pair invariant, and we define

$$(\psi * \varphi)(x, y) = \int \psi(x, z)\varphi(z, y)\, d\mu(z),$$

then $\psi * \varphi$ is a point pair invariant (assuming the integral absolutely convergent).

Proof. For $a \in G$, we get

$$\begin{aligned}\int \psi(ax, z)\varphi(z, ay)\, d\mu(z) &= \int \psi(x, a^{-1}z)\varphi(a^{-1}z, y)\, d\mu(z)\\ &= \int \psi(x, z)\varphi(z, y)\, d\mu(z)\end{aligned}$$

by the invariance of Haar measure, as was to be shown.

CONV 3. Under conditions of absolute convergence, we have associativity, that is

$$C_\psi \circ C_\varphi = C_{\psi * \varphi} \qquad \text{or} \qquad \psi * (\varphi * f) = (\psi * \varphi) * f.$$

Proof. This is a simple application of Fubini's theorem, under conditions of absolute convergence. Note that point pair invariance is not used here; the relation depends only on general measure theory and Fubini.

Theorem 2.3. *Let X be a weak symmetric space. Under conditions of absolute convergence, the algebra of convolutions of point pair invariant operators is commutative.*

Proof. This is proved exactly like Theorem 1.8, because a convolution operator commutes with the averaging operation $\varphi \mapsto \varphi_2$ of §1. In other words, let $C = C_\psi$ be convolution with a point pair invariant ψ. Then

$$(C_1\varphi_2)(x, x_0) = (C\varphi)_2(x, x_0)$$

in the notation of Lemma 1.7. This time we use Fubini's theorem to justify the formula. Instead of differentiating under the integral sign, we integrate under the integral sign. Since one of the integrals is over a compact set (the isotropy group at x_0) one needs only that the convolution integral $\psi * \varphi$ is absolutely convergent. This concludes the proof.

IV, §3. EXAMPLE: THE GAMMA FUNCTION

In the present book, we deal almost entirely with invariant differential operators rather than integral operators. However, for other approaches or applications, integral operators dominate. We take this opportunity to point to the most basic of the integral operators. Cf. for instance, Maass [Maa 56], [Maa 71], Gindikin [Gin 64], Bengtson [Ben 83], and many others. We shall give an exposition of some of these integral operators elsewhere. Here we limit ourselves to the definition of only one of them, on $\mathrm{Pos}_n = \mathrm{Pos}_n(\mathbf{R})$, the space of real positive definite symmetric matrices.

We note that the function

$$(Z, Y) \mapsto \mathrm{tr}(YZ^{-1})$$

is a point pair invariant on Pos_n. It follows that the function

$$(Z, Y) \mapsto e^{-\mathrm{tr}(YZ^{-1})} = \exp(-\mathrm{tr}(YZ^{-1}))$$

is a point pair invariant, which goes to zero rapidly as $\mathrm{tr}(YZ^{-1})$ goes to infinity. We define the **gamma kernel** or **gamma point pair invariant** to be

$$\varphi(Z, Y) = e^{-\mathrm{tr}(Z^{-1}Y)} = e^{-\mathrm{tr}(YZ^{-1})}.$$

On the other hand, we define the **gamma function** Γ_n on the set of characters χ on Pos_n, to be

$$\Gamma_n(\chi) = \int_{\mathrm{Pos}_n} e^{-\mathrm{tr}(Y)}\chi(Y)\,d\mu_n(Y),$$

where μ_n is the Haar measure on Pos_n, normalized by

$$d\mu_n(Y) = |Y|^{-(n+1)/2}d\mu_{\mathrm{euc}}(Y),$$

and

$$d\mu_{\text{euc}}(Y) = \prod_{1 \leqq i \leqq j \leqq n} dy_{ij}.$$

For $n = 1$, the definition coincides with the classical gamma function

$$\Gamma(s) = \int_0^\infty e^{-y} y^s \frac{dy}{y}.$$

Here characters are parametrized by a complex variable s, and the integral has a half plane of convergence. A similar situation prevails in the higher dimensional case. If we express

$$Y = T^t T$$

with an upper triangular matrix

$$T = \begin{pmatrix} t_{11} & t_{12} & \cdots & t_{1n} \\ 0 & t_{22} & & t_{2n} \\ \vdots & \vdots & \ddots & \vdots \\ 0 & 0 & \cdots & t_{nn} \end{pmatrix} \qquad \begin{matrix} t_{ii} \in \mathbf{R}^+, \\ t_{ij} \in \mathbf{R} \text{ for } i < j, \end{matrix}$$

and $s_1, \ldots, s_n$ are n complex variables, we define

$$\chi_s(Y) = t_{11}^{s_1} \cdots t_{nn}^{s_n}.$$

Then a computation shows that for $\text{Re}(s_i) > (n - i)$,

$$(1) \qquad \Gamma_n(\chi_s) = \sqrt{\pi}^{\,n(n-1)/2} \prod_{i=1}^{n} \Gamma(\tfrac{1}{2}(s_i - (n - i))).$$

Thus the higher dimensional gamma function can be factored in terms of the ordinary one.

By Proposition 2.2, we know that a character χ on Pos_n is an eigenfunction of convolution with a point pair invariant, and in particular of convolution with the gamma kernel. Proposition 2.2 also shows that the eigenvalue is the gamma function, in other words, we have the formula

$$(2) \qquad (\varphi * \chi)(Z) = \Gamma_n(\chi)\chi(Z).$$

We shall not use the gamma function in this book, but one can take different directions where it plays a primary role, so we thought it valuable to point to its existence, as well as other integral transforms on symmetric spaces.

IV, §4. K-INVARIANCE OR BI-INVARIANCE AND EIGENFUNCTIONS OF CONVOLUTIONS

Let G be a connected unimodular Lie group, K a compact subgroup, and $X \approx G/K$ a reductive homogeneous space. We assume that the Haar measure on K is normalized to have total measure 1.

For a function f on X, we defined its **K-invariant projection** (or average) $\mathrm{Av}_K f = {}^K f$ by the integral

$$\text{(1)} \qquad \mathrm{Av}_K f(x) = {}^K f(x) = \int_K f(kx)\, dk.$$

For any continuous function f, we have $\mathrm{Av}_K f(I) = f(I)$, where I is the chosen origin on X, corresponding to the unit coset in G/K. We are faced with various operators:

— Invariant differential operators $D \in \mathrm{DO}(X)^G$.

— The averaging operator Av_K.

— Convolution operators C_φ, with point pair invariants φ.

We shall see that these operators commute with each other (for convolutions, absolute convergence of the relevant integrals needs to be assumed). Hence they leave stable each other's eigenspaces. Theorem 5.2 of Chapter III can then be applied to see that the eigenspace of eigenfunctions with a given character has dimension 1. We proceed systematically. We let:

$\mathrm{Fu}(\mathrm{DO}(X)^G, \lambda)$ = λ-eigenspace of $\mathrm{Fu}(X)$ for the operation of $\mathrm{DO}(X)^G$.

$\mathrm{Fu}^K(\mathrm{DO}(X)^G, \lambda)$ = subspace of K-invariant functions.

Proposition 4.1. *The K-invariant projection* Av_K *gives a projection*

$$\mathrm{Av}_K : \mathrm{Fu}(\mathrm{DO}(X)^G, \lambda) \to \mathrm{Fu}^K(\mathrm{DO}(X)^G, \lambda).$$

Proof. We differentiate (1) by a differential operator D_x (with respect to the variable x), which we may bring under the integral sign. By the left invariance of D_x, we have

$$D_x f(kx) = D(f \circ L_k)(x) = (Df)(kx).$$

Then $(Df)(kx) = \lambda f(kx)$, where $\lambda = \lambda(D, f)$ is the eigenvalue of D on f. We integrate over K to get

$$D_x {}^K f(x) = \int_K D_x f(kx)\, dk = \int_K \lambda f(kx)\, dk = \lambda {}^K f(x).$$

Hence Av_K maps the λ-eigenspace into the λ-eigenspace of K-invariant functions, and it obviously maps such a function to itself, so it is a projection, as was to be shown.

Note that the above proof is based on the additional commutation of D with left translation by elements of G on X.

Next, consider Av_K and C_φ. *We suppose the convolution integral*

$$C_\varphi f(x) = \int_X \varphi(x, y) f(y)\, d\mu(y)$$

to be absolutely convergent, with a Haar measure μ on X. Since K is compact, we meet no further convergence question when using Fubini's theorem together with integration over K. Hence we conclude that Av_K and C_φ commute, that is

$$C_\varphi \circ \mathrm{Av}_K = \mathrm{Av}_K \circ C_\varphi$$

on the space of functions for which the convolution with φ is absolutely convergent. Again, we get:

Proposition 4.2. *Let φ be a point pair invariant. Then Av_K gives a projection of a C_φ-eigenspace onto the subspace of K-invariants.*

Thirdly, we come to the commutation of differential operators and C_φ. This is more serious, and we go to Pos_n to avoid technical complications.

Theorem 4.3. *Let $X = \mathrm{Pos}_n$. Let φ be a point pair invariant. Under conditions of absolute convergence, and in particular on $C_c^\infty(X)$, elements of $\mathrm{DO}(X)^G$ commute with convolution C_φ.*

Proof. The formal proof is short, but it uses two serious past results as follows. Let $D \in \mathrm{DO}(X)^G$ be an invariant differential operator. Then

$$\begin{aligned} D_x \int_X \varphi(x, y) f(y)\, d\mu(y) \\ &= \int_X D_1\varphi(x, y) f(y)\, d\mu(y) \\ &= \int_X \tilde{D}_2\varphi(x, y) f(y)\, d\mu(y) \quad \text{by Lemma 1.4} \\ &= \int_X \varphi(x, y) Df(y)\, d\mu(y) \quad \text{by Proposition 3.3} \end{aligned}$$

so $DC_\varphi f = C_\varphi Df$, which proves the theorem.

Although using the space $C_c^\infty(X)$ is formally efficient, it is mathematically deficient, because we actually want to apply the commutation to functions which do not have compact support, such as characters and their K-projections to spherical functions. Hence, in practice, one has to rely on a point pair invariant which has sufficiently rapid decay at infinity, compared to the growth of the function and its derivatives to which the operators are applied. In practice, the functions have polynomial growth, and the point pair invariant has exponential decay, so all is well. The gamma kernel is just a typical example of this situation.

Next we consider the analogous notions on the group G rather than the homogeneous space X. There is just enough difference to warrant tabulating certain results separately on the group, which we now do.

In defining spherical functions previously, the essential conditions was that of eigenfunctions for the algebra of invariant differential operators. We are now concerned with the analogous property with convolution operators. Let as usual:

$C(K\backslash G/K)$ = vector space of continuous K-bi-invariant functions on G;

$C_c(K\backslash G/K)$ = vector space of such functions with compact support.

For continuous functions it is harmless to use the above notation, but for C^∞ functions, we emphasize and repeat that the notation means functions on G which are K-bi-invariant. If we wish to emphasize this in the notation, we write

$$C^\infty(G)^{K,K} \qquad \text{and} \qquad C_c^\infty(G)^{K,K},$$

the superscripts indicating that the functions are on G, and are fixed under the action of K on the left and on the right.

Let f, g be functions on G. We define their **convolution** $f * g$ by the integral

$$(f * g)(x) = \int_G f(xy^{-1})g(y)\,dy,$$

with respect to a fixed Haar measure dy on G. Of course, the definition is meaningful only under absolute convergence of the integral. This absolute convergence will be immediately satisfied in the applications.

Routinely from standard measure theory, $L^1(G)$ is an associative algebra under convolution. Observe that *the K-bi-invariant functions form a subalgebra*. Indeed, if f, g are K-bi-invariant, then trivially

$$(f * g)(kx) = (f * g)(x)$$

directly from the K-left invariance f. For the right invariance, let $k \in K$. Then

$$(f * g)(xk) = \int_G f(xky^{-1})g(y)\,dy = \int_G f(xy^{-1})g(y)\,dy = (f * g)(x)$$

using the right invariance of the measure under $y \mapsto yk$ and the right K-invariance of g.

We shall look at the situation unsymmetrically. We let R_φ denote convolution with φ on the right, so by definition

$$(R_\varphi f)(x) = (f * \varphi)(x).$$

Theorem 4.4. *Let G be a unimodular Lie group and K a compact subgroup. Let $f \in C(K\backslash G/K)$ be a continuous K-bi-invariant function on G, not identically zero. If f is an eigenfunction of all right convolutions R_φ with $\varphi \in C_c^\infty(K\backslash G/K)$, then f is C^∞ and is an eigenfunction of $\mathrm{IDO}(G)^K$.*

Proof. By definition and unimodularity,

$$(f * \varphi)(x) = \int_G f(y^{-1})\varphi(yx)\,dy.$$

Using a C^∞ partition of unity on φ, differentiating under the integral sign shows that $f * \varphi$ is C^∞. By hypothesis, $f * \varphi = \lambda_\varphi f$ for some number λ_φ, the eigenvalue. We can pick φ such that $\lambda_\varphi \neq 0$. Indeed, let $\{\psi_n\}$ be a C^∞ Dirac sequence on G, and let $\varphi_n = {}^K\psi_n^K$. Then $f * \varphi_n$ converges pointwise to f, so

$$(f * \varphi_n)(x) = \lambda_{\varphi_n} f(x) \to f(x) \qquad \text{as} \quad n \to \infty.$$

Let x be a point such that $f(x) \neq 0$. We conclude that $\lambda_{\varphi_n} \neq 0$ for n large. It follows that f itself is C^∞, because $f = (f * \varphi)/\lambda_\varphi$. Let $D \in \mathrm{IDO}(G)^K$. Differentiating under the integral sign and using the G-invariance shows that

$$D(f * \varphi) = f * D\varphi = \lambda_{D\varphi} f.$$

But $D(f * \varphi) = D(\lambda_\varphi f) = \lambda_\varphi Df$. Hence f is an eigenfunction, with eigenvalue

$$\lambda(D, f) = \lambda_{D\varphi}/\lambda_\varphi$$

for all φ such that $\lambda_\varphi \neq 0$. This concludes the proof.

Note that Theorem 4.4 is basically a piece of abstract nonsense, with little structure in the hypotheses. We may view it as some sort of converse to considerations of spherical functions in Chapter III,

Theorem 6.3. To give the converse argument differentiating under the integral sign, what comes up naturally is left convolution $L_\varphi(f) = \varphi * f$ as in the next lemma.

Lemma 4.5. *Let G be a Lie group and K a compact Lie subgroup. Suppose that every eigenspace of* $\mathrm{IDO}(G)^K$ *in* $C^\infty(K\backslash G/K)$ *is one-dimensional (if not zero). Let f be a* C^∞ *eigenfunction of* $\mathrm{IDO}(G)^K$, $f \neq 0$. *Then f is an eigenfunction for left convolution* L_φ *for all* $\varphi \in C_c(K\backslash G/K)$.

Proof. The left convolution integral is given by

$$(\varphi * f)(x) = \int \varphi(xy^{-1}) f(y)\, dy = \int \varphi(y^{-1}) f(yx)\, dy.$$

We can then differentiate under the integral sign, and for

$$\varphi \in C_c(K\backslash G/K), \quad D \in \mathrm{IDO}(G)^K,$$

we find

$$D(\varphi * f) = \varphi * Df = \lambda(D, f)\varphi * f.$$

Hence $\varphi * f$ is a K-bi-invariant eigenfunction of $\mathrm{IDO}(G)^K$, with the same eigencharacter as f, whence by hypothesis it is a scalar multiple of f itself, thereby proving the lemma.

The next theorem of Gelfand [Gel 50] gives a condition under which the left convolution is equal to right convolution. This statement is the analogue of Selberg's condition for weak symmetric spaces.

Theorem 4.6. *Let G be locally compact unimodular, and let K be a compact subgroup. Let* τ *be an anti-automorphism of G of order* 2 *such that given* $x \in G$ *there exist* $k_1, k_2 \in K$ *satisfying*

$${}^\tau x = x^\tau = k_1 x k_2.$$

Then the algebra $C_c(K\backslash G/K)$ *is commutative. In fact, one has* $f * g = g * f$ *if both* f, g *are K-bi-invariant, one of them is in* $C_c(K\backslash G/K)$ *and the other is in* $C(K\backslash G/K)$.

Proof. Haar measure is invariant under $k \mapsto {}^\tau x$ because the measure dilation factor $\Delta(\tau)$ satisfies

$$1 = \Delta(\tau^2) = \Delta(\tau)\Delta(\tau),$$

so $\Delta(\tau) = 1$. Also $f(x) = f({}^{\tau}x)$ for any K-bi-invariant f. Then

$$\begin{aligned}(f * g)(x) = (f * g)({}^{\tau}x) &= \int f({}^{\tau}xy^{-1})g(y)\,dy \\ &= \int f({}^{\tau}x{}^{\tau}y^{-1})g({}^{\tau}y)\,dy \quad \text{by } y \mapsto {}^{\tau}y \\ &= \int f({}^{\tau}y^{-1})g({}^{\tau}x{}^{\tau}y)\,dy \quad \text{by } y \mapsto yx \\ &= \int f(y)g({}^{\tau}xy^{-1})\,dy \quad \text{by } y \mapsto {}^{\tau}y^{-1} \\ &= (g * f)({}^{\tau}x) = (g * f)(x),\end{aligned}$$

which concludes the proof.

Example. The hypotheses of Theorem 4.6 are of course satisfied for $G = \mathrm{SL}_n(\mathbf{R})$ or $\mathrm{GL}_n^+(\mathbf{R})$ (connected component of $\mathrm{GL}_n(\mathbf{R})$), or $\mathrm{GL}_n(\mathbf{R})$, and K the unitary subgroup. We take τ to be the transpose. The decomposition of a matrix $x = sk$ into a product of a symmetric matrix and an element $k \in K$ immediately shows that ${}^t x = {}^t ks = k_1 x k_2$ because ${}^t k = k^{-1}$.

Theorem 4.7. *Let $G = \mathrm{GL}_n(\mathbf{R})$ or $\mathrm{SL}_n(\mathbf{R})$. Let f be a K-bi-invariant function on G, eigenfunction of $\mathrm{IDO}(G)^K$. Then f is an eigenfunction for right convolution R_φ with $\varphi \in C_c(K\backslash G/K)$. Anticipating the definition from §5, on G, a function is spherical if and only if it is convolution spherical.*

Proof. Theorem 6.1 of Chapter III proves the one-dimensionality of the $\mathrm{IDO}(G)^K$ eigenspaces. Lemma 4.5 then shows that f is an eigenfunction for left convolutions, and Theorem 4.6 then gives it for right convolutions. This proves the first statement. The if and only if assertion comes by combining this latest implication with Theorem 4.4. We are done.

IV, §5. CONVOLUTION SPHERICALITY

The theorems at the end of the last section show that in the concrete cases at hand, the conditions of being an eigenfunction for $\mathrm{IDO}(G)^K$ or for the convolution operation of $C_c(K\backslash G/K)$ are equivalent. Some properties which we shall now list depend directly on the convolution operation in a Haar measure theoretic setting, and have their source in Gelfand [Gel 50]. Thus we give an exposition based just on that setting. For the convenience of the reader, we now reproduce [Lan 75/85], Chapter IV, §3, giving such properties, and following the above reference.

We let G be a locally compact group, unimodular, with a compact subgroup K. We say that a function f on G is **convolution spherical**, or C-**spherical** for short, if it satisfies the following properties:

SPH$_C$ 1. f is bi-invariant and continuous.

SPH$_C$ 2. f is an eigenfunction of $C_c(K\backslash G/K)$ on the right, i.e.

$$f * \psi = \lambda(f, \psi) f$$

for $\psi \in C_c(K\backslash G/K)$ and some complex number $\lambda(f, \psi)$.

SPH$_C$ 3. $f(e) = 1$, where e is the unit element of G.

The third condition is a normalization. A function satisfying the first two properties, and such that $f(e) \neq 0$, can be divided by $f(e)$ to yield a function satisfying all three properties.

Note that the eigenvalue $\lambda(f, \psi)$ is

$$\lambda(f, \psi) = (f * \psi)(e),$$

which we see from conditions **SPH$_C$ 2** and **SPH$_C$**, evaluating at e.

The next theorem gives a fundamental example of spherical functions.

Theorem 5.1. *Assume that $G = PK$, where P is a closed subgroup, and $P \times K \to PK = G$ is a topological isomorphism. Let*

$$\chi : P \to \mathbf{C}^*$$

be a character (continuous homomorphism), which we extend to a function on G by setting $\chi(pk) = \chi(p)$. Then χ is a right eigenvector of $C_c(K\backslash G/K)$, i.e.

$$\chi * \psi(x) = \lambda(\rho, \psi)\chi(x),$$

and the function f such that

$$f(x) = \int_K \chi(kx)\, dk$$

is C-spherical with eigenvalues $\lambda(\chi, \psi) = (\chi * \psi)(e) = \lambda(f, \psi)$.

Proof. Write $x = p_1 k_1$. Then for $\psi \in C_c(K\backslash G/K)$ we get

$$\begin{aligned} \chi * \psi(x) &= \int_G \chi(xy^{-1})\psi(y)\, dy \\ &= \int_G \chi(p_1 y)\psi(y^{-1})\, dy. \end{aligned}$$

Writing $y = pk$ we have $\chi(p_1 y) = \chi(p_1 p) = \chi(p_1)\rho(p) = \chi(p_1)\rho(y)$, so our last expression is

$$\begin{aligned} &= \chi(p_1) \int_G \chi(y)\psi(y^{-1})\,dy \\ &= \lambda(\chi, \psi)\chi(x). \end{aligned}$$

This proves that χ is an eigenvector, and also gives us the explicit expression for the eigenvalue $\lambda(\chi, \psi)$. For f, we now have

$$\begin{aligned} f * \psi(x) &= \int_G f(xy^{-1})\psi(y)\,dy \\ &= \int_K \int_G \chi(kxy^{-1})\psi(y)\,dy\,dk \\ &= \int_K \lambda(\chi, \psi)\chi(kx)\,dk \\ &= \lambda(\chi, \psi) f(x), \end{aligned}$$

so that f is also an eigenvector, with the same eigenvalue as χ. Clearly $f(e) = 1$, and f is bi-invariant since χ is invariant on the right, while the integral takes care of left invariance. This proves Theorem 5.1.

The approach to spherical functions via the equation of the next theorem, and the subsequent theorem giving an equivalent condition in terms of the characters of the L^1-algebra stem from Gelfand [Gel 50] and Godement [God 52b]. See also [God 57].

Theorem 5.2. *Let f be a continuous function on G, not identically 0. Then f is C-spherical if and only if for all $x, y \in G$ we have*

$$\int_K f(xky)\,dk = f(x)f(y).$$

Proof. Assume that f is C-spherical. For each x let

$$F_x(y) = \int_K f(xky)\,dk.$$

Let $\varphi \in C_c(K\backslash G/K)$. Then

$$\begin{aligned} F_x * \varphi(y) &= \int_G F_x(yz^{-1})\varphi(z)\,dz \\ &= \int_G \int_K f(xkyz^{-1})\varphi(z)\,dk\,dz. \end{aligned}$$

Interchange the integrals, let $z \mapsto zy$, then let $z \mapsto zk$. We see that the last expression is

$$\begin{aligned} &= \int_K \int_G f(xz^{-1})\varphi(zky)\,dz\,dk \\ &= (f * \varphi'_y)(x) \end{aligned}$$

where

$$\varphi'_y(z) = \int_K \varphi(zky)\,dk.$$

Since φ'_y is bi-invariant, we finally obtain

$$F_x * \varphi(y) = \lambda(f, \varphi'_y) f(x),$$

where $\lambda(f, \varphi'_y)$ is the eigenvalue. Let $x = e$. Then

$$F_e(y) = f(y),$$

so $F_e = f$. We get

$$(f * \varphi)(y) = \lambda(f, \varphi'_y) f(e) = \lambda(f, \varphi'_y),$$

so that

$$F_x * \varphi(y) = (f * \varphi)(y) f(x).$$

On the other hand, let $\{\varphi_n\}$ be a Dirac sequence, and apply what we just obtained to $\varphi = \varphi_n$. We know that

$$F_x * \varphi_n \to F_x \quad \text{and} \quad f * \varphi_n \to f.$$

Since F_x, f are both bi-invariant, we can replace φ_n by ${}^K\varphi_n^K$. Hence

$$\int_K f(xky)\,dk = F_x(y) = f(y)f(x).$$

This proves half of our theorem.

Conversely, assume that f satisfies the stated functional equation. Let x_0 be such that $f(x_0) \neq 0$. Then

$$f(x_0)f(y) = \int_K f(x_0kk_1y)\,dk = f(x_0)f(k_1y),$$

so $f(y) = f(k_1y)$ for all $k_1 \in K$, and f is left invariant. A similar argument shows that f is right invariant, so f is bi-invariant. Then

$$f(x_0) = \int_K f(x_0k)\,dk = f(x_0)f(e),$$

so that $f(e) = 1$. Finally, let $\varphi \in C_c(K\backslash G/K)$. By definition,

$$f * \varphi(x) = \int_G f(xy^{-1})\varphi(y)\,dy.$$

Integrate over K on the outside, let $y \mapsto yk^{-1}$, change the order of integration, to get this expression

$$\begin{aligned} &= \int_G \int_K f(xky^{-1})\varphi(y)\,dk\,dy \\ &= \int_G f(x)f(y^{-1})\varphi(y)\,dy \\ &= (f * \varphi(e))f(x). \end{aligned}$$

Therefore f is an eigenvector of $C_c(K\backslash G/K)$, thereby proving the second half of our theorem.

Theorem 5.3. *Let $f \in C(K\backslash G/K)$. Then f is C-spherical if and only if the map*

$$T_f = T: \varphi \mapsto \int_G \varphi(x)f(x)\,dx$$

is an algebra homomorphism of $C_c(K\backslash G/K)$ into $\mathbf{C}$.

Proof. By definition,

$$T(\varphi * \psi) = \int_G \int_G \varphi(xy^{-1})\psi(y)f(x)\,dy\,dx.$$

Interchange $dy\,dx$ to $dx\,dy$, let $x \mapsto xy$, get the right side

$$= \int_G \int_G \varphi(x)\psi(y)f(xy)\,dx\,dy.$$

Integrate with respect to K on the outside, let $x \mapsto xk$, move the integral with respect to K inside, getting

$$T(\varphi * \psi) = \int_G \int_G \varphi(x)\psi(y) \int_K f(xky)\,dk\,dx\,dy. \tag{1}$$

On the other hand,

$$T(\varphi)T(\psi) = \int_G \int_G \varphi(x)\psi(y)f(x)f(y)\,dx\,dy, \tag{2}$$

so the implications $\Rightarrow$ in Theorem 5.3 is clear from one implication in Theorem 5.2.

Conversely, assume that T is an algebra homomorphism, i.e.

$$T(\varphi * \psi) = T(\varphi)T(\psi)$$

for all $\varphi, \psi \in C_c(K\backslash G/K)$. Then the functional equation for f follows at once from the equality between (1) and (2), and we can apply the reverse implication in Theorem 5.2 to conclude the proof.

Note. If we assume f is bounded in Theorem 5.3, then

$$T_f: \varphi \mapsto \int_G \varphi(x) f(x)\, dx$$

extends to an algebra homomorphism of

$$L^1(K\backslash G/K)$$

into $\mathbf{C}$.

Theorem 5.4. *Any continuous algebra homomorphism of* $L^1(K\backslash G/K)$ *into* $\mathbf{C}$ *is of the form*

$$\varphi \mapsto (f * \varphi)(e)$$

for some bounded C-spherical function f.

Proof. By measure theory, given a character $T \neq 0$ of the algebra $L^1(K\backslash G/K)$, there exists a bounded measurable function f such that

$$T(\varphi) = \int_G \varphi(x) f(x)\, dx, \qquad \text{all } \varphi \in L^1(K\backslash G/K).$$

Replace $\varphi(x)$ by $\varphi(k_1 x k_2)$, integrate with respect to K, let $x \mapsto k_1^{-1} x k_2^{-1}$. This shows that we can replace f by

$$\int_K \int_K f(k_1 x k_2)\, dk_1\, dk_2,$$

i.e. we may assume that f is bi-invariant. From (1) and (2) we get

$$\int_K f(xky)\, dk = f(x) f(y)$$

for almost all $(x, y) \in G \times G$. To show that f can be replaced by a continuous function, let $\psi \in C_c(G)$ be such that

$$\int_G \psi(y) f(y)\, dy \neq 0,$$

and assume without loss of generality that this last integral is equal to 1 (after multiplying ψ by a constant if necessary). Then

$$\int_G \varphi(x) f(x)\,dx \int_G \psi(y) f(y)\,dy = \int_G \int_G \varphi(x)\psi(y) \int_K f(xky)\,dk\,dy\,dx$$

which, after using Fubini and letting $y \mapsto k^{-1}y$ and $y \mapsto x^{-1}y$ is

$$\begin{aligned} &= \int_G \int_K \int_G \varphi(x)\psi(k^{-1}x^{-1}y) f(y)\,dy\,dk\,dx \\ &= \int_G \int_G \int_K \varphi(x)\psi(kx^{-1}y) f(y)\,dk\,dy\,dx. \end{aligned}$$

We can replace f by

$$g(x) = \int_G \int_K \psi(kx^{-1}y) f(y)\,dk\,dy,$$

which is continuous. This proves our theorem.

As a final application of the functional equation in Theorem 5.2, we show how convolution with a spherical function behaves vis-à-vis translations. For a function $f \in C(K\backslash G/K)$ and $x \in G$ we let f^{xK} be defined by

$$f^{xK}(y) = \int_K f(xky)\,dk.$$

Then $f^{xK} \in C(K\backslash G/K)$ and $f^{xK}(e) = f(x)$. If φ is a C-spherical function, then Theorem 5.2 tells us that $\varphi^{xK}(y) = \varphi(x)\varphi(y)$.

Theorem 5.5. *Let $\varphi \in C(K\backslash G/K)$ be C-spherical. For*

$$f \in C_c(K\backslash G/K),$$

let

$$T_\varphi(f) = \int_G f(y)\varphi(y)\,dy.$$

Then

$$T_\varphi(f^{xK}) = \varphi(x^{-1})T_\varphi(f).$$

Proof. We have

$$T_\varphi(f^{xK}) = \int_G f^{xK}(y)\varphi(y)\,dy = \int_K \int_G f(xky)\varphi(y)\,dy\,dk$$

$$= \int_G f(xy)\varphi(y)\,dy \quad \text{[by Fubini, and letting } y \mapsto k^{-1}y]$$

$$= \int_G f(y)\varphi(x^{-1}y)\,dy = \int_G f(y)\varphi(x^{-1}ky)\,dy$$

for every $k \in K$. Integrating over K and using the functional equation of Theorem 5.2, we obtain

$$T_\varphi(f^{xK}) = \int_G f(y)\varphi(x^{-1})\varphi(y)\,dy,$$

which proves the theorem.

Corollary 5.6. *Let G be a Lie group with an Iwasawa decomposition $G = UAK$, and K compact. Let $\varphi = \varphi(\chi, x)$ be the spherical kernel, that is, as in Chapter III, §1,*

$$\varphi(\chi, x) = \int_K (\chi\delta^{1/2})(\mathrm{Iw}_A(kx))\,dk.$$

Let $\mathbf{S}$ be the spherical transform defined by

$$(\mathbf{S}f)(\chi) = \int_G \varphi(\chi, y)f(y)\,dy.$$

Then for $f \in C_c(K\backslash G/K)$ we have

$$(\mathbf{S}f^{xK})(\chi) = \varphi(\chi, x^{-1})(\mathbf{S}f)(\chi).$$

Proof. Let $\varphi = \varphi_\chi$ in Theorem 5.5, with a character χ, and use Theorem 5.1 to see that φ_χ is C-spherical. This concludes the proof.

Remark. In Chapter VI, §2, we shall describe how various formulas such as the above can be pulled back to the Lie algebra of A and its dual.

IV, §6. THE SPHERICAL TRANSFORM AS MULTIPLICATIVE HOMOMORPHISM

This section is another one having only to do with measure theory. So we assume that G is a Lie group, unimodular, with a weak Iwasawa decomposition

$$G = UAK.$$

We let δ be the modular function defined by the relation

$$\int_U f(ua)\,du = \delta(a)\int_U f(au)\,du.$$

No other property will be used. The **Harish transform** is then defined for $f \in C_c(K\backslash G/K)$ by the formula

$$\mathbf{H}f(a) = \delta(a)^{1/2}\int_U f(au)\,du = a^\rho \int_U f(au)\,du.$$

We recall the **convolution product** on G, defined by

$$f * g(x) = \int_G f(xy^{-1})g(y)\,dy.$$

Proposition 6.1. *For $f, g \in C_c(K\backslash G/K)$ we have*

$$\mathbf{H}(f * g) = \mathbf{H}f * \mathbf{H}g.$$

In other words, the Harish transform is a homomorphism for the convolution.

Proof. We reproduce the proof as in [La 75/85] for the convenience of the reader. We have:

$$\begin{aligned}\mathbf{H}f * \mathbf{H}g(a) &= \int_A \mathbf{H}f(ab^{-1})\mathbf{H}g(b)\,db \\ &= \int\int\int (ab^{-1})^\rho f(ab^{-1}v)b^\rho g(bu)\,dv\,du\,db \\ &= a^\rho \int\int\int f(ab^{-1}v)g(bu)\,dv\,du\,db.\end{aligned}$$

On the other hand,

$$\begin{aligned}&\mathbf{H}(f * g)(a) \\ &\quad = a^\rho \int_U (f * g)(au)\,du \\ &\quad = a^\rho \int_U\int_G f(auy)g(y^{-1})\,dy\,du \quad (\text{by } y \mapsto y^{-1}) \\ &\quad = a^\rho \int_U\int_G f(ay)g(y^{-1}u)\,dy\,du \quad (\text{by } y \mapsto u^{-1}y) \\ &\quad = a^\rho \int_U\int_A\int_U f(abv)g(v^{-1}b^{-1}u)\,db\,dv\,du \quad (\text{by } y = bvk) \\ &\quad = a^\rho \int_U\int_A\int_U f(ab^{-1}v)g(v^{-1}bu)\,db\,du\,dv \quad (\text{by } b \mapsto b^{-1})\end{aligned}$$

$$= a^{\rho} \int\int\int f(ab^{-1}v)g(v^{-1}ub)\delta(b)^{-1}\, db\, du\, dv \quad (\text{by } bu \mapsto ub)$$

$$= a^{\rho} \int\int\int f(ab^{-1}v)g(ub)\delta(b)^{-1}\, db\, du\, dv \quad (\text{by } u \mapsto vu)$$

$$= a^{\rho} \int\int\int f(ab^{-1}v)g(bu)\, db\, du\, dv \quad (\text{by } ub \mapsto bu)$$

which is the same as the last expression obtained above, thus proving our theorem.

Next we come to the **Mellin transform**, defined for $\varphi \in C_c(A)$ by the integral

$$\mathbf{M}\varphi(\chi) = \int_A \varphi(a)\chi(a)\, da,$$

where $da = d^*a$ is the Haar measure on A, and $\chi: A \to \mathbf{C}^*$ is a character, namely a continuous homomorphism.

Proposition 6.2. *The Mellin transform is a multiplicative homomorphism, in the sense that for* $\varphi, \psi \in C_c(A)$ *we have*

$$\mathbf{M}(\varphi * \psi) = \mathbf{M}\varphi\mathbf{M}\psi,$$

so it transforms the convolution product to the ordinary product.

Proof. Directly from the definitions,

$$\begin{aligned} \mathbf{M}(\varphi * \psi)(\chi) &= \int_A (\varphi * \psi)(a)\chi(a)\, da \\ &= \int_A\int_A \varphi(ab^{-1})\psi(b)\chi(a)\, db\, da. \end{aligned}$$

We change the order of integration, $db\, da = da\, db$, and we let $a \mapsto ab$, to get

$$\begin{aligned} &= \int_A\int_A \varphi(a)\psi(b)\chi(a)\chi(b)\, db\, da \\ &= \mathbf{M}\varphi(\chi)\mathbf{M}\psi(\chi), \end{aligned}$$

thus concluding the proof.

Theorem 6.3. *On* $C_c(K\backslash G/K)$, *we have* $\mathbf{S}(f * g) = \mathbf{S}f\mathbf{S}g$, *that is, the spherical transform is a multiplicative homomorphism transforming the convolution product to the ordinary product.*

Proof. Propositions 6.1 and 6.2 put together with Proposition 5.1 of Chapter III, $\mathbf{S} = \mathbf{MH}$.

Later we shall study Harish-Chandra inversion. It is then useful to keep in mind a general formalism on the group, which we extract in a completely general lemma, which shows how punctual inversion at the origin implies an L^2-metric preservation property.

Let G be a unimodular locally compact group with Haar measure dx. If f is a function on G, we define as usual

$$f^*(x) = \overline{f(x^{-1})}.$$

We note the formula for the convolution quadratic form at 0, namely if $f * f^* \in L^1(G)$,

$$(1) \qquad (f * f^*)(e) = \|f\|_2^2,$$

where the L^2-norm on the right is with respect to the Haar measure defining the convolution product on the left. This is immediate from the definition

$$(f * f^*)(e) = \int_G f(ey^{-1})\overline{f(y^{-1})}\, dy,$$

after letting $y \mapsto y^{-1}$ and using the unimodularity.

Recall that in Chapter III, Corollary 6.4, we showed that the spherical transform with unitary characters is a homomorphism for the star operation $f \mapsto f^*$. The next result is concerned with both stars, and lays the first stone for inversion theory on G.

Proposition 6.4. *Let G be a unimodular locally compact group with a given Haar measure. Let E, F be subspaces of $L^1(G)$. Assume E closed under convolution and under $f \mapsto f^*$. Assume F closed under multiplication and complex conjugation. Let $S: E \to F$ be a linear map which is also a star homomorphism for both stars, that is, for all $f, g \in E$,*

$$S(f * g) = (Sf)(Sg) \qquad \textit{and} \qquad S(f^*) = \overline{Sf}.$$

Let $\eta = \eta_G$ be a measurable function on G such that $\eta g \in L^1(G)$ for all $g \in F$. Define

$$Tg = \int_G g(y)\eta(y)\, dy \qquad \textit{for } g \in F.$$

Suppose that

$$T \circ S = C\delta_e \quad \textit{(Dirac delta)}$$

for some constant C. Then for all $f \in E$,

$$\|Sf\|_{2,\eta}^2 = C\|f\|_2^2.$$

Proof. Routine. For $f \in E$,

$$\begin{aligned}
\|Sf\|_{2,\eta}^2 &= \int_G Sf \cdot \overline{Sf}(y)\eta(y)\,dy \quad \text{(by definition, if the integral exists)}\\
&= \int_G S(f * f^*)(y)\eta(y)\,dy \quad \text{(exists by assumption)}\\
&= T(S(f * f^*))\\
&= C(f * f^*)(e) = C\|f\|_2^2
\end{aligned}$$

by (1). This proves the proposition.

We recall that when $S = \mathbf{S}$ is the spherical transform, then the formula

$$\mathbf{S}(f^*) = \overline{\mathbf{S}f} \quad \text{on unitary characters}$$

was proved in Theorem 7.4 of Chapter III.

We shall apply Proposition 6.4 to the spherical transform. The problem is therefore to determine the function η to which the proposition will apply.

By using a polar decomposition in Theorem 3.2 of Chapter VI we shall see how the multiplicative homomorphic property can be formulated entirely on A, on which one can define a twisted convolution product.

IV, §7. THE MELLIN TRANSFORM AND THE PALEY–WIENER SPACE

In this section, we derive the most classical Paley–Wiener theorem in terms of classical notation for the Mellin transform. We do this so that readers can compare both the framework of the ordinary coordinates and the more invariant and further reaching context of the convex sets which will be carried out in full in the next section, at which time we shall make appropriate historical comments.

Let A be a multiplicative Lie group with an isomorphism $A \xrightarrow{\approx} \mathbf{R}^{+(r)}$. The **Mellin transform** is defined by

$$\begin{aligned}
\mathbf{M}f(s) &= \int_A f(a)a^s d^*a\\
&= \int_0^\infty \cdots \int_0^\infty f(a_1, \ldots, a_r) a_1^{s_1} \cdots a_r^{s_r} \prod \frac{da_i}{a_i},
\end{aligned}$$

where

$s = (s_1, \ldots, s_r)$ are complex variables;

$a = (a_1, \ldots, a_r)$ and $d^*a = \prod da_i/a_i \quad (i = 1, \ldots, r)$.

The transform is defined on functions f for which the integral is absolutely convergent. We wish to characterize its image for $f \in C_c^\infty(A)$. The notation becomes simpler if we deal with an additive rather than multiplicative group, so we let $\mathfrak{a} = \mathbf{R}^r$ and by the change of variables $a_i = e^{x_i}$ $(i = 1, \ldots, r)$ or $x_i = \log a_i$, $x = \log a$, the Mellin transform of a function f on $\mathfrak{a}$ becomes

$$(\mathbf{M}_\mathfrak{a} f)(s) = \int_\mathfrak{a} f(x) e^{sx}\, dx = \int_\mathfrak{a} f(x_1, \ldots, x_r) e^{s_1 x_1} \cdots e^{s_r x_r}\, dx_1 \cdots dx_r.$$

Here we write sx for $s \cdot x$ which is the ordinary dot product $\sum s_i x_i$. Note that restricting s to

$$s = -\mathbf{i}\lambda \quad (\text{with } \lambda \in \mathbf{R}^r)$$

gives a function of λ which is none other than the **Fourier transform**.

There are two ways to define the **Paley–Wiener space** $\mathrm{PW}(\mathbf{C}^r)$, depending on how a basic inequality is formulated. At first, let us define $\mathrm{PW}(\mathbf{C}^r)$ to consist of those entire functions h for which there exists a positive number c such that for all positive integers N there is a constant C_N for which

$$|h(\sigma + \mathbf{i}t)| \leqq C_N \frac{e^{c|\sigma|}}{(1 + |t|)^N} \qquad \text{for all } \sigma, t \in \mathbf{R}^r,$$

where:

$s = \sigma + \mathbf{i}t$, $\quad \sigma = (\sigma_1, \ldots, \sigma_r) = \mathrm{Re}(s)$, $\quad t = (t_1, \ldots, t_r) = \mathrm{Im}(s)$;

$|\sigma|$ and $|t|$ are the scalar product norms, say;

$|\zeta|$ is the positive definite hermitian norm for $\zeta \in \mathbf{C}^r$.

We may call the above inequality the **weak PW inequality**. The second definition is to replace this inequality by the **strong PW inequality**

$$|h(\zeta)| \leqq C'_N \frac{e^{c|\sigma|}}{(1 + |\zeta|)^N} \qquad \text{for all } \zeta \in \mathbf{C}^r.$$

It turns out that the two definitions are equivalent. The proof will come from the next results. In Theorem 7.1, we prove that the Mellin transform maps $C_c^\infty(\mathfrak{a})$ into the Paley–Wiener space of functions satisfying the strong inequality. On the other hand, the proof of Theorem 7.2 will show that the inverse Mellin transform of a function satisfying the weak PW inequality has compact support, i.e. is in $C_c^\infty(\mathfrak{a})$. Fourier inversion concludes the proof that the two definitions are equivalent.

Then the growth condition for h may be expressed by saying that h has at most exponential growth with respect to the real part σ, and is super polynomially decreasing uniformly in every strip of finite width (r-dimensional strip, that is).

We now state and prove the two theorems giving the required estimates and the inversion.

Theorem 7.1. *The Mellin transform maps $C_c^\infty(\mathbf{R}^r)$ into the Paley–Wiener space, and*

$$\mathbf{M}_\mathfrak{a}: C_c^\infty(\mathfrak{a}) = C_c^\infty(\mathbf{R}^r) \to \mathrm{PW}(\mathbf{C}^r)$$

is a linear isomorphism.

Proof. The proof is basically routine, using integration by parts. We carry it out. Note that for the first statement, we show that $\mathbf{M}_\mathfrak{a}$ maps $C_c^\infty(\mathbf{R}^r)$ into the Paley–Wiener space defined by the strong PW inequality. For each coordinate s_i, we have

$$\begin{aligned} s_i(\mathbf{M}_\mathfrak{a} f)(s) &= \int f(x) s_i e^{s_1 x_1} \cdots e^{s_r x_r}\, dx_1 \cdots dx_r \\ &= f(x)e^{sx}\big|_{-\infty}^{\infty} - \int (\partial_i f)(x) e^{sx}\, dx \\ &= -\int (\partial_i f)(x) e^{sx}\, dx = -(\mathbf{M}_\mathfrak{a}(\partial_i f))(s) \end{aligned}$$

because the boundary term is 0 by the compact support of f. Thus multiplication by a variable corresponds to minus the partial derivative under the transform $\mathbf{M}_\mathfrak{a}$. In particular, we obtain the estimate

$$|s_i|\,|\mathbf{M}_\mathfrak{a} f(s)| \leqq \|\partial_i f\|_1 e^{c|\sigma|}.$$

Then by iterating, for any polynomial P,

$$(1) \qquad |P(s)|\,|\mathbf{M}_\mathfrak{a} f(s)| = \mathbf{M}_\mathfrak{a}(P^-(\partial) f)(s),$$

where $P^-(\partial) = P(-\partial) = P(-\partial_1, \ldots, -\partial_r)$. Then

$$|P(s)|\,|\mathbf{M}_\mathfrak{a} f(s)| \leqq \|P^-(\partial) f\|_1 e^{c|\sigma|}.$$

This proves that $\mathbf{M}_\mathfrak{a} f$ is in the Paley–Wiener space defined by the strong PW inequality.

For the inversion we assume the ordinary Fourier inversion theorem. Given a function $h \in \mathrm{PW}(\mathbf{C}^r)$, we consider the restriction of h to the imaginary axis $\mathbf{i}\mathbf{R}^r = \mathbf{i}\mathfrak{a}$. The integral

$$\int_{\mathfrak{a}} h(\mathbf{i}t)e^{-\mathbf{i}tx}\,dt$$

is the Fourier transform of the function $h^{\#}(t) = h(\mathbf{i}t)$, and up to a constant factor (equal to $(2\pi)^r$) inverts the transform $\mathbf{M}_{\mathfrak{a}}$. We shall shift the vertical space of integration appropriately.

Given $\sigma = (\sigma_1, \ldots, \sigma_r) \in \mathbf{R}^r$ we define the integral of some function $h(s)$,

$$\int_{\mathrm{Re}(s)=\sigma} h(s)\,ds = \int_{\mathrm{Re}(s_1)=\sigma_1} \cdots \int_{\mathrm{Re}(s_r)=\sigma_r} h(s)\,ds_1 \cdots ds_r$$

to be the repeated integral over the variables on the vertical lines $\mathrm{Re}(s_i) = \sigma_i$ for $i = 1, \ldots, r$. Of course, this repeated integral must be absolutely convergent, which will be trivially the case in what follows. For h in the Paley–Wiener space, we claim that the integral

$$(2) \qquad \int_{\mathrm{Re}(s)=\sigma} h(s)e^{sx}\,ds$$

is independent of the real part σ. This amounts to the analogous statement in each variable, shifting the vertical line of integration one variable at a time. So let s for a moment denote one complex variable, and consider the rectangle $\mathrm{Rec}_{\sigma,T}$ as shown on the figure.

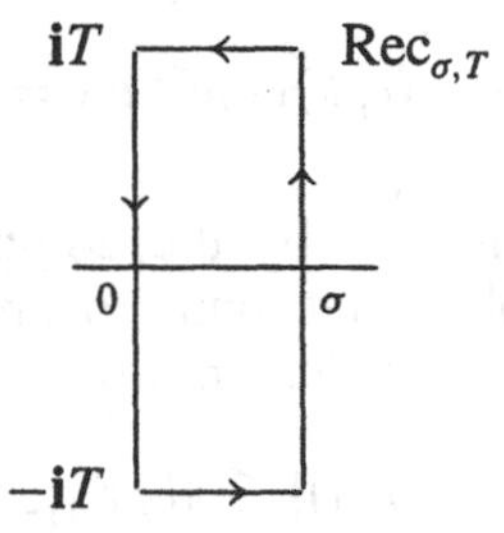

Then

$$\int_{\mathrm{Rec}_{\sigma,T}} h(s)e^{sx}\,ds = 0.$$

By the Paley–Wiener estimate, the function has arbitrary large polynomial decay on the top and bottom segments of the rectangle, say for the top segment,

$$\int_{\mathrm{top}} h(s)e^{sx}\,ds = O(T^{-N}) \qquad \text{for } T \to \infty,$$

because the length of the segment is bounded, and the term e^{sx} is bounded on the segment. This shows that the integral (2) is independent of σ.

Theorem 7.2. *Define*

$$ {}^t\mathbf{M}_{\mathfrak{a}}^{-}(h) = \int_{\mathrm{Re}(s)=\sigma} h(s)e^{-sx}\,ds. $$

Then

$$ {}^t\mathbf{M}_{\mathfrak{a}}^{-} \circ \mathbf{M}_{\mathfrak{a}} = (2\pi)^r \,\mathrm{id} \qquad \textit{and} \qquad \mathbf{M}_{\mathfrak{a}} \circ {}^t\mathbf{M}_{\mathfrak{a}}^{-} = (2\pi)^r \,\mathrm{id}. $$

Proof. The constant is the one thing coming from Fourier inversion, which we assume.

All that remains to be proved is that for $h \in \mathrm{PW}(\mathbf{C}^r)$ the function f defined by

$$ f(x) = \int_{\mathrm{Re}(s)=0} h(s)e^{-sx}\,ds $$

has compact support. So we fix x, and shift the space of integration with an appropriate σ. We need only the weak PW inequality

$$ |h(\sigma + it)| \leqq C_N \frac{e^{c|\sigma|}}{(1+|t|)^N}. $$

We take N so that the vertical integral is finite, for instance, $N = r+1$, and let

$$ \int_{\mathfrak{a}} (1+|t|)^{-N}\,dt = C' < \infty. $$

Then

$$ |f(x)| \leqq C_N C' e^{-\sigma x} e^{c|\sigma|}. $$

Suppose $|x| > c$. Let $\sigma = ux$ with u real $\to \infty$. Then

$$ c|\sigma| - \sigma x = cu|x| - ux^2 \to -\infty \qquad \text{as } u \to \infty. $$

Hence $f(x) = 0$, which concludes the proof of Theorem 7.1.

So far, we have given arguments with coordinates to fit the usual formulation of Cauchy's theorem for functions of one variable, as a tool to shift the vertical line of integration. For this purpose, any convenient norm was adequate. However, we want more precise information on the support of test functions, and on the order of growth of their images in the Paley–Wiener space. Hence we shall have to use a more invariant way of formulating these notions, which we now carry out. The application to support questions will be given in the next section.

Invariant formulation and intertwining relations

The second part of this section goes into the invariant formulation for the Fourier–Mellin transform, and some standard formulas showing how multiplication by polynomials and differentiation with constant coefficient partial differential operators correspond to each other under the Fourier–Mellin transform.

First we start with a finite dimensional vector space $\mathfrak{a}$ *over* $\mathbf{R}$, *and define*

$$\mathbf{M}_{\mathfrak{a}}\colon C_c^{\infty}(\mathfrak{a}) \to \mathrm{PW}(\mathfrak{a}_{\mathbf{C}}^{\vee})$$

by the integral

$$(\mathbf{M}_{\mathfrak{a}}g)(\zeta) = \int_{\mathfrak{a}} g(H)e^{\zeta(H)}\,dH \qquad \text{with } \zeta \in \mathfrak{a}_{\mathbf{C}}^{\vee}.$$

With a factor $-\mathbf{i}$ in the exponent, the corresponding integral would be called the **Fourier transform F**, so by definition

$$(\mathbf{F}g)(\zeta) = \int_{\mathfrak{a}} g(H)e^{-\mathbf{i}\zeta(H)}\,dH \qquad \text{for } g \in C_c^{\infty}(\mathfrak{a}).$$

Of course, we can define the **Mellin transform** of a function f on A by the integral

$$(\mathbf{M}f)(\zeta) = \int_A f(a)a^{\zeta}\,da.$$

We stick to $\mathbf{M}_{\mathfrak{a}}$ for the rest of the chapter. Writing $\zeta = \sigma + \mathbf{i}\lambda$ with $\sigma, \lambda \in \mathfrak{a}^{\vee}$, we see that

$$|e^{\zeta(H)}| = e^{\mathrm{Re}(\zeta)(H)} = e^{\sigma(H)}.$$

The Haar measures dH *on* $\mathfrak{a}$ *and* $d\lambda$ *on* $\mathfrak{a}^{\vee}$ *are now normalized so that Fourier inversion holds.* Thus letting $h \in \mathrm{PW}(\mathfrak{a}_{\mathbf{C}}^{\vee})$ and

$$g(H) = ({}^t\mathbf{M}_{\mathfrak{a}}^{-}h)(H) = \int_{\mathfrak{a}^{\vee}} h(\mathbf{i}\lambda)e^{-\mathbf{i}\lambda(H)}\,d\lambda$$

we have:

M 1. $\mathbf{M}_{\mathfrak{a}}^{-1} = {}^t\mathbf{M}_{\mathfrak{a}}^{-}$ on $\mathrm{PW}(\mathfrak{a}_{\mathbf{C}}^{\vee})$.

For every $\sigma \in \mathfrak{a}^\vee$ and $h \in \mathrm{PW}(\mathfrak{a}_{\mathbf{C}}^\vee)$ we can shift the vertical space of integration, and we have

$$\begin{aligned}({}^t\mathbf{M}_{\mathfrak{a}}^{-}h)(H) &= \int_{\mathrm{Re}(\zeta)=\sigma} h(\zeta)e^{-\zeta(H)}\,d\zeta = \int_{\mathfrak{a}^\vee} h(\sigma+\mathbf{i}\lambda)e^{-(\sigma+\mathbf{i}\lambda)(H)}\,d\lambda \\ &= e^{-\sigma(H)}\int_{\mathfrak{a}^\vee} h(\sigma+\mathbf{i}\lambda)e^{-\mathbf{i}\lambda(H)}\,d\lambda.\end{aligned}$$

We can rewrite this equality in the form

M 2. $$e^{\sigma(H)}(\mathbf{M}_{\mathfrak{a}}^{-1}h)(H) = \int_{\mathfrak{a}^\vee} h(\sigma+\mathbf{i}\lambda)e^{-\mathbf{i}\lambda(H)}\,d\lambda.$$

Of course, for $g \in C_c^\infty(\mathfrak{a})$ and $h \in \mathbf{M}_{\mathfrak{a}}g$ we can rewrite **M 2** in the form

$$e^{\sigma(H)}g(H) = \int_{\mathfrak{a}^\vee} (\mathbf{M}_{\mathfrak{a}}g)(\sigma+\mathbf{i}\lambda)e^{-\mathbf{i}\lambda(H)}\,d\lambda.$$

Next we deal with the differential operators. Let $Q \in S(\mathfrak{a})$, and $h = \mathbf{M}_{\mathfrak{a}}g$. Then

M 3. $$(\mathcal{D}(Q)g)(H) = \int_{\mathfrak{a}^\vee} (Q^- h)(\mathbf{i}\lambda)e^{-\mathbf{i}\lambda(H)}\,d\lambda,$$

or alternatively

$$\mathcal{D}(Q)\mathbf{M}_{\mathfrak{a}}^{-1} = \mathbf{M}_{\mathfrak{a}}^{-1} \circ Q^-.$$

Thus multiplication by Q^- on the Paley–Wiener side corresponds to partial differentiation on the C_c^∞ side. This is just differentiation under the integral sign of the relation $g = {}^t\mathbf{M}_{\mathfrak{a}}^{-}h$ using **M 1**. Thus

$$\begin{aligned}(\mathcal{D}(Q)g)(H) &= \int_{\mathfrak{a}^\vee} h(\mathbf{i}\lambda)\mathcal{D}_H(Q)e^{-\mathbf{i}\lambda(H)}\,d\lambda \\ &= \int_{\mathfrak{a}^\vee} h(\mathbf{i}\lambda)\,Q(-\mathbf{i}\lambda)e^{-\mathbf{i}\lambda(H)}\,d\lambda,\end{aligned}$$

which proves the formula.

Finally let $P \in S(\mathbf{i}\mathfrak{a}^\vee)$. Note that $\mathbf{i}\mathfrak{a}^\vee$ is a vector space over $\mathbf{R}$, and monomials in $S(\mathbf{i}\mathfrak{a}^\vee)$ are just formal commutative products

$$P = (\mathbf{i}\lambda_i)\cdots(\mathbf{i}\lambda_N) \qquad \text{with } \lambda_1,\ldots,\lambda_N \in \mathfrak{a}^\vee.$$

The value of such a monomial at an element $H \in \mathfrak{a}$ is

$$P(H) = \mathbf{i}^N \lambda_1(H) \cdots \lambda_N(H).$$

This is just one way to normalize the notation. [Alternatively, one could use a polynomial $P \in S(\mathfrak{a}^\vee)$, and then use the polynomial $P^{(\mathbf{i})}$ in $S(\mathbf{i}\mathfrak{a}^\vee)$ defined by $P^{(\mathbf{i})}(H) = P(\mathbf{i}H)$. Our chosen convention goes along with the conventions in [GaV 88], but not with those of some other authors, e.g. Anker [Ank 91].] With a polynomial $P \in S(\mathbf{i}\mathfrak{a}^\vee)$ as agreed upon, we have the commutation rule:

M 4.

$$P(H)g(H) = \int_{\mathfrak{a}^\vee} h(\mathbf{i}\lambda)\mathcal{D}(P^-)e^{-\mathbf{i}\lambda(H)}\, d\lambda = \int_{\mathfrak{a}^\vee} \mathcal{D}(P)h(\mathbf{i}\lambda)e^{-\mathbf{i}\lambda(H)}\, d\lambda,$$

or alternatively

$$P\,{}^t\mathbf{M}_{\mathfrak{a}}^- = {}^t\mathbf{M}_{\mathfrak{a}}^- \circ \mathcal{D}(P) = {}^t\mathbf{M}_{\mathfrak{a}}^- \circ \tilde{P}$$

with our old notation $\tilde{P}$ to denote the differential operator associated to P. The relation is immediate, by differentiating the exponential and using the formula for the transpose of a differential operator, which introduces a minus sign, under conditions of absolute convergence, and h together with its derivatives vanishing at infinity.

For the application to the next section, we need still one more formula, which will be formulated and proved separately to keep the next section as self-contained as possible. Let $P \in S(\mathfrak{a})$ and $g \in C_c^\infty(\mathfrak{a})$ as before. Then

M 5.

$$P(\zeta)(\mathbf{M}_{\mathfrak{a}}g)(\zeta) = \int_{\mathfrak{a}} (\mathcal{D}(P^-)g)(H)e^{\zeta(H)}\, dH$$

or alternatively

$$P\mathbf{M}_{\mathfrak{a}} = \mathbf{M}_{\mathfrak{a}} \circ \mathcal{D}(P^-).$$

See Lemma 8.1 below.

Remark. We carried out the above formalism in the context of C_c^∞ and the Paley–Wiener space. However, all that is needed to insure the validity of the formulas is that the arguments differentiating under the integral sign and integrating by parts be valid. For this all that is needed is that the functions be super polynomially decreasing. We shall apply the formulas in this context to functions in the Schwartz space in Chapter X.

IV, §8. BEHAVIOR OF THE SUPPORT

We stated Theorem 7.1 without any specification of the support. Similarly, in Chapters II and III we did not specify anything about the support of the Harish transform. We shall now make this support more precise. It is worth while to develop the theory more generally to include not only support in balls of radius R as Helgason did [Hel 66], following earlier work of Ehrenpreis–Mauntner on $SL_2(\mathbf{R})$ [EhM 55], but also support in certain convex sets as in Anker [Ank 91], Proposition 9, for application to the Schwartz space. It takes very little extra space to deal with those convex sets, and there is a big pay-off. So we start *ab ovo*.

Let $\zeta \in \mathfrak{a}_{\mathbf{C}}^{\vee}$ and $\sigma = \mathrm{Re}(\zeta)$. We shall estimate by using the formula

$$|e^{\zeta(H)}| = e^{\mathrm{Re}(\zeta)(H)} = e^{\sigma(H)}.$$

Let C be a compact convex subset of $\mathfrak{a}$. If f has support in C, then

$$(1) \qquad |(\mathbf{M}_{\mathfrak{a}} f)(\zeta)| \leqq \|f\|_1 \max_{H \in C} e^{\sigma(H)},$$

where $\|f\|_1$ is the L^1-norm of f on $\mathfrak{a}$ with respect to the chosen Lebesgue (Haar) measure.

We define the **quasi seminorm** associated to C to be the function on $\mathfrak{a}^{\vee}$ given by

$$q^C(\sigma) = \max_{H \in C} \sigma(H).$$

Then (1) can be rewritten

$$(2) \qquad |(\mathbf{M}_{\mathfrak{a}} f)(\zeta)| \leqq \|f\|_1 e^{q^C(\sigma)}.$$

Remarks. Suppose $0 \in C$. Since $\sigma(0) = 0$ for all $\sigma \in \mathfrak{a}^{\vee}$, it follows that $q^C(\sigma) \geqq 0$, and q^C behaves like a seminorm, namely for all $\sigma, \lambda \in \mathfrak{a}^{\vee}$,

$$q^C(\sigma + \lambda) \leqq q^C(\sigma) + q^C(\lambda).$$

For $t \geqq 0$,

$$q^C(t\sigma) = t q^C(\sigma).$$

The one difference with a seminorm is that we consider only $t > 0$ for the above homogeneity condition. *Suppose in addition that C is a neighborhood of* 0, i.e. *contains an open set containing* 0. *Then $q^C(\sigma) > 0$ if $\sigma \neq 0$*, because C contains an open ball (for some norm), which is symmetric about the origin, so given $\sigma \neq 0$, there exists v in this ball such that $\sigma(v) \neq 0$ and $q^C(\sigma) \neq 0$. Hence q^C behaves like a norm.

Actually, suppose that the convex set C is symmetric and contains a neighborhood of 0. Then q^C is a norm. An L^2-version of the Paley–Wiener theorem for such non-degenerate convex sets is given in Stein–Weiss [StW 71], Chapter III, Theorem 4.9. For the example of those convex sets which will be especially relevant for us later, see the Anker convex sets at the end of the section.

We return to general considerations. Let $f \in C_c^\infty(\mathfrak{a})$. By Theorem 7.1, we know of course that $\mathbf{M}_\mathfrak{a} f$ is in the Paley–Wiener space. However, we now have an alternative way of filtering this space. For each compact convex set C in $\mathfrak{a}$, we define

$\mathrm{PW}_C(\mathfrak{a}_\mathbf{C}^\vee) =$ set (actually complex vector space) of entire functions h on $\mathfrak{a}_\mathbf{C}^\vee$ such that for each positive integer N there is a constant C_N having the property that for all $\zeta \in \mathfrak{a}_\mathbf{C}^\vee$,

$$|h(\zeta)| \leqq C_N \frac{1}{(1+|\zeta|)^N} e^{q^C(\mathrm{Re}\,\zeta)}.$$

We call q^C the **exponential order** of the above function h. Also let:

$C_C^\infty(\mathfrak{a}) =$ vector space of C^∞ functions on $\mathfrak{a}$ with support in C.

We shall relate the above spaces. We need the fact that multiplication by polynomials corresponds to differential operators as in the next lemma.

Lemma 8.1. *Let $g \in C_c^\infty(\mathfrak{a})$. Let P be a polynomial function on $\mathfrak{a}^\vee$, so P is in the symmetric algebra of $\mathfrak{a}$. Then for $\zeta \in \mathfrak{a}_\mathbf{C}^\vee$ we have*

$$P(\zeta)(\mathbf{M}_\mathfrak{a} g)(\zeta) = \int_\mathfrak{a} (\mathcal{D}(P^-)g)(H) e^{\zeta(H)}\, dH = \mathbf{M}_\mathfrak{a}(\mathcal{D}(P^-)g)(\zeta),$$

where $P^-(\zeta) = P(-\zeta)$ (cf. Proposition 6.3 *of Chapter* II).

Proof. The routine is

$$\begin{aligned} P(\zeta)(\mathbf{M}_\mathfrak{a} g)(\zeta) &= \int_\mathfrak{a} g(H) P(\zeta) e^{\zeta(H)}\, dH \\ &= \int_\mathfrak{a} g(H) \mathcal{D}_H(P) e^{\zeta(H)}\, dH \\ &= \int_\mathfrak{a} (\mathcal{D}(P^-)g)(H) e^{\zeta(H)}\, dH \end{aligned}$$

by the above mentioned proposition, thus proving the lemma.

Theorem 8.2. *Let C be a compact convex set in $\mathfrak{a}$. The Mellin transform*

$$\mathbf{M}_{\mathfrak{a}}: C_C^\infty(\mathfrak{a}) \to \mathrm{PW}_C(\mathfrak{a}_{\mathbf{C}}^\vee)$$

maps $C_C^\infty(\mathfrak{a})$ into $\mathrm{PW}_C(\mathfrak{a}_{\mathbf{C}}^\vee)$.

Proof. To show that the image of $\mathbf{M}_{\mathfrak{a}}$ is contained in $\mathrm{PW}_C(\mathfrak{a}_{\mathbf{C}}^\vee)$, the inequality (2) is a first step. To get the polynomial decay, we apply Lemma 8.1. The function $\mathcal{D}(P^-)g$ has compact support, we can apply (2) to get the exponential growth, and we can divide by the absolute value of the polynomial $|P(\zeta)|$ for $\|\zeta\|$ (any fixed norm) sufficiently large to get the arbitrary large polynomial decay, thus concluding the proof that the image has exponential order q^C.

Under the pull-back of the Mellin transform to $\mathfrak{a}$, we have the transpose operator, defined by

$$({}^t\mathbf{M}_{\mathfrak{a}}h)(H) = \int_{\mathrm{Re}(\zeta)=0} h(\zeta)e^{\zeta(H)}\,d\zeta,$$

and with the usual minus sign,

$$({}^t\mathbf{M}_{\mathfrak{a}}^-h)(H) = \int_{\mathrm{Re}(\zeta)=0} h(\zeta)e^{-\zeta(H)}\,d\zeta.$$

Theorem 8.3. *Let C be a compact convex subset of $\mathfrak{a}$. Then ${}^t\mathbf{M}_{\mathfrak{a}}^-$ and ${}^t\mathbf{M}_{\mathfrak{a}}$ map $\mathrm{PW}_C(\mathfrak{a}_{\mathbf{C}}^\vee)$ into $C_C^\infty(\mathfrak{a})$, that is, if $H \notin C$, then*

$$\int_{\mathrm{Re}(\zeta)=0} h(\zeta)e^{\zeta(H)}\,d\zeta = 0 \qquad \textit{for } h \in \mathrm{PW}_C(\mathfrak{a}_{\mathbf{C}}^\vee).$$

The Mellin transform induces a linear isomorphism

$$\mathbf{M}_{\mathfrak{a}}: C_C^\infty(\mathfrak{a}) \to \mathrm{PW}_C(\mathfrak{a}_{\mathbf{C}}^\vee)$$

whose inverse is ${}^t\mathbf{M}_{\mathfrak{a}}^-$, up to a constant factor depending on the choice of Haar measures.

Proof. By Theorem 7.2 and of course Theorem 8.2, the first statement implies the second, so it suffices to prove that

$${}^t\mathbf{M}_{\mathfrak{a}}^-: \mathrm{PW}_C(\mathfrak{a}_{\mathbf{C}}^\vee) \to C_C^\infty(\mathfrak{a})$$

maps $\mathrm{PW}_C(\mathfrak{a}_{\mathbf{C}}^\vee)$ into $C_C^\infty(\mathfrak{a})$. As in Theorem 7.1, we shift the space of integration. Let h be in the Paley–Wiener space, with exponential order q^C. Let $H \notin C$. We have to show that $({}^t\mathbf{M}_{\mathfrak{a}}h)(H) = 0$. By convexity, there exists a functional $\sigma \in \mathfrak{a}^\vee$, and a real number c defining the hyperplane $\sigma(X) = c$ (for X variable in $\mathfrak{a}$), such that $\sigma(H) > c$ and $\sigma(C) \leqq c$. (This is the standard separation theorem for convex sets, already recalled in Chapter I, before Proposition 5.1). By definition,

$$q^C(\sigma) = \max_{x \in C} \sigma(X).$$

By Cauchy's theorem and the arbitrary fast polynomial decay, the values $h(\zeta)$ tend to 0 on horizontal segments at heights tending to infinity, and hence

$$\int_{\mathrm{Re}(\zeta)=0} h(\zeta)e^{-\zeta(H)}\,d\zeta = \int_{\mathrm{Re}(\zeta)=\sigma} h(\zeta)e^{-\zeta(H)}\,d\zeta.$$

We replace σ by $t\sigma$ with t positive tending to infinity. For $\mathrm{Re}(\zeta) = t\sigma$ we have with N large, say $N \geqq \dim \mathfrak{a} + 1$ to make the integral absolutely convergent,

$$|h(\zeta)e^{-\zeta(H)}| \leqq C_{C,N} \frac{1}{(1+|\zeta|)^N} e^{q^C(t\sigma) - t\sigma(H)}$$

and if we put $c_1 = \sigma(H)$, then

$$q^C(t\sigma) - t\sigma(H) \leqq t(c - c_1),$$

which tends to $-\infty$ as $t \to \infty$. Hence the integral on the right is equal to 0, which proves the theorem.

Up to now in this section, we have dealt only with a general finite dimensional vector space $\mathfrak{a}$ over the reals, and its complexification $\mathfrak{a}_{\mathbf{C}}$.

We turn to the situation when we have a positive definite scalar product B on $\mathfrak{a}$, arising in practice, and when we have a fixed basis $\{\alpha_1, \ldots, \alpha_r\}$ which gives rise to the notion of semipositivity and positivity as in Chapter I, §4.

Example of a convex set. The scalar product norm gives rise to the most basic example of a convex set, namely the ball of radius R which we denote by $\bar{\mathbf{B}}_R(0)$. Let $H \in \mathfrak{a}$ and $H > 0$ (with respect to the given basis). Let σ be the functional such that H_σ is the unit vector in the direction of H. Then $\sigma(H) = |H|$, and $H_\sigma > 0$, so $\sigma > 0$. Furthermore, if $X \in \mathfrak{a}$, $|X| \leqq R$, that is $X \in \bar{\mathbf{B}}_R(0)$, then $\sigma(X) \leqq R$. Thus σ and R constitute an example for the proof, with $c = R$. In the case of balls of radius R, we also call R the **exponential order,** because if $C = \bar{\mathbf{B}}_R(0)$ then

$$q^C(\sigma) = R|\sigma|.$$

There will be other important convex sets, see the Anker convex sets below.

In some applications, we want to multiply h by a meromorphic function which does not have poles in, say, the left half space including the imaginary axis $\mathbf{i}\mathfrak{a}^\vee$. See the end of Chapter VI, §8. Thus it is convenient to formulate the half space version of Paley–Wiener as follows. Let h be a function on the half space $\operatorname{Re}(\zeta) \leqq 0$, $\zeta \in \mathfrak{a}^\vee_{\mathbf{C}}$. Let C be a compact convex subset of $\mathfrak{a}$. We define h to be a **Paley–Wiener function** of exponential order q^C **on the half space** if h is holomorphic on this half space, and for each positive integer N there is a constant C_N such that for $\operatorname{Re}(\zeta) \leqq 0$, we have

$$|h(\zeta)| \leqq C_N(1+|\zeta|)^{-N} e^{q^C(\operatorname{Re}\zeta)}.$$

Corollary 8.4. *Let $h(\zeta)$ be a Paley–Wiener function on the half space $\operatorname{Re}(\zeta) \leqq 0$, of exponential order q^C on the half space. Let $H \notin C$. Then*

$$\int_{\operatorname{Re}(\zeta)=0} h(\zeta) e^{\zeta(H)}\, d\zeta = 0.$$

Proof. We have

$$\int_{\operatorname{Re}(\zeta)=0} h(\zeta) e^{\zeta(H)}\, d\zeta = \int_{\operatorname{Re}(\zeta)=0} h(-\zeta) e^{-\zeta(H)}\, d\zeta.$$

The proof of Theorem 8.3 now applies without change, integrating over $\operatorname{Re}(\zeta) = t\sigma$.

We make the special case of balls explicit.

Corollary 8.5. *Let $h(\zeta)$ be a Paley–Wiener function on the half space $\operatorname{Re}(\zeta) \leqq 0$, of exponential order $\leqq R$ (taking $C = \bar{\mathbf{B}}_R(0)$). Let $H \in \mathfrak{a}^+$ and $R < |H|$ (B-scalar product norm). Then*

$$\int_{\operatorname{Re}(\zeta)=0} h(\zeta) e^{\zeta(H)}\, d\zeta = 0.$$

Proof. This is a special case of Corollary 8.4, but explicitly one could take σ such that H_σ is the unit vector in the direction of H, so $\sigma > 0$.

In the applications we have to take the Weyl group into account. More generally, we note that *the transform $\mathbf{M}_\mathfrak{a}$ is equivariant with respect to linear automorphisms of $\mathfrak{a}$ with determinant* ± 1, or even with respect to differential automorphisms whose absolute Jacobian determinant is

1. This is immediate from the definition and the change of variables $H \mapsto wH$ in the integral

$$(\mathbf{M}_{\mathfrak{a}} wf)(\zeta) = \int_{\mathfrak{a}} f(w^{-1}H)e^{\zeta(H)}\,dH = (w\mathbf{M}_{\mathfrak{a}} f)(\zeta).$$

Thus we obtain:

Proposition 8.6. *Let W be a finite group of linear automorphisms of $\mathfrak{a}$. Let C be a compact convex subset of $\mathfrak{a}$, and assume that C is stable under the action of W. Then $C_C^\infty(\mathfrak{a})^W$ and $\mathrm{PW}_C(\mathfrak{a}_{\mathbf{C}}^\vee)^W$ correspond to each other under the Mellin transform isomorphism.*

The Harish-Chandra and spherical transform

So far in this section, we have dealt exclusively with the additive theory, on $\mathfrak{a}$ or $\mathfrak{a}^\vee$, and the Mellin transform in its additive version. We finally turn to the Harish transform, which occurs multiplicatively, and we let $G = \mathrm{SL}_n(\mathbf{R})$ so $\mathfrak{a}$ is the space of diagonal $n \times n$ matrices with trace 0. We now look at the support in the context of the Harish-Chandra transform, as in Anker [Ank 91], Corollary 12.

Let $f \in C_c(K\backslash G/K)$ be K-bi-invariant. Its Harish transform is given by the integral

$$\mathbf{H}f(a) = \delta^{1/2}(a) \int_U f(au)\,du.$$

Of course, the factor $\delta^{1/2}(a)$ does not affect the support. The function f pulls back to a function $f_{\mathfrak{a}}$ on $\mathfrak{a}$ via the exponential map, because

$$f(k_1 a k_2) = f(a) \qquad \text{for all} \quad k_1, k_2 \in K.$$

Furthermore, we now use Chapter I, §4. If $a \in A^+$, or equivalently $a = \exp H$ with $H \in \mathfrak{a}^+$, then

$$f(a) = f(a^+) \qquad \text{and} \qquad f_{\mathfrak{a}}(H) = f_{\mathfrak{a}}(wH) \qquad \text{for all } w \in W.$$

Let C be a compact, convex, W-stable subset of $\mathfrak{a}$. We define

$$A_C = \exp C \qquad \text{and} \qquad G_C = K(\exp C)K,$$

so G_C is the subset of elements whose polar representative in $\mathfrak{a}$ lies in C. We let:

$$\begin{aligned} C_C(K\backslash G/K) &= \text{vector space of continuous functions on } G,\ K\text{-bi-} \\ &\quad\ \text{invariant, whose pull-back to } \mathfrak{a} \text{ have support in } C \\ &= \text{vector space of elements } f \in C_c(K\backslash G/K) \text{ with} \\ &\quad\ \text{support in } G_C. \end{aligned}$$

We define $C_C(A)$ and $C_C^\infty(A)$, $C_C^\infty(K\backslash G/K)$ in the same way.

We say that C is **U-polar expanding** if given $a \in A$, $a \notin A_C$ and $u \in U$, then $au \notin G_C$.

Example. By Theorem 6.2 of Chapter I, the ball $\bar{\mathbf{B}}_R(0)$ is U-polar expanding. We shall give other examples shortly.

Theorem 8.7. *Suppose C is U-polar expanding. Then the Harish-Chandra transform induces linear maps*

$$\mathbf{H}\colon C_C(K\backslash G/K) \to C_C(A)^W \quad \textit{and} \quad C_C^\infty(K\backslash G/K) \to C_C^\infty(A)^W.$$

Proof. This is immediate from the integral definition of the Harish transform, and the definition of U-polar expanding.

Corollary 8.8. *Suppose C is U-polar expanding. Then the spherical transform induces a linear map*

$$\mathbf{S}\colon C_C^\infty(K\backslash G/K) \to \mathrm{PW}_C(\mathfrak{a}_C^\vee)^W.$$

Proof. All we have to do is put together Theorem 8.3 for the Mellin transform, Theorem 8.7 for the Harish transform, the fact that their composite is the spherical transform, and Proposition 8.6 to take W into account.

The spherical inversion theorem to complement Corollary 8.8 requires additional properties of spherical functions, and will be carried out in Chapters VIII and IX, following Anker [Ank 91]. We assume the reader is familiar with Chapter I, §4.

Examples. Anker convex sets. Let $\Lambda \in \mathfrak{a}^\vee$, $\Lambda >_{\mathcal{A}'} 0$. We define the **Anker convex set**:

$$C_\Lambda = C_{\Lambda,1} = \text{set of elements } H \in \mathfrak{a} \text{ such that}$$

$$\Lambda(wH) \leqq 1 \qquad \text{for all } w \in W.$$

The convexity and W-stability of C_Λ are immediate. *Furthermore, C_Λ is compact.* To prove this, write $\Lambda = \sum c_i\alpha_i$ with $c_i > 0$ cf. Chapter I, §4). Let $H \in C_\Lambda$. Write $H^+ = \sum y_i H_{\alpha_i'}$ with the dual basis $\{\alpha_1', \ldots, \alpha_r'\}$.

By hypothesis,

$$1 \geqq \Lambda(H^+) = \sum c_i y_i.$$

Since $H^+ \geqq 0$ we have $y_i \geqq 0$ for all i, so the coefficients y_i are bounded, thus proving H^+ bounded, and hence $\mathcal{C}_\Lambda$ is bounded, being in the image of a bounded set by W. This proves $\mathcal{C}_\Lambda$ compact.

For $R > 0$ we let:

$\mathcal{C}_{\Lambda,R}$ = subset of $\mathfrak{a}$ consisting of the elements H such that

$$\Lambda(wH) \leqq R \quad \text{for all } w \in W.$$

Note that $\mathcal{C}_{\Lambda,R} = \mathcal{C}_{R^{-1}\Lambda,1}$. These sets $\mathcal{C}_{\Lambda,R}$ are analogues of balls of radius R. They play a fundamental role in Anker's extension of Paley–Wiener. They satisfy some important classical properties, which we summarize by making a definition. Let V be a finite dimensional vector space over $\mathbf{R}$. Let $\mathcal{C}$ be a compact convex subset of V. We say that $\mathcal{C}$ is **Minkowskian**, or a **Minkowski convex set**, if it satisfies the additional conditions:

MINK 1. $\mathcal{C}$ is stable under multiplication by real t with $0 \leqq t \leqq 1$, that is $t\mathcal{C} \subset \mathcal{C}$ for such t. In particular, $0 \in \mathcal{C}$.

MINK 2. $\mathcal{C}$ is a compact neighborhood of 0.

Proposition 8.9. *The Anker convex sets $\mathcal{C}_\Lambda$ are Minkowskian.*

Proof. First $0 \in \mathcal{C}_\Lambda$, because as in Chapter I, §4, the element

$$|W|^{-1} = \sum_{w \in W} wH \qquad \text{for } H \in \mathcal{C}_\Lambda$$

is fixed under W, all its diagonal components are equal, the trace is 0, so the element is 0. The homogeneity condition **MINK 1** is then immediate, and so is the fact that $\mathcal{C}_\Lambda$ contains a basis of $\mathfrak{a}$, so **MINK 2** is also satisfied. Minkowskian convex sets are routinely used by functional analysts, e.g., [Yos 65/88], p. 24.

To simplify the notation, we shall abbreviate:

$$G_{\mathcal{C}_\Lambda} = G_\Lambda, \qquad A_{\mathcal{C}_\Lambda} = A_\Lambda, \qquad C^\infty_{\mathcal{C}_\Lambda}(A) = C^\infty_\Lambda(A),$$
$$C^\infty_{\mathcal{C}_\Lambda}(K\backslash G/K) = C^\infty_\Lambda(K\backslash G/K),$$

and similarly with R, that is, $G_{\Lambda,R}$, $C^\infty_{\Lambda,R}(K\backslash G/K)$, etc.

Proposition 8.10. *Let $\Lambda >_{\mathcal{A}'} 0$. The Anker convex set C_Λ is U-polar expanding.*

Proof. This is merely Theorem 6.2 of Chapter I, put there for use here, because $a \notin G_\Lambda$ implies $au \notin G_\Lambda$ for $u \in U$.

In particular, Theorem 8.7 and Corollary 8.8 apply to these convex sets.

Remark. In other theories of convex sets, the symmetric ones play an important role (symmetry with respect to the origin), but symmetry is irrelevant for the properties we develop at first. It happens to be satisfied in the most important case used by Anker, when $\Lambda = \rho$ or a multiple of ρ, because $-\rho$ is in the orbit $W\rho$, so *the convex closure* $\mathrm{Co}(W\rho)$ *is symmetric*. Cf. Chapter X, §1, (5).

CHAPTER V

Gelfand–Naimark Decomposition and the Harish-Chandra c-Function

This chapter continues the integral representation of spherical functions. Most of it is due to Harish-Chandra [Har 58a], who recognized and developed the role of $\bar{U} = {}^t U$ in his theory of spherical inversion. This chapter develops the information which arises from the Gelfand–Naimark decomposition.

In this decomposition, K is essentially replaced by $\bar{U}$, and integration over K is replaced by integration over $\bar{U}$. Harish-Chandra's so-called c-function arises naturally in two contexts: the present context of integral formulas, and as a function giving the coefficients of a certain linear combination of certain series representing spherical functions, to be handled in Chapter VIII as an application of the polar decomposition. Here we take Harish-Chandra's integral formula for the c-function as a definition, which we immediately connect with the main asymptotic behavior of the spherical functions on $SL_n(\mathbf{R})$. We then derive the product formula for the c-function due to Bhanu-Murty [Bha 60]. For this formula, the characters are parametrized by $n-1$ complex variables in the case of SL_n. Our presentation uses natural coordinates of $SL_n(\mathbf{R})$, and in this way a basis for the A-characters giving rise to the so-called Selberg power function.

For a much more elaborate continuation of the integral theory of the c-function, see [Har 68]. At the present level, we are really developing a basic pattern whereby we squeeze whatever can be squeezed out of three or four decompositions, namely Iwasawa, Cartan, Gelfand–Naimark, polar. In this chapter, we deal with the immediate consequences of the Gelfand–Naimark decomposition. The next three chapters will describe what can be squeezed out of the polar decomposition, and are logically independent of the present chapter. However, one has to project the exposition in a totally ordered way on the page axis, and we have chosen to present the Harish-Chandra integral for the $\boldsymbol{c}$-function first. Readers can do as they please. Both [Hel 84] and [GaV 88] follow the order of discovery by Harish-Chandra.

Of course, at the end of the independent exposition of the effect of the Gelfand–Naimark and polar decompositions, following Harish-Chandra one puts the two aspects together, so Theorem 3.6 of Chapter VIII will show that certain objects which came up independently in the theory of the Harish-Chandra series are equal to the objects which come up in the theory of the Harish-Chandra integral. A loop closes, but the closure of the loop should not obscure the fact that the integral theory of the $\boldsymbol{c}$-function is independent of the Harish-Chandra series.

The determination of the $\boldsymbol{c}$-function will be done by induction on a repeated integral. Three natural bases of $\mathfrak{a}$ or its dual space $\mathfrak{a}^\vee$ intervene. The inductive step will be carried out with respect to one of them. In §7, readers will find a systematic discussion of these bases and of the way the $\boldsymbol{c}$-function can be expressed both invariantly and in terms of the coordinatization with one of them. Readers may find Chapter I, §4, illuminating in the present context.

V, §1. THE GELFAND–NAIMARK DECOMPOSITION AND THE HARISH-CHANDRA MAPPING OF $\bar{U}$ INTO $M\backslash K$

We let $G = \mathrm{SL}_n(\mathbf{R})$. Then G has the Iwasawa decomposition

$$G = UAK,$$

We shall use Chapter III, §3, giving basic commutation properties and a description of the Weyl group. By Corollary 3.2 of Chapter III, we know that the centralizer M of A in K consists of all diagonal matrices in G having ± 1 as diagonal elements. We let:

B = **Borel subgroup** of the Iwasawa decomposition, that is by definition,

$$B = UAM = MAU,$$

so B is just the group of upper triangular matrices with arbitrary diagonal elements having determinant 1. We let

$$\mathfrak{b} = \text{Lie algebra of } B = \mathfrak{a} + \mathfrak{n}.$$

We shall be especially concerned with

$$\bar{U} = {}^tU,$$

which is the group of unipotent lower diagonal matrices, with Lie algebra $\bar{\mathfrak{n}}$. Thus

$$\bar{\mathfrak{n}} = {}^t\mathfrak{n}.$$

We now come to still another product decomposition of G, concerning the role of the group $\bar{U}$. An axiomatization will be given in §3.

Theorem 1.1 ([GeN 50]). *The product map*

$$B \times \bar{U} \to B\bar{U}$$

is a differential isomorphism of $B \times \bar{U}$ on an open subset of G. The complement consists of finitely many manifolds of lower dimension.

Proof. First we prove the injectivity of the product map. Suppose

$$T_1\bar{u}_1 = T_2\bar{u}_2 \qquad \text{with } u_1, u_2 \in U \text{ and } T_1, T_2 \in B.$$

Then $\bar{u}_2\bar{u}_1^{-1} = T_2^{-1}T_1$. The left side is lower triangular while the right side is upper triangular. Hence both are diagonal. But the left side has its diagonal elements equal to 1, and hence so does the right side. Hence finally $u_1 = u_2$ and $T_1 = T_2$, thus proving the injectivity. Observe that the proof is completely algebraic, and applies to complex matrices as well. In particular, it follows that the map

$$B(\mathbf{C}) \times \bar{U}(\mathbf{C}) \to B(\mathbf{C})\bar{U}(\mathbf{C}) \subset G(\mathbf{C})$$

is a complex analytic embedding. Hence $B \times \bar{U} \to B\bar{U} \subset G$ is a real differential embedding. (See also Lemmas 3.5 and 3.6.) Since

$$\dim \bar{U} + \dim B = \dim G,$$

and $\bar{U}, B$ are algebraically irreducible, it follows that $B\bar{U}$ is open in G. Finally, the map $B \times \bar{U} \to B\bar{U}$ is birational (because of the injectivity), whence the complement of the image consists of a proper algebraic subset, and is therefore of lower dimension, thus concluding the proof.

Remark 1. Theorem 1.1 is all that we shall need for the applications to the Harish-Chandra $\boldsymbol{c}$-function and integration. There is of course some independent interest about what the complement of $B\bar{U}$ looks like. We shall therefore interpolate another section dealing with this question in §2, but we won't need it subsequently. It may, however, help the reader get into Harish-Chandra's works, and anyhow provides insight into the algebraic structure of G. A purely differential argument will be given in §3.

We consider the Iwasawa decomposition

$$G = UAK,$$

with the Iwasawa projections

$$\mathrm{Iw}_U(uak) = u, \qquad \mathrm{Iw}_A(uak) = a, \qquad \mathrm{Iw}_K(uak) = k.$$

We let the **Harish-Chandra mapping**

$$\psi_{M\backslash K}\colon \bar{U} \to M\backslash K \qquad \text{be the map such that} \qquad \bar{u} \mapsto M\,\mathrm{Iw}_K(\bar{u}).$$

Theorem 1.2. *The map $\psi_{M\backslash K}$ is a differential isomorphism of $\bar{U}$ onto an open subset of $M\backslash K$. The complement of the image of $\psi_{M\backslash K}$ is a finite union of subsets of lower dimension.*

Proof. This is immediate from Theorem 1.1, since we have a differential isomorphism from the Iwasawa projection

$$UA\backslash G \to K,$$

whence a differential isomorphism

$$UAM\backslash G \to M\backslash K.$$

Then we apply Theorem 1.1 to get $\bar{U}$ as an open subset of $AUM\backslash G$, whence an open subset of $M\backslash K$ under $\psi_{M\backslash K}$.

Remark 2. Readers may now skip §2 and go directly to §3 where we determine the Jacobian determinant of the map $\psi_{M\backslash K}$, and various applications to integral formulas.

Remark 3. Previously, and especially Chapter III, §7, we used only the Iwasawa decomposition $G = UAK$. One may define an **anti-Iwasawa decomposition** $G = KAU$, with its three projections on K, A, U respectively which we could denote by $\mathrm{Iw}'_K, \mathrm{Iw}'_A, \mathrm{Iw}'_U$. Harish-Chandra used this anti-Iwasawa decomposition in connection with $\bar{U}$. However, Wallach in [Wal 88] manages by using only the Iwasawa decomposition all the way through, and he drew our attention to this possibility, which we adopt. We have an immediate relation between the Iwasawa and anti-Iwasawa projections, namely for $x \in G$:

$$\mathrm{Iw}'_U(x^{-1}) = \mathrm{Iw}_U(x)^{-1}, \qquad \mathrm{Iw}'_A(x^{-1}) = \mathrm{Iw}_A(x)^{-1},$$
$$\mathrm{Iw}'_K(x^{-1}) = \mathrm{Iw}_K(x)^{-1}.$$

These are obvious from the definitions. But we shall not use the anti-Iwasawa decomposition.

Remark 4. The above decompositions hold as well for $\mathrm{GL}_n(\mathbf{R})$, in exactly the same way.

Axiomatization of the Gelfand–Naimark decomposition

In line with our general practice, we give a useful set of axioms.

Let $G = PK = UAK$, $P = UA$, be a weak Iwasawa decomposition with K compact, G, U, A unimodular. By a **Gelfand–Naimark decomposition** of such G, we mean the additional data of a closed Lie subgroup M of K and a closed unimodular subgroup $\bar{U}$ of G such that, if we put $B = PM$, then:

GN 1. $P \times M \to PM = B$ is a differential isomorphism.

GN 2. $B \times \bar{U} \to B\bar{U}$ is a differential isomorphism of $B \times \bar{U}$ with an open subset of G, whose complement has measure 0.

GN 3. M normalizes $U, \bar{U}$ and centralizes A.

GN 4. The map $\psi_{M\backslash K}\colon \bar{U} \to M\backslash K$ defined by $\psi_{M\backslash K}(\bar{u}) = M\,\mathrm{Iw}_K(\bar{u})$ gives a differential isomorphism of $\bar{U}$ with an open subset of $M\backslash K$ whose complement has measure 0. We call $\psi_{M\backslash K}$ the **Harish-Chandra mapping.**

Note that the first condition is automatic, since the product Iwasawa decomposition $P \times K \to PK$ induces a differential isomorphism

$$P \times M \to PM$$

because M splits locally in K.

A Gelfand–Naimark decomposition was proved above for $\mathrm{SL}_n(\mathbf{R})$, and a similar one is valid for $\mathrm{GL}_n(\mathbf{R})$, additionally with M finite. The general conditions we have listed here are all that is needed to make valid a number of results, including the integral formulas of §3. We have kept the notation $\bar{U}$ to fit the application we have in mind, instead of a more neutral letter (say V), to fit specific applications as they arise in this book, but the general results of §3 are valid with $\bar{U}$ unrelated to U.

V, §2. THE BRUHAT DECOMPOSITION

We continue with $G = \mathrm{SL}_n(\mathbf{R})$. *This section will not be used elsewhere in the book.*

We shall get further insight on the Gelfand–Naimark decomposition, in terms of a double coset of $B = UAM = MAU$. It turns out that there is one main double coset of B in G, which we look at first. The following element of GL_n will play an important role:

$$\omega_0 = \begin{pmatrix} 0 & \cdots & 0 & 1 \\ 0 & \cdots & 1 & 0 \\ & \cdots & & \\ 1 & \cdots & 0 & 0 \end{pmatrix}.$$

To obtain an element in $\mathrm{SL}_n(\mathbf{R})$ playing the same role, we have to distinguish cases depending on n. If $n = 2m$, $2m + 1$ with m even, we let $\omega_n = \omega_0$ above. If m is odd, let

$$\omega_n = \begin{pmatrix} 0 & \cdots & 0 & -1 \\ 0 & \cdots & 1 & 0 \\ \vdots & \iddots & \vdots & \vdots \\ 1 & \cdots & 0 & 0 \end{pmatrix}.$$

so we just change the upper right component from 1 to -1 to insure that $\det(\omega) = 1$ so $\omega \in \mathrm{SL}_n(\mathbf{R})$. We call ω_n the **special Weyl element**. We abbreviate ω_n by ω. We then have

$$\omega^2 \in M,$$

i.e. ω^2 is diagonal with ± 1 on the diagonal. We shall write $[\omega]$ for the conjugation action by ω. We note that $\omega \in K$, so ${}^t\omega = \omega^{-1}$, and our notation is consistent with past notation. If S is a subset of G, we also use the notation

$$[\omega]S = S^{\omega^{-1}} = S^{\omega}.$$

In all cases, we have the following action on the two groups A and U:

$$A^{\omega} = A \quad \text{and} \quad U^{\omega} = \bar{U},$$

where $\bar{U}$ is the group of lower diagonal unipotent matrices. In particular, $\omega \in M'$. Actually, if D is a diagonal matrix, possibly singular so that some components may be 0, we still have D^{ω} diagonal, so in fact

$$\mathfrak{a}^{\omega} = \mathfrak{a},$$

where $\mathfrak{a}$ is the Lie algebra of diagonal matrices. More precisely:

$$\text{If} \quad X = \begin{pmatrix} x_1 & 0 & \cdots & 0 \\ \vdots & \vdots & \ddots & \vdots \\ 0 & 0 & \cdots & x_n \end{pmatrix} \quad \text{then} \quad X^{\omega} = \begin{pmatrix} x_n & 0 & \cdots & 0 \\ \vdots & \vdots & \ddots & \vdots \\ 0 & 0 & \cdots & x_1 \end{pmatrix},$$

that is, conjugation by ω reverses the order of the diagonal components. These statements are proved by direct computation with the matrices, and make explicit on SL_n general facts from semisimple Lie groups.

Proposition 2.1. *We have*

$$B\bar{U}\omega = B\omega U = B\omega B.$$

So by Theorem 1.1, *$B\omega B$ is open in G, and its complement has lower dimension in G.*

Proof. We just write $B\bar{U}\omega = B\omega U\omega\omega = B\omega U = B\omega B$, this last equality because $A\omega = \omega A$ and $M\omega = \omega M$, $B = MAU = UAM$, $A = AA$, $M = MM$, so we can move one factor A or M from the left to the right of ω. This concludes the proof.

Next we shall describe very precisely the complement of $B\omega B$. We are headed to the Bruhat decomposition of G into double cosets of B, that is $B \times B$ with $x \in G$. Observe that if $x = uak$ is the Iwasawa decomposition of an element $x \in G$, then

$$BxB = BkB,$$

so double cosets are determined by elements of K. But only special elements are needed to get a double coset decomposition of G with respect to B, as in the next theorem of Bruhat.

Theorem 2.2. *The map $w \mapsto BwB$ is a bijection of the Weyl group W and the set of double cosets of B in G.*

The proof will take the rest of the section.

We shall follow Harish-Chandra's general formation of Bruhat's decomposition [Har 56], see also Chevalley's completely algebraic treatment [Che 55].

We start with an analogue of Chapter III, Theorem 4.1, on the Lie algebras. Dealing with the case of SL_n simplifies the visualization. For convenience, we use the following terminology. Let X be strictly upper triangular. We define the **grading** of X to be the earliest diagonal on which X contains a non-zero component, provided $X \neq 0$.

If H is regular in $\mathfrak{a}$, that is, $H \in \mathfrak{a}'$, then $HX - XH$ has the same grading as X. This is another way of seeing Proposition 3.1(i) of Chapter III.

Proposition 2.3. *Let $H \in \mathfrak{a}'$ be a regular element of $\mathfrak{a}$. Then the map*

$$\Phi_H : u \mapsto H^u - H \qquad \text{of } U \to \mathfrak{n}$$

is a differential, actually algebraic, isomorphism. Here we have denoted conjugation exponentially, that is, $H^u = u^{-1}Hu$. The *derivative is given by*

$$\Phi'_H(u) = -\mathbf{c}(u^{-1}) \circ H_{\mathfrak{n}}, \tag{1}$$

where $\mathbf{c}(u^{-1})$ *is conjugation on the left by* u^{-1}, *and* $H_{\mathfrak{n}}$ *is the action of* H *on* $\mathfrak{n}$ *in the regular representation of the Lie algebra.*

Proof. As for injectivity, suppose $H^u - H = H^v - H$. Then $H^u = H^v$ so uv^{-1} commutes with H. But uv^{-1} is strictly upper triangular, centralizes H, and so is equal to the identity, thus proving injectivity.

Next we show surjectivity. We give the proof in the same style as the corresponding proof in Chapter III. Let $S = H + \mathfrak{n}$. Then U acts on S by conjugation, and the action is algebraic, polynomial.

Proposition 2.4 ([Har 54], Lemma 8). *The conjugation action by* U *on* $S = H + \mathfrak{n}$ *makes* S *a principal homogeneous space. In particular, the isotropy group at each point is trivial.*

Proof. Actually, we prove the same statement for the complexification $U(\mathbf{C})$, $\mathfrak{n}(\mathbf{C})$, $H \in \mathfrak{a}(\mathbf{C})$ having distinct components. The statement is purely algebraic over a field. We write an element v of S in the form $v = H + Z$ with Z strictly upper triangular. We first prove the last statement, i.e. that the isotropy group of v in U is trivial. This amounts to showing that if $Z \in \mathfrak{n}$, and

$$u^{-1}Hu - H = u^{-1}Zu - Z$$

then u is the identity. We write $u = I + X$ with $X \in \mathfrak{n}$. Then the hypothesis means that

$$HX - XH \equiv ZX - XZ \mod X^2,$$

where $\operatorname{mod} X^2$ means mod the image of some combination XHX and XZX, which have the effect of strictly increasing the grading of these linear maps. Similarly, the grading of $ZX - XZ$ is strictly greater than the grading of X. However, since H is regular, the grading of $HX - XH$ is the same as the grading of X, which is a contradiction unless $X = 0$. Thus we have proved that the isotropy group at each point is trivial.

Since U and S have the same dimension, it follows that every orbit of U in S contains a generic point of S. Now given $Z_0 \in \mathfrak{n}(\mathbf{C})$ there exists $u \in U(\mathbf{C})$ such that $H^u = H + Z_0 + Z$ with Z generic; there exists $v \in U(\mathbf{C})$ such that $(H + Z_0)^v = H + Z_0 + Z$. Then

$$H^{uv^{-1}} = H + Z_0,$$

which proves the homogeneity, and also proves the surjectivity in Proposition 2.3, at least over the algebraic closure **C**. But when we pick H, Z real, we already know that there is a unique solution u such that $\Phi_H(u) = Z$, so this u must be real, thus proving that Φ_H is bijective.

Being algebraic (or complex analytic) it follows that Φ_H is a differential isomorphism. One can also compute its tangent map at a given $u \in U$, in a routine matter to see that this map is a linear isomorphism. Namely with the congruence modulo terms containing X at least twice,

$$\begin{aligned}\Phi_H(u(I+X)) &= (I+X)^{-1}u^{-1}Hu(I+X) - H \\ &\equiv \Phi_H(u) + (XH - HX)^u.\end{aligned}$$

Hence we find the derivative $\Phi'_H(u) = -\mathbf{c}(u^{-1}) \circ H_{\mathfrak{n}}$ as asserted. In Chapter III Proposition 4.2, we already noted that $H_{\mathfrak{n}}$ is a linear isomorphism, so $\Phi'_H(u)$ is a linear isomorphism, and we get a differential proof that Φ_H is a differential isomorphism (taking the bijectivity into account). This concludes the proof of Propositions 2.3 and 2.4.

Next we shall apply these propositions to the proof of Theorem 2.2, which will occur through a lemma.

Lemma 2.5. *Let $x \in G$ and let k be its K-Iwasawa component. Then $\mathfrak{b}^x = \mathfrak{b}^k$. Furthermore:*

(i) *We have $\mathfrak{b} = (\mathfrak{b} \cap \mathfrak{b}^k) + \mathfrak{n}$.*

(ii) *Let $H \in \mathfrak{a}$ be regular. There exist $u, v \in U$ and $w \in W$ such that*

$$H^{wvk} = H^u.$$

Proof. That $\mathfrak{b}^x = \mathfrak{b}^k$ is immediate since $\mathfrak{b}$ is stable under conjugation by elements of A and U. Next, observe that

$$\mathfrak{b}^k \cap \mathfrak{n} = \mathfrak{n}^k \cap \mathfrak{n}, \tag{2}$$

because $\mathfrak{b} \cap \mathfrak{n} = \mathfrak{n}$; and if $x \in \mathfrak{b}$, $X = H + Z$ with $H \in \mathfrak{a}$, $Z \in \mathfrak{n}$ then X^k has the same eigenvalues as X, so if $X^k \in \mathfrak{n}$ then $(H+Z)^k \in \mathfrak{n}$ and hence

$$H + Z \in \mathfrak{n}^{k^{-1}},$$

so $H + Z$ is nilpotent, and finally $H = 0$ so $X \in \mathfrak{n}$. This gives one inclusion for (2), and the other inclusion is trivial, so we get (2).

Next we are ready to prove (i). By the standard isomorphism theorem for vector spaces, using (2) we get:

$$(\mathfrak{b}^k \cap \mathfrak{b})/(\mathfrak{n}^k \cap \mathfrak{n}) \approx (\mathfrak{b} \cap \mathfrak{b}^k)/(\mathfrak{b} \cap \mathfrak{b}^k \cap \mathfrak{n}) \approx ((\mathfrak{b} \cap \mathfrak{b}^k) + \mathfrak{n}) \subset \mathfrak{b}/\mathfrak{n} \approx \mathfrak{a}.$$

It will suffice to prove that the dimensions of the two extreme vector spaces are equal, that is to prove

(3) $$[(\mathfrak{b}^k \cap \mathfrak{b}) : (\mathfrak{n}^k \cap \mathfrak{n})] = \dim \mathfrak{a}.$$

We use an orthogonality relation in $\mathfrak{g}$ with respect to the trace form, namely

$$\mathfrak{n}^\perp = \mathfrak{a} + \bar{\mathfrak{n}} \qquad \text{whence} \qquad (\mathfrak{n}^k)^\perp = (\mathfrak{a} + \bar{\mathfrak{n}})^k = \bar{\mathfrak{b}}^k \qquad \text{with } \bar{\mathfrak{b}} = \mathfrak{a} + \bar{\mathfrak{n}},$$

because conjugation by k preserves the trace form. Hence

(4) $$(\mathfrak{b} \cap \mathfrak{b}^k \cap \mathfrak{n})^\perp = (\mathfrak{n}^k \cap \mathfrak{n})^\perp = \bar{\mathfrak{b}}^k + \bar{\mathfrak{b}}.$$

Let $V = \bar{\mathfrak{b}}^k + \bar{\mathfrak{b}}$. Then ${}^tV = \mathfrak{b}^k + \mathfrak{b}$ and

(5) $$\dim V = \dim {}^tV = 2 \dim \mathfrak{b} - \dim(\mathfrak{b}^k \cap \mathfrak{b}).$$

Hence by (4),

(6) $$\dim(\mathfrak{n}^k \cap \mathfrak{n}) = \dim \mathfrak{b} + \dim \mathfrak{n} - 2 \dim \mathfrak{b} + \dim(\mathfrak{b}^k \cap \mathfrak{b}),$$

which proves (3), and concludes the proof of the first part of the lemma.

As to the second part, let H be regular in $\mathfrak{a}$. Given $k \in K$ we shall prove that there exist $u, v \in U$ and $w \in M'$ such that

(7) $$H^u = H^{wvk}.$$

By (i) there exist $Z \in \mathfrak{n}$ such that $H + Z \in \mathfrak{b} \cap \mathfrak{b}^k$. By Proposition 2.3, there exists $u \in U$ such that $H + Z = H^u$. Since $(H + Z)^{k^{-1}} \in \mathfrak{b}$ and its eigenvalues are the same as those of H (up to a permutation), there exists $w \in W$ (actually $w \in M'$) and $Z_1 \in \mathfrak{n}$ such that

$$(H + Z)^{k^{-1}} = H^w + Z_1 = H^{wv},$$

this last equality by using Proposition 2.3 again. Then $H + Z = H^{wvk}$, which we combine with $H + Z = H^u$ to conclude the proof of the lemma.

We now come to the proof of Theorem 2.2 proper. We consider the map

$$k' \mapsto Bk'B \qquad \text{with } k' \in M',$$

and we want to prove the map is a bijection. We start with the surjectivity. A double coset BxB is actually equal to BkB with the k-Iwasawa component $k \in K$ of x. Let H be regular in $\mathfrak{a}$. By Lemma 2.5(ii) there exist $u, v \in U$ and $w \in M'$ such that

$$H = H^{wvku^{-1}},$$

so $wvku^{-1}$ centralizes H, and hence by Chapter III, Proposition 4.2, $wvku^{-1} = d$ for some diagonal matrix d. Hence finally $k = v^{-1}w^{-1}du$, and

$$BkB = Bw^{-1}duB = Bw^{-1}B,$$

thus proving the surjectivity.

As to injectivity, suppose $BkB = Bk'B$ with $k, k' \in M'$. We have to show that $k^{-1}k' \in M$. There exist $b_1, b_2 \in B$ such that $b_1k = k'b_2$. We use the following immediate facts: For $b \in B$ and $\mathbf{c}(b)$ the conjugation action, we have $\mathbf{c}(b)H \equiv H \bmod \mathfrak{n}$. Furthermore, for $k \in M'$, the conjugation $\mathbf{c}(k)$ permutes the eigenspaces for the Lie-action of $\mathfrak{a}$ (directly from the definitions), and so induces an automorphism of $\mathfrak{n} + \bar{\mathfrak{n}}$. Hence

$$\mathbf{c}(b_1k)H = \mathbf{c}(b_1)\mathbf{c}(k)H = \mathbf{c}(b_1)H_1 \equiv H_1 \quad \bmod \mathfrak{n}$$

with H_1 regular in $\mathfrak{a}$, uniquely representing $\mathbf{c}(k)H \bmod \mathfrak{n}$. But also

$$\mathbf{c}(k'b_2)H = \mathbf{c}(k')\mathbf{c}(b_2)H \equiv H' \quad \bmod \mathfrak{n} + \bar{\mathfrak{n}}$$

with H' regular in $\mathfrak{a}$, uniquely representing $\mathbf{c}(k')H \bmod \mathfrak{n} + \bar{\mathfrak{n}}$. It follows that $H_1 = H'$, so $\mathbf{c}(k)H = \mathbf{c}(k')H$. By Chapter III, Corollary 3.3, it follows that $k^{-1}k' \in M$, which proves the injectivity, and concludes the proof of Theorem 2.2.

V, §3. JACOBIAN FORMULAS

The Harish-Chandra results of this section continue the tabulation of Jacobian determinants from Chapter I, Chapter III, §7. Jacobian determinants for the polar decomposition will be given in Chapter VI, §1. At the cost of repetition, we give a fairly systematic tabulation for the convenience of the reader. We continue to use pre-Iwasawa decompositions $G = PK$. We shall also deal with a decomposition involving $\bar{U}$ following Harish-Chandra. We found Wallach's exposition useful [Wal 73], 7.6, and [Wal 88], 2.4, as well as Helgason [Hel 68], A-2, and [Hel 84], Chapter I, §3. As Wallach writes: "The trick of using all the 'built in invariances' to compute integral formulas is taken from Harish-Chandra's recent lectures on harmonic analysis on p-adic groups."

We shall meet various product decompositions of G, such as

$$G = UAK = PK \supset UAM\bar{U} = B\bar{U} \qquad \text{with} \qquad B = UAM = AUM,$$

which we have encountered previously. We now give a lemma which pins down the Jacobian determinant in each case. The lemma will be applied to all of these situations.

We denote by δ_Q the modular function on a Lie group Q. Note that assuming we are dealing with Lie groups is not really necessary, the computations are valid in the context of Haar measure on locally compact groups, with Radon–Nikodym derivative in lieu of the Jacobian. However, we preferred to use a language which will fit the context of the applications directly. Note that p-adic groups are usually p-adic analytic, so the notion of Jacobian of analytic maps between such groups applies as well.

Let G be a Lie group with a Haar measure dx. Let Q, H be closed Lie subgroups. Suppose the product map

$$\eta_{Q,H}: Q \times H \to QH \subset G$$

is a differential isomorphism with an open subset of G whose complement has measure 0. We let dq, dh denote Haar measures on Q, H, respectively.

By the basic property of the Jacobian of a map, and its effect locally on a Haar differential form, abbreviating $J\eta_{Q,H}$ by J, we have for all $f \in C_c(G)$,

$$\int_G f(x)\,dx = \int_Q \int_H f(qh)J(q,h)\,dq\,dh. \tag{1}$$

Thus J denotes the absolute value of the Jacobian determinant of the map.

Lemma 3.1. *Notation as above, the function J depends only on the variable $h \in H$, and may therefore be denoted by $J(h)$.*

Proof. Starting with (1), we find for $q_1 \in Q$:

$$\begin{aligned}\int_G f(x)\,dx &= \int_Q \int_H f(qh)J(q,h)\,dq\,dh \\ &= \int_G f(q_1x)\,dx = \int_Q \int_H f(q_1qh)J(q,h)\,dq\,dh \\ &= \int_H \int_Q f(qh)J(q_1^{-1}q,h)\,dq\,dh \quad \text{by } q \mapsto q_1^{-1}q.\end{aligned}$$

Hence we obtain $J(q_1^{-1}q,h) = J(q,h)$ for all q_1, q, h so $J(q,h)$ is independent of q, thus proving the lemma.

We are headed toward Lemma 3.5 and Theorem 3.6, with $Q = B$ and $H = \bar{U}$, when both G and $\bar{U}$ are unimodular. However, there are other applications, so we proceed step by step with minimal assumptions at each step. The next lemma assumes G unimodular.

Lemma 3.2. *Let G be a unimodular Lie group. Let Q, H be two closed Lie subgroups. Suppose the product map*

$$\eta_{Q,H}: Q \times H \to QH \subset G$$

is a differential isomorphism with an open subset of G whose complement has measure 0. Then for $f \in C_c(G)$, the integral

$$\int_Q \int_H f(qh)\delta_H^{-1}(h)\, dq\, dh$$

is a Haar integral on G. More precisely, if $J = J\eta_{Q,H}$ denotes the absolute value of the Jacobian determinant with respect to a given Haar measure dx on G, then

$$J(q, h) = J(h) = c\delta_H^{-1}(h),$$

with the constant $c = J(e)$.

Proof. Let $h_1 \in H$. By the unimodularity of G we now get

$$\begin{aligned}\int_G f(x)\, dx &= \int_G f(xh_1)\, dx \\ &= \int_Q \int_H f(qhh_1)J(h)\, dq\, dh \quad \text{[by Lemma 3.1]} \\ &= \int_Q \int_H f(qh)J(hh_1^{-1})\delta_H^{-1}(h_1)\, dh\, dq\end{aligned}$$

by the definition of the modular function δ_H on H, as we make the right translation $h \mapsto hh_1^{-1}$. Therefore

$$J(hh_1^{-1}) = \delta_H(h_1)J(h),$$

from which it follows that $J(h) = J(e)\delta_H^{-1}(h)$, as was to be shown.

Remark on automorphisms. In the forthcoming proofs we shall use the following standard property. Let G be a locally compact group, and let θ be a topological group automorphism of G. Then there exists a constant $\Delta(\theta)$ such that for any Haar measure dx on G and $f \in C_c(G)$, we have

$$\int_G f(x^\theta)\, dx = \Delta(\theta) \int_G f(x)\, dx.$$

For example, suppose Q is a closed subgroup of G and θ is an element of the normalizer of Q in G. Then $q \mapsto q^\theta = \theta^{-1}q\theta$ is an automorphism of Q, and so

$$\int_Q f(q^\theta)\,dq = \Delta(\theta)\int_Q f(q)\,dq.$$

The map $\theta \mapsto \Delta(\theta)$ is a continuous homomorphism of the normalizer of Q into $\mathbf{R}^+$, i.e. a character of this normalizer into $\mathbf{R}^+$.

Instead of unimodularity of G, the next lemma makes assumptions on H, especially that H normalizes Q.

Lemma 3.3. *Let G be a Lie group and Q, H closed subgroups such that H normalizes Q, and the product map*

$$Q \times H \to QH = G$$

is a differential isomorphism. Then:

(i) *For $f \in C_c(G)$, the functional*

$$f \mapsto \int_Q\int_H f(hq)\,dh\,dq$$

is a Haar (left invariant) functional on G.

(ii) *If all characters $H \to \mathbf{R}^+$ are trivial* (*i.e.* $\mathrm{conthom}(H, \mathbf{R}^+) = \{1\}$), *then for all f as above,*

$$\int_Q\int_H f(hq)\,dh\,dq = \int_Q\int_H f(qh)\,dh\,dq.$$

(iii) *Again if* $\mathrm{conthom}(H, \mathbf{R}^+) = \{1\}$, *we have equality of the modular functions*

$$\delta_G(hq) = \delta_G(qh) = \delta_Q(q).$$

Proof. This is really a variation of Proposition 1.1 in Chapter I, but we reproduce the steps for the convenience of the reader and completeness of tabulation of Jacobian computations, with the present notation. Note first that the functional

$$f \mapsto \iint f(hq)\,dq\,dh \tag{2}$$

is trivially left H-invariant. For left Q-invariance, let $q_1 \in Q$. Then

$$\int_Q f(q_1hq)\,dq = \int_Q f(hq_1^hq)\,dq = \int_Q f(hq)\,dq,$$

by the left invariance of Haar measure on Q. This proves that the functional in (2) is a Haar functional on G.

For (ii), we write

$$(3)\qquad \iint f(hq)\,dh\,dq = \iint f(hqh^{-1}h)\,dq\,dh = \iint f(qh)\,dq\,dh$$

by the hypothesis that characters of H into $\mathbf{R}^+$ are trivial. This proves (ii).

For (iii) we argue similarly. Let $h_1 \in H$ and $q_1 \in Q$. Then

$$(4)\qquad \iint f(hqh_1q_1)\,dh\,dq = \iint f(hh_1q^{h_1}q_1)\,dh\,dq$$

$$= \iint f(hq^{h_1}q_1)\,dh\,dq \quad [\text{by } h \mapsto hh_1^{-1}]$$

$$= \iint f(hqq_1)\,dq\,dh$$

[by the hypothesis that characters of H into $\mathbf{R}^+$ are trivial]

$$= \iint f(hq)\delta_Q(q_1)\,dq\,dh$$

[by the definition of the modular function on Q]. This proves that

$$\delta_G(h_1q_1) = \delta_Q(q_1).$$

Similarly,

$$\iint f(hqq_1h_1)\,dq\,dh = \iint f(hqh_1)\delta_Q(q_1)\,dq\,dh$$

$$= \iint f(hh_1q^{h_1})\delta_Q(q_1)\,dq\,dh$$

$$= \iint f(hh_1q)\delta_Q(q_1)\,dq\,dh$$

$$= \iint f(hq)\delta_Q(q_1)\,dq\,dh,$$

which proves

$$\delta_G(q_1h_1) = \delta_Q(q_1).$$

This concludes the proof of Lemma 3.3.

Example. The condition that conthom$(H, \mathbf{R}^+)$ is trivial is satisfied when H is compact, or $H = \bar{U}$, or H is a group such as $SL_n(\mathbf{R})$. Of course, Lemma 3.3 is under the additional condition that H normalizes Q. In particular, if $H = M$ is finite, the condition is satisfied, but not in a very interesting way. Also note that if $Q = P = AU$ as in an Iwasawa decomposition, $p = au$, then

$$\delta_P(p) = \delta_P(a).$$

Next we apply the above measure computations to the Gelfand–Naimark decompositions axiomatized in §1.

Lemma 3.4. *Let $G = UAK = PK$ be a weak Iwasawa decomposition, complemented by a Gelfand–Naimark decomposition, with $B = PM$ and $\bar{U}$. Let*

$$\eta_{B,\bar{U}} : B \times \bar{U} \to B\bar{U}$$

be the product map. Then

$$J\eta_{B,\bar{U}} = constant.$$

In other words, there is a choice of Haar measure dx such that for all $f \in C_c(G)$ we have

$$\int_G f(x)\, dx = \int_U \int_A \int_M \int_{\bar{U}} f(uam\bar{u})\delta_P^{-1}(a)\, du\, da\, dm\, d\bar{u}.$$

Proof. This is a special case of Lemma 3.2 with $Q = B$ and $H = \bar{U}$. Note the order of the variables ua rather than au, so the Haar measure db on B with that order of the variables in terms of $du\, da$ needs the modular function $\delta_P^{-1}(a)$ in applying Lemma 3.2. We are thus using Proposition 1.3 of Chapter I. On the other hand, $\delta_M = \delta_H = 1$ because M is compact and $\bar{U} = H$ is unimodular by assumption.

In line with Lemma 3.4, there is a commutative diagram

$$
\begin{array}{ccc}
U \times A \times M \times \bar{U} & \xrightarrow{\ \eta\ } & UAM\bar{U} \subset G \\
& \searrow{\scriptstyle\psi_{\text{Har}}} \quad \nearrow{\scriptstyle\text{Iw}} & \\
& U \times A \times K &
\end{array}
$$

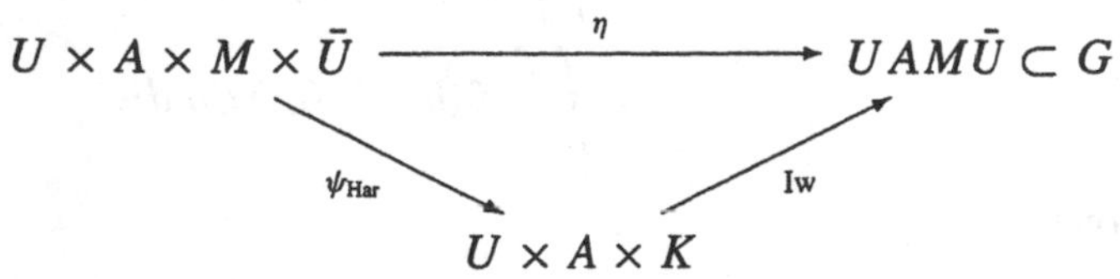

where the top map is the product map of Lemma 3.4; the right map is the product map of the Iwasawa decomposition; and the left map is given as follows. Write $m\bar{u} = u'a'k'$ in its Iwasawa decomposition. Then

(5a) $$\psi_{\mathrm{Har}}(u, a, m, \bar{u}) = (u'', aa', k')$$

where $u''aa'k' = uam\bar{u}$ is the Iwasawa decomposition of $uam\bar{u}$, and trivially

$$u'' = uau'a^{-1} = \mathrm{Iw}_U(uam\bar{u})$$

because A normalizes U.

We also remark that for all $x \in G$ and $m \in M$, we have

(5b) $$\mathrm{Iw}_A(mx) = \mathrm{Iw}_A(x).$$

This is immediate, writing $x = u'a'k'$, and using the condition that m normalizes U and centralizes A. *In particular, the function*

$$k \mapsto \mathrm{Iw}_A(mkx)$$

is really defined on $M\backslash K$, a fact which is often used.

Lemma 3.5. *The map* ψ_{Har} *has absolute Jacobian*

$$J\psi_{\mathrm{Har}}(u, a, m, \bar{u}) = \delta_P(\mathrm{Iw}_A(\bar{u})),$$

up to the normalizing constant factor.

Proof. Keeping the letters as above, we see that up to constant factors, the absolute Jacobians are $\delta_P^{-1}(a)$ for η by Lemma 3.4, and $\delta_P^{-1}(aa')$ for the Iwasawa map Iw on the right. But

$$\delta_P^{-1}(aa') = \delta_P^{-1}(a)\delta_P^{-1}(a').$$

By the chain rule (multiplicativity of the Jacobian for $\eta = \mathrm{Iw} \circ \psi_{\mathrm{Har}}$), it follows that

$$J\psi_{\mathrm{Har}}(u, a, m, \bar{u}) = \delta_P(a') = \delta_P(\mathrm{Iw}_A(\bar{u})),$$

which proves the lemma.

Of course, there is the constant factor, which comes from choosing the Haar measures on G giving the Jacobians for η and for Iw$'$ respectively. The formula of the next theorem determines the constant factor in the main relation which is of concern to us. Among other things, it determines the absolute Jacobian of the Harish-Chandra boundary map $\psi_{K/M}$. We get it essentially as a corollary of the determination of $J\psi_{\mathrm{Har}}$ by a projection argument for integration, amounting to Fubini's theorem.

Theorem 3.6. *Let $G = UAK$ be a weak Iwasawa decomposition with K compact, also having a Gelfand–Naimark decomposition with subgroups $M, \bar{U}$. Normalize the measures so that K, M have measure 1. The function $\bar{u} \mapsto \delta_P(\mathrm{Iw}_A(\bar{u}))$ is in $L^1(\bar{U})$, and we can then normalize $d\bar{u}$ so that*

$$\int_{\bar{U}} \delta_P(\mathrm{Iw}_A(\bar{u}))\, d\bar{u} = 1.$$

Then for $f \in C(K)$, we have

$$\int_K f(k)\, dk = \int_M \int_{\bar{U}} f(m\, \mathrm{Iw}_K(\bar{u}))\delta_P(\mathrm{Iw}_A(\bar{u}))\, d\bar{u}\, dm.$$

For $M\backslash K$ we then have

$$J\psi_{M\backslash K}(\bar{u}) = \delta_P(\mathrm{Iw}_A(\bar{u})),$$

and for $f \in C(M\backslash K)$,

$$\int_{M\backslash K} f(k_M)\, dk_M = \int_{\bar{U}} f(M\, \mathrm{Iw}_K(\bar{u}))\delta_P(\mathrm{Iw}_A(\bar{u}))\, d\bar{u}.$$

Proof. The proof is forced by Lemma 3.5 and Fubini's theorem, as follows. Let $F \in C_c(U \times A \times K)$ be of the form

$$F(u, a, k) = f_1(u) f_2(a) f(k),$$

where $f \in C(K)$, $f_1 \in C_c(U)$, $f_2 \in C_c(A)$, and f_1, f_2 have total integral 1 on U and A respectively. Then with some constant c, Lemma 3.5 gives

$$\int_K f(k)\, dk = \int_U \int_A \int_K F(u, a, k)\, du\, da\, dk$$

$$= c \int_U \int_A \int_M \int_{\bar{U}} f_1(ua\, \mathrm{Iw}_U(m\bar{u})a^{-1}) f_2(a\, \mathrm{Iw}_A(m\bar{u})) f(m\, \mathrm{Iw}_K(\bar{u})) \\ \delta_P(\mathrm{Iw}_A(m\bar{u}))\, du\, da\, dm\, d\bar{u}.$$

Using Fubini, we integrate over A and U first. The right invariance of da and du then allows us to cancel $a\, \mathrm{Iw}_U(m\bar{u})a^{-1}$ in the U-integral, and $\mathrm{Iw}_A(m\bar{u})$ in the A-integral. Therefore our integral is

$$= c \int_M \int_{\bar{U}} f(m\, \mathrm{Iw}_K(\bar{u})\delta_P(\mathrm{Iw}_A(m\bar{u}))\, dm\, d\bar{u}$$

$$= c \int_M \int_{\bar{U}} f(m\, \mathrm{Iw}_K(\bar{u}))\delta_P(\mathrm{Iw}_A(\bar{u}))\, dm\, d\bar{u}$$

because $\text{Iw}_A(m\bar{u}) = \text{Iw}_A(\bar{u})$. Now, if $f \in C(M\backslash K)$, then we obtain the formula of the theorem with the constant c in front. Choosing f to be the constant function 1 shows that under the normalization as stated in the theorem, we have $c = 1$.

The above arguments also show that the absolute Jacobian $\delta_P \circ \text{Iw}_A$ is absolutely integrable on $\bar{U}$, because the constant function 1 is integrable on K or $M\backslash K$, and it is general L^1-theory of the chain rule that one then gets the integrability of the absolute Jacobian. See, for instance, [Lan 93], Chapter XXI, Theorem 2.6. This concludes the proof of Theorem 3.6.

Remark. In the next chapter, the polar Haar measure will be normalized by a local condition at the origin. Here, the key Haar measure is normalized by a global integral condition starting with Theorem 3.6.

V, §4. INTEGRAL FORMULAS FOR SPHERICAL FUNCTIONS

This section derives Harish-Chandra expressions for the spherical functions as integrals over $\bar{U}$. The matter has only to do with continued abstract nonsense of Haar integrals, so:

Throughout this section, we let $G = UAK$ be an Iwasawa decomposition, with K compact, having measure 1. *We also assume that there is a Gelfand–Naimark decomposition, with subgroups $M, \bar{U}$. The measures are normalized as in Lemma* 3.6, *that is, K and K/M have total measure* 1, *and*

$$\int_{\bar{U}} \delta(\text{Iw}_A(\bar{u}))\, d\bar{u} = 1.$$

Here we abbreviate $\delta = \delta_P$ for the Iwasawa character.

We don't need any further assumptions for the time being. For a character

$$\chi: A \to \mathbf{C}^*,$$

we let φ_χ be the corresponding Harish-Chandra spherical function. By definition, for $x \in G$,

$$(1) \qquad \varphi_\chi(x) = \varphi(\chi, x) = \int_K (\chi\delta^{1/2})(\text{Iw}_A(kx))\, dk,$$

and therefore also

$$(2) \qquad = \int_{M\backslash K} (\chi\delta^{1/2})(\mathrm{Iw}_A(k_M x))\,dk_M,$$

where k_M is the variable in $M\backslash K$.

Proposition 4.1. *Under the Iwasawa and Gelfand–Naimark decompositions, and the prescribed Haar measure normalizations, we have*

$$\varphi(\chi, x) = \int_{\bar{U}} (\chi\delta^{1/2})(\mathrm{Iw}_A(\mathrm{Iw}_K(\bar{u})x))\delta_P(\mathrm{Iw}_A(\bar{u}))\,d\bar{u}.$$

Proof. We apply Theorem 3.6 to the function f_x on K given by

$$f_x(k) = (\chi\delta^{1/2})(\mathrm{Iw}_A(kx)).$$

The desired expression falls out.

For the next expressions, we need the formulas for $x, y \in G$, $a \in A$:

$$(3) \qquad \mathrm{Iw}_A(\mathrm{Iw}_K(x)y) = \mathrm{Iw}_A(x)^{-1}\,\mathrm{Iw}_A(xy)$$

$$\text{or } \mathrm{Iw}_A(xy) = \mathrm{Iw}_A(x)\,\mathrm{Iw}_A(\mathrm{Iw}_K(x)y),$$

$$(4) \qquad \mathrm{Iw}_A(xa) = a\,\mathrm{Iw}_A(a^{-1}xa).$$

(3) comes from writing the Iwasawa decomposition of x and applying the definitions. Cf. formula (4) of Chapter III, §7. The other formula is immediate.

Proposition 4.2. *Under those same conditions, let $a \in A$. Then*

$$\text{(i) } \varphi(\chi, a) = \int_{\bar{U}} (\chi\delta^{1/2})(\mathrm{Iw}_A(\bar{u}a))(\chi^{-1}\delta^{1/2})(\mathrm{Iw}_A(\bar{u}))\,d\bar{u}$$

$$\text{(ii) } \qquad = (\chi\delta^{1/2})(a)\int_{\bar{U}} (\chi\delta^{1/2})(\mathrm{Iw}_A(a^{-1}\bar{u}a))(\chi^{-1}\delta^{1/2})(\mathrm{Iw}_A(\bar{u}))\,d\bar{u}$$

$$\text{(iii) } \qquad = (\chi\delta^{-1/2})(a)\int_{\bar{U}} (\chi^{-1}\delta^{1/2})(\mathrm{Iw}_A(a\bar{u}a^{-1})(\chi\delta^{1/2})(\mathrm{Iw}_A(\bar{u}))\,d\bar{u}.$$

Proof. We start with the formula in Proposition 4.1. We then apply (3) to get (i) directly, taking $x = \bar{u}$. Next, we apply (4). Then (ii) drops out. For (iii), we make the change of variables $\bar{v} = a^{-1}\bar{u}a$ with Jacobian $\delta(a)^{-1}$.

For Harish-Chandra's own version of the proofs, see [Har 58a], Corollary 1 of Lemma 44. In any case, specializing to the trivial character χ, we obtain as in Harish:

Corollary 4.3. *Let* $\bar{u}^a = a^{-1}\bar{u}a$. *Then*

$$\delta^{-1/2}(a)\int_K \delta^{1/2}(\mathrm{Iw}_A(ka))\,dk = \int_{\bar{U}} \delta^{1/2}(\mathrm{Iw}_A(\bar{u}^a))\delta^{1/2}(\mathrm{Iw}_A(\bar{u}))\,d\bar{u},$$

or also

$$\delta^{1/2}(a)\int_K \delta^{1/2}(\mathrm{Iw}_A(ka))\,dk = \int_{\bar{U}} \delta^{1/2}(\mathrm{Iw}_A(a\bar{u}a^{-1}))\delta^{1/2}(\mathrm{Iw}_A(\bar{u}))\,d\bar{u}.$$

Remark on notation. Up to now, in computing Jacobians, it has been convenient to use a symbol Iw_A for the projection on the A-Iwasawa component, in order to fit the chain rule. However, in some forthcoming applications, we are dealing with expressions such as those in Propositions 4.2 and 4.3 where it will be convenient to eliminate some letters. Thus we shall abbreviate the notation for the A-Iwasawa projection of an element $x \in G$, by writing

$$\mathrm{Iw}_A(x) = x_A \qquad \text{so for instance,} \qquad \mathrm{Iw}_A(a\bar{u}a^{-1}) = (a\bar{u}a^{-1})_A.$$

Then the second formula of Corollary 4.3 reads

$$\delta^{1/2}(a)\int_K \delta^{1/2}((ka)_A)\,dk = \int_{\bar{U}} \delta^{1/2}((a\bar{u}a^{-1})_A)\delta^{1/2}(\bar{u}_A)\,d\bar{u}.$$

The Lie industry writes $a(x)$ for the A-component, but we do not follow this practice, because in an expression $a(a\bar{u}a^{-1})$ we find too many a's with different meanings.

V, §5. THE c-FUNCTION AND THE FIRST SPHERICAL ASYMPTOTICS

For this section and the next, we suppose that $G = \mathrm{SL}_n(\mathbf{R})$. *Cf. the remark preceding Lemma* 5.1. *Also* $\bar{U} = {}^tU =$ *lower triangular unipotent matrices.*

The reason for going to the specific group is that we shall need the eigenspace decomposition of $\mathfrak{n}$ for the regular representation of A, whose properties are obvious for concrete groups, with no need for general Lie theory. Our approach in this section is an expansion of Bhanu-Murty's approach [Bha 60]. We are indebted to Nolan Wallach for his suggestions on the use of the wedge product.

We consider the integral of Theorem 4.2. The integrand consists of two factors, one of which has the conjugation $a\bar{u}a^{-1}$. For an element $\bar{u}$ the value of the eigencharacter α_{ji} with $i < j$ on an element $a \in A$ is

$$a^{\alpha_{ji}} = a_j a_i^{-1} = a^{-\alpha_{ij}}.$$

An element $\bar{u}$ has the form

$$\bar{u} = \begin{pmatrix} 1 & & & 0 \\ & 1 & & \\ & & \ddots & \\ & x_{ji} & & 1 \end{pmatrix},$$

so that

$$a\bar{u}a^{-1} = \begin{pmatrix} 1 & & & 0 \\ & 1 & & \\ & & \ddots & \\ a_j x_{ji} a_i^{-1} & & & 1 \end{pmatrix}.$$

We **define** $a \mapsto \infty$ to mean $a^{\alpha_{ij}} \to \infty$ for all pairs $i < j$. Then for each $\bar{u}$,

$$\lim_{a\to\infty} a\bar{u}a^{-1} = 1.$$

We shall use the formula from Proposition 4.2(iii) for the Harish-Chandra spherical function, namely

$$\boxed{\varphi(\chi, a) = (\chi\delta^{-1/2})(a) \int_{\bar{U}} (\chi^{-1}\delta^{1/2})((a\bar{u}a^{-1})_A)(\chi\delta^{1/2})(\bar{u}_A)\, d\bar{u}.}$$

Taking the limit under the integral sign formally, and defining the **Harish-Chandra *c*-function** by

$$\text{(1)} \qquad c_{\text{Har}}(\chi) = \int_{\bar{U}} (\chi\delta^{1/2})(\bar{u}_A)\, d\bar{u},$$

we then have formally the asymptotic relation

$$\text{(2)} \qquad \varphi_\chi(a) \sim c_{\text{Har}}(\chi)(\chi\delta^{-1/2})(a) \qquad \text{for } a \to \infty.$$

To justify all this, we need:

– a range of validity for the absolute convergence of the integral defining the c-function in (1);

– justification of an application of the dominated convergence theorem to take the limit under the integral sign for the asymptotics (2).

We shall specify a suitable half space for the character in terms of the complex parametrization of the character. We shall also describe the analytic continuation going beyond the domain of absolute convergence. The problem is to solve for $\bar{u}_A$ in terms of $\bar{u}$ with as little pain as possible.

Our definition of the **c**-function is in line with the Harish-Chandra integral formula for the spherical functions, and thus continues the theory in a natural fashion, with the added information which comes from the Gelfand–Naimark decomposition. In Chapter VIII we shall give a completely different construction for spherical functions in terms of a Harish-Chandra series, and certain functions will appear as the coefficients of a linear combination. It will then be a theorem that these coefficients are the same functions which arise in the present section. Basic properties depending on the two definitions are proved independently of each other. Only at the end are the two definitions proved equivalent.

We begin with lemmas of Harish-Chandra [Har 58a]. See also [Hel 84], Chapter IV, Lemma 6.5 and Corollary 6.6; and Wallach [Wal 88], 3.A.2.3.

We recall that A^+ is the subset of A consisting of those elements a such that $a^\alpha > 1$ for all relevant characters α, that is, $a_i/a_j > 1$ for all $i < j$. There is a dual notion. For $d = 1, \ldots, n$ we define the **product character** p_d on A as in Chapter I, §1 and §6, by

$$p_d(a) = a_1 \cdots a_d,$$

which is just the product of the first d coordinates. Then we define on $\mathrm{SL}_n(\mathbf{R})$:

$A_{p>1}$ = *subset of elements* $a \in A$ *such that* $p_d(a) > 1$ *for all* $d = 1, \ldots, n-1$ *(of course,* $p_n(a) = 1$*).*

$A_{p<1}$ = *subset of elements* $a \in A$ *such that* $p_d(a) < 1$ *for all* d *as above.*

Remark. One can also give the definitions on $\mathrm{GL}_n(\mathbf{R})$, replacing $p_d(a)$ by $p_d(a)/p_n(a)^{d/n}$, which is homogeneous of degree 0 in a (i.e. invariant under scalar multiplication). Since it's a pain to carry the full product $p_n(a)^{d/n}$ throughout, we assume $G = \mathrm{SL}_n(\mathbf{R})$ unless otherwise specified, e.g. the next lemma, where the reader can see just once how $p_n(a)$ intervenes.

Lemma 5.1. *For* $a \in A$ *on* GL_n, *we have the relation*

$$\delta(a) = p_n(a)^{n+1} \prod_{i=1}^{n} a_i^{-2i} = \prod_{j=1}^{n-1} \frac{p_j(a)^2}{p_n(a)}$$

or equivalently

$$\delta^{1/2}(a) = p_n(a)^{-(n-1)/2} \prod_{j=1}^{n-1} p_j(a)$$

$$= \prod_{j=1}^{n-1} p_j(a) \qquad \text{if } a \in A \cap \mathrm{SL}_n(\mathbf{R}).$$

In particular, if $G = \mathrm{SL}_n(\mathbf{R})$ *so* $p_n(a) = 1$, *and* $p_d(a) < 1$ *for* $d = 1, \ldots, n-1$, *then*

$$\delta(a) < 1.$$

Proof. The identity is a simple piece of algebra. The first equality is just the standard expression for $\delta(a)$, already computed in Chapter I, §1. For the second equality, one divides $\prod a_i^i$ by $(a_1 \cdots a_n)^n = p_n(a)^n$. The application to diagonal $a \in \mathrm{SL}_n(\mathbf{R})$ comes from $p_n(a) = 1$.

In the next lemmas, we shall deal with elements $\bar{u}$, $a\bar{u}a^{-1}$ and we note that they lie in SL_n, so their A-projections are naturally in SL_n. The problem is to have reasonably explicit expressions for the A-projections in terms of $\bar{u}$ and a. For this, we require some identities which come from a natural representation of G, namely in the present case, the standard representation of G on $\mathbf{R}^n$ with its standard scalar product, and its wedge products. In effect, the following computations are expressed in terms of explicit determinants in Bhanu-Murty [Bha 60]. We thank Wallach, who pointed out to us that using the wedge product makes them more transparent.

For $g \in G$, the action of G on $\mathbf{R}^n$ extends to the alternating products $\bigwedge^d g$ on $\bigwedge^d \mathbf{R}^n$. We let $e_1, \ldots, e_n$ be the standard unit column vectors. Note first that for $u \in U$, we have

$$\left(\bigwedge^d u\right)(e_1 \wedge \cdots \wedge e_d) = ue_1 \wedge \cdots \wedge ue_d = e_1 \wedge \cdots \wedge e_d,$$

because for each i, $ue_i = e_i +$ linear combination of e_j with $j < i$, and the extra linear combination is killed in the wedge product.

Now let $g = ubk$ be the Iwasawa decomposition of g. Then $g^{-1} = k^{-1}b^{-1}u^{-1}$, and therefore

$$g^{-1}e_1 \wedge \cdots \wedge g^{-1}e_d = b_1^{-1} \cdots b_d^{-1} k^{-1}e_1 \wedge \cdots \wedge k^{-1}e_d. \tag{3}$$

Thus u disappears on the right side. We take the natural positive definite scalar product on $\bigwedge^d \mathbf{R}^n$, for which the elements $e_{j_1} \wedge \cdots \wedge e_{j_d}$ with

$j_1 < \cdots < j_d$ form an orthonormal basis. Then for the norm of this scalar product, using the fact that k is unitary, we conclude:

$$\|g^{-1}e_1 \wedge \cdots \wedge g^{-1}e_d\| = b_1^{-1} \cdots b_d^{-1} = p_d^{-1}(g_A). \tag{4}$$

We apply this to $g = \bar{u} \in \bar{U}$. Write

$$\bar{u}^{-1}e_1 \wedge \cdots \wedge \bar{u}^{-1}e_d = e_1 \wedge \cdots \wedge e_d + \sum_{(j)\neq(1,\ldots,d)} c_{(j)}e_{j_1} \wedge \cdots \wedge e_{j_d}, \tag{5}$$

where $c_{(j)} = c_{(j)}(\bar{u})$. Then

$$p_d^{-2}(\bar{u}_A) = \|\bar{u}^{-1}e_1 \wedge \cdots \wedge \bar{u}^{-1}e_d\|^2 = 1 + \sum_{(j)\neq(1,\ldots,d)} c_{(j)}^2. \tag{6}$$

Next we carry out the same computation for $g = a\bar{u}a^{-1}$ with $a \in A$. We get

$$\begin{aligned}
&(a\bar{u}a^{-1})^{-1}e_1 \wedge \cdots \wedge (a\bar{u}a^{-1})^{-1}e_d \\
&\qquad = (a_1 \cdots a_d)^{-1}a\bar{u}^{-1}e_1 \wedge \cdots \wedge a\bar{u}^{-1}e_d \\
&\qquad = p_d(a)^{-1} \bigwedge\nolimits^d a(\bar{u}^{-1}e_1 \wedge \cdots \wedge \bar{u}^{-1}e_d) \\
\text{[by (5)]} &\qquad = p_d(a)^{-1}\Big[\Big(\bigwedge\nolimits^d a\Big)(e_1 \wedge \cdots \wedge e_d) \\
&\qquad\qquad + \sum_{(j)} c_{(j)}a_{j_1} \cdots a_{j_d}e_{j_1} \wedge \cdots \wedge e_{j_d}\Big],
\end{aligned}$$

where $c_{(j)} = c_{(j)}(\bar{u})$ as before, and the sum is taken as before for $(j) \neq (1, \ldots, d)$. The norm square then gives

$$\begin{aligned}
p_d^{-2}((a\bar{u}a^{-1})_A) &= \|(a\bar{u}a^{-1})^{-1}e_1 \wedge \cdots \wedge (a\bar{u}a^{-1})^{-1}e_d\|^2 \\
&= 1 + \sum_{(j)} p_d(a)^{-2}(a_{j_1} \cdots a_{j_d})^2 c_{(j)}^2.
\end{aligned} \tag{7}$$

The two identities (6) and (7) are the basic ones solving the implicit determination of the A-component for elements of $\bar{U}$. As a first application, we shall obtain Harish-Chandra inequalities.

Lemma 5.2 ([Har 58a], Lemma 43). *Let $\bar{u} \in \bar{U}$. Then for all d:*

(i) $$p_d(\bar{u}_A) \leqq 1.$$

(ii) *If $a \in A^+$ then $p_d((a\bar{u}a^{-1})_A/\bar{u}_A) \geqq 1$.*

Proof. For (i) we use (6), which immediately gives $p_d(\bar{u}_A) \leqq 1$, as desired. For (ii) we use (7). Let $a \in A^+$. Then $p_d(a) > a_{j_1} \cdots a_{j_d}$ for all $(j) \neq (1, \dots, d)$. Therefore the coefficient of $c_{(j)}^2$ in the sum on the right of (7) is < 1 for each (j), and so

$$p_d^{-2}((a\bar{u}a^{-1})_A) \leqq 1 + \sum_{(j)} c_{(j)}^2 = p_d^{-2}(\bar{u}_A).$$

Hence

$$1 \leqq p_d((a\bar{u}a^{-1})_A)/p_d(\bar{u}_A) = p_d((a\bar{u}a^{-1})_A/\bar{u}_A),$$

which proves (ii) and concludes the proof of Lemma 5.2.

Remark. In the notation of Chapter I, §4, we can write the two inequalities in the form

$$\log \bar{u}_A \leqq_{A'} 0 \qquad \text{and} \qquad \log \bar{u}_A \leqq_{A'} \log (a\bar{u}a^{-1})_A.$$

We let p^s be the **Selberg character**, formed with the products $p_1, \dots, p_{n-1}$, that is, for $s = (s_1, \dots, s_{n-1}) \in \mathbf{C}^{n-1}$, by definition on $A \subset \mathrm{SL}_n(\mathbf{R})$,

$$p^s(a) = p_1(a)^{s_1} \cdots p_{n-1}(a)^{s_{n-1}}.$$

Note that here, we use the character (power function) as Selberg did. With our notation, we then have the expression

$$\delta^{1/2} = p^1 = p^{(1,\dots,1)} \qquad \text{with } s = (1, \dots, 1). \tag{8}$$

For the pull-back to the Lie algebra, see §7.

In whatever domain of absolute convergence it turns out to be valid, we then have the definition of the Harish $\boldsymbol{c}$-function in terms of the variables $s_1, \dots, s_{n-1}$:

$$\boldsymbol{c}_{\mathrm{Har}}(p^s) = \int_{\bar{U}} p^{s+1}(\bar{u}_A)\, d\bar{u}, \tag{9}$$

with the normalized measure $d\bar{u}$, which will also be compared with the euclidean measure $d_{\mathrm{euc}}\bar{u}$, another natural Haar measure. In terms of the x_{ji}-coordinates at the beginning of this section, we have

$$d_{\mathrm{euc}}\bar{u} = \prod_{i<j} dx_{ji}.$$

If $d\bar{u} = c\,d_{\mathrm{euc}}\bar{u}$, then directly from the definitions we find that $\delta = p^2$, so

$$c = \int_{\bar{U}} \delta(\bar{u}_A) d_{\mathrm{euc}}\bar{u} = \int_{\bar{U}} p^2(\bar{u}_A) d_{\mathrm{euc}}\bar{u}.$$

Computations will be carried out with the euclidean measure. Hence we **define**

$$(10) \qquad C_n(s) = \int_{\bar{U}} p^{s+1}(\bar{u}_A) d_{\text{euc}}\bar{u}.$$

Then

$$(11) \qquad d\bar{u} = \frac{1}{C_n(1)} d_{\text{euc}}\bar{u} \quad \text{with} \quad C_n(1) = \int_{\bar{U}} p^2(\bar{u}_A) d_{\text{euc}}\bar{u},$$

and the Harish-Chandra **c**-function is the quotient

$$(12) \qquad \boldsymbol{c}_{\text{Har}}(p^s) = \frac{C_n(s)}{C_n(1)}.$$

Lemma 5.3. *The integrals of* (9), (10), (11), (12) *are absolutely convergent for* $\text{Re}(s_j) \geqq 1$ $(j = 1, \ldots, n-1)$.

Proof. This follows because $\delta \circ \text{Iw}_A \in L^1(\bar{U})$ and $p_j(\bar{u}_A) \leqq 1$ by Lemmas 5.1 and 5.2.

In particular, the integral defining $C_n(1)$ is absolutely convergent. It will be proved by an explicit induction in Theorem 6.2 that actually, the range of absolute convergence for all these integrals is $\text{Re}(s_i) > 0$ for $i = 1, \ldots, n-1$.

Lemma 5.2 is sufficient to give us asymptotics for the spherical functions in a certain range. We are now parametrizing these functions by the Selberg power function. It will be convenient to define

$$\text{Re}(s) > r \quad \text{(for a real number } r\text{)}$$

to mean $\text{Re}(s_i) > r$ for all $i = 1, \ldots, n-1$, and similarly for $\text{Re}(s) \geqq r$.

Theorem 5.4. *Suppose* $\text{Re}(s) \geqq 1$. *Then*

$$\begin{aligned} \varphi(p^s, a) &= p^{s-1}(a) \int_{\bar{U}} p^{-s+1}((a\bar{u}a^{-1})_A) p^{s+1}(\bar{u}_A)\, d\bar{u} \\ &\sim p^{s-1}(a)\, \boldsymbol{c}_{\text{Har}}(p^s) \qquad \textit{for } a \in A^+, a \to \infty. \end{aligned}$$

Proof. The first formula is a rewriting of Proposition 4.2(iii) with the Selberg character. We then rewrite the integrand to get

$$(13) \qquad \frac{\varphi(p^s, a)}{p^{s-1}(a)} = \int_{\bar{U}} p^{1-s}((a\bar{u}a^{-1})_A/\bar{u}_A) p^2(\bar{u}_A)\, d\bar{u}.$$

But $p^2 = \delta$, and $\delta \circ \mathrm{Iw}_A \in L^1(\bar{U})$. By Lemma 5.2(ii) and the hypothesis $\mathrm{Re}(1-s) \leqq 0$, we conclude that

$$|p^{1-s}((a\bar{u}a^{-1})_A/\bar{u}_A)| \leqq 1,$$

and so the left side is bounded. Thus the dominated convergence theorem applies and we can take the limit under the integral sign as $a \to \infty$ to conclude the proof.

Remark 1. The integral formula in Theorem 5.3 amounts to Theorem 1 in [Bha 60a], taking into account the order in which one writes various things, e.g. Iwasawa or anti-Iwasawa decompositions, etc.

Remark 2. An asymptotic relation in a larger range will follow from a more precise range of absolute convergence for the integrals to be given in the next section. We gave the limited version first so readers can see right away what's going on, and see the general idea of the argument independently of the Bhanu-Murty inductive computation. Thus Theorem 5.4 will be immediately extended. However, it is also useful to have it in the above range, because we can then concentrate on the more delicate range in Theorem 6.5. In the proof of Theorem 5.4, we used Lemma 5.2(ii). In the proof of Theorem 6.5, we shall use Lemma 5.2(i) instead.

V, §6. THE BHANU-MURTY FORMULA FOR THE c-FUNCTION

As Wallach pointed out to us, Bhanu-Murty's inductive computation of the c-function can be made more transparent by using subgroups as follows. (Compare with [Ter 88], p. 104.) We let:

$\bar{U}_{n,1}$ = unipotent group of first columns, consisting of the matrices

$$\bar{u}_{n,1}(x) = \begin{pmatrix} 1 & 0 & \cdots & 0 \\ x_{21} & 1 & \cdots & 0 \\ \vdots & \vdots & \ddots & \vdots \\ x_{n1} & 0 & \cdots & 1 \end{pmatrix} = \begin{pmatrix} 1 & 0 \\ x & I_{n-1} \end{pmatrix} \qquad \text{with } x \in \mathbf{R}^{n-1}.$$

We also embed and identify $\bar{U}_{n-1}$ in $\bar{U}_n$ on the lower right, so we identify

$$\bar{U}_{n-1} = \begin{pmatrix} 1 & 0 \\ 0 & \bar{U}_{n-1} \end{pmatrix} \qquad \text{or more generally} \qquad \mathrm{GL}_{n-1} = \begin{pmatrix} 1 & 0 \\ 0 & \mathrm{GL}_{n-1} \end{pmatrix}.$$

The euclidean variables on $\bar{U}_{n-1}$ are therefore x_{ji} with $j = 3, \ldots, n$ and $2 \leqq i < j$. Note that GL_{n-1} (and so $\bar{U}_{n-1}$) leaves e_1 fixed.

Also note that GL_{n-1} *normalizes* $\bar{U}_{n,1}$. This is verified by direct computation. For instance let $e_1, \ldots, e_n$ be the standard unit vectors. We may write $\mathbf{R}^n$ as a direct sum

$$\mathbf{R}^n = (e_1) \oplus (e_2, \ldots, e_n).$$

Then $\bar{U}_n$ leaves $(e_2, \ldots, e_n)$ stable, and $\bar{U}_{n,1}$ is the kernel of the restriction of $\bar{U}_n$ to $(e_2, \ldots, e_n)$. Let $g \in \mathrm{GL}_{n-1}$ and $v \in (e_2, \ldots, e_n)$. Then $gv \in (e_2, \ldots, e_n)$, so

$$g^{-1}\bar{u}_{n,1}gv = g^{-1}gv = v,$$

whence $g^{-1}\bar{u}_{n,1}g \in \bar{U}_{n,1}$. This proves the normalization assertion.

We have a product decomposition

$$\bar{U}_{n-1} \times \bar{U}_{n,1} \to \bar{U}_{n-1}\bar{U}_{n,1} = \bar{U}_n.$$

The Jacobian determinant is trivially equal to 1. We write an element $\bar{u}_n \in \bar{U}_n$ in terms of its two components,

$$\bar{u}_n = \bar{u}_{n-1}\bar{u}_{n,1}.$$

We define

$$(1) \quad F_n(s) = \int_{\bar{U}_{n,1}} p^{s+1}((\bar{u}_{n,1})_A) d_{\mathrm{euc}}\bar{u}_{n,1} = \int_{\mathbf{R}^{n-1}} p^{s+1}(\bar{u}_{n,1}(x)_A)\, dx.$$

Fubini's theorem in the next lemma will show immediately the absolute convergence for $\mathrm{Re}(s_i) \geqq 1$ ($i = 1, \ldots, n-1$), and the explicit inductive computation in Theorem 6.2 will show it for $\mathrm{Re}(s_i) > 0$.

Lemma 6.1. *In the domain of absolute convergence, we have for* $n \geqq 3$:

$$C_n(s) = F_n(s)C_{n-1}(s_2, \ldots, s_{n-1}).$$

For $n = 2$, *we have* $C_2(s_1) = F_2(s_1)$.

Proof. We have by formula (3) of §4,

$$C_n(s) = \int_{\bar{U}} p^{s+1}(\bar{u}_A) d_{\mathrm{euc}}\bar{u} = \int_{\bar{U}_{n-1}} \int_{\bar{U}_{n,1}} p^{s+1}((\bar{u}_{n-1})_A) d_{\mathrm{euc}}\bar{u}_{n-1} d_{\mathrm{euc}}\bar{u}_{n,1}$$

$$= \int_{\bar{U}_{n-1}} \int_{\bar{U}_{n,1}} p^{s+1}((\bar{u}_{n-1})_A) p^{s+1}((\bar{u}_{n-1})_K \bar{u}_{n,1})_A) d_{\mathrm{euc}}\bar{u}_{n-1} d_{\mathrm{euc}}\bar{u}_{n,1}.$$

We can use Fubini's theorem and integrate with respect to $d_{\mathrm{euc}}\bar{u}_{n,1}$ first. Let $k = (\bar{u}_{n-1})_K$, so that $k \in K_{n-1}$ and conjugation by k normalizes $\bar{U}_{n,1}$. Then

$$\begin{aligned}\int_{\bar{U}_{n,1}} p^{s+1}((k\bar{u}_{n,1})_A)d_{\mathrm{euc}}\bar{u}_{n,1} &= \int_{\bar{U}_{n,1}} p^{s+1}((k\bar{u}_{n,1}k^{-1})_A)d_{\mathrm{euc}}\bar{u}_{n,1}\\ &= \int_{\bar{U}_{n,1}} p^{s+1}((\bar{u}_{n,1})_A)d_{\mathrm{euc}}\bar{u}_{n,1}\\ &= F_n(s).\end{aligned}$$

In other words, we can cancel the conjugation by k, because K_{n-1} is compact, and the Haar measure transforms by a continuous homomorphism of K_{n-1} into $\mathbf{R}^+$, which is trivial. On the other hand, the restriction of p^{s+1} to A_{n-1} again is the Selberg power function in terms of one fewer variable, namely in terms of $s_2, \ldots, s_{n-1}$. We may write this formally as

$$p^{s+1}((\bar{u}_{n-1})_A) = p_2^{s_2+1} \cdots p_{n-1}^{s_{n-1}+1}((\bar{u}_{n-1})_A), \tag{2}$$

or in terms of the diagonal elements $1, a_2, \ldots, a_{n-1}$ we have

$$p^{s+1}(1, a_2, \ldots, a_{n-1}) = a_2^{s_2+1}(a_2a_3)^{s_3+1} \cdots (a_2 \cdots a_{n-1})^{s_{n-1}+1}.$$

Hence we may write

$$\int_{\bar{U}_{n-1}} p^{s+1}((\bar{u}_{n-1})_A)\, d\bar{u} = C_{n-1}(s_2, \ldots, s_{n-1}).$$

This proves Lemma 6.1, so far with $\mathrm{Re}(s_i) \geqq 1$.

However, computing the integral explicitly will give us a larger domain of absolute convergence. Let $\bar{u} \in \bar{U}_{n-1}$ so $\bar{u} = \bar{u}(x)$ with $x \in \mathbf{R}^{n-1}$, that is

$$\bar{u} = \begin{pmatrix} 1 & & & 0 \\ x_2 & 1 & & \\ \vdots & \vdots & \ddots & \\ x_n & 0 & \cdots & 1 \end{pmatrix}.$$

Then

$$\begin{aligned}\bar{u}^{-1}e_1 &= e_1 - x_2e_2 - \cdots - x_ne_n,\\ \bar{u}^{-1}e_j &= e_j \qquad \text{for} \quad j \geqq 2.\end{aligned}$$

Hence

$$\bar{u}^{-1}e_1 \wedge \cdots \wedge \bar{u}^{-1}e_d = e_1 \wedge \cdots \wedge e_d - \sum_{j=d+1}^{n} x_je_j \wedge e_2 \wedge \cdots \wedge e_d.$$

Taking the norm as in Lemma 5.2 now gives

$$p_d(\bar{u}_A)^{-2} = 1 + \sum_{j=d+1}^{n} x_j^2. \tag{3}$$

Therefore we find:

$$p^{s+1}((\bar{u}_{n,1})_A) = \prod_{d=2}^{n}\left(1 + \sum_{j=d}^{n} x_j^2\right)^{-(s_{d-1}+1)/2} \tag{4}$$

and

$$\begin{aligned} F_n(s) &= F_n(s_1, \ldots, s_{n-1}) \\ &= \int_{\mathbf{R}^{n-1}} \prod_{d=2}^{n}\left(1 + \sum_{j=d}^{n} x_j^2\right)^{-(s_{d-1}+1)/2} dx_2 \cdots dx_n. \end{aligned} \tag{5}$$

In particular, things start with the basic integral and $s \in \mathbf{C}$, $\mathrm{Re}(s) > 0$:

$$\begin{aligned} C_2(s) = F_2(s) &= \int_{-\infty}^{\infty} \frac{1}{(1+x^2)^{(s+1)/2}}\, dx \\ &= \sqrt{\pi}\frac{\Gamma\left(\frac{s}{2}\right)}{\Gamma\left(\frac{s+1}{2}\right)} = \mathbf{B}\left(\frac{s}{2}, \frac{1}{2}\right), \end{aligned} \tag{6}$$

where $\mathbf{B}$ is the standard beta function, $\mathbf{B}(x, y) = \Gamma(x)\Gamma(y)/\Gamma(x+y)$. Cf. [Lan 75/85], p. 85.

We are now ready to evaluate as in [Bha 60a], Theorem 3:

Theorem 6.2. *The integrals defining $C_n(s)$, $F_n(s)$, and $\mathbf{c}_{\mathrm{Har}}(p^s)$ are absolutely convergent for* $\mathrm{Re}(s) > 0$. *We have the explicit expressions for $n \geqq 3$:*

(i) $F_n(s) = F_n(s_1, \ldots, s_{n-1}) = F_2(s_1 + \cdots + s_{n-1})F_{n-1}(s_1, \ldots, s_{n-2})$

(ii) $\displaystyle = \prod_{j=2}^{n} F_2(s_1 + \cdots + s_{j-1})$

(iii) $\displaystyle = \prod_{j=2}^{n} \mathbf{B}\left(\frac{1}{2}(s_1 + \cdots + s_{j-1}), \tfrac{1}{2}\right)$

In particular,

$$F_n(1^{(n-1)}) = \prod_{i=1}^{n-1} \mathbf{B}\left(\frac{i}{2}, \frac{1}{2}\right).$$

Proof. We separate out the variable x_n in (5) by letting

$$x_n = y_n,$$
$$x_i = y_i(1+x_n^2)^{1/2} \qquad \text{for } i = 2, \ldots, n-1.$$

Then

$$dx_2 \wedge \cdots \wedge dx_n = (1+y_n^2)^{(n-2)/2}\, dy_2 \wedge \cdots \wedge dy_n.$$

The x_n-integral becomes

$$\int_{-\infty}^{\infty} (1+y_n^2)^{(-1/2)(s_1+\cdots+s_{n-1}+1)} dy_n = F_2(s_1+\cdots+s_{n-1}).$$

The remaining integral is just $F_{n-1}(s_1, \ldots, s_{n-2})$, which proves the theorem.

Corollary 6.3. $C_n(1) = \prod_{i=1}^{n-1} \mathbf{B}\left(\frac{i}{2}, \frac{1}{2}\right)^{n-i}$.

Proof. Immediate from Lemma 6.1 and Theorem 6.2(iii).

Lemma 6.1 and Theorem 6.2 give us a product decomposition for $\mathbf{c}_{\mathrm{Har}}(p^s)$ inductively. We see the sum of the variables $s_i + \cdots + s_{j-1}$ appearing for $i < j$. Just to state more easily the product decomposition, we introduce new variables

$$s_{ij} = s_i + \cdots + s_{j-1} \qquad \text{for } i < j \leqq n.$$

They correspond to coordinates for a different choice of basis for the characters. The expression of the product in terms of the notation of relevant characters on $\mathfrak{a}$ will be explained later. Right now, we want to give the product expression in as neat a form as possible.

Theorem 6.4. *For $n = 2$, with $s = s_1$, we have*

$$\mathbf{c}_{\mathrm{Har}}(p^s) = \frac{1}{\sqrt{\pi}} \frac{\Gamma\left(\frac{s}{2}\right)}{\Gamma\left(\frac{s+1}{2}\right)} = \mathbf{B}\left(\frac{s}{2}, \frac{1}{2}\right) / \mathbf{B}\left(\frac{1}{2}, \frac{1}{2}\right).$$

For $n \geqq 2$,

$$\begin{aligned} \mathbf{c}_{\mathrm{Har}}(p^s) &= C_n(1)^{-1} \prod_{i<j\leqq n} \mathbf{B}\left(\frac{1}{2}s_{ij}, \frac{1}{2}\right) \\ &= \prod_{i<j\leqq n} \mathbf{B}\left(\frac{1}{2}s_{ij}, \frac{1}{2}\right) / \mathbf{B}\left(\frac{i}{2}, \frac{1}{2}\right). \end{aligned}$$

We note that the poles of each factor on the right are on the hyperplanes

$$\tfrac{1}{2}s_{ij} = \text{seminegative integer.}$$

As an application of Theorems 6.2 and 6.4, we may now give the stronger version of the asymptotics in Theorem 5.4.

Theorem 6.5. *Let* $\mathrm{Re}(s) > 0$. *Then*

$$\begin{aligned}\varphi(p^s, a) &= p^{s-1}(a)\int_{\bar{U}} p^{1-s}((a\bar{u}a^{-1})_A)p^{1+s}(\bar{u}_A)\,d\bar{u}\\ &\sim p^{s-1}(a)c_{\mathrm{Har}}(s) \qquad \textit{for } a \in A^+, a \to \infty.\end{aligned}$$

Proof. Without loss of generality, by Theorem 5.4, we may suppose

$$0 < \mathrm{Re}(s) \leqq 1.$$

By Theorem 6.2, we know the absolute convergence of the integral

$$c_{\mathrm{Har}}(s) = \int_{\bar{U}} p^{1+s}(\bar{u}_A)\,d\bar{u}.$$

By Lemma 5.2(i) we know that $p_d((a\bar{u}a^{-1})_A) \leqq 1$ for all d, so for $0 < \mathrm{Re}(s) \leqq 1$,

$$\left|p^{1-s}((a\bar{u}a^{-1})_A)\right| \leqq 1.$$

We may therefore apply the dominated convergence theorem in taking the limit as $a \to \infty$. This concludes the proof.

Theorem 6.5 was one of our main goals. Readers may now look how this result is translated in more invariant terms in Theorem 7.2 of the next section.

V, §7. INVARIANT FORMULATION ON $\mathfrak{a}_{\mathbf{C}}^{\vee}$

We shall pull back the multiplicative expressions on A to additive expressions on $\mathfrak{a}$, and also on its complexification $\mathfrak{a}_{\mathbf{C}} = \mathbf{C} \otimes_{\mathbf{R}} \mathfrak{a}$. We can then obtain an expression for the c-function which coincides with the product formula due in general to Gindikin–Karpelevic [GiK 62], and reproduced in various books, e.g., [Hel 84], Chapter IV, Theorems 6.13, 6.14; and [GaV 88], Theorem 4.7.5.

Let $\mathfrak{g} = \mathfrak{sl}_n(\mathbf{R})$ = Lie algebras of $n \times n$ real matrices with trace 0. Let $\mathfrak{a}$ be the subspace of diagonal matrices

$$H = \begin{pmatrix} h_1 & & \\ & \ddots & \\ & & h_n \end{pmatrix} \quad \text{so} \quad \sum_{i=1}^{n} h_i = 0.$$

Let $\{\alpha\} = \{\alpha_{ij}\}_{i<j}$ be the set of $(\mathfrak{a}, \mathfrak{n})$-characters, which for our purposes we may define by

$$\alpha_{ij}(H) = h_i - h_j.$$

No other property will be needed for the present discussion. On $\mathfrak{a}$ we have a positive definite scalar product defined by the **trace form**, namely

$$\langle H, H' \rangle = \operatorname{tr}(HH').$$

Let E_{ii} be the diagonal matrix with i-th component 1 and other components 0. If $\alpha = \alpha_{ij}$, let

$$H_\alpha = H_{ij} = E_{ii} - E_{jj}.$$

Then

$$\langle H_\alpha, H_\alpha \rangle = 2 \qquad \text{for all } \alpha \in R^+.$$

Let $\alpha_i = \alpha_{i,i+1}$ for $i = 1, \ldots, r = n - 1$. *Warning*: It is not so that $\{\alpha_1, \ldots, \alpha_r\}$ is the dual basis of $\{H_{12}, H_{23}, \ldots, H_{r,r+1}\}$. The dual basis $\{\lambda_1, \ldots, \lambda_r\}$ will be described shortly.

The positive definite scalar product extends to a **non-degenerate symmetric bilinear form B** on the complexification $\mathfrak{a}_{\mathbf{C}} = \mathbf{C} \otimes_{\mathbf{R}} \mathfrak{a}$, so again

$$B(H, H') = \operatorname{tr}(HH') \qquad \text{for } H, H' \in \mathfrak{a}_{\mathbf{C}}.$$

Then B induces a $\mathbf{C}$-isomorphism $\mathfrak{a}_{\mathbf{C}} \xrightarrow{\approx} \mathfrak{a}_{\mathbf{C}}^\vee$ of $\mathfrak{a}_{\mathbf{C}}$ with its complex dual space $\mathfrak{a}_{\mathbf{C}}^\vee$, and of course, over the reals, we get an isomorphism $\mathfrak{a} \xrightarrow{\approx} \mathfrak{a}^\vee$. We may then view $\{\alpha_1, \ldots, \alpha_r\}$ as a $\mathbf{C}$-basis for $\mathfrak{a}_{\mathbf{C}}^\vee$.

Let λ_i be the functional defined by

$$\lambda_i(H) = h_1 + \cdots + h_i \qquad \text{for } i = 1, \ldots, r.$$

Thus $\lambda_i \in \mathfrak{a}^\vee$ or $\mathfrak{a}_{\mathbf{C}}^\vee$ depending whether we agree to deal with the real Lie algebra or its complexification. Note that directly from the definition,

$$\lambda_i(H_{j,j+1}) = \delta_{ij},$$

and therefore $\{\lambda_1, \ldots, \lambda_r\}$ is the dual basis.

Under the isomorphism $\mathfrak{a}_{\mathbf{C}} \to \mathfrak{a}_{\mathbf{C}}^{\vee}$ the symmetric bilinear scalar product is defined on the dual space. Let $\lambda \in \mathfrak{a}_{\mathbf{C}}^{\vee}$. Let $H_\lambda \in \mathfrak{a}_{\mathbf{C}}$ be the vector such that

$$\langle H_\lambda, H\rangle = \lambda(H) \qquad \text{for all } H \in \mathfrak{a}, \text{ or } H \in \mathfrak{a}_{\mathbf{C}}.$$

This general definition is compatible with our previous definition of H_α when α is an $(\mathfrak{a}, \mathfrak{n})$-character, namely

$$H_{\alpha_{ij}} = H_{ij} = E_{ii} - E_{jj} \qquad \text{for } i < j.$$

Then by definition, for $\lambda, \mu \in \mathfrak{a}_{\mathbf{C}}^{\vee}$ we have

$$\langle \lambda, \mu\rangle = \langle H_\lambda, H_\mu\rangle.$$

Thus $\{\lambda_1, \ldots, \lambda_r\}$ and $\{\alpha_1, \ldots, \alpha_r\}$ are B-dual bases of each other, that is

$$\langle \lambda_i, \alpha_j\rangle = \delta_{ij}.$$

With the basis $\{\lambda_1, \ldots, \lambda_r\}$ we can coordinatize the characters. Let $s = (s_1, \ldots, s_r)$ be an r-tuple of complex numbers, and let

$$\zeta = \zeta_s = s_1\lambda_1 + \cdots + s_r\lambda_r.$$

Then $\zeta_s \in \mathfrak{a}_{\mathbf{C}}^{\vee}$. For $H \in \mathfrak{a}$ we have

$$\exp(\zeta_s(H)) = p^s(\exp H),$$

so *ζ_s is the pull back of the power character to the Lie algebra.*

We note in summary that there are three natural bases:

$\{H_{12}, H_{23}, \ldots, H_{r,r+1}\}$ for $\mathfrak{a}$ or $\mathfrak{a}_{\mathbf{C}}$;

$\{\alpha_{12}, \alpha_{23}, \ldots, \alpha_{r,r+1}\} = \{\alpha_1, \ldots, \alpha_r\}$ for $\mathfrak{a}^{\vee}$ or $\mathfrak{a}_{\mathbf{C}}^{\vee}$;

$\{\lambda_1, \ldots, \lambda_r\}$ the dual basis of the first and B-dual of the second.

We have

$$\boxed{\rho = \lambda_1 + \cdots + \lambda_r \qquad \text{and} \qquad \tau = 2\rho = 2\lambda_1 + \cdots + 2\lambda_r.}$$

These additive relations correspond to the multiplicative relation $\delta^{1/2} = p^{(1,\ldots,1)}$ from §5. This terminates the tabulation of useful bases, and we now turn to expressing the c-function more invariantly in terms of elements of $\mathfrak{a}_{\mathbf{C}}$ and $\mathfrak{a}_{\mathbf{C}}^{\vee}$.

Proposition 7.1. *Let $s = (s_1, \ldots, s_r)$ be real. Then $\langle \zeta_s, \alpha\rangle > 0$ for all $\alpha \in \mathcal{R}(\mathfrak{n})$ (or equivalently for all $\alpha = \alpha_1, \ldots, \alpha_r$) if and only if*

$$s_i > 0 \qquad \textit{for all } i.$$

Proof. Immediate from the duality of the respective bases above.

We define:

$$\lambda > 0 \iff \lambda \in \mathfrak{a}^\vee \text{ and } \langle \alpha, \lambda \rangle > 0 \text{ for all } \alpha \in \mathcal{R}(\mathfrak{n}),$$
$$\text{or equivalently, } \langle \alpha_i, \lambda \rangle > 0 \text{ for } i = 1, \ldots, r.$$

$\mathfrak{a}^{\vee +} = \text{set of } \lambda \in \mathfrak{a}^\vee,\ \lambda > 0.$

$\chi_\zeta(a) = a^\zeta$ for $\zeta \in \mathfrak{a}^\vee_{\mathbf{C}}$.

$c_{\mathrm{Har}}(\zeta) = c_{\mathrm{Har}}(\chi_\zeta)$ for $\zeta \in \mathfrak{a}^\vee_{\mathbf{C}}$.

Thus we may view c_{Har} as a function on $\mathfrak{a}^\vee_{\mathbf{C}}$. As usual, stricter notation would require that we index the function to indicate going to the complexified Lie algebra dual, but we usually omit such index, just as we write both $c_{\mathrm{Har}}(\chi)$ and $c_{\mathrm{Har}}(s)$ with a complex r-tuple $s = (s_1, \ldots, s_r)$, given a coordinatization of the additive variable. We define $\lambda \geqq 0$, $\lambda < 0$, $\lambda \leqq 0$ similarly.

We can decompose an arbitrary element $\zeta \in \mathfrak{a}^\vee_{\mathbf{C}}$ into its real and imaginary part, namely

$$\zeta = \mathrm{Re}(\zeta) + \mathbf{i}\,\mathrm{Im}(\zeta) \qquad \text{with} \qquad \mathrm{Re}(\zeta), \mathrm{Im}(\zeta) \in \mathfrak{a}^\vee.$$

The following conditions are equivalent, directly from the definitions:

$$\begin{aligned} \mathrm{Re}(\zeta) > 0 \quad &\iff \quad \langle \alpha, \mathrm{Re}(\zeta) \rangle > 0 \qquad \text{for all } \alpha \in \mathcal{R}(\mathfrak{n}), \\ &\iff \quad \langle \alpha_i, \mathrm{Re}(\zeta) \rangle > 0 \qquad \text{for } i = 1, \ldots, r. \end{aligned}$$

We can now state Theorem 6.5 more invariantly.

Theorem 7.2. *Let $\zeta \in \mathfrak{a}^\vee_{\mathbf{C}}$ be such that* $\mathrm{Re}(\zeta) > 0$. *Then we have the asymptotic relation*

$$\begin{aligned} c_{\mathrm{Har}}(\zeta) &= \lim_{\substack{H \to \infty \\ H \in \mathfrak{a}^+}} e^{-(\zeta - \rho)(H)} \varphi_\zeta(\exp H) \\ &= \lim_{\substack{a \to \infty \\ a \in A^+}} (\chi_\zeta^{-1} \delta^{1/2})(a) \varphi_\zeta(a). \end{aligned}$$

In other words,

$$\varphi_\zeta(a) \sim c_{\mathrm{Har}}(\zeta) a^{\zeta - \rho} \qquad \textit{for } a \to \infty,\ a \in A^+.$$

Proof. This is merely a translation of Theorem 6.5, taking Proposition 7.1 into account.

Notational Remark. We defined the character χ_ζ without $\mathbf{i}$ in the exponent. The literature is split on this. Harish-Chandra started by putting $\mathbf{i}$ in [Har 58], but later did not as in [Har 68]. Helgason [Hel 84] and Anker [Ank 91] put the $\mathbf{i}$, but Gangolli–Varadarajan do not [GaV 88]. The conflict arises from two incompatible desiderata: fit the (x, y)-symmetry of the classical Fourier analysis notation $e^{\mathbf{i}xy}$, or use a formalism which is algebraically as natural as possible. It turns out that the convention without the $\mathbf{i}$ gives rise to the simplest formalism in the long run, especially in the theory of $\Gamma\backslash G$ as in [Har 68] (Langlands). It is true that pure imaginary characters play the fundamental role in classical Fourier analysis, and in the K-bi-invariant Harish-Chandra inversion. However, the distinguished aspect of the pure imaginary axis does not persist in more general contexts for spectral inversion.

Next we express the Bhanu-Murty formula in a more invariant way. We start with a lemma.

Lemma 7.3. *For all $\alpha \in \mathcal{R}(\mathfrak{n})$ we have $\langle H_\alpha, H_\alpha \rangle = 2 = \langle \alpha, \alpha \rangle$.*

Proof. This is obvious by direct verification, since an element $\alpha \in \mathcal{R}(\mathfrak{n})$ is of type α_{ij} $(i < j)$ represented by the matrix with 1 and -1 in the i-th resp. j-th diagonal position.

In §6 we used the notation $s_{ij} = s_i + \cdots + s_{j-1}$. We now write

$$s_\alpha = s_{ij} \qquad \text{if } \alpha = \alpha_{ij} \text{ with } i < j,$$

$$z_\alpha = \frac{\langle \zeta, \alpha \rangle}{\langle \alpha, \alpha \rangle} = \frac{s_\alpha}{2}.$$

We let ζ_α be the restriction of ζ to the one-dimensional space generated by α, so

$$\zeta_\alpha(\alpha) = \langle \zeta, \alpha \rangle.$$

Then Bhanu-Murty's theorem can be stated in the form:

Theorem 7.4. *Let c_n be the constant*

$$c_n = \prod_{i<j} \mathbf{B}\left(\frac{1}{2}, \frac{1}{2}\right) / \mathbf{B}\left(\frac{i}{2}, \frac{1}{2}\right).$$

Then

$$c_{\mathrm{Har},n}(\zeta) = c_n \prod_\alpha \mathbf{B}\left(\frac{s_\alpha}{2}, \frac{1}{2}\right) / \mathbf{B}\left(\frac{1}{2}, \frac{1}{2}\right) = c_n \prod_\alpha c_{\mathrm{Har},2}(\zeta_\alpha).$$

The poles of $c_{\mathrm{Har},n}$ *are at the points such that for some* α,

$$z_\alpha \in \text{seminegative integer}.$$

The above product formula is accompanied by a geometric interpretation. For each $(\mathfrak{a}, \mathfrak{n})$-character $\alpha \in \mathcal{R}(\mathfrak{n})$ there is an embedding of SL_2 in SL_n in the α-place. More concretely, let $\alpha = \alpha_{ij}$ with $i < j$. Then the Lie algebra of SL_2 is embedded in $\mathfrak{sl}_n$ on the four corners of the square with coordinates

$$(i,i), (i,j), (j,j), (j,i),$$

illustrated on the figure, with $x_{ii} + x_{jj} = 0$.

$$\begin{pmatrix} 0 & & & & & & & & \\ & \ddots & & & & & & & \\ & & 0 & & & & & & \\ & & & x_{ii} & & \cdots & x_{ij} & & \\ & & & & 0 & & & & \\ & & & \vdots & & \ddots & \vdots & & \\ & & & & & 0 & & & \\ & & & x_{ji} & & \cdots & x_{jj} & 0 & \\ & & & & & & & \ddots & \\ & & & & & & & & 0 \end{pmatrix}.$$

The exponential of this Lie algebra is the embedded SL_2 in SL_n. Thus the Harish-Chandra c-function on SL_n is reduced to the c-function on SL_2.

Bhanu-Murty's was extended to a more general Gindikin–Karpelevic product formula [GiK 62], reproduced in [Hel 84] and [GaV 88].

V, §8. COROLLARIES ON THE ANALYTIC BEHAVIOR OF c_{Har}

We can read off some analytic or continuity properties of the c-function immediately from Theorem 6.4 or its translation Theorem 7.4. These properties will be needed later in proving the spherical inversion theorem, so we tabulate them in a section designed for convenient reference. We shall consider the regions $\mathrm{Re}(\zeta) \geqq 0$, $\mathrm{Re}(\zeta) = 0$, and $\mathrm{Re}(\zeta) \leqq 0$.

Lemma 8.1.

(i) *For* $\lambda \in \mathfrak{a}^\vee$, *we have* $c_{\mathrm{Har}}(-i\lambda) = \overline{c_{\mathrm{Har}}(i\lambda)}$.

(ii) *For* $\zeta \in \mathfrak{a}_{\mathbf{C}}^{\vee}$ *the function* $c_{\mathrm{Har}}(\zeta)c_{\mathrm{Har}}(-\zeta)$ *is invariant under* W.

(iii) *The functions* $|c_{\mathrm{Har}}(\mathbf{i}\lambda)|$ *and* $|c_{\mathrm{Har}}(\mathbf{i}\lambda)|^2$ *(for* $\lambda \in \mathfrak{a}^{\vee}$*) are invariant under* W, *that is*

$$|c_{\mathrm{Har}}(\mathbf{i}\lambda)| = |c_{\mathrm{Har}}(\mathbf{i}w\lambda)| \qquad \textit{for } \lambda \in \mathfrak{a}^{\vee}.$$

Proof. The first assertion is merely due to the fact that the gamma function has the stated property. For (ii), we consider both sets of characters $\mathcal{R}(\mathfrak{n})$ and $\mathcal{R}('\mathfrak{n}) = -\mathcal{R}(\mathfrak{n})$. Theorem 7.4 tells us that up to a constant factor, $c(\zeta)$ is a product over all $\alpha \in \mathcal{R}(\mathfrak{n})$. Hence

$$\text{(1)} \quad c(\zeta)c(-\zeta) = \text{constant} \cdot \prod_{\alpha \in \mathcal{R}(\mathfrak{n})} c_2\left(\frac{\langle w\zeta, \alpha\rangle}{2}\right) c_2\left(\frac{\langle -w\zeta, \alpha\rangle}{2}\right)$$

$$= \text{constant} \cdot \prod_{\alpha \in \mathcal{R}(\mathfrak{n}) \cup \mathcal{R}('\mathfrak{n})} c_2\left(\frac{\langle \zeta, w^{-1}\alpha\rangle}{2}\right)$$

because $\langle w\zeta, \alpha\rangle = \langle \zeta, w^{-1}\alpha\rangle$ by the G-invariance of the form $B = \langle\ ,\ \rangle$. An element $w \in W$ permutes $\mathcal{R}(\mathfrak{n}) \cup \mathcal{R}('\mathfrak{n})$, and (ii) follows. Then (iii) is immediate from (ii) and (i). This concludes the proof.

Of course, Lemma 8.1 is to be interpreted in the natural way at the poles of c_{Har}, so the set of poles is stable under W. We want a convenient way of expressing the divisor of poles, and we use Harish-Chandra's notation. The polar divisor can actually be defined by a simple equation. We let

$$\Pi_+ = \prod_{\alpha \in \mathcal{R}(\mathfrak{n})} \alpha \qquad \text{and} \qquad \Pi_+^{\vee}(\zeta) = \prod_{\alpha \in \mathcal{R}(\mathfrak{n})} \langle \zeta, \alpha\rangle.$$

Note that $\langle \zeta, \alpha\rangle = 0$ (with variable $\zeta \in \mathfrak{a}_{\mathbf{C}}^{\vee}$) defines a hyperplane. We let

$$\mathbf{b} = \Pi_+^{\vee} c_{\mathrm{Har}},$$

so $\mathbf{b}$ is a meromorphic function on $\mathfrak{a}_{\mathbf{C}}^{\vee}$. We are interested in its poles in the right half plane, but we may go a bit to the left of the imaginary half space.

Lemma 8.2.

(i) *The function* $\zeta \mapsto c_{\mathrm{Har}}^{-1}(\zeta)$ *is holomorphic for* $\mathrm{Re}(\zeta) > -1$. *Its divisor of zeros in this region is defined by the equation*

$$\Pi_+^{\vee}(\zeta) = 0 = \prod_{\alpha \in \mathcal{R}(\mathfrak{n})} \langle \zeta, \alpha\rangle.$$

(ii) *Alternatively, the functions* $\mathbf{b}, \mathbf{b}^{-1}$ *are holomorphic on the domain* $\mathrm{Re}(\zeta) > -1$, *so* $\mathbf{b}$ *is a unit in the ring of holomorphic functions on that domain.*

Proof. We recall that the gamma function has its poles at the negative integers and 0. The function $1/\Gamma$ is entire. The product formula for c^{-1} has factors

$$\frac{\Gamma\left(\frac{s}{2}+\frac{1}{2}\right)}{\Gamma\left(\frac{s}{2}\right)},$$

up to a constant factor. For $\mathrm{Re}(s) > -1$, the numerator has no zero or pole because it is translated to the right by $\frac{1}{2}$. For s in a neighborhood of 0, the function

$$\frac{s}{2}\Gamma\left(\frac{s}{2}\right)$$

is invertible holomorphic, because the gamma function has a simple pole at the origin with residue 1. This concludes the proof of (ii), and of the lemma.

Remark. See Chapter XII, Theorem 4.3, where $c = \Pi_+^\vee(\zeta)/\Pi_+^\vee(\zeta)$ for $\mathrm{SL}_n(\mathbf{C})$.

Lemma 8.3. *For each* $w \in W$, *the function* $\zeta \mapsto c_{\mathrm{Har}}(w\zeta)/c_{\mathrm{Har}}(\zeta)$ *has a holomorphic extension to the imaginary axis* $\mathbf{i}\mathfrak{a}^\vee$, *that is,* $\mathrm{Re}(\zeta) = 0$. *We denote this extension by* $\gamma(w, \zeta)$, *so we may write*

$$\gamma(w,\zeta) = c_{\mathrm{Har}}(w\zeta)/c_{\mathrm{Har}}(\zeta) \qquad \text{for } \zeta \in \mathbf{i}\mathfrak{a}^\vee.$$

Proof. For $w \in W$ we have

$$\Pi_+^\vee(w\zeta) = \prod_{\alpha\in\mathcal{R}(\mathfrak{n})} \langle w\zeta, \alpha\rangle = \prod_{\alpha\in\mathcal{R}(\mathfrak{n})} \langle \zeta, w^{-1}\alpha\rangle.$$

Note that w permutes pairs $\{(\alpha, -\alpha)\}$ with $\alpha \in \mathcal{R}(\mathfrak{n})$. Hence the divisors of poles of $c_{\mathrm{Har}}(w\zeta)$ and $c_{\mathrm{Har}}(\zeta)$ are the same in a neighborhood of 0 by Lemma 8.2. Hence the quotient is a holomorphic unit on the imaginary axis, as was to be shown.

We identify $\mathfrak{a}_{\mathbf{C}}^\vee$ with $\mathbf{C}^r$ by means of the basis $\{\lambda_1, \ldots, \lambda_r\}$. To make subsequent estimates, we can sometimes use any norm on $\mathfrak{a}_{\mathbf{C}}^\vee = \mathbf{C}^r$, and we denote such a norm by $|\zeta|$ in the next lemma. If we need a specific norm for a more refined result, we shall specify it, namely the B-scalar product norm (trace norm) in Theorem 8.5.

Lemma 8.4. *Let* $M = \dim \mathfrak{n}$. *Then uniformly for* ζ *in the half plane* $\mathrm{Re}(\zeta) \geqq 0$,

$$|c_{\mathrm{Har}}^{-1}(\zeta)| = O(|\zeta|^{M/2} \qquad \textit{for } |\zeta| \to \infty.$$

Proof. This is a standard, routine consequence of Stirling's formula, which we recall. Cf. texts in complex analysis, e.g., [Lan 95/99]. Stirling's formula states that for a complex variable z,

$$\log \Gamma(z) = \left(z - \tfrac{1}{2}\right) \log z - z + \tfrac{1}{2} \log 2\pi + O\left(\frac{1}{|z|}\right)$$

for $|z| \to \infty$, uniformly in each region

$$|\arg z| \leqq \pi - \delta \qquad \text{with } \delta > 0.$$

For a fixed complex number a, we obtain

$$\log \Gamma(z + a) = (z + a - \tfrac{1}{2}) \log z - z + \tfrac{1}{2} \log 2\pi + O\left(\frac{1}{|z|}\right)$$

uniformly in the above region. In particular, if a is real and if we set $r = |z|$ and $\theta = \arg z$, $z = x + iy$ then

$$|\Gamma(z + a)| = r^{x+a-1/2} e^{-x} (2\pi)^{1/2} e^{O(1/r)}$$

uniformly in the above region. Taking a quotient of two such expressions, the factors with exponential growth cancel. Even more precisely,

$$\left| \frac{\Gamma(z + \frac{1}{2})}{\Gamma(z)} \right| \ll |z|^{1/2} \qquad \text{for } |z| \to \infty, \quad \text{in } \mathrm{Re}(z) \geqq 0.$$

We apply this estimate with each variable z_α. The number of variables is the cardinality of $\mathcal{R}(\mathfrak{n})$, which is $M = \dim \mathfrak{n}$. The estimate for $|c_{\mathrm{Har}}|^{-2}$ follows as asserted.

We are now ready to connect with Chapter IV, §8, where we defined the exponential order of a function in the Paley–Wiener space, with respect to a convex set $\mathcal{C}$. We denoted this order by $q^{\mathcal{C}}$. We now refer to Chapter IV, Corollaries 8.4 and 8.5, and give an example taking the c-function into account. In those corollaries, we considered the half space $\mathrm{Re}(\zeta) \leqq 0$, and a Paley–Wiener function on this left half space. Let h be a holomorphic function on this half space such that h has polynomial growth, that is there is some constant k such that

$$|h(\zeta)| = O(|\zeta|^k) \qquad \text{for} \quad |\zeta| \to \infty, \quad \mathrm{Re}(\zeta) \leqq 0.$$

Then Fh is again a Paley–Wiener function on the half space, with the same exponential order. Applying the above mentioned corollaries to the function Fh with

$$h(\zeta) = c(-\zeta)^{-1},$$

and using Lemma 8.4, we get:

Proposition 8.5. *Let C be a W-invariant, compact convex set in $\mathfrak{a}$. Let $F(\zeta)$ be a Paley–Wiener function on the half space* $\mathrm{Re}(\zeta) \leqq 0$, *of exponential order $\leqq q^C$. Let $H \in \mathfrak{a}$ and $H \notin C$. Then*

$$\int_{\mathrm{Re}(\zeta)=0} F(\zeta) e^{\zeta(H)} c(-\zeta)^{-1} \, d\zeta = 0.$$

Example. Let $C = \bar{\mathbf{B}}_R(0)$ be the ball of radius R, let the exponential order of F be $\leqq R$, and $R < |H|$ (B-scalar product norm).

CHAPTER VI

Polar Decomposition

So far we have dealt mostly with the Iwasawa decomposition from various points of view: projection on invariant differential operators, spherical functions, etc. Already in studying spherical functions, the action of K on the left was crucial. In the present chapter, and the next two, we deal with an alternate continuation of Chapter III, independent of the Gelfand–Naimark decomposition, namely we deal with the polar decomposition $G = KAK$. We also complement Chapter IV, §6 in §3.

In §1, we start with the computation of the Jacobian formula for the polar decomposition.

In §2 we give the C^∞ version of Chevalley's theorem. This is part of the program, further developed in §3, to show how the polar decomposition allows us to transfer certain questions concerning the spherical transform on the group to questions of essentially euclidean Fourier analysis (somewhat jazzed up) on the Lie algebra. Thus we show that axioms which will form the basis of the inversion theory in Chapter IX are satisfied in the context of Iwasawa and polar decompositions.

In the rest of the chapter we give other results of Harish-Chandra concerning polar decompositions using regular elements. The transpose group $\bar{U}$ appears briefly, but not in the fundamental way of Chapter V.

Most of this chapter is due to Harish-Chandra, who developed systematically analysis on the group and on the Lie algebra of A in the case of K-bi-invariant functions [Har 58].

VI, §1. THE JACOBIAN OF THE POLAR MAP

We let $G = \mathrm{SL}_n(\mathbf{R})$.

We have the Iwasawa decomposition

$$G = UAK,$$

and as in Chapter III, §2, the direct sum decomposition

$$\mathfrak{n} = \bigoplus_{\alpha \in \mathcal{R}(\mathfrak{n})} \mathfrak{n}_\alpha,$$

where $\mathfrak{n}_\alpha$ is the α-eigenspace for the relevant character α of $\mathfrak{a}$. We defined the **regular elements** A' to consist of those $a \in A$ such that $a^\alpha \neq 1$ for all $\alpha \in \mathcal{R}(\mathfrak{n})$. Now let:

$A^+ =$ open subset of A consisting of all $a \in A$ such that $a^\alpha > 1$ for all $\alpha \in \mathcal{R}(\mathfrak{n})$.

We recall the Weyl group W from Chapter III, §3,

$$W = M'/M.$$

As previously explained, we may view W as the group of permutations of the coordinates (induced from the embedding in the $n \times n$ diagonal matrices).

Proposition 1.1. *The set A^+ is a fundamental domain for the action of W on the set of regular elements A'.*

Proof. This is trivially seen on $\mathrm{SL}_n(\mathbf{R})$. First a permutation of the coordinates can be found to change a given element of A to another element a having the property that $a_i > a_{i+1}$ for all $i = 1, \ldots, n-1$. Then such an element a lies in A^+. If $a \in A^+$ and $w \in W$, $w \neq \mathrm{id}$, then $[w]a = b$ cannot lie in A^+, i.e. we have $b_i < b_j$ for some $j > i$, which proves the proposition.

We note that if $a \in A^+$ and $c \in \mathbf{R}^+ = \mathbf{R}_{>0}$, then $ca \in A^+$. In other words, *A^+ is a positive cone.*

We define the **polar coordinate map** or simply the **polar map**

$$\mathbf{p} = \mathbf{p}_G \colon K \times A \times K \to G \quad \text{by} \quad \mathbf{p}(k, a, k') = kak'.$$

The surjectivity is proved in Chapter I, Theorem 5.5. Letting $X = G/K$, we also have the polar map into X by composing $\mathbf{p}_G$ with the natural map $G \to G/K$, namely

$$\mathbf{p}_X \colon K \times A \to X \quad \text{given by} \quad \mathbf{p}_X(k, a) = kaK = kak^{-1}K.$$

Let $X = \mathrm{SPos}_n$. Under the quadratic map

$$G/K \to \mathrm{SPos}_n \quad \text{induced by} \quad g \mapsto g\,{}^t g,$$

we get still another "polar map." $K \times A \to X \approx X$. Note that every element of Pos_n can be diagonalized, so every element of Pos_n is of the form

$$x = kak^{-1} \quad \text{for some positive diagonal matrix } a.$$

This diagonalization then also holds for SPos_n. For those matrices with distinct eigenvalues (the regular elements) the decomposition

$$x = kak^{-1}$$

is unique up to a permutation of the diagonal elements, and up to elements of K which are diagonal and preserve orthonormality, in other words, diagonal matrices whose diagonal elements consist of ± 1. Hence the map

$$\mathbf{p}: K \times A \to \mathrm{SPos}_n \qquad \text{given by} \qquad (k, a) \mapsto kak^{-1}$$

is a covering of degree $|M|\,|W| = |M|n!$ over the set of regular elements, that is the set of elements with distinct eigenvalues. The Jacobian computation to verify that the map is a local isomorphism above a regular element will be carried out in Theorem 1.5.

The non-regular elements of A are those elements a such that $a^\alpha = 1$ for some α. In the vector space $\mathfrak{a}$, this condition is equivalent to $\alpha(\log a) = 0$. Thus the complement of $\mathfrak{a}'$ in $\mathfrak{a}$ consists of a finite number of hyperplanes. By abuse of language, we also say that the complement of A' in A consists of a finite number of hyperplanes $A_{(\alpha)} = \mathrm{Ker}\,\alpha$ in A, indexed by the relevant characters. We may then summarize the situation as follows.

Proposition 1.2. *The polar map $\mathbf{p}_G$ induces a covering of degree $|M|\,|W|$*

$$\mathbf{p}': K \times A' \times K \to KA'K = G'$$

over the set of regular elements. The complement of the image is a finite union of sets $KA_{(\alpha)}K$, with hyperplanes $A_{(\alpha)}$, and is therefore of lower dimension. We then get a covering of degree 1, i.e. a differential isomorphism

$$\mathbf{p}^+: K/M \times A^+ \times K \to KA^+K = G^+.$$

Here K/M is the coset space (homogeneous space, not a group).

For simplicity, we shall omit the superscript and write $\mathbf{p}$ instead of $\mathbf{p}'$ or $\mathbf{p}^+$. The context will specify the domain where we consider the polar coordinate map.

Haar measure computation

Both in this chapter and the next we shall determine the Jacobian determinants of certain maps. So we make here some general remarks on manifolds.

Let $p: X \to Y$ be a morphism of differential manifolds having the same dimension (everything is C^∞). Let Ω be a volume form on Y and Ψ a volume form on X. For each $x \in X$ we have the tangent linear map

$$T_x p: T_x X \to T_{p(x)} Y.$$

We assume that X, Y have the same finite dimension. Then we have the top degree induced map

$$\bigwedge^{\max} T_x p: \bigwedge^{\max} T_x X \to \bigwedge^{\max} T_{p(x)} Y.$$

This map gives rise to the contravariant map on the differential forms by composition,

$$p^*: \bigwedge^{\max} T_{p(x)} Y \to \bigwedge^{\max} T_x X$$

such that for $v_i \in T_x X$,

$$(p^*\Omega)(v_1, \ldots, v_N) = \Omega(p_* v_1, \ldots, p_* v_N).$$

Then there is a unique function h on X such that

$$p^*\Omega = h\Psi.$$

Each volume form gives rise to a measure, which we write

$$|\Omega| = \mu_\Omega,$$

obtained by integrating a function against the absolute value of the form, say locally in a chart. Then using the change of variables formula we see that this integral is actually defined on the manifold. Cf. [Lan 99], Chapter XVI, §4. Suppose p is a local isomorphism. We use the notation $h = Jp$, so

$$p^*|\Omega| = (Jp)|\Psi|.$$

We shall determine Jp for various mappings p. When f is fixed throughout a discussion, we may delete it and write simply J for the **absolute value of the Jacobian determinant**, which we call the **absolute Jacobian**, so by definition,

$$p^*|\Omega| = J|\Psi|.$$

Theorem 1.3. *Suppose p is a covering except possibly on a closed subset of measure 0. Then for all $f \in C_c(Y)$ we have*

$$(\deg p) \int_Y f(y)\, d\mu_\Omega(y) = \int_X f(p(x)) J(x)\, d\mu_\Psi(x).$$

Proof. Immediate from the change of variables formula locally in charts, as in the above mentioned reference.

We return to the polar map $\mathbf{p}\colon \tilde{G} \to G$, where $\tilde{G} = K \times A \times K$. We apply the theorem to $\mathbf{p}$ restricted to $K \times A' \times K$, or possibly to $K \times A^+ \times K$. We conclude that there exists a C^∞ function $J = J\mathbf{p}$ on $K \times A' \times K$ such that for all $f \in C_c(G)$ we have

$$(1) \quad (\deg \mathbf{p}) \int_G f(x)\, dx = \int_K \int_{A'} \int_K f(kak') J(k, a, k')\, dk\, da\, dk'.$$

Theorem 1.4. *The Jacobian $J = J\mathbf{p}$ depends only on the A-variable a. In particular, for $f \in C_c(K\backslash G/K)$, assuming K has total measure 1, we have*

$$(\deg \mathbf{p}) \int_G f(x)\, dx = \int_A f(a) J\mathbf{p}(a)\, da.$$

Proof. We use only the fact that G is unimodular. Let $k_1 \in K$. From left invariance of Haar measure dx, adjusted so that $c = 1$, we get

$$\begin{aligned}(2) \qquad \int_G f(x)\, dx &= \int_G f(k_1 x)\, dx \\ &= \int_K \int_{A'} \int_K f(k_1 kak') J(k, a, k')\, dk\, da\, dk' \\ [\text{by } k \mapsto k_1^{-1}k] \quad &= \int_K \int_{A'} \int_K f(kak') J(k_1^{-1}k, a, k')\, dk\, da\, dk'.\end{aligned}$$

This being true for all $f \in C_c(G)$, comparing (1) and (2), we conclude that

$$J(k_1^{-1}k, a, k') = J(k, a, k')$$

for all k, a, k', whence J is independent of k in its first variable. Similarly, using the right invariance of Haar measure dx, we conclude the independence of J from the third variable k' to conclude the proof.

The above computations were general Haar measure computations. We now want to determine $J(a)$, and for this we have to have more information about the actual effect of the tangent linear map to get at its determinant. For a Lie group G and $x \in G$ we let τ_x be left translation by x. For our purposes here the exponential at the origin is the usual one, and otherwise

$$\exp_x((\tau_x)_* u) = \tau_x(\exp_e(u)) = x \cdot \exp_e(u). \tag{3}$$

Furthermore, by the chain rule,

$$\frac{d}{dt}(x \cdot \exp_e(tu))\bigg|_{t=0} = (\tau_x)_* u. \tag{4}$$

Since $J\mathbf{p}(k, a, k') = J\mathbf{p}(a)$ depends only on a, we have to determine the tangent map only at (e, a, e), in other words we have to determine $T_{e,a,e}\mathbf{p}$ on $T_{e,a,e}\tilde{G}$, and then the determinant with respect to specific Haar volume forms. Before doing this, we make more general remarks on the Lie algebra. Cf. $SL_2(\mathbf{R})$ [Lan 75/85], Chapter VII, §2, which we essentially follow except for replacing 2 by n.

Let $G, \tilde{G}$ be Lie groups and $\mathfrak{g}, \tilde{\mathfrak{g}}$ their Lie algebras. Let

$$p: \tilde{G} \to G$$

be a differential morphism. Let $\tilde{x} \in \tilde{G}$ and $\tilde{u} \in T_{\tilde{e}}\tilde{G}$. Then by (3) and (4),

$$(T_{\tilde{x}}p)((\tau_{\tilde{x}})_*(\tilde{u})) = \frac{d}{dt}p(\tilde{x} \cdot \exp_{\tilde{e}}(t\tilde{u}))\bigg|_{t=0}. \tag{5}$$

Suppose $G, \tilde{G}$ have the same dimension. Let

$$\{X_1, \dots, X_N\} \quad \text{and} \quad \{\tilde{X}_1, \dots, \tilde{X}_N\}$$

be bases of $\mathfrak{g}, \tilde{\mathfrak{g}}$ respectively. The exterior products

$$X_1 \wedge \cdots \wedge X_N \quad \text{and} \quad \tilde{X}_1 \wedge \cdots \wedge \tilde{X}_N$$

are bases of $\bigwedge^N \mathfrak{g}$ and $\bigwedge^N \tilde{\mathfrak{g}}$ respectively. There is a unique N-form Ω_e on $\mathfrak{g}$ such that $\Omega_e(X_1, \dots, X_N) = 1$, or equivalent for all $Y_1, \dots, Y_N \in \mathfrak{g}$,

$$Y_1 \wedge \cdots \wedge Y_N = \Omega_e(Y_1, \dots, Y_N) X_1 \wedge \cdots \wedge X_N.$$

We call Ω_e the **form dual to the chosen basis**.

Under left translation, $Y_1, \dots, Y_N$, determine N left invariant vector fields on G, and Ω_e determines a G-invariant volume form on G, whose value at $x \in G$ is denoted by Ω_x or $\Omega(x)$. If $\eta_j(x) = (\tau_x)_* Y_j$ for $j = 1, \dots, N$, then

$$\Omega_x(\eta_1(x), \dots, \eta_N(x)) = \Omega_e(Y_1, \dots, Y_N).$$

The form Ω is a **Haar form** on G, i.e. a left invariant volume form, and we have a similar Haar form $\tilde{\Omega}$ on $\tilde{G}$, determined by the choice of basis elements. Actually, Ω is determined by the wedge product $X_1 \wedge \cdots \wedge X_N$, and similarly for $\tilde{\Omega}$. If we change the order of the basis elements, then we change the forms by ± 1.

Assume that p is a local differential isomorphism at each point of an open set of $\tilde{G}$. Write $x = p(\tilde{x})$. Then we get what we call the **determinant map**

$$\bigwedge^N (T_{\tilde{x}} p) \colon (\tau_{\tilde{x}})_* \left(\bigwedge^N \tilde{\mathfrak{g}}\right) \to (\tau_x)_* \left(\bigwedge^N \mathfrak{g}\right),$$

where the index $*$ denotes the functorially induced map on whatever, in the present instance,

$$(\tau_{\tilde{x}})_* = \bigwedge^N (T_{\tilde{e}} \tau_{\tilde{x}}).$$

We also get

$$(p^*\Omega)(\tilde{x}) = \pm (Jp)(\tilde{x}) \tilde{\Omega}(\tilde{x}) \tag{6}$$

with a function $Jp > 0$. We call the function Jp the **absolute Jacobian of p, relative to the given forms $\Omega, \tilde{\Omega}$.**

Change of Variables Formula. *On an open set $\tilde{V}$ where p is a differential isomorphism, for $f \in C_c(\tilde{V})$ we have*

$$\int_{p(\tilde{V})} f(x)\, d\mu_\Omega(x) = \int_{\tilde{V}} f(p(\tilde{x})) Jp(\tilde{x})\, d\mu_{\tilde{\Omega}}(\tilde{x}).$$

Having recalled this formalism of the change of variables in terms of the Lie algebra, we are ready to get down to our concrete application. So far, we have not used the semisimple Lie decomposition of $\mathfrak{n}$ under the action of A, except for defining A^+. We shall now use it for Jacobian computations. The next theorems are due to Harish-Chandra in the general context of semisimple Lie groups. We first state a weak version, leaving out the determination of a constant factor.

Theorem 1.5. *Let $\mathbf{p} \colon K \times A' \times K \to G = \mathrm{SL}_n(\mathbf{R})$ be the polar decomposition. Let $J = J\mathbf{p}$. Then up to a constant factor c, with respect to the Haar measures on both sides, for a regular in A', we have*

$$J\mathbf{p}(a) = c \prod_{\alpha \in \mathcal{R}(\mathfrak{n})} |a^\alpha - a^{-\alpha}| \neq 0;$$

or letting $m = \dim \mathfrak{n}$,

$$= 2^m c \prod_{\alpha \in \mathcal{R}(\mathfrak{n})} \sinh \alpha(\log a).$$

The constant factor is left unspecified in the above formulation. We shall specify it by defining more precisely certain choices of Haar measures relative to the Iwasawa decomposition $G = UAK$ and its Lie algebra decomposition

$$\mathfrak{g} = \mathfrak{n} \oplus \mathfrak{a} \oplus \mathfrak{k} = \mathfrak{n} \oplus \mathfrak{a} \oplus {}^t\mathfrak{n}.$$

By a **semisimple Iwasawa basis** for $\mathfrak{g}$ we mean a basis consisting of:

a basis $\{H_1, \dots, H_r\}$ of $\mathfrak{a}$;
a basis of $\mathfrak{a}_\mathfrak{n}$-eigenvectors $\{v_\alpha\}$ $(\alpha \in \mathcal{R}(\mathfrak{n}))$ for $\mathfrak{n}$;
the basis $\{{}^t v_\alpha\}$ for ${}^t\mathfrak{n}$.

We are using notation which will go over to the general case. Of course, on $\mathrm{SL}_n(\mathbf{R})$, we can take v_α to be a matrix E_{ij} $(i < j)$. Note that each ${}^t v_\alpha$ is an eigenvector for the regular action $\mathfrak{a}_\mathfrak{g}$ of $\mathfrak{a}$ on $\mathfrak{g}$, with eigencharacter $-\alpha$.

A semisimple basis for $\mathfrak{g}$ then determines a basis for $\mathfrak{k}$ as follows. We let

$$z_\alpha = v_\alpha - {}^t v_\alpha.$$

Then $z_\alpha \in \mathfrak{k}$, and it is immediately verified that z_α is an eigenvector for the action of $H_\mathfrak{g}^2$ $(H \in \mathfrak{a})$ with eigencharacter $\alpha^2(H)$. To understand this properly, however, one must put it in context of the larger structure, whereby putting

$$w = \tfrac{1}{2}(v - {}^t v) \qquad \text{and} \qquad u = \tfrac{1}{2}(v + {}^t v)$$

decomposes $\mathfrak{g}$ into a direct sum of the skew-symmetric and symmetric elements

$$\mathfrak{g} = \mathfrak{g}_{\mathrm{Sk}} \oplus \mathfrak{g}_{\mathrm{Sym}} = \mathfrak{k} \oplus \mathfrak{p} = \mathfrak{k} \oplus \mathfrak{a} \oplus \mathfrak{p}^{(0)}$$

with respect to the transpose. By $\mathfrak{p}^{(0)}$ we mean the subspace of elements in $\mathfrak{p}$ with 0 diagonal. We define

$$u_\alpha = \tfrac{1}{2}(v_\alpha + {}^t v_\alpha),$$

and then for each regular element $H \in \mathfrak{a}$, the regular representation $H_\mathfrak{g}$ induces an isomorphism

$$H_\mathfrak{g} \colon \mathfrak{k} \to \mathfrak{p}^{(0)}.$$

Let:

$\mathfrak{k}_\alpha =$ image of $\mathfrak{n}_\alpha$ under the skew-symmetrizing map $v \mapsto v - {}^t v$;

$\mathfrak{p}_\alpha^{(0)} =$ image of $\mathfrak{n}_\alpha$ under the symmetrizing map $v \mapsto v + {}^t v$.

Then each $H \in \mathfrak{a}$ induces linear homomorphisms

$$H_{\mathfrak{g},\alpha} \colon \mathfrak{k}_\alpha \to \mathfrak{p}_\alpha^{(0)} \quad \text{and} \quad \mathfrak{p}_\alpha^{(0)} \to \mathfrak{k}_\alpha.$$

It is immediately verified that if two relevant characters have the same square, then they are equal (it just comes out from the α_{ij} representation). Thus:

$\mathfrak{k}_\alpha = \alpha^2$-eigenspace for the action of all operators $H_\mathfrak{g}^2$ on $\mathfrak{k}$, $H \in \mathfrak{a}$.

Then

$$\mathfrak{k} = \bigoplus_{\alpha \in \mathcal{R}(\mathfrak{n})} \mathfrak{k}_\alpha.$$

The map

$$v_\alpha \mapsto z_\alpha$$

induces an isomorphism of $\mathfrak{n}_\alpha$ with $\mathfrak{k}_\alpha$.

Given a semisimple basis for $\mathfrak{g}$, we define the **adapted polar basis** for $\mathrm{Lie}(K \times A \times K) = \tilde{\mathfrak{k}} \oplus \tilde{\mathfrak{a}} \oplus \tilde{\tilde{k}}$, to be the basis consisting of:

the basis $\{\tilde{z}_\alpha\}$ $(\alpha \in \mathcal{R}(\mathfrak{n}))$ *of* $\tilde{\mathfrak{k}}_\alpha$, *letting* $\tilde{w} = (w, 0, 0)$ *for* $w \in \mathfrak{k}$;
the basis $\{\tilde{H}_1, \ldots, \tilde{H}_r\}$ *of* $\tilde{\mathfrak{a}}$ *as above, letting* $\tilde{H} = (0, H, 0)$ *for* $H \in \mathfrak{a}$;
the basis $\{\tilde{\tilde{z}}_\alpha\}$ $(\alpha \in \mathcal{R}(\mathfrak{n}))$ *for* $\tilde{\tilde{\mathfrak{k}}}$ *such that* $\tilde{\tilde{w}} = (0, 0, w)$.

We let Ω_e be the form on $\mathfrak{g}$ dual to

$$\bigwedge_{\alpha \in \mathcal{R}(\mathfrak{n})} v_\alpha \wedge \bigwedge_j H_j \wedge \bigwedge_{\alpha \in \mathcal{R}(\mathfrak{n})} {}^t v_\alpha.$$

Here we take the wedge products in a given order. Choosing another order changes the dual form by a sign, which disappears when we take the associated measures.

Similarly, we let $\tilde{\Omega}_{\tilde{e}}$ be the form on $\mathrm{Lie}(K \times A \times K) = \tilde{\mathfrak{k}} \oplus \tilde{\mathfrak{a}} \oplus \tilde{\tilde{\mathfrak{k}}}$ dual to

$$\bigwedge_{\alpha \in \mathcal{R}(\mathfrak{n})} \tilde{z}_\alpha \wedge \bigwedge_j \tilde{H}_j \wedge \bigwedge_{\alpha \in \mathcal{R}(\mathfrak{n})} \tilde{\tilde{z}}_\alpha.$$

We let Ω and $\tilde{\Omega}$ be the forms on G and $K \times A \times K$ obtained by left translation of Ω_e and $\tilde{\Omega}_{\tilde{e}}$ respectively. We call such a pair of forms **adapted**, and also say that $\tilde{\Omega}$ is **adapted** to Ω.

We note that the forms Ω and $\tilde{\Omega}$ are Haar forms on $G = UAK$ and $\tilde{G} = K \times A \times K$ respectively. In particular, any other Haar forms (or reading things mod ± 1, corresponding Haar measures) differ from these two by constant factors. Thus the Jacobian determinant of $\mathbf{p}$ can be pinned down more precisely by using adapted Haar forms. Also note that if we multiply the semisimple basis elements for $\mathfrak{g}$ by scalars, then the corresponding elements of the adapted basis for $\tilde{G}$ get multiplied by the same scalars, and the absolute Jacobian is unchanged. The next theorem gives its value.

Theorem 1.6. *Let Ω and $\tilde{\Omega}$ be adapted Haar forms on G and $\tilde{G} = K \times A \times K$ respectively. Let $J\mathbf{p}$ be the absolute Jacobian with respect to these forms. Then for $a \in A'$, we have*

$$J\mathbf{p}(a) = \prod_{\alpha \in \mathcal{R}(\mathfrak{n})} |a^{\alpha} - a^{-\alpha}|.$$

Proof. As already pointed out after Theorem 1.4, $J\mathbf{p}(a)$ is the absolute Jacobian of $\mathbf{p}$ at (e, a, e) on $\tilde{G}$. Thus we have to compute the effect of the tangent map $T_{e,a,e}\mathbf{p}$ on elements of the adapted basis of $\tilde{\mathfrak{g}}$, and then take the wedge product of the images of the basis elements.

We start with a basis element $\tilde{z}_{\alpha}$. We write $(e, a, e) = \tilde{a}$. We claim that

$$(1.6.1) \qquad (T_{e,a,e}\mathbf{p})((\tau_{\tilde{a}})_* \tilde{z}_{\alpha}) = (\tau_a)_*(a^{-\alpha} v_{\alpha} - a^{\alpha}\, {}^t v_{\alpha}).$$

Indeed, for $w \in \mathfrak{k}$, $\mathbf{p}(\exp(tw), a, e)) = (\exp(tw))a$, and directly from (5), the left side of (1.6.1) is

$$\frac{d}{dt}(\exp(tz_{\alpha}))a \Big|_{t=0}.$$

We write

$$\exp(tz_{\alpha})a = aa^{-1}\exp(tz_{\alpha})a = a \cdot \exp(tz_{\alpha}^{a}),$$

and observe that z_{α}^{a} is precisely the vector expression on the right of (1.6.1), that is

$$z_{\alpha}^{a} = a^{-\alpha} v_{\alpha} - a^{\alpha}\, {}^t v_{\alpha}.$$

Then (1.6.1) follows from (4), with $x = a$.

Next, we have for $H \in \mathfrak{a}$,

$$(1.6.2) \qquad (T_{e,a,e}\mathbf{p})((\tau_{\tilde{a}})_*\tilde{H}) = (\tau_a)_*H,$$

which is immediate from the analogous argument.

Thirdly, we claim that for all $\tilde{\tilde{w}} = (0, 0, w)$ with $w \in \mathfrak{k}$, we have

$$(1.6.3) \qquad (T_{e,a,e}\mathbf{p})(\tau_{\tilde{a}})_*\tilde{\tilde{w}}) = (\tau_a)_*(w).$$

But

$$\begin{aligned}(T_{e,a,e}\mathbf{p})((\tau_{\tilde{a}})_*\tilde{\tilde{w}}) &= \frac{d}{dt}\mathbf{p}(e, a, \exp(tw))\bigg|_{t=0} \\ &= \frac{d}{dt}a \cdot \exp_e(tw)\bigg|_{t=0} \\ &= (\tau_a)_*w,\end{aligned}$$

as was to be shown. This time there was no need to commute factors. Thus we have determined the tangent linear map $T_{e,a,e}\mathbf{p}$ on the basis elements.

Then for the volume form, we first note that

$$(a^{-\alpha}v_\alpha - a^{\alpha}\,{}^tv_\alpha) \wedge (v_\alpha - {}^tv_\alpha) = (a^\alpha - a^{-\alpha})v_\alpha \wedge {}^tv_\alpha.$$

Therefore taking the wedge product we get

$$\begin{aligned}&\bigwedge_\alpha \mathbf{p}_*(\tau_{\tilde{a}})_*\tilde{z}_\alpha \wedge \bigwedge_j \mathbf{p}_*(\tau_{\tilde{a}})_*\tilde{H}_j \wedge \bigwedge_\alpha \mathbf{p}_*(\tau_{\tilde{a}})_*\tilde{\tilde{z}}_\alpha \\ &= \text{image under } (\tau_a)_* \text{ of } \pm\bigwedge_\alpha(a^{-\alpha}v_\alpha - a^\alpha\,{}^tv_\alpha) \wedge \bigwedge_j H_j \wedge \bigwedge_\alpha(v_\alpha - {}^tv_\alpha) \\ &= \bigwedge_\alpha(a^\alpha - a^{-\alpha})v_\alpha \wedge {}^tv_\alpha \wedge \bigwedge_j H_j.\end{aligned}$$

This proves that

$$\pm(\mathbf{p}^*\Omega)(e, a, e) = \prod_{\alpha\in\mathcal{R}(\mathfrak{n})}(a^\alpha - a^{-\alpha})\tilde{\Omega}(e, a, e),$$

which concludes the proof of the theorem.

Remark. Choosing $z_\alpha = v_\alpha - {}^tv_\alpha$ without the factor one-half has the effect of eliminating powers of 2 in the final answer. Sometimes one wants this normalization, and sometimes one wants $w_\alpha = z_\alpha/2$. They both occur in the literature.

The general situation and the polar SSk decomposition

As readers have no doubt noticed, the above proofs are rather formal, and it is now appropriate to axiomatize what we have used. So we let G be a Lie group with an Iwasawa decomposition $G = UAK = AUK$, with K compact. *We suppose that the Lie algebra is Lie-semisimple, as defined in Chapter* III, §2, *so* $\mathfrak{n}$ *has a direct sum decomposition into* $\mathfrak{a}_{\mathfrak{n}}$ *eigenspaces*

$$\mathfrak{n} = \bigoplus_{\alpha} \mathfrak{n}_{\alpha},$$

and $\dim \mathfrak{n}_{\alpha} = m(\alpha)$ may be greater than 1. By a **polar decomposition** of G **associated to the Iwasawa decomposition** we mean that the product map

$$\mathbf{p}: K \times A \times K \to KAK = G$$

is a surjection on G, and satisfies the following conditions. These conditions depend on the definitions of the set of regular elements A'; the Weyl group, as in Chapter IV, §3; and the set A^+ defined at the beginning of this section. Those definitions are valid completely generally. Furthermore, the conditions just axiomatize Propositions 1.1 and 1.2.

POL 1. The set A^+ is a fundamental domain for the action of W on A'.

POL 2. The polar map

$$K/M \times A' \times K \to KA'K$$

induces a covering of degree $|W|$ over the set of regular elements, and the complement of the image has measure 0. Thus we get a differential isomorphism

$$K/M \times A^+ \times K \to KA^+K.$$

If M is finite, which it is in the case of $\mathrm{SL}_n(\mathbf{R})$, then the only other structure which was relevant in defining adapted volume forms was the transpose. Hence to make the proof go through we axiomatize the transpose as follows.

We suppose given an **anti-involution** of G, meaning a Lie group anti-automorphism of order 2, which we denote by $\mathbf{t}$. Defining θ by $\theta x = \mathbf{t}x^{-1}$ we also assume that θ is an automorphism of order 2. Note that $\mathbf{t}x = \theta x^{-1}$ for all $x \in G$. The symbol $\mathbf{t}$ suggests the "transpose" operation on matrices, and the symbol θ is used by the Lie industry for what is called the Cartan involution. By the chain rule, the induced linear maps on the Lie algebra satisfy

$$T_e\mathbf{t} = -T_e\theta.$$

We say that $\mathbf{t}$ or θ is **adapted to the Iwasawa decomposition**, and that θ is a **Cartan involution**, if they satisfy **TR 1–TR 5** below.

TR 1. The Lie algebra $\mathfrak{g} = \mathrm{Lie}(G)$ has the direct sum decomposition

$$\mathfrak{g} = \mathfrak{a} \oplus \mathfrak{n} \oplus \mathbf{t}\mathfrak{n}, \qquad (\text{writing } \mathbf{t} \text{ for } T_e\mathbf{t})$$

and $T_e\mathbf{t}$ preserves the eigenspace decomposition, that is $\mathbf{t}$ induces an isomorphism

$$\mathbf{t}_\alpha : \mathfrak{n}_\alpha \to (\mathbf{t}\mathfrak{n})_{-\alpha} \qquad \text{for each } \alpha \in \mathcal{R}(\mathfrak{n}).$$

As a matter of notation, we may write $\mathbf{t}\mathfrak{n} = (T_e\mathbf{t})\mathfrak{n} = \bar{\mathfrak{n}} = \mathfrak{n}^-$.

Since $\mathfrak{n}+\mathbf{t}\mathfrak{n}$ is stable under $\mathbf{t}$, we may then define the **skew-symmetric** (resp. **symmetric**) subspaces of $\mathfrak{n}+\mathbf{t}\mathfrak{n}$ in the natural way, and denote them by Sk (resp. $\mathrm{Sym}^{(0)}$). In other words:

Sk = subspace of all elements in $\mathfrak{n}+\mathbf{t}\mathfrak{n}$ of the form $v - \mathbf{t}v$, $v \in \mathfrak{n}$;

$\mathrm{Sym}^{(0)}$ = subspace of all elements in $\mathfrak{n}+\mathbf{t}\mathfrak{n}$ of the form $v + \mathbf{t}v$, $v \in \mathfrak{n}$.

TR 2. We have $T_e\mathbf{t} = \mathrm{id}$ on $\mathfrak{a}$.

Then we define

$$\mathrm{Sym} = \mathfrak{a} + \mathrm{Sym}^{(0)}.$$

Thus we can express the direct sum decomposition of the Lie algebra in the form

$$\mathfrak{g} = \mathrm{Sym} + \mathrm{Sk}.$$

Letting $\mathfrak{k} = \mathrm{Lie}(K)$, we assume further:

TR 3. $\mathrm{Sym} = \mathbf{c}(K)\mathfrak{a}$.

TR 4. $\mathrm{Sk} = \mathfrak{k}$.

Letting $\mathfrak{p} = \mathrm{Sym}$, we recover a Cartan Lie decomposition, with Cartan Lie adapted θ as in Chapter II, §6.

Since $A = \exp \mathfrak{a}$ is connected (isomorphic to $\mathbf{R}^+)^r$) it follows at once from **TR 2** that A is pointwise fixed by $\mathbf{t}$, that is

$$\mathbf{t}a = a \qquad \text{or} \qquad \theta a = a^{-1} \qquad \text{for all } a \in A.$$

Similarly, the connected component of K is pointwise fixed, but we impose the global condition:

TR 5. For all $k \in K$ we have $\mathbf{t}k = k^{-1}$ or $\theta k = k$.

Finally we formulate a global group decomposition of G. The axiom can also be expressed in the context of a Cartan decomposition.

Pol SSk or Global Cartan Decomposition. Let $\mathbf{P} = \exp \mathfrak{p}$. Then $G = \mathbf{P}K$, and

$$\mathbf{P}_{\text{SSk}} \text{ or } \mathbf{P}_{\text{CA}}: \mathbf{P} \times K \to \mathbf{P}K = G$$

is a differential isomorphism.

For $G = \mathrm{SL}_n(\mathbf{R})$, this axiom is routinely satisfied by elementary algebra, and is even true in Hilbert space. Indeed, given $x \in G$, let

$$p = (x\,{}^t x)^{1/2},$$

and define k by $pk = x$, so

$$k = p^{-1}x.$$

It is immediate that ${}^t p = p$, ${}^t k = k^{-1}$ because $k\,{}^t k = p^{-1}x\,{}^t x p^{-1} = I$, and p, k are defined in a C^∞ (actually real analytic) way from x itself. They provide the inverse of the product map $\mathbf{P} \times K \to G$.

Under Conditions **TR 1–TR 5**, Theorem 1.6 essentially goes through. We have hedged with the word "essentially" because the general case presents one added complication. As already evident in **POL 2**, one has to consider the homogeneous space K/M on the left in the polar decomposition. If M is finite, then there is no added complication. One selects a basis for each eigenspace $\mathfrak{n}_\alpha$. That such a basis has more than one element is immaterial for what follows, except for introducing the multiplicity $m(\alpha)$. One defines **adapted bases** as we did for $\mathrm{SL}_n(\mathbf{R})$, and the Jacobian formula then has the shape

$$J\mathbf{p}(a) = \prod_{\alpha \in \mathcal{R}(\mathfrak{n})} |a^\alpha - a^{-\alpha}|^{m(\alpha)}.$$

The proof is the same.

In some cases of the general theory of semisimple Lie groups, M is not necessarily finite, but is compact, so one has to be a little careful about the way K/M is related to $\bar{U}$, or at the Lie algebra level, $\mathfrak{k}/\mathfrak{m}$ is related to $\bar{\mathfrak{n}} = \mathbf{t}\mathfrak{n}$. The situation here meets that of the Gelfand–Naimark decomposition studied in Chapter V, involving the Harish-Chandra mapping

$$\psi_{M\backslash K}: \bar{U} \to M\backslash K.$$

At this point, we want to give priority to working out $\mathrm{SL}_n(\mathbf{R})$ over giving a full exposition of the general theory, so we stop here with the general comments.

VI, §2. FROM *K*-BI-INVARIANT FUNCTIONS ON *G* TO *W*-INVARIANT FUNCTIONS ON $\mathfrak{a}$

Let $G = \mathrm{SL}_n(\mathbf{R})$, although the exposition will be such that it is valid for a Lie group with an Iwasawa decomposition $G = UAK$ (K compact), and a Cartan and polar decomposition satisfying one more condition made explicit below.

The restriction map

$$\mathrm{res}_A : C(K\backslash G/K) \to C(A)$$

on the space of continuous functions actually maps a K-bi-invariant function f to a function f_A which is invariant under the Weyl group W, as follows directly from the definitions, i.e., $f_A \in C(A)^W$. We shall mostly view the restriction map on the usual spaces of test functions, so in particular

$$f \mapsto f_A \quad \text{gives a linear map} \quad C_c(K\backslash G/K) \to C_c(A)^W$$

or also a linear map on the K-bi-invariant C^∞ functions

$$\mathrm{res}_A : C_c^\infty(G)^{K,K} \to C_c^\infty(A)^W.$$

Note that the restriction map is injective, actually on the space of all K-bi-invariant functions, because if $f_A = 0$ then $f = 0$ since $G = KAK$.

As for the surjectivity, it is satisfied in the standard situation of semisimple Lie groups, and $\mathrm{SL}_n(\mathbf{R})$ in particular, on the spaces where we want it. For clarity, we formulate one more axiom, defining the polar decomposition to be **smooth**.

POL 3. The restriction map $C^\infty(G)^{K,K} \to C^\infty(A)^W$ is surjective, and so gives a linear isomorphism between these spaces.

Since K is assumed compact, this formulation is equivalent with the condition that the restriction gives a linear isomorphism

$$C_c^\infty(G)^{K,K} \to C_c^\infty(A)^W,$$

because functions with compact support correspond to each other. Furthermore, being C^∞ is a local condition, so multiplying a function by a K-bi-invariant C^∞ bump function in the neighborhood of a point, equal to 1 in some neighborhood of the point, and using a K-bi-invariant partition of unity, shows that the two versions of **POL 3**, with respect to all C^∞ functions, or with respect to those with compact support, are actually equivalent.

Theorem 2.1. *Let $G = \mathrm{SL}_n(\mathbf{R})$. The restriction map*

$$\mathrm{Fu}(G) \to \mathrm{Fu}(A)$$

induces a linear isomorphism

$$C^\infty(G)^{K,K} \xrightarrow{\approx} C^\infty(A)^W,$$

and similarly with C_c^∞ instead of C^∞.

Proof. For definiteness, let us deal with C^∞. The injectivity is clear. For the surjectivity, let $f \in C^\infty(A)^W$ be a smooth function invariant under W. Given an element k_1ak_2 of G, we want to define

$$F(k_1ak_2) = f(a).$$

The first problem is to show that F is well defined. Suppose $k_1ak_2 = b \in A$. It will suffice to prove that $f(a) = f(b)$, and for this it suffices to prove that a, b differ by conjugation with an element W. But taking $b\,{}^tb$, we find

$$k_1a^2k_1^{-1} = b^2,$$

so a^2, b^2 have the same eigenvalues, whence a^2, b^2 differ by conjugation with some $w \in W$, i.e. a permutation of the eigenvalues $(a_1, \ldots, a_n)$, thus proving that $f(a) = f(b)$, and F is well defined. It is essentially obvious that F is continuous, see (1) below, but the differentiability presents a problem.

Theorem 2.1 is an immediate consequence of a theorem of Glaeser [Gla 63]:

Theorem 2.2. *A function $f \in C^\infty(\mathfrak{a})^W$ is a C^∞ function of the elementary W-invariant polynomials.*

Dadok [Dad 82] gave two proofs of Theorem 2.1 (actually Theorem 2.3 below), one of which is reproduced in [Hel 84], Chapter II, Theorem 5.8. The second proof in Part 2 of his paper is done by approximation and skew invariant trigonometric polynomials, and proves Glaeser's theorem. We shall prove only Theorem 2.1, and we won't use Glaeser's theorem.

Furthermore, Coifman told us how to prove the C^∞ property still another way, by using a theorem of Bernstein characterizing C^∞ functions by close polynomial approximations. According to Coifman, this approximation theorem, actually a calculus lemma, and variants of it, are used to prove regularity theorems in PDE. In order not to interrupt the flow of the present section, this calculus result will be dealt with in an appendix to the section. It really belongs in basic texts in analysis. Here we give the details how the proof of Theorem 2.1 follows from the calculus lemma.

In the first place, let us return to the continuity property. By general topology, there is a quotient space

$$A \to A/W,$$

identifying points of A under the action of W, and this quotient is universal for continuous maps invariant under the action of W. Then for continuous functions, we have an isomorphism

$$C(A)^W \xrightarrow{\approx} C(A/W).$$

In particular, every continuous function on A/W lifts to a W-invariant continuous function on A, and also to a K-bi-invariant continuous function on G.

We have a topological isomorphism of coset spaces

(1) $$K\backslash G/K \to A/W.$$

Proof. The natural map is clearly continuous. The arguments at the beginning of the proof of Theorem 2.1 show that it is bijective, and the local compactness shows that it is bicontinuous, as was to be shown.

In particular, we obtain two linear isomorphisms

(2) $$C(K\backslash G/K) \xrightarrow{\approx} C(A)^W \xrightarrow{\approx} C(A/W).$$

Note that we can identify $C(K\backslash G/K)$ with $C(G)^{K,K}$ (K-bi-invariant continuous functions on G) because of the bicontinuous map in (1). However, $C^\infty(K\backslash G/K)$ really means $C^\infty(G)^{K,K}$, i.e. K-bi-invariant C^∞ functions on G, which we cannot identify with C^∞ functions on $(K\backslash G/K)$. The latter are not even defined because the space $(K\backslash G/K)$ has singularities. This entire section shows how to get around that complication. We shall prove a serviceable theorem, although not quite all of Glaeser's theorem.

We reduce Theorem 2.1 to a theorem on euclidean space as follows. Let $\mathbf{P} = \exp \mathfrak{p}$, with $\mathfrak{p} = \mathrm{Sym}_0$ (symmetric matrices with trace 0), so $\mathbf{P} = \mathrm{SPos}_n$. The map

(3) $$\mathbf{P} \times K \to \mathbf{P}K = G$$

is a differential isomorphism, as recalled in §1, **POL SSk**. Let $\mathfrak{a} = \mathrm{Lie}(A)$ be the space of diagonal matrices with trace 0, so $A = \exp \mathfrak{a}$. Let

$f \in C^\infty(A)^W$. Let F be the function on $\mathfrak{a}$ corresponding to f under the exponential map, that is

$$F = f \circ \exp.$$

Then $F \in C(\mathfrak{a})^W$, and we thereby have a linear isomorphism

$$C^\infty(A)^W \to C^\infty(\mathfrak{a})^W. \tag{4}$$

We also have a commutative diagram

$$\begin{array}{ccc} C^\infty(\mathbf{P})^{\mathfrak{c}(K)} & \xrightarrow{\circ\exp} & C^\infty(\mathfrak{p})^{\mathfrak{c}(K)} \\ {\scriptstyle \mathrm{res}}\downarrow & & \downarrow{\scriptstyle \mathrm{res}} \\ C^\infty(A)^W & \xrightarrow[\circ\exp]{} & C^\infty(\mathfrak{a})^W \end{array} \tag{5}$$

the vertical maps being restriction, and the horizontal maps being composition with the exponential. Both horizontal maps are linear isomorphisms since the exponential map is a differential isomorphism on $\mathfrak{p}$ and on $\mathfrak{a}$ (with their images). But from the **Pol SSk** decomposition (3), we have immediately a linear isomorphism

$$\mathrm{res}\colon C^\infty(G)^{K,K} \to C^\infty(\mathbf{P})^{\mathfrak{c}(K)}. \tag{6}$$

Therefore Theorem 2.1 is reduced to proving:

Theorem 2.3 (C^∞ version of Chevalley's Theorem). *The restriction map*

$$\mathrm{res}\colon C^\infty(\mathfrak{p})^{\mathfrak{c}(K)} \to C^\infty(\mathfrak{a})^W \tag{7}$$

is an isomorphism, and similarly for C_c^∞ instead of C^∞.

Proof. Of course the injectivity of the restriction map is obvious, so the content of the theorem lies in the surjectivity. We note that $\mathfrak{a}, \mathfrak{p}$ are finite dimensional vector spaces over $\mathbf{R}$, so Theorem 2.1 is reduced to a euclidean space property.

We follow Coifman's suggestion to apply the calculus lemma of Bernstein. Given $f \in C_c^\infty(\mathfrak{a})^W$, fix a positive integer d. Let S be a cube containing the support of f. There exists a constant $B(d, S)$ such that for all positive integers N there is a polynomial P_N of degree $\leqq N$ such that

$$\|f - P_N\|_{\infty,S} \leqq \frac{B(d,S)}{N^d}.$$

Since f is W-invariant, we can average over W and use the triangle inequality to get

$$\|f - \mathrm{Av}_W(P_N)\|_{\infty,S} \leqq \frac{B(d,S)}{N^d}.$$

The polynomial $\mathrm{Av}_W(P_N)$ is symmetric, and is therefore a polynomial in the elementary symmetric polynomials $I_1, \ldots, I_r$, of degree differing from $\deg P_N$ only by some constant factor. Let $f_{\mathfrak{p}}$ be the pull-back of f to $\mathfrak{p}$ and let $Q_{N,\mathfrak{p}}$ be the polynomial pulled back from $\mathrm{Av}_W(P_N)$. Then $Q_{N,\mathfrak{p}}$ is a polynomial in the coordinates of a symmetric matrix. In fact, if $\sigma_1, \ldots, \sigma_r$ are the coefficients of the characteristic polynomial $\det(tI - M)$ for a symmetric matrix M, then

$$Q_{N,\mathfrak{p}} = P_N(\sigma_1, \ldots, \sigma_r).$$

Let $S_{\mathfrak{p}}$ be the inverse image of S in $\mathfrak{p}$, so $S_{\mathfrak{p}}$ is compact. Then

$$\|f_{\mathfrak{p}} - Q_{N,\mathfrak{p}}\|_{\infty,S_{\mathfrak{p}}} \leqq \frac{B(d,S)}{N^d},$$

in other words, $\|f_{\mathfrak{p}} - Q_{N,\mathfrak{p}}\|_\infty$ satisfies the same sort of inequality on its support for all d. We can then apply the Bernstein calculus lemma in reverse direction to conclude that $f_{\mathfrak{p}}$ is C^∞, thus concluding the proof of the theorem.

Appendix. The Bernstein calculus lemma

We follow Coifman's exposition for this entire appendix. First we make some remarks in the analogous case of Fourier series, whose properties are covered in a standard way in advanced calculus courses. Suppose a function is C^∞ with compact support. Say the support is contained in the interval $(-\pi, \pi)$ so we can use the standard normalization of Fourier series. Say we are in one variable first to get the idea of the proof.

Lemma A1. *Let f be a continuous function on* **R** *with period 2π. The following conditions are equivalent.*

C^∞ 1. *The function f is C^∞.*

C^∞ 2. *Given a positive integer p, there is a constant $B(p) = B(p, f)$ such that for every positive integer N there is a trigonometric polynomial T_N of degree $\leqq N$ such that*

$$\|f - T_N\|_\infty \leqq \frac{B(p)}{N^p}.$$

C^∞ 3. *Same as C^∞ 2, except with the L^2-norm instead of the sup norm, with some constant $B'(p)$,*

$$\|f - T_N\|_2 \leqq \frac{B'(p)}{N^p}.$$

C^∞ 4. *Let $c_k = c_k(f)$ be the Fourier coefficient of f ($k \in \mathbf{Z}$). For every positive integer d, we have*

$$|c_k| = O(1/|k|^d) \qquad \textit{for } |k| \to \infty.$$

Proof. This is routine from the elementary theory of Fourier series. Let

$$\sum_{k \in \mathbf{Z}} c_k e^{ikx} = \sum_{k \in \mathbf{Z}} c_k \chi^k(x)$$

be the Fourier series of f, with Fourier coefficients c_k. Assume first that f is C^∞ Then given an integer p, integrating by parts p times, we have

$$|c_k| = O(k^{-p}) \qquad \text{for } k \to \infty.$$

Hence for each positive integer N there exists a trigonometric polynomial T_N (in the character χ and χ^{-1}, $\chi(x) = e^{ix}$) such that

$$\|f - T_N\|_\infty \leqq \frac{B(p)}{N^p}.$$

Furthermore, this inequality implies the L^2-inequality

$$\|f - T_N\|_2 \leqq \frac{B'(p)}{N^p},$$

with constants $B(p)$ and $B'(p)$.

Conversely suppose that the L^2-inequality is satisfied with some integer $p \geqq 3$, but the function f is only assumed continuous, say. The partial sums of the Fourier series give the best L^2-approximation to the function, so if $S_{f,N}$ denotes the N-th partial sum of the Fourier series of f, then

$$\|f - S_{f,N}\|_2 \ll \frac{1}{N^p} \qquad \text{so} \qquad \sum_{k=N+1}^{\infty} |c_k|^2 \ll \frac{1}{N^p}.$$

Therefore

$$|c_k| \ll |k|^{-p}.$$

In particular, the Fourier series can be differentiated term by term $p-2$ times, so the function f is of class C^{p-2}. This being true for every p, we conclude that f is C^∞. This concludes the proof.

The above argument is valid in arbitrary dimensions, except for a slightly more complicated notation. The Fourier series

$$\sum c_{(k)}\chi^{(k)} \qquad \text{with} \quad \chi^{(k)}(x) = e^{ik_1x_1}\cdots e^{ik_rx_r}$$

is indexed by r-tuples $(k_1, \ldots, k_r)$, with an r-tuple of variables $(x_1, \ldots, x_r)$. We have a bound for the number of monomials of total degree d, namely $\ll d^{r-1}$. Hence

$$\text{number of monomials of total degree} \ll d \text{ is } \ll d^r.$$

Then the above argument works since we may lose r derivatives, but letting p tend to infinity still shows that a continuous function with rapidly decreasing Fourier coefficients is C^∞ and conversely. For the record, we state the theorem in several variables.

Lemma A2. *Let f be a continuous function on $\mathbf{R}^r/2\pi\mathbf{Z}^r$. Then f is C^∞ if and only if it satisfies any one of the following equivalent conditions.*

(a) *For each positive integer d there is a constant $B(d) = B(d, f)$ such that for every positive integer N there is a trigonometric polynomial T_N of degree $\leqq N$ such that*

$$\|f - T_N\|_\infty \leqq \frac{B(d)}{N^d}.$$

(b) *Same as* (a) *except for the final inequality, which now reads for the L^2-norm,*

$$\|f - T_N\|_2 \leqq \frac{B'(d)}{N^d}.$$

(c) *Let $c_{(k)}$ be the Fourier coefficients of f, and let $|(k)| = \max|k_i|$ for $(k) \in \mathbf{Z}^r$. Then for every positive integer d we have*

$$|c_{(k)}| = O(1/|(k)|^d) \qquad \textit{for } |(k)| \to \infty.$$

Now we have to formulate and prove the above for ordinary polynomials.

Lemma A3 (Bernstein). *Let $f \in C(\mathbf{R}^r)$ be a continuous function on euclidean space. The following conditions are equivalent.*

(a) *The function f is C^∞.*

(b) *Given a finite cube S, and a positive integer d, there exists a constant $B(d, S) = B(d, S, f)$ having the following property. For every positive integer N there exists a polynomial P_N of degree $\leqq N$, such that*

$$\|f - P_N\|_{\infty,S} \leqq \frac{B(d, S)}{N^d}.$$

(c) *Same as in* (b), *except for the L^2-norm instead of the sup norm, that is, for some constant $B'(d, S)$,*

$$\|f - P_N\|_{2,S} \leqq \frac{B'(d, S)}{N^d}.$$

Proof. We begin with some general remarks. The property of being C^∞ is local. In particular, f is C^∞ if and only if for every function $h \in C_c^\infty(\mathbf{R}^r)$ the function fh is C^∞. Furthermore, it is sufficient that fh be C^∞ for all C^∞ bump functions h, i.e. functions with compact support, equal to 1 on a small ball centered at a point, and decreasing to 0 outside a ball of slightly larger radius.

After this remark, we come to the proof proper. The implication (b) $\Rightarrow$ (c) is trivial. We show that (a) $\Rightarrow$ (b) by using the result for the Fourier approximation. Given a cube S, by a dilation and a translation, we may reduce the desired inequality for a function on the ordinary cube of sides 2π, centered at the origin, and we may assume the function is periodic C^∞. By the Fourier series Lemma A2, for each positive integer M there is a trigonometric polynomial T_M such that

$$\|f - T_M\|_\infty \leqq \frac{B(2d)}{M^{2d}}.$$

We write

$$T_M = \sum_{|(k)| \leqq M} c_{(k)} \chi^{(k)}.$$

We use the Taylor series for e^{ix}, in one variable x, say with some constant C_0,

$$e^{ix} = E_M(x) + O(C_0^M/M!) \qquad \text{for } M \to \infty,$$

with an ordinary polynomial $E_M(x)$ of degree M. The estimate $O(C_0^M/M!)$ is of course for x in the fixed interval of length 2π. Then dealing with an r-tuple $x = (x_1, \ldots, x_r)$ we have

$$\begin{aligned}\chi^{(k)}(x) &= E(\pm x_1)^{|k_1|} \cdots E(\pm x_r)^{|k_r|} + O(C_1^M/M!) \\ &= E_{(k)}(x) + O(C_1^M/M!),\end{aligned}$$

where $\deg E_{(k)} \leqq M|k_1| + \cdots + M|k_r| \leqq rM^2$. Then

$$T_M = \sum_{|(k)| \leqq M} c_{(k)} E_{(k)} + \sum_{|(k)| \leqq M} c_{(k)} O\left(\frac{C^M}{M!}\right)$$

with a suitable constant C,

$$= P_{rM^2} + R(M)$$

where P_{rM^2} is a polynomial of degree $\leqq rM^2$, and $R(M)$ is the remainder sum on the right. But $|c_{(k)}| \ll 1/|(k)|^d$, and for d sufficiently large compared to r, the series

$$\sum \frac{1}{|(k)|^d}$$

is absolutely convergent, so

$$R(M) = O\left(\frac{C^M}{M!}\right) \qquad \text{for } M \to \infty.$$

Thus finally for each positive integer M we get a polynomial P_{rM^2} of degree $\leqq rM^2$ such that

$$\|f - P_{rM^2}\|_{\infty,S} \leqq \frac{B''(2d)}{M^{2d}}.$$

Given a positive integer N, we let M be the largest integer such that $rM^2 \leqq N$. Then we see that inequality (b) is satisfied, thus concluding the proof that (a) $\Rightarrow$ (b).

Finally, we have to show that (c) $\Rightarrow$ (a) Thus we have to go back from an L^2-estimate to some sort of Fourier series estimate. We have to understand better the relation between Fourier approximation and polynomial approximation.

For this, making a dilation, we may assume without loss of generality that the cube is $[-1, 1]^r$. For simplicity, let us discuss first the case $r = 1$. Instead of using the ordinary Fourier series, we want to use an orthonormalization of the polynomials

$$1, x, x^2, \ldots, x^n, \ldots$$

over this interval, resulting in a sequence of polynomials ψ_n of degree n. For a continuous function f, suppose (c) is satisfied, so we have bounds on the approximation by polynomials. Let

$$\mathrm{pr}_N(f) = \sum_{n=0}^{N} \langle f, \psi_n \rangle \psi_n = \sum_{n=0}^{N} c_n \psi_n$$

be the partial sum of the polynomial L^2-orthonormal series, i.e. the L^2-expansion of f in terms of the above orthonormal sequence. Since the orthonormal expansion gives the closest L^2-approximation to the function, we conclude from (c) that

$$\| \mathrm{pr}_N(f) - \mathrm{pr}_{N-1}(f) \|_{2,s} \leqq 2\frac{B(d)}{N^d}.$$

Therefore the Fourier coefficients are rapidly decreasing, that is

$$(*) \qquad |c_n| = |\langle f, \psi_n \rangle| \leqq 2\frac{B(d)}{n^d}.$$

We now want to conclude that f is C^∞. To do this we choose the orthogonal polynomials in a way which relates directly to the ordinary Fourier series expansion. The polynomials we get are the Tchebychev polynomials, but no previous knowledge of these polynomials will be necessary.

For simplicity, we deal first with one variable. By a dilation (rescaling) and a translation, we may assume that the support of f is in a closed subinterval of $(0, 1)$, say $[\frac{1}{2}, \frac{3}{4}]$, and then we make f even on $[-1, 1]$ so we can use the Fourier approximation in terms of $\cos n\theta$ with integers $n \geqq 0$. Hypothesis (c) is satisfied with $[-A, A] = [-1, 1]$, and the usual polynomial variable x.

Now the substitution $x = \cos\theta$ is a C^∞ change of charts on the interval, except at the end points which don't matter because for our purposes we have a function with support in $[-\frac{3}{4}, \frac{3}{4}]$. We then have

$$d\theta = \frac{dx}{(1-x^2)^{1/2}}.$$

Trivially, there exists a polynomial R_n of degree n such that

$$\cos n\theta = R_n(\cos\theta) = R_n(x).$$

Quick proof: Let $u = e^{i\theta}$ so $\cos\theta = \frac{1}{2}(u + u^{-1})$. Let $v = u + u^{-1}$ and $w = u - u^{-1}$. Note that $w^2 = v^2 - 4$ so even powers of w are polynomials in v. Hence all polynomials in u, u^{-1} are of the form $P_1(v) + P_2(v)w$, whence a polynomial $P(u, u^{-1})$ invariant under the interchange $u \mapsto u^{-1}$ is actually a polynomial in v, of the same degree.

The functions $\cos n\theta$ $(n \geqq 0)$ form a complete orthogonal system for even functions on $[-\pi, \pi]$, and hence the polynomials $R_n(x)$ form a complete orthogonal system for even functions on $[-1, 1]$ with respect to the measure

$$d\mu(x) = \frac{dx}{(1-x^2)^{1/2}}.$$

Furthermore, if

$$f(\cos\theta) = F(\theta) = \sum_{n=0}^{\infty} c_n \cos n\theta$$

is the (not necessarily convergent) Fourier series of F, then the polynomial Fourier series of f has the same Fourier coefficients, that is

$$f(x) = \sum_{n=0}^{\infty} c_n R_n(x).$$

Indeed, with the constant factor π for $n \neq 0$ and 2π for $n = 0$, we have

$$(\pi \text{ or } 2\pi)c_n = \int_{-\pi}^{\pi} F(\theta)\cos(n\theta)\,d\theta = \int_{-1}^{1} f(x)R_n(x)\,d\mu(x).$$

Therefore condition (c) implies that in terms of the θ-variable, the ordinary Fourier series has ordinary Fourier coefficients tending rapidly to 0, and hence the function is C^∞ by Lemma A1.

The proof in several variables follows in exactly the same way, with multi-indices, using Lemma A2.

VI, §3. PULLING BACK CHARACTERS AND SPHERICAL FUNCTIONS TO $\mathfrak{a}$

The smooth polar decomposition of G allows us to pull back systematically K-bi-invariant functions on G and W-invariant functions on A to $\mathfrak{a}$, by means of the exponential map

$$\exp\colon \mathfrak{a} \to A,$$

which is a differential isomorphism. If f is a K-bi-invariant function on G or a function on A, we let $f_{\mathfrak{a}}$ be its pull-back to $\mathfrak{a}$, we shall be principally concerned with pulling back characters and spherical functions, especially the Harish-Chandra spherical kernel. Ultimately we want to show that the inversion axioms of Chapter IX are satisfied.

Given a character ζ in $\mathfrak{a}_{\mathbf{C}}^{\vee}$ so an element of the dual space of $\mathfrak{a}_{\mathbf{C}}$, let χ_ζ be the character of A defined by

$$\chi_\zeta(a) = a^\zeta.$$

Then the spherical function $\varphi(\chi, x)$ pulls back on $\mathfrak{a}$ to a function $\Phi(\zeta, H)$ defined by

$$\Phi(\zeta, H) = \varphi(\chi_\zeta, \exp H).$$

A given Haar measure dx on G decomposes according to Theorem 1.5. We let $J(a)$ be its Jacobian factor on A, and **define** $\nu(H)$ by

$$\nu(H)\,dH = J(a)\,d^*a$$

where d^*a is our chosen Haar measure on A, sometimes denoted da. By Theorem 1.6 on $\mathfrak{a}^+$ we have

$$\nu(H) = \prod_\alpha (e^{\alpha(H)} - e^{-\alpha(H)}).$$

We may now define the **Φ_ν-transform** of a function $f \in C_c^\infty(\mathfrak{a})$ by

$$(\Phi_\nu f)(\zeta) = \int_{\mathfrak{a}} \Phi(\zeta, H) f(H) \nu(H)\,dH.$$

We are making this definition by integrating over all of $\mathfrak{a}$. Since the polar decomposition of G is a covering of degree $|W|$ outside a set of measure 0, it follows directly from the definitions that for $f \in C_c(K\backslash G/K)$, the spherical transform and the Φ_ν-transform are related by the formula

$$(1) \qquad (\mathbf{S}f)_{\mathfrak{a}}(\zeta) = |W|^{-1}\Phi_\nu f_{\mathfrak{a}}(\zeta) = \int_{\mathfrak{a}^+} \Phi(\zeta, H) f(H) \nu(H)\,dH.$$

All the formulas proved for $\varphi(\chi, x)$ hold for $\Phi(\zeta, H)$, *mutatis mutandis.*

For instance, $\Phi(\zeta, H)$ is W-invariant in H because $\varphi(\chi, a)$ is K-bi-invariant in a, and $\Phi(\zeta, H)$ is W-invariant in ζ by Chapter III, Theorem 5.2. By Chapter III, Proposition 7.3, we get that Φ is even:

$$\Phi(-\zeta, -H) = \Phi(\zeta, H).$$

We define

$$\Phi^-(\zeta, H) = \Phi(-\zeta, H) = \Phi(\zeta, -H).$$

Translating Theorem 5.6 of Chapter III in the present context, we get:

Theorem 3.1. *Let ζ, ζ' be elements of $\mathfrak{a}^\vee_{\mathbf{C}}$. Then $H \mapsto \Phi(\zeta, H)$ and $H \mapsto \Phi(\zeta', H)$ are equal if and only if there exists $w \in W$ such that $\zeta = w\zeta'$.*

Finally we deal with the multiplicative properties of the Φ_ν-operator. We continue with a smooth polar decomposition $G = KAK$. The Φ_ν-operator is not a multiplicative homomorphism with respect to the convolution product on $\mathfrak{a}$, but we can define another product, which we call the **G-product** on $C_c^\infty(\mathfrak{a})^W$ as follows. By Theorem 2.1, a function in this space is of the form $f_{\mathfrak{a}}$ with

$$f \in C_c^\infty(K\backslash G/K) = C_c^\infty(G)^{K,K}.$$

Let f^{xK} be the function already encountered in Chapter VII, §5, namely

$$f^{xK}(y) = \int_K f(xky)\,dk. \tag{2}$$

We first write down a product on A rather than $\mathfrak{a}$. For

$$f, g \in C_c^\infty(K\backslash G/K)$$

and $a \in A$, we let

$$(f_A \# g_A)(a) = \int_A f^{aK}(b^{-1})g(b)\,db. \tag{3}$$

Thus the inner integral over K gives a twist to the convolution product on A, to define the new product on A. Of course, we then can pull back further to the Lie algebra $\mathfrak{a}$. Let $H \in \mathfrak{a}$ and $a = \exp H$. We **define**

$$(f_{\mathfrak{a}} \# g_{\mathfrak{a}})(H) = (f_A \# g_A)(\exp H) = \int_{\mathfrak{a}} f_{\mathfrak{a}}^{aK}(-Z)g(Z)\nu(Z)\,dZ. \tag{4}$$

Theorem 3.2. *The Φ_ν and Φ_ν^- transforms are multiplicative homomorphisms from the G-product # to the ordinary product of functions, that is*

$$\Phi_\nu^-(f_{\mathfrak{a}} \# g_{\mathfrak{a}}) = \Phi_\nu^-(f_{\mathfrak{a}})\Phi_\nu^-(g_{\mathfrak{a}}).$$

In particular, the Φ_ν^--transform of $C_c^\infty(\mathfrak{a})^W$ is an algebra. Furthermore, putting $f_{\mathfrak{a}}^*(H) = \overline{f_{\mathfrak{a}}(-H)}$, on pure imaginary characters (in $\mathbf{i}\mathfrak{a}^\vee$) we have

$$\Phi_\nu^-(f_{\mathfrak{a}}^*) = \overline{\Phi_\nu^-(f_{\mathfrak{a}})}.$$

Proof. The first formula is immediate from Theorem 6.3 of Chapter IV, and the definitions. Note that we do not divide by $|W|$ in (3) and (4), and so

$$(f * g)(a) = |W|^{-1}(f_A \# g_A)(a).$$

Using (1) and the above formula, we see that a factor $|W|^{-1}$ occurs twice, and this factor occurs twice in the product $\mathbf{S}f \cdot \mathbf{S}g$, thus giving the multiplicativity formula for Φ_ν. The second formula comes from the pull-back to $\mathfrak{a}_{\mathbf{C}}^\vee$ of Chapter III, Corollary 7.4.

From Chapter IV, Theorem 5.5 and Corollary 5.6, applied to the pull-back to the Lie algebra that we have just gone through, we obtain directly the behavior of the Φ_ν-transform under translation.

Theorem 3.3. *For $f \in C_c^\infty(\mathfrak{a})^W$ and $H \in \mathfrak{a}$, let τ_H be translation by H, $\tau_H(v) = v + H$, so $(f \circ \tau_H)(v) = f(v + H)$. Then for $\zeta \in \mathfrak{a}_{\mathbf{C}}^\vee$,*

$$(\Phi_\nu^-(f \circ \tau_H))(\zeta) = \Phi(\zeta, H)(\Phi_\nu^- f)(\zeta).$$

Finally, on $\mathfrak{a}$ we have both $L^1(dH)$ and $L^1(\nu\, dH)$ which we abbreviate by $L^1(\nu)$, as well as $L^2(\nu)$. The next theorem describes the quadratic form of the #-product at the origin.

Theorem 3.4. *For* $f_\mathfrak{a} \in C_c^\infty(\mathfrak{a})^W$ *we have*

$$(f_\mathfrak{a} \# f_\mathfrak{a}^*)(0) = \|f_\mathfrak{a}\|_{2,\nu}^2.$$

Proof. From the definitions,

$$\begin{aligned}(f_\mathfrak{a} \# f_\mathfrak{a}^*)(0) = (f * f^*)(e) &= \int_G f(y^{-1})\overline{f(y^{-1})}\, dy \\ &= \int_G f(y)\overline{f(y)}\, dy = \|f\|_2^2.\end{aligned}$$

Pulling back to $\mathfrak{a}$ with the Jacobian gives precisely $\|f_\mathfrak{a}\|_{2,\nu}^2$ on the right side (cf. (3)), thus concluding the proof.

VI, §4. LEMMAS USING THE SEMISIMPLE LIE IWASAWA DECOMPOSITION

For this section and the next we essentially follow [Har 58a], p. 266 et seq.

We continue to work with $G = \mathrm{SL}_n(\mathbf{R})$. Readers will note that the proofs work under the axiomatization of the polar decomposition and the Cartan involution. We use as before the eigenspace decomposition

$$\mathfrak{n} = \bigoplus_{\alpha \in \mathcal{R}(\mathfrak{n})} \mathfrak{n}_\alpha$$

which is a direct sum of the non-zero α-eigenspaces with a set of non-zero characters $\mathcal{R}(\mathfrak{n})$ on $\mathfrak{a}$, called the $(\mathfrak{a}, \mathfrak{n})$-**characters**. Cf. Chapter III, §2.

We let A' be the subgroup of regular elements in A, so the subgroup of elements b such that $b^\alpha \neq 1$ for all $\alpha \in \mathcal{R}(\mathfrak{n})$.

We shall need some formulas. We write an element $v \in \mathfrak{n}$ as a sum of its components v_α. Any $v \in \mathfrak{g}$ can be written as

$$v = u + w \qquad \text{with } u \in \mathrm{Sym} \text{ and } w \in \mathrm{Sk}$$

in the usual manner, with

$$u = \tfrac{1}{2}(v + {}^t v) \qquad \text{and} \qquad w = \tfrac{1}{2}(v - {}^t v).$$

These will be applied especially to

$$v_\alpha = u_\alpha + w_\alpha \in \mathfrak{n}_\alpha.$$

We shall use Chapter II, Propositions 6.4 and 6.7. Recall first especially that

$$\theta_{\mathfrak{g}} v = -{}^t v \quad \text{and} \quad \theta_* \tilde{v} = (\theta_{\mathfrak{g}} v)^{\sim} \quad \text{for all } v \in \mathfrak{g}.$$

For $v \in \mathfrak{g}$ and $b \in A$ we use the notation for conjugation

(1) $$\mathbf{c}(b)v = bvb^{-1} = v^{b^{-1}} \quad \text{so} \quad \mathbf{c}(b^{-1})v = v^b.$$

For $v_\alpha \in \mathfrak{n}_\alpha$ and $b \in A'$, we have

(2) $$\mathbf{c}(b)v_\alpha = bv_\alpha b^{-1} = b^\alpha v_\alpha.$$

We now get two relations:

(3) $$2w_\alpha = v_\alpha - {}^t v_\alpha$$

(4) $$2w_\alpha^b = b^{-\alpha} v_\alpha - b^\alpha\, {}^t v_\alpha.$$

Eliminating from (3) and (4), and defining

$$S_\alpha(b) = \tfrac{1}{2}(b^\alpha - b^{-\alpha}) \quad \text{and} \quad C_\alpha(b) = \tfrac{1}{2}(b^\alpha + b^{-\alpha})$$

$$g_\alpha(b) = \frac{b^{-\alpha}}{S_\alpha(b)}$$

we find

(5) $${}^t v_\alpha = g_\alpha(b) w_\alpha - S_\alpha(b)^{-1} w_\alpha^b,$$

Note that g_α is defined as an analytic function on A', but is not defined on the boundary. Also, our g_α is 2 times the g_α of Harish-Chandra. Formula (5) is significant by itself in a precise way, but it also shows more generally how ${}^t\mathfrak{n}$ can be expressed in terms of $\mathfrak{k}$ and conjugation, namely:

Lemma 4.1. *Let $b \in A'$. Then*

$$\theta_{\mathfrak{g}} \mathfrak{n} = {}^t\mathfrak{n} \subset \mathfrak{k}^b + \mathfrak{k} = \mathbf{c}(\mathbf{b}^{-1})\mathfrak{k} + \mathfrak{k},$$

$$\mathfrak{g} = \mathfrak{k}^b \oplus \mathfrak{a} \oplus \mathfrak{k} \quad \textit{because} \quad \mathfrak{g} = {}^t\mathfrak{n} \oplus \mathfrak{a} \oplus \mathfrak{k} = \theta_{\mathfrak{g}}\mathfrak{n} \oplus \mathfrak{a} \oplus \mathfrak{k}.$$

Thus we see that the $\mathfrak{n}$ in an Iwasawa decomposition of $\mathfrak{g}$ can be replaced by a conjugate of $\mathfrak{k}$. We may then rewrite (5) in the weaker form

(6) $$-\theta_{\mathfrak{g}} v_\alpha = {}^t v_\alpha \equiv g_\alpha(b) w_\alpha \mod \mathfrak{k}^b.$$

Lemma 4.2. *Let $v_\alpha \in \mathfrak{n}_\alpha$. Let $\theta x = {}^t x^{-1}$. Then for $b \in A'$,*

$$\theta_* \tilde{v}_\alpha = -({}^t v_\alpha)^{\sim} \equiv -g_\alpha(b) \tilde{w}_\alpha \mod \tilde{\mathfrak{k}}^b.$$

Proof. The first equality comes from Proposition 6.4(2) of Chapter II. Then using (6) above concludes the proof.

We let

$$\mathfrak{R} = \mathbf{R}[\ldots, g_\alpha, \ldots] = \text{ring generated over the reals by the functions } g_\alpha,\ \alpha \in \mathcal{R}(\mathfrak{n});$$

$$\mathfrak{R}_+ = \text{ideal generated by the functions } g_\alpha.$$

Lemma 4.3. *The ring* $\mathfrak{R}$ *and the ideal* $\mathfrak{R}_+$ *are stable under the action of* IDO(A).

Proof. For $v \in \mathfrak{a}$ we have by direct computation

$$\mathcal{D}(v)g_\alpha = -\alpha(v)(g_\alpha + \tfrac{1}{2}g_\alpha^2), \tag{7}$$

which gives both an explicit formula and a proof of the lemma.

Remark. We use the word "ring" to include the unit element, following current standard usage. Harish-Chandra's convention is different, and his $\mathfrak{R}$ is our $\mathfrak{R}_+$.

VI, §5. THE TRANSPOSE IWASAWA DECOMPOSITION AND POLAR DIRECT IMAGE

We shall first go through a formalism analogous to that of Chapter V. We get decompositions where the transpose ${}^t\mathfrak{n}$ or $\theta_\mathfrak{g}\mathfrak{n}$ plays a role instead of $\mathfrak{n}$, and also a conjugation of $\mathfrak{k}$ instead of $\mathfrak{n}$.

Following our general policy not to spend time on generalities if it takes time, we assume that $G = \mathrm{SL}_n(\mathbf{R})$. However, we basically follow Harish-Chandra [Har 58a], pp. 265–268.

We shall give various proofs by induction again, and we recall the notation that IDO$(G)_q$ is the space of invariant differential operators of degree $\leqq q$, so q is the filtration index. We continue with the notation where

$$\mathcal{D}(v) = \tilde{v}$$

is the invariant differential operator associated to a vector $v \in \mathfrak{g}$. For $b \in A$, pursuing notation as in Chapter II, we let:

$$\mathcal{J}(\mathfrak{k}^b, \mathfrak{k}) = \tilde{\mathfrak{k}}^b\, \mathrm{IDO}(G) + \mathrm{IDO}(G)\tilde{\mathfrak{k}}.$$

$$\mathcal{J}({}^t\mathfrak{n}, \mathfrak{k}) = ({}^t\mathfrak{n})^\sim\, \mathrm{IDO}(G) + \mathrm{IDO}(G)\tilde{\mathfrak{k}} = \mathcal{D}({}^t\mathfrak{n})\, \mathrm{IDO}(G) + \mathrm{IDO}(G)\mathcal{D}(\mathfrak{k}).$$

$$\mathcal{J}({}^t\mathfrak{n}, \mathfrak{k}) = \mathcal{J}(\theta_\mathfrak{g}\mathfrak{n}, \mathfrak{k}) = \theta_*(\tilde{\mathfrak{n}})\, \mathrm{IDO}(G) + \mathrm{IDO}(G)\tilde{\mathfrak{k}}.$$

Cf. Chapter II, §6(1), and Proposition 6.4(2). Because we want to apply these previous results, it will be more relevant to express ourselves in terms of θ rather than in terms of $\mathbf{t}$.

Theorem 5.1. *We have a direct sum decomposition*

$$\mathrm{IDO}(G) = \mathcal{D}_G(S(\mathfrak{a})) \oplus \mathcal{J}({}^t\mathfrak{n}, \mathfrak{k}) = \mathcal{D}_G(S(\mathfrak{a})) + \mathcal{J}(\theta_{\mathfrak{g}}\mathfrak{n}, \mathfrak{k}).$$

For every $b \in A'$,

$$\mathcal{J}(\theta_{\mathfrak{g}}\mathfrak{n}, \mathfrak{k}) = \mathcal{J}({}^t\mathfrak{n}, \mathfrak{k}) \subset \mathcal{D}_G(S(\mathfrak{a})) + \mathcal{J}(\mathfrak{k}^b, \mathfrak{k}),$$

and also

$$\mathrm{IDO}(G) = \mathcal{D}_G(S(\mathfrak{a})) \oplus \mathcal{J}(\mathfrak{k}^b, \mathfrak{k}).$$

If $D \in \mathcal{J}(\mathfrak{k}^b, \mathfrak{k})$ and $f \in C_c^\infty(K\backslash G/K)$ is K-bi-invariant, then $Df(b) = 0$.

Proof. The first decomposition follows from Corollary 1.3 of Chapter V, using a basis of $\mathfrak{g}$ coming from bases of ${}^t\mathfrak{n}$, $\mathfrak{a}$, and $\mathfrak{k}$.

As to the second, we use the eigenspace decomposition $\mathfrak{n} = \sum \mathfrak{n}_\alpha$. By (6) of §4, each $v_\alpha \in \mathfrak{n}_\alpha$ is such that

$$\mathcal{D}({}^t v_\alpha) \equiv g_\alpha(b)\tilde{w}_\alpha \mod \tilde{\mathfrak{k}}^b. \tag{1}$$

Let $D_\alpha \in \mathrm{IDO}(G)$ have degree q. Then

$$\begin{aligned} \mathcal{D}({}^t v_\alpha)D_\alpha &\equiv g_\alpha(b)\tilde{w}_\alpha D_\alpha \mod \tilde{\mathfrak{k}}^b\,\mathrm{IDO}(G) \\ &\equiv g_\alpha(b)[\tilde{w}_\alpha, D_\alpha] \mod \mathcal{J}(\mathfrak{k}^b, \mathfrak{k}) \end{aligned} \tag{2}$$

with $g_\alpha(b)[\tilde{w}_\alpha, D_\alpha] \in \mathrm{IDO}(G)_{q-1}$. Inductively,

$$\mathcal{J}({}^t\mathfrak{n}, \mathfrak{k}) \subset \mathcal{D}_G(S(\mathfrak{a})) + \mathcal{J}(\mathfrak{k}^b, \mathfrak{k}),$$

which proves the inclusion $\mathrm{IDO}(G) \subset \mathcal{D}_G(S(\mathfrak{a})) + \mathcal{J}(\mathfrak{k}^b, \mathfrak{k})$, and hence equality.

Let $f \in C_c^\infty(K\backslash G/K)$. If $D \in \mathrm{IDO}(G)\tilde{\mathfrak{k}}$ then trivially $Df = 0$. Next, suppose $D = \tilde{w}^b E$ for some $E \in \mathrm{IDO}(G)$, $w \in \mathfrak{k}$. Write $\tilde{w}^b = \mathbf{c}(b^{-1})\tilde{w}$. Then

$$\begin{aligned} (\mathbf{c}(b^{-1})\tilde{w})(Ef) &= (L(b^{-1})R(b)\tilde{w})Ef \\ &= (R(b)\tilde{w})Ef \\ &= R(b)(\tilde{w}(R(b^{-1})(Ef))), \end{aligned}$$

which evaluated at b is equal to $(\tilde{w}(R(b^{-1})(Ef)))(e) = 0$, because $R(b^{-1})(Ef)$ is left K-invariant, so we get 0 at e whence $(Df)(b) = 0$.

The direct sum property then follows, because if a differential operator D is in the intersection $\mathcal{D}_G(S(\mathfrak{a})) \cap \mathcal{J}(\mathfrak{k}^b, \mathfrak{k})$ and $Df(b) = 0$ for all $f \in C_c^\infty(K\backslash G/K)$, then $D = 0$ since there exist functions f having arbitrarily prescribed Taylor series at b. This concludes the proof of the theorem.

We stated the theorem with a dependence on the chosen element b. However, the proof actually yields a stronger inductive result uniformly on A', emphasizing another feature of the situation. The ring $\mathfrak{R}$ has a direct sum decomposition

$$\mathfrak{R} = \mathbf{R} \oplus \mathfrak{R}_+,$$

of the constants and the ideal $\mathfrak{R}_+$. We then get a direct sum decomposition of the polynomial algebra with extended coefficients,

$$\mathfrak{R}S(\mathfrak{a}) = S(\mathfrak{a}) \oplus \mathfrak{R}_+ S(\mathfrak{a}).$$

We may then apply the Harish-Chandra maps $\mathcal{D}_G$ or $\mathcal{D}_A$. For instance, we obtain a direct sum decomposition

$$\mathfrak{R}\mathrm{IDO}(A) = \mathrm{IDO}(A) \oplus \mathfrak{R}_+\mathrm{IDO}(A),$$

about which we say more later. For the moment, we stay on G.

Given $P \in \mathfrak{R}S(\mathfrak{a})$ with extended coefficients, let $E = \mathcal{D}_G(P)$. We can write

$$P = P_0 + P_+$$

where P_0 has constant coefficients, and P_+ has coefficients in $\mathfrak{R}_+$, so

$$P_+ = \sum_{j \geqq 1} \psi_j P_j \qquad \text{with } \psi_j \in \mathfrak{R}_+ \text{ and } P_j \in S(\mathfrak{a}).$$

Putting $E = \mathcal{D}_G(P)$, say (the analogous statements hold also for $\mathcal{D}_A(P)$), we then let

$$E_+(b) = \sum_{j \geqq 1} \psi_j(b) \mathcal{D}_G(P_j) \qquad \text{and} \qquad E(b) = \mathcal{D}_G(P_0) + E_+(b)$$

be the differential operator with constant coefficients, evaluating the functions ψ_j at b. This evaluation is independent of the representation of P_+ as a sum $\sum \psi_j P_j$, and is called the **(polynomial) expression of** E at b.

Theorem 5.2. *Given $D \in \mathrm{IDO}(G)$, there exists a unique $P \in \mathfrak{R}S(\mathfrak{a})$ such that putting $E = \mathcal{D}_G(P)$, for all $b \in A'$, we have*

$$D \equiv E(b) \quad \mathrm{mod}\, \mathcal{J}(\mathfrak{k}^b, \mathfrak{k}).$$

Proof. Existence will be proved in Lemma 5.3. The uniqueness of $E(b)$ for each $b \in A'$ comes from the direct sum decomposition of Theorem 5.1. The uniqueness of the element E is then due to the analyticity.

Actually, we have a two-tier decomposition. From Chapter II, Proposition 6.8, there is a unique $P_0 \in S(\mathfrak{a})$ such that

(3) $$D \equiv \mathcal{D}_G(P_0) \quad \mathrm{mod}\, \theta_*(\tilde{\mathfrak{n}})\, \mathrm{IDO}(G) + \mathrm{IDO}(G)\tilde{\mathfrak{k}},$$

so we have already the constant coefficient part of the desired operator. We now want to catch the part with coefficients in $\mathfrak{R}_+$.

Lemma 5.3. *Let $D' \in \theta_*(\tilde{\mathfrak{n}})\, \mathrm{IDO}(G)_d$. Then there exists a unique $P_+ \in \mathfrak{R}_+ S(\mathfrak{a})_{d-1}$ such that for all $b \in A'$,*

$$D' \equiv \mathcal{D}_G(P_+)(b) \quad \mathrm{mod}\, \mathcal{J}(\mathfrak{k}^b, \mathfrak{k}).$$

Proof. We return to formulas (1) and (2) of Theorem 5.1. We select a basis v_α for each $\mathfrak{n}_\alpha$. We may write

$$D' \equiv \sum \mathcal{D}(\theta_{\mathrm{g}} v_\alpha) D_\alpha \quad \mathrm{mod}\, \mathcal{J}(\mathfrak{k})$$

with some $D_\alpha \in \mathrm{IDO}(G)_{d-1}$. Hence by Theorem 5.1, (1) and (2),

(4) $$D \equiv \sum_\alpha -g_\alpha(b)[\tilde{w}_\alpha, D_\alpha] \quad \mathrm{mod}\, \mathcal{J}(\mathfrak{k}^b, \mathfrak{k}).$$

Each $[\tilde{w}_\alpha, D_\alpha]$ has degree $\leqq d-1$, and we can apply induction to the operators $[\tilde{w}_\alpha, D_\alpha]$ to conclude the proof of the lemma, which also proves Theorem 5.2.

We then obtain an additional projection from the direct sum decompositions of Theorems 5.1 and 5.2. Indeed, we define the **polar projection**

$$\mathbf{p}_{A'} \colon \mathrm{IDO}(G) \to \mathfrak{R}\mathrm{IDO}(A) \subset \mathrm{DO}^{\mathrm{an}}(A')$$

into the algebra of differential operators with analytic coefficients on A', by the formula

$$\mathbf{p}_{A'}(D) = \mathcal{D}_A(P) = \mathcal{D}_A(P_0) + \mathcal{D}_A(P_+)$$

where P_+ is the polynomial of Lemma 5.3. In terms of the more precise degree description of Lemma 5.3, we have

$$(5) \qquad \mathbf{p}_{A'}(D) = E_0 + \sum_{j \leqq 1} \psi_j E_j \quad \text{with } \psi_j \in \mathfrak{R}_+ \text{ and } \deg E_j < \deg D.$$

Here we put $E_0 = \mathcal{D}_A(P_0)$, and $E_j = \mathcal{D}_A(P_j)$ in the notation before Theorem 5.2. From §1 we have the polar decomposition, and thus its projection

$$\mathrm{Pol}_{A^+} \colon G^+ = KA^+K \to A^+.$$

Then we get the direct image on differential operators

$$(\mathrm{Pol}_{A^+})_* \colon \mathbb{D}\mathrm{O}(G)^K \to \mathrm{DO}^{\mathrm{an}}(A^+)$$

as defined in Chapter II. This direct image is sometimes called the **polar radial component** of the differential operator.

Theorem 5.4. *For $D \in \mathbb{D}\mathrm{O}(G)^K$ we have*

$$(\mathrm{Pol}_{A^+})_*(D) = \mathbf{p}_{A^+}(D).$$

In other words, $\mathbf{p}_{A^+}$ is the polar direct image of the differential operator on A^+.

Proof. This is immediate from the definition of the direct image, the annihilation statement of Theorem 5.1, and Lemma 5.3.

Finally we relate the polar direct image with the Harish-Chandra mapping, involving conjugation by $\delta^{-1/2}$. This is Lemma 16 of [Har 58a].

Proposition 5.5. *Let $D \in \mathbb{D}\mathrm{O}(G)^K$ have degree d. Then*

$$\mathbf{p}_{A'}(D) \equiv \delta^{-1/2}\mathbf{h}(D) \circ \delta^{1/2} \mod \mathfrak{R}_+ \mathbb{D}\mathrm{O}(A)_{d-1}.$$

More precisely, there exists $E_+ \in \mathfrak{R}_+ \mathbb{D}\mathrm{O}(A)_{d-1}$ such that

$$\mathbf{p}_{A'}(D) = \delta^{-1/2}\mathbf{h}(D) \circ \delta^{1/2} + E_+ \quad \text{on } A'.$$

Proof. This comes directly from Proposition 6.8 of Chapter II, which tells us that

$$\mathcal{D}_G(P_0) = \delta^{-1/2}\mathbf{h}(D) \circ \delta^{1/2}$$

The definition of $\mathbf{p}_{A'}(D)$ as $\mathcal{D}_A(P_0) + \mathcal{D}_A(P_+)$ concludes the proof.

VI, §6. *W*-INVARIANTS

Although there is no simple characterization of the image of $\mathbf{p}_A$, known at present, still Proposition 5.5 combined with the known image of $\mathbf{h}$ itself, namely $\mathrm{IDO}(A)^W$, is enough to prove some facts about this image, by reducing the proofs to $\mathbf{h}$, and we now list some of these, as in [Har 58a], Corollary of Theorem 2, §6. The main result here is the upper bound for an eigenspace of $\mathbf{p}_{A'}$ on some open connected subset of A', which in practice will be in A^+. The upper bound will be combined with Theorem 3.1 in Chapter VIII to get a basis for this space.

We recall that $S(\mathfrak{a})$ is free of dimension $|W|$ over the Weyl group invariants $S(\mathfrak{a})^W$. Let $\{E_i\}$ $(i = 1, \ldots, |W|)$ be a basis of $\mathrm{IDO}(A)$ over $\mathrm{IDO}(A)^W$, so we have

$$\text{(1)} \qquad \mathrm{IDO}(A) = \sum_{i=1}^{|W|} \mathrm{IDO}(A)^W E_i \quad \text{(direct).}$$

We remind the reader that $\mathfrak{R}$ is the ring generated by the functions g_α (cf. §1). By definition, the ring has the unit element.

Recall that conjugation

$$E \mapsto \delta^{-1/2} E \circ \delta^{1/2} \quad \text{which we abbreviate } E^{\delta^{1/2}}$$

is an automorphism of $\mathrm{IDO}(A)$. Since $\mathrm{IDO}(A)^W$ is the image of $\mathbf{h}$ by Theorem 4.3 of Chapter VI, we get

$$\text{(2)} \qquad \mathrm{IDO}(A) = \sum_{i=1}^{|W|} (\mathrm{Im}\,\mathbf{h})^{\delta^{1/2}} E_i^{\delta^{1/2}}.$$

It is slightly advantageous to use a basis which is lifted from a vector space $\mathcal{H}$ over the constants, and this vector space is stable under the Harish conjugation.

Example as in Helgason [Hel 84], Chapter IV, Proposition 5.1. Let Har_W be the vector space of harmonic polynomials, and

$$\mathrm{HDO}(A) = \mathcal{D}_A(\mathrm{Har}_W) = \widetilde{\mathrm{Har}}_W$$

the corresponding space of invariant differential operators, called the **harmonic differential operators**. Then $\mathrm{HDO}(A)$ is stable under the Harish conjugation.

Let $\mathcal{H}$ be a vector space as above and let $\{E_i\}$ $(i = 1, \ldots, |W|)$ be a basis of $\mathcal{H}$ over $\mathbf{R}$. Then the linear combination of §2 yields

$$\text{(3)} \qquad \mathrm{IDO}(A) = \sum (\mathrm{Im}\,\mathbf{h})^{\delta^{1/2}} E_i,$$

or in other words,

$$\mathbb{D}\mathrm{O}(A) \approx (\mathrm{Im}\,\mathbf{h}) \otimes \mathcal{H}.$$

Theorem 6.1. *Let $\mathcal{H} = \mathrm{HDO}(A)$, or more generally a subspace of $\mathbb{D}\mathrm{O}(A)$ such that $\mathbb{D}\mathrm{O}(A) = \mathbb{D}\mathrm{O}(A)^W \otimes \mathcal{H}$ and $\mathcal{H}$ is stable under $\delta^{1/2}$-conjugation. Then*

$$\mathbb{D}\mathrm{O}(A) \subset \mathfrak{R}(\mathrm{Im}\,\mathbf{p}_{A'})\mathcal{H}.$$

Proof. Letting $\{E_i\}$ be as above, for any $E \in \mathbb{D}\mathrm{O}(A)$ there exist $D_i \in \mathbb{D}\mathrm{O}(G)^K$ such that by using Proposition 5.5,

$$E = \sum \mathbf{h}(D_i)^{\delta^{1/2}} E_i \equiv \sum E_i \mathbf{p}_{A'}(D_i) \mod \mathfrak{R}_+ \mathbb{D}\mathrm{O}(A)_{d-1}$$

if $d = \deg E$. One completes the proof by induction.

Let γ be a character of $\mathbb{D}\mathrm{O}(G)^K$. Let V be an open non-empty connected subset of A', and consider the γ-eigenspace:

$\mathrm{An}_V(\mathbf{p}_{A'}, \gamma)$ = vector space of analytic functions on V having eigencharacter γ.

By definition, $\mathrm{An}(\mathbf{p}_{A'}, \gamma)$ consists of those analytic functions φ such that

$$\mathbf{p}_{A'}(D)\varphi = \gamma(D)\varphi \qquad \text{for all } D \in \mathbb{D}\mathrm{O}(G)^K.$$

Theorem 6.2. *We have* $\dim \mathrm{An}_V(\mathbf{p}_{A'}, \gamma) \leqq |W|$.

Proof. Let $b \in V$. Suppose the dimension of the eigenspace is $> |W|$. Then there exists a function $\varphi \neq 0$ in $\mathrm{An}_V(\mathbf{p}_{A'}, \gamma)$ such that

$$(E_i\varphi)(b) = 0 \qquad \text{for } i = 1, \ldots, |W|.$$

By Theorem 6.1, we conclude that $(E\varphi)(b) = 0$ for all $E \in \mathbb{D}\mathrm{O}(A)$, whence $\varphi = 0$, thus proving the theorem.

Remark. Harish-Chandra developed the "ramified case" much further than the above. For a book exposition, see [GaV 88], Theorem 4.4.7, Corollaries 4.4.8 and 4.4.9. Because of various simplifications, this additional theory is not needed for the current development of the spherical inversion theorem, whether on $C_c^\infty(K\backslash G/K)$ or on the Harish-Chandra Schwartz space. Cf. Remark 1 following Theorem 2.8 of Chapter X.

CHAPTER VII

The Casimir Operator

To begin with, this chapter provides an example for an especially important invariant differential operator, and its direct image on A with respect to the Iwasawa decomposition and the polar decomposition. It is independent of Chapter V and is a direct continuation of Chapter VI.

For the first time in this book, duality on Lie(G) becomes important. We introduce a symmetric non-singular bilinear form which is G-invariant. In §1, we shall describe the two special realizations on $SL_n(\mathbf{R})$, namely the trace form and the Killing form. However, subsequent arguments do not depend on such special realizations, so we list clearly the two conditions under which all arguments eventually go through. It is sometimes easier to work with these conditions than with their specific ad hoc realizations, but we also don't want to spend extra space verifying various properties, so we shift to $SL_n(\mathbf{R})$ right away to avoid any lengthening of the exposition.

Harish-Chandra worked systematically with the Casimir operator, see especially [Har 58a], §6 and §7 which we cover in the present chapter. He used this operator to get a series expression for the spherical functions [Har 58a], to be treated in the next chapter. Helgason has preferred to deal with the Laplace operator [Hel 84], which *ipso facto* requires more foundational material on Riemannian geometry. Dealing with Casimir eliminates the need for such material, and we therefore prefer it here. Actually, the direct images to G/K of the Casimir operator and the Laplacian are equal, so readers can connect as they wish with the literature.

VII, §1. BILINEAR FORMS OF CARTAN TYPE

In Chapter VI, §1, we discussed a possible axiomatization of some conditions related to a semisimple Iwasawa decomposition, conditions **TR 1**–**TR 5**, involving either an anti-involution or an involution. Let us assume that we are in this situation, and as in Chapter VI, §1, use the symbols $\mathbf{t}$ and θ for these operations. We now put an additional structure of scalar products. A bilinear form B on $\mathfrak{g}$ will be said to be of **Cartan type** if it satisfies the following conditions.

BIL 1. The form B is non-singular, symmetric, $\mathbf{c}(G)$-invariant and $\mathbf{t}$-invariant.

This means that if $\mathbf{c}_x$ denotes conjugation by $x \in G$, and $v, w \in \mathfrak{g}$ then

$$B(\mathbf{c}_x v, \mathbf{c}_x w) = B(v, w)$$
$$\text{and } B(\mathbf{t}v, \mathbf{t}w) = B(v, w) \text{ or } B(\theta v, \theta w) = B(v, w).$$

One can use either $\mathbf{t}$ or θ, since on the Lie algebra, we have $\mathbf{t}_{\text{Lie}} = -\theta_{\text{Lie}}$ (putting the functorial index in the notation).

BIL 2. B is positive definite on $\mathfrak{p} = \text{Sym}$ and negative definite on $\mathfrak{k} = \text{Sk}$.

In the general case, Sym and Sk refer to the symmetric and skew-symmetric elements respectively, with respect to the operation $\mathbf{t}$ on $\mathfrak{g}$. Here we deal with the transpose, and describe the specific example of $\text{SL}_n(\mathbf{R})$. We let $\mathfrak{g}$ = Lie algebra of real $n \times n$ real matrices with trace 0. Define the **invariant trace form** $B = B_{\text{itr}}$ by the formula

$$B(v, w) = B_{\text{itr}}(v, w) = \text{tr}(vw).$$

where tr is the trace of a matrix (trace in the representation on $\mathbf{R}^n$). Define the **Killing form**

$$B_{\text{Ki}}(v, w) = \text{tr}(\text{Lie}(v)\,\text{Lie}(w)).$$

The trace is the usual trace of an endomorphism of a finite dimensional vector space, in this case $\mathfrak{g}$, and $\text{Lie}(v)$ is the regular representation of v in the Lie algebra, denoted by $\text{ad}(v)$ in the Lie industry.

Proposition 1.1. *The invariant trace form and the Killing form are of Cartan type, i.e. satisfy* **BIL 1** *and* **BIL 2**.

Proof. This is immediate, because the trace is invariant under conjugation.

Let $\mathfrak{g}_{\mathbf{C}}$ be the complexification of $\mathfrak{g}$, that is, $\mathbf{C} \otimes_{\mathbf{R}} \mathfrak{g}$. Then the trace form and Killing form can be defined by the same formulas on $\mathfrak{g}_{\mathbf{C}}$ and Proposition 1.1 remains true for the action of $G(\mathbf{C})$ by conjugation. In other words, we can take $x \in \mathrm{SL}_n(\mathbf{C})$ instead of $\mathrm{SL}_n(\mathbf{R})$. Going to $\mathbf{C}$ has the technical advantage that every non-singular $n \times n$ matrix with distinct eigenvalues can be diagonalized by conjugation with an element of $\mathrm{SL}_n(\mathbf{C})$. Following standard practice, we write $\mathfrak{sl}_n(\mathbf{R})$ and $\mathfrak{sl}_n(\mathbf{C})$ for the Lie algebras of $\mathrm{SL}_n(\mathbf{R})$ and $\mathrm{SL}_n(\mathbf{C})$ respectively.

Proposition 1.2. *On $\mathfrak{sl}_n(\mathbf{C})$, the invariant trace form and Killing form differ by a constant factor. In fact, for $v, w \in \mathfrak{sl}_n(\mathbf{C})$ and $n \geqq 2$ we have*

$$B_{\mathrm{Ki}}(v, w) = 2nB(v, w) = 2n\,\mathrm{tr}(vw).$$

Proof. Let E_{ij} be the $n \times n$ matrix with ij-component 1 and all other components 0. Let $\mathfrak{a} = \mathrm{Lie}(A)$ be the space (Lie algebra) of diagonal matrices with trace 0. Let

$$u_i = E_{ii} - E_{i+1,i+1} \qquad \text{for } i = 1, \ldots, n-1.$$

Then $\mathfrak{a}$ (resp. $\mathfrak{a}_{\mathbf{C}}$) is the space generated by $u_1, \ldots, u_{n-1}$ over $\mathbf{R}$ (resp. $\mathbf{C}$). We have

$$\mathfrak{g} = \mathfrak{a} + \sum_{i \neq j} \mathbf{R}E_{ij} \qquad \text{and} \qquad \mathfrak{g}_{\mathbf{C}} = \mathfrak{a}_{\mathbf{C}} + \sum_{i \neq j} \mathbf{C}E_{ij}.$$

For $u \in \mathfrak{a}_{\mathbf{C}}$ let u_{ii} ($i = 1, \ldots, n$) be its diagonal components. Then

$$[u, E_{ij}] = (u_{ii} - u_{jj})E_{ij} = \alpha_{ij}(u)E_{ij}$$

with the usual characters α_{ij}. Then

$$\begin{aligned} B_{\mathrm{Ki}}(u, u) &= \mathrm{tr}((\mathrm{Lie}\, u)^2) = \sum_{i,j} (u_{ii} - u_{jj})^2 \\ &= 2n\,\mathrm{tr}(u^2) \quad \left[\text{because } \sum u_{ii} = 0\right] \\ &= 2nB(u, u). \end{aligned}$$

This proves the formula

$$B_{\mathrm{Ki}}(u, u) = 2nB(u, u) \tag{1}$$

for all diagonal matrices with trace 0. Let $\mathfrak{g}'_{\mathbf{C}}$ be the set of matrices in $\mathfrak{g}_{\mathbf{C}}$ with distinct eigenvalues. Then any element of $\mathfrak{g}'_{\mathbf{C}}$ is conjugate to a diagonal matrix, and by the invariance of the two forms under conjugation, this proves formula (1) for the matrices $v \in \mathfrak{g}'_{\mathbf{C}}$ and distinct diagonal elements. Since $\mathfrak{g}'_{\mathbf{C}}$ is dense in $\mathfrak{g}_{\mathbf{C}}$, the formula holds for all elements $v \in \mathfrak{g}_{\mathbf{C}}$. Then the proposition follows by polarization, i.e. applying formula (1) to $v + w$ instead of v, thereby concluding the proof.

We define the **twisted trace form** $B_{\mathfrak{t}}$ by the formula

$$B_{\mathfrak{t}}(v, w) = B(\mathbf{t}v, w) = B(v, \mathbf{t}w) = B_{\mathfrak{t}}(w, v),$$

so $B_{\mathfrak{t}}$ is also symmetric, and it is positive definite on all of $\mathfrak{g}$.

For the record, we tabulate the behavior of the matrices E_{ij} with respect to the twisted trace form (the positive definite form on $\mathfrak{g}$). We let:

$\mathfrak{g}^{(0)}$ = matrices with 0 diagonal components.

Proposition 1.3.

(a) $\mathfrak{g}^{(0)} = \mathfrak{a}^{\perp}$ *is the orthogonal complement of* $\mathfrak{a}$.

(b) *The* E_{ij} $(i \neq j)$ *form an orthonormal basis of* $\mathfrak{g}^{(0)}$.

Proof. Matrix multiplication, in particular for all indices i, j, k, l:

$$E_{ij}E_{kl} = \delta_{jk}E_{il}.$$

Warning. *The element* $u_i = E_{ii} - E_{i+1,i+1}$ *is not orthogonal to* u_{i+1} *for* $i = 1, \ldots, n-2$. Note here that SL_n introduces a complication over GL_n. Of course, one can pick orthogonal bases for $\mathfrak{sl}_n(\mathbf{R})$ in various ways, but no particular way seems more useful than others.

Finally, we note that the above specific computations are covered by much more general facts concerning Lie algebras. For instance, the uniqueness of the bilinear form up to a constant factor belongs to the following general context.

Let G be a group acting on a finite dimensional vector space V, so we have a representation $G \to \mathrm{Aut}(V)$. Fix a non-singular bilinear form

$$B_0 = \langle\ ,\ \rangle$$

on V. For any other bilinear form B there is a unique linear map $A: V \to V$ such that

$$B(v, w) = \langle v, Aw\rangle \qquad \text{for all } v, w \in V.$$

If B_0, B are G-invariant, then A is in $\mathrm{End}_G(V)$. Actually, B_0 induces a linear isomorphism of the vector space of bilinear forms on V with $\mathrm{End}(V)$,

$$\mathrm{Bil}(V) \to \mathrm{End}(V) \qquad \text{by } B \mapsto A,$$

and an isomorphism of the G-invariant subspaces

$$\mathrm{Bil}_G(V) \to \mathrm{End}_G(V).$$

Proposition 1.4. *If the representation $G \to \mathrm{End}(V)$ is irreducible, and if there is one non-zero G-invariant bilinear form on V, then it is uniquely determined up to a scalar multiple, that is*

$$\dim \mathrm{Bil}_G(V) = 1.$$

Proof. By Schur's lemma, $\mathrm{End}_G(V)$ consists of the scalars, so the proposition drops out.

In the case of $G = \mathrm{SL}_n(\mathbf{R})$, Proposition 1.6 applies to the conjugation representation on the tangent space $\mathfrak{g} = \mathrm{Lie}(G)$ at the origin because of the standard theorem:

Theorem 1.5. *The conjugation representation of $\mathrm{SL}_n(\mathbf{R})$ on $\mathfrak{sl}_n(\mathbf{R})$ is irreducible.*

Within the present context, we make no use of such a result because we work concretely with a choice for the invariant form, and Proposition 1.4 computes the specific constant which relates the two possible natural choices in this concrete case. For the reader's convenience, we sketch a (computational) proof. We have formulas

$$E_{ij}E_{kl} = \delta_{jk}E_{il} \tag{2}$$

and so

$$[E_{ij}, E_{kl}] = \delta_{jk}E_{il} - \delta_{li}E_{kj}. \tag{3}$$

In particular, for $i \neq 1$,

$$[E_{ij}, E_{jl}] = E_{il}.$$

Let V be a G-conjugation invariant subspace of $\mathfrak{g}$. Decomposing with respect to the action of A (or the regular representation of $\mathfrak{a}$) and using the semisimplicity of $\mathfrak{sl}_n(\mathbf{R})$, we conclude that V is a direct sum of spaces generated by some E_{ij} and some diagonal matrices. As to the former, if $E_{ij} \in V$, then so is $xE_{ij}x^{-1}$ for all $x \in G$. Take

$$x = x(t) = \exp(tE_{jl}), \qquad \text{with } i \neq l,$$

take d/dt, and evaluate at $t = 0$. By (7) we get $\pm E_{il}$, which therefore lies in V. A similar argument shows that $E_{kj} \in V$ for all k. Thus V contains all the spaces $\mathfrak{n}_\alpha$ for all α. One argues similarly to get diagonal matrices, namely for $i \neq j$,

$$[E_{ij}, E_{ji}] = H_{ij}$$

(cf. Lemma 3.2 below).

VII, §2. THE CASIMIR DIFFERENTIAL OPERATOR

We start with some multilinear algebra. Let V be a finite dimensional vector space over a field of characteristic $\neq 2$ (because we are going to use symmetric bilinear forms and quadratic forms and we don't want to keep track of denominators). We then have the symmetric algebra $S(V)$ as well as for the dual space $S(V^\vee)$. Let B be a non-singular symmetric bilinear form on V. Then B induces an isomorphism

$$V \xrightarrow{\approx} V^\vee \quad \text{and therefore an isomorphism} \quad S(V) \xrightarrow{\approx} S(V^\vee) = \mathrm{Pol}(V).$$

Let $\{v_1, \ldots, v_N\}$ be a basis of V and let $\{\lambda_1, \ldots, \lambda_N\}$ be the dual basis in $V^\vee$. Then we can define the **B-dual basis of** V, namely $\{v'_1, \ldots, v'_N\}$ such that

$$B(v_i, v'_j) = \delta_{ij}.$$

There is a unique linear isomorphism

$$V \otimes V^\vee \to \mathrm{End}(V) \tag{1}$$

such that for $v \in V$ and $\lambda \in V^\vee$, the corresponding element $h_{v,\lambda} \in \mathrm{End}(V)$ is given by

$$h_{v,\lambda}(w) = \lambda(w)v.$$

In the tensor product $V \otimes V^\vee \approx \mathrm{End}(V)$, the element corresponding to the identity in $\mathrm{End}(V)$ is given in terms of dual bases $\{v_1, \ldots, v_N\}$ and $\{\lambda_1, \ldots, \lambda_N\}$ by

$$\sum v_i \otimes \lambda_i.$$

Thus all such elements are equal in the tensor product $V \otimes V^\vee$. Using the linear isomorphism $V \to V^\vee$ induced by B, it follows that if $\{v_i\}, \{v'_i\}$ and $\{w_i\}, \{w'_i\}$ are pairs consisting of a basis and its B-dual basis, then in the tensor product $V \otimes V$ we have

$$\sum v_i \otimes v'_i = \sum w_i \otimes w'_i.$$

A basis determines its B-dual basis uniquely. Furthermore, we can pick $w_i = v_i'$ and $w_i' = v_i$ (interchanging v_i, v_i') so that in particular

$$\sum v_i \otimes v_i' = \sum v_i' \otimes v_i.$$

Proposition 2.1. *The element in the tensor algebra*

$$\sum v_i \otimes v_i'$$

is independent of the choice of basis $\{v_1, \ldots, v_N\}$. *The element of the symmetric algebra*

$$Q_B = \sum_{i=1}^{N} v_i v_i'$$

is independent of the choice of basis.

Proof. The first statement has been proved above. The second statement is then immediate in the symmetric algebra, which is a homomorphic image of the tensor algebra.

Remark. The element Q_B represents the quadratic form associated with B under the isomorphism of V and $V^\vee$ induced by B. But this interpretation will not be needed. In any case, one may call Q_B in Proposition 2.1 the **Casimir polynomial.**

We return to the Lie algebra of a Lie group G, with a bilinear form B of Cartan type on the Lie algebra $\mathfrak{g}$, which we take as our vector space V. From Chapter V we recall the **Harish-Chandra linear isomorphism**

$$\mathcal{D}: S(\mathfrak{g}) \to \mathbb{D}\mathrm{O}(G)$$

which to each vector $v \in \mathfrak{g}$ associates the Lie derivative $\tilde{v} = \mathcal{D}(v)$, then extends to the tensor algebra, and finally to the symmetric algebra by symmetrization. We define the **Casimir operator** to be

$$\omega = \omega_B = \mathcal{D}(Q_B) = \tilde{Q}_B \in \mathbb{D}\mathrm{O}(G).$$

As we shall see next, the Casimir operator actually lies in $\mathbb{D}\mathrm{O}(G)^G$.

Proposition 2.2.

(a) *We have* $Q_B \in S(\mathfrak{g})^{c(G)}$, *in other words,* Q_B *is invariant under conjugation by* G, *and* $\omega_B \in \mathbb{D}\mathrm{O}(G)^G$.

(b) *For any basis* $\{v_i\}$ *of* $\mathfrak{g}$ *with* B*-dual basis* $\{v_i'\}$, *we have*

$$\omega_B = \sum \tilde{v}_i \tilde{v}_i' = \sum \mathcal{D}(v_i)\mathcal{D}(v_i').$$

Proof. That Q_B is $\mathbf{c}(G)$-invariant follows from the assumption that B is $\mathbf{c}(G)$-invariant and the natural isomorphism $V \xrightarrow{\approx} V^\vee$ induced by B. The invariance of the Casimir operator then follows from Proposition 1.5 of Chapter V, namely the fact that the Harish isomorphism commutes with the conjugation action of G. This proves (a). As to (b), first observe that the equality in the tensor product in Proposition 2.1 implies the equality in the algebra of differential operators

$$\sum \tilde{v}_i \tilde{v}_i' = \sum \tilde{v}_i' \tilde{v}_i.$$

Therefore the Harish symmetrization applied to the polynomial

$$Q_B \in S_2(V)$$

yields

$$\begin{aligned} \omega_B = \mathcal{D}(Q_B) = \tilde{Q}_B &= \frac{1}{2}\sum (\tilde{v}_i \tilde{v}_i' + \tilde{v}_i' \tilde{v}_i) \\ &= \sum \tilde{v}_i \tilde{v}_i'. \end{aligned}$$

This concludes the proof.

VII, §3. THE *A*-IWASAWA AND HARISH-CHANDRA DIRECT IMAGES

In this section, we apply Proposition 2.2. We use a semisimple basis of the usual type, adapted to the semisimple decomposition of the Lie algebra under $\mathfrak{a}$-action. To avoid any lengthening of the essential features, we remain concrete, and so we let:

$G = \mathrm{SL}_n(\mathbf{R})$;

$B =$ invariant trace form.

We shall compute systematically direct images of the Casimir operator. In this section we deal with the Iwasawa and Harish-Chandra direct images, thus providing examples of the general considerations in Chapters II and III. The computations will be used later in connection with the Harish-Chandra series for spherical functions. We start with a basis for $\mathfrak{g}$ consisting of $\mathfrak{a}$-eigenvectors.

We let:

$\{H_i\}$ $(i = 1, \ldots, r)$ be an orthonormal basis of $\mathfrak{a}$ with respect to B,

so

$$r = \dim_{\mathbf{R}} \mathfrak{a}.$$

For each $\alpha \in \mathcal{R}(\mathfrak{n})$ we let E_α be the natural basis of $\mathfrak{n}_\alpha$ (remember, we are on $\mathrm{SL}_n(\mathbf{R})$), so $E_\alpha = E_{ij}$ $(i < j)$. We let

$$E_{-\alpha} = {}^tE_\alpha = \mathbf{t}E_\alpha$$

be the transpose. Then E_α has eigencharacter α on $\mathfrak{a}$, and $E_{-\alpha}$ has eigencharacter $-\alpha$ on $\mathfrak{a}$. The elements ($\mathfrak{a}$-eigenvectors)

(1) $\{H_i\}_{i=1,\ldots,r}$, $\{E_\alpha\}_{\alpha\in\mathcal{R}(\mathfrak{n})}$, $\{E_{-\alpha}\} = \{{}^tE_\alpha\}_{\alpha\in\mathcal{R}(\mathfrak{n})}$,

form an $\mathfrak{a}$-semisimple basis of $\mathfrak{g}$. If $\alpha \neq \pm\beta$, then $\mathfrak{n}_\alpha$ and $\mathfrak{n}_\beta$ are orthogonal for B, and we have

$$B(E_\alpha, E_{-\alpha}) = \mathrm{tr}(E_\alpha {}^tE_\alpha) = 1.$$

Thus the dual basis of (1) is simply

(1′) $\{H_i\}_{i=1,\ldots,r}$, $\{E_{-\alpha}\}_{\alpha\in\mathcal{R}(\mathfrak{n})}$, $\{E_\alpha\}_{\alpha\in\mathcal{R}(\mathfrak{n})}$.

The bases (1) and (1′) will be said to be **the standard B-dual bases of the Lie algebra $\mathfrak{g} = \mathfrak{sl}_n(\mathbf{R})$.**

Remark. The above notation fits the general case. The only added complications of the general case are that the eigenspaces may not have dimension 1, so one has to pick a basis of each eigenspace; and also there is a possibly non-trivial Lie algebra $\mathfrak{m}$ of dimension > 0 which has to be taken into account. The proofs for the subsequent result apply equally well to the general case.

Applying Proposition 2.1 we obtain as a special example:

Proposition 3.1. *On $\mathfrak{sl}_n(\mathbf{R})$, with the standard B-dual bases for the invariant trace form B, we have (with $r = n - 1$):*

$$Q_B = \sum_{i=1}^{r} H_i^2 + \sum_{\alpha\in\mathcal{R}(\mathfrak{n})} (E_\alpha E_{-\alpha} + E_{-\alpha} E_\alpha),$$

$$\omega_B = \sum_{i=1}^{r} \tilde{H}_i^2 + \sum_{\alpha\in\mathcal{R}(\mathfrak{n})} (\tilde{E}_\alpha \tilde{E}_{-\alpha} + \tilde{E}_{-\alpha} \tilde{E}_\alpha).$$

Warning. In the expression for Q_B, note that the product $E_\alpha E_{-\alpha}$ is taken in the symmetric algebra $S(\mathfrak{g})$, and is of course not the matrix product. The context will (should) always make clear which product is intended. The notational problem does not arise in the product expression for the differential operators in the sum for ω_B, since the product is just composition of differential operators. Also note that $E_\alpha E_{-\alpha} = E_{-\alpha} E_\alpha$ in the commutative algebra $S(\mathfrak{g})$.

Next we establish some notation for the study of the direct image of the Casimir operator to A. We let:

$$\boxed{\omega_{B,\mathfrak{a}} = \tilde{H}_1^2 + \cdots + \tilde{H}_r^2 = \mathcal{D}(H_1^2 + \cdots + H_r^2).}$$

The element $\omega_{B,\mathfrak{a}}$ is not the direct image to A of ω. The direct image will be determined shortly in Theorem 3.3. Rather, $\omega_{B,\mathfrak{a}}$ is the **Casimir operator on A determined by the restriction of B to $\mathfrak{a}$** (positive definite). The sum of squares defining $\omega_{B,\mathfrak{a}}$ is equal to its own symmetrization, independent of the choice of orthonormal basis of $\mathfrak{a}$ because A is abelian. Thus $\omega_{B,\mathfrak{a}}$ is the ordinary Laplacian of calculus in euclidean space.

For any character λ on $\mathfrak{a}$, there is a unique vector $H_\lambda \in \mathfrak{a}$ such that

$$\lambda(H) = B(H, H_\lambda) \qquad \text{for all } H \in \mathfrak{a}.$$

The positive definite form B on $\mathfrak{a}$ induces a similar form on the dual space $\mathfrak{a}^\vee$ by the formula

$$\langle \lambda, \mu \rangle = B(H_\lambda, H_\mu) \qquad \text{for } \lambda, \mu \in \mathfrak{a}^\vee.$$

If $\alpha = \alpha_{ij}$ with $i < j$, then

$$H_\alpha = E_{ii} - E_{jj} = \begin{pmatrix} 0 & & & & & & & & \\ & \ddots & & & & & & & \\ & & 0 & & & & & & \\ & & & 1 & & & & & \\ & & & & 0 & & & & \\ & & & & & \ddots & & & \\ & & & & & & 0 & & \\ & & & & & & & -1 & \\ & & & & & & & & 0 \\ & & & & & & & & & \ddots \\ & & & & & & & & & & 0 \end{pmatrix}.$$

This is immediate from the definitions.

Lemma 3.2. *Let $Z_\alpha = E_\alpha - {}^tE_\alpha \in \mathfrak{k}$. Then in the Lie algebra $\mathfrak{sl}_n(\mathbf{R})$, we have*

$$[E_\alpha, Z_\alpha] = -H_\alpha \in \mathfrak{a} \quad \text{or equivalently} \quad [E_\alpha, E_{-\alpha}] = H_\alpha.$$

Proof. This follows by direct matrix multiplication. Note that in taking the bracket product $[E_\alpha, Z_\alpha]$ we are now using the linear representation, so that in the expression

$$[v, w] = vw - wv$$

the products vw and wv on the right are the ordinary matrix products. Using the fact that $[v, v] = 0$, if $\alpha = \alpha_{ij}$ direct matrix multiplication yields

$$[E_\alpha, E_{-\alpha}] = E_{ij}E_{ji} - E_{ji}E_{ij} = E_{ii} - E_{jj},$$

as desired. The proof in the general case takes somewhat longer, see [Hel 78], Chapter III, Theorem 5.5 or [GaV 88], 2.6.62 and 4.2.1.

Theorem 3.3. *The Iwasawa direct image of Casimir is given by*

$$(\mathrm{Iw}_A)_*(\omega_B) = \omega_{B,\mathfrak{a}} - \sum_{\alpha\in\mathcal{R}(\mathfrak{n})} \tilde{H}_\alpha.$$

Proof. The general method for obtaining the direct image is given in Theorem 3.3 of Chapter V. We apply it in the present case. We write

$$E_{-\alpha} = {}^tE_\alpha = E_\alpha + {}^tE_\alpha - E_\alpha = E_\alpha - Z_\alpha,$$

so that taking the corresponding differential operators, we get

$$\tilde{E}_{-\alpha} = \tilde{E}_\alpha - \tilde{Z}_\alpha,$$

and therefore, recalling $\mathcal{J}(\tilde{\mathfrak{n}}, \tilde{\mathfrak{k}}) = \tilde{\mathfrak{n}}\,\mathrm{IDO}(G) + \mathrm{IDO}(G)\tilde{\mathfrak{k}}$, we have

$$\tilde{E}_\alpha \tilde{E}_{-\alpha} = \tilde{E}_\alpha^2 - \tilde{E}_\alpha \tilde{Z}_\alpha \equiv 0 \mod \mathcal{J}(\mathfrak{n}, \mathfrak{k}).$$

On the other hand,

$$\tilde{E}_{-\alpha}\tilde{E}_\alpha = (\tilde{E}_\alpha - \tilde{E}_\alpha)\tilde{E}_\alpha = \tilde{E}_\alpha^2 - \tilde{Z}_\alpha \tilde{E}_\alpha.$$

Therefore

$$\begin{aligned}\sum_{\alpha\in\mathcal{R}(\mathfrak{n})} (\tilde{E}_\alpha \tilde{E}_{-\alpha} + \tilde{E}_{-\alpha}\tilde{E}_\alpha) &\equiv \sum_{\alpha\in\mathcal{R}(\mathfrak{n})} -\tilde{Z}_\alpha \tilde{E}_\alpha \mod \mathcal{J}(\mathfrak{n}, \mathfrak{k}) \\ &\equiv \sum_{\alpha\in\mathcal{R}(\mathfrak{n})} [\tilde{E}_\alpha, \tilde{Z}_\alpha] \mod \mathcal{J}(\mathfrak{n}, \mathfrak{k})\end{aligned}$$

$$\text{[by Chapter V, §1(2)]} \quad \equiv \sum_{\alpha \in \mathcal{R}(\mathfrak{n})} [E_\alpha, Z_\alpha]^{\sim} \quad \operatorname{mod} \mathcal{J}(\mathfrak{n}, \mathfrak{k})$$

$$\text{[by Lemma 3.2]} \quad \equiv \sum_{\alpha \in \mathcal{R}(\mathfrak{n})} -\tilde{H}_\alpha \quad \operatorname{mod} \mathcal{J}(\mathfrak{n}, \mathfrak{k}).$$

Hence from Proposition 3.1 we now see that the pure $\mathfrak{a}$-component of Casimir is precisely that given in the statement of the theorem. This concludes the proof.

Anders Karlsson has pointed out to us that there is an easy way to describe an orthonormal basis $\{H_i\}$ $(i = 1, \ldots, r)$, as follows. We let

$$C_1 = \begin{pmatrix} 1 & & & & \\ & -1 & & & \\ & & 0 & & \\ & & & \ddots & \\ & & & & 0 \end{pmatrix},$$

$$C_2 = \begin{pmatrix} 1 & & & & & \\ & 1 & & & & \\ & & -2 & & & \\ & & & 0 & & \\ & & & & \ddots & \\ & & & & & 0 \end{pmatrix} \quad \cdots \quad C_r = \begin{pmatrix} 1 & & & & \\ & 1 & & & \\ & & \ddots & & \\ & & & 1 & \\ & & & & -r \end{pmatrix}$$

so $C_1, \ldots, C_r$ are diagonal matrices with trace 0. They are orthogonal, but not orthonormal, so they form an orthogonal basis of $\mathfrak{a}$ for the trace form. Let

$$\tau = \sum_{\alpha \in \mathcal{R}(\mathfrak{n})} \alpha = \log \delta.$$

Then

$$H_\tau = \sum_{\alpha \in \mathcal{R}(\mathfrak{n})} H_\alpha = C_1 + \cdots + C_r.$$

We have

$$\langle C_i, C_i \rangle = i^2 + i.$$

We can then define an orthonormal basis by letting

$$H_i = \frac{1}{(i^2 + i)^{1/2}} C_i,$$

making everything explicit in Theorem 3.3. Unless otherwise specified, however, we don't need any property except those of a general orthonormal basis.

Next as usual we let

$$\rho = \frac{1}{2} \sum_{\alpha \in \mathcal{R}(\mathfrak{n})} \alpha = \tfrac{1}{2}\tau,$$

and we give some formulas about ρ, which come up in the Harish-Chandra conjugation. By definition, $\langle \rho, \rho \rangle = B(H_\rho, H_\rho)$. Then:

$$(2) \qquad H_\rho = \frac{1}{2} \sum_{\alpha \in \mathcal{R}(\mathfrak{n})} H_\alpha = \sum_{i=1}^{r} \langle H_\rho, H_i \rangle H_i = \sum_{i=1}^{r} \rho(H_i) H_i,$$

$$(3) \qquad \langle \rho, \rho \rangle = \frac{1}{2} \sum_{\alpha \in \mathcal{R}(\mathfrak{n})} \rho(H_\alpha) = \sum_{i=1}^{r} \rho(H_i)^2.$$

Proofs. The first formula (2) is just the expression of an element in a vector space with a positive definite scalar product, in terms of an orthonormal basis, namely $\{H_i\}$ in the present case. The second formula is obtained by taking the value of ρ on the elements of the first formula, thus concluding the proof.

We recall from Chapter II, §5, the definition of the **Harish-Chandra image** of a differential operator $D \in \mathrm{IDO}(G)^K$, namely

$$\mathbf{h}(D) = \delta^{-1/2} (\mathrm{Iw}_A)_*(D) \circ \delta^{1/2}.$$

No use so far has been made of Chapter III, Proposition 2.1, which tells us that for a polynomial $P_\mathfrak{a} \in S(\mathfrak{a})$ we have

$$(4) \qquad \mathcal{D}(P_\mathfrak{a}^{S(\rho)}) = \delta^{-1/2} \mathcal{D}(P_\mathfrak{a}) \circ \delta^{1/2}.$$

This formula will now come to the fore. Let

$$P_\mathfrak{a} = \sum_{i=1}^{r} H_i^2 - \sum_{\alpha \in \mathcal{R}(\mathfrak{n})} H_\alpha .$$

Theorem 3.3 tells us that

$$(\mathrm{Iw}_A)_*(\omega_B) = \mathcal{D}(P_\mathfrak{a}).$$

Thus we have developed all the tools to get the next theorem, whose effect is to get rid of first-order terms.

Theorem 3.4. *The Harish-Chandra image of Casimir is given by*

$$\mathbf{h}(\omega_B) = \omega_{B,\mathfrak{a}} - \langle \rho, \rho \rangle .$$

Proof. By definition,

$$P_{\mathfrak{a}}^{S(\rho)} = \sum_{i=1}^{r}(H_i + \rho(H_i))^2 - \sum_{\alpha\in\mathcal{R}(\mathfrak{n})}(H_\alpha + \rho(H_\alpha)).$$

Hence expanding the square and applying the Harish-Chandra map

$$\mathcal{D}_A\colon S(\mathfrak{a}) \to \mathbb{D}\mathrm{O}(A) \qquad \text{also written} \qquad \mathcal{D}_A(P) = \tilde{P},$$

we obtain

$$\begin{aligned}\mathcal{D}_A(P_{\mathfrak{a}}^{S(\rho)}) &= \sum_{i=1}^{r}\tilde{H}_i^2 + \sum_{i=1}^{r}2\rho(H_i)\tilde{H}_i + \sum_{i=1}^{r}\rho(H_i)^2 - \sum_{\alpha\in\mathcal{R}(\mathfrak{n})}(\tilde{H}_\alpha + \rho(H_\alpha))\\ &= \omega_{B,\mathfrak{a}} + \tilde{H}_\tau + \langle\rho,\rho\rangle - \tilde{H}_\tau - 2\langle\rho,\rho\rangle\end{aligned}$$

by using formula (2) (with a tilde) and formula (3). Then (4) and Theorem 3.3 conclude the proof.

Remark. The above proof essentially follows that of the general case as in Harish-Chandra [Har 58a], Corollary 2, p. 271, see also [GaV 88], Lemma 2.6.10. The conjugation by the square root stems from a general formalism, which will recur, and which will be considered in a slightly more general case in the next section. We also remark that Gangolli [Gan 71] first recognized the significance of getting rid of first-order terms to make stronger and needed estimates than were made up to that time, for instance to complete the proof of the Paley–Wiener theorem set up by Helgason in [Hel 66].

Casimir and the Laplacian. We also note that Theorem 3.4 proves for the Casimir operator the same formula for the direct image as occurs in [Hel 84], Chapter II, Proposition 3.8, for the Laplacian. Since the direct image on invariant differential operators from G/K to A is a linear isomorphism, those interested in the connection with the differential geometric aspects can now conclude:

Corollary 3.5. *The direct image of Casimir to G/K is equal to the direct image of the Laplacian to G/K.*

Since we shall be concerned with the homogeneous space G/K, readers will appreciate that differential geometry is not needed to develop the Harish-Chandra theory. However, it may be needed for other purposes, and Corollary 3.5 allows readers to connect with various parts of the literature. [Hel 84] relegates Casimir to Exercise 4 of Chapter II.

The Casimir operator on G is not elliptic, but its direct image to G/K or A is elliptic. In the special case of $\mathrm{SL}_2(\mathbf{R})$ and G/K = upper half plane with coordinates (x, y), $y > 0$, and the θ coordinate on K, one has

$$\omega = 4y^2\left(\frac{\partial^2}{\partial x^2} + \frac{\partial^2}{\partial y^2}\right) - 4y\frac{\partial^2}{\partial x \partial \theta}.$$

Thus the $\partial/\partial\theta$ annihilates a function which is independent of θ (i.e. right K-invariant on the group), and deleting the term containing θ, one sees that what is left is the Laplacian on the upper half plane. Cf. [Lan 75/85], Chapter X, §2, p. 198.

Scaling the form B. For definiteness, we picked the trace form B right from the start. However, the Lie industry mostly uses the Killing form because it is the only one available canonically, independently of any special representation, when dealing with an arbitrary Lie group G and its Lie algebra. So the question arises, how do the formulas we have derived change when we scale the form B. The answer is that essentially they don't. Let us make this more precise.

Let $B_1 = cB$ with some number $c > 0$. This is the most general change allowable in practice since we require the form to be conjugation invariant (and also θ-invariant). Let $\{Z_1, \ldots, Z_N\}$ be a basis of $\mathfrak{g}$ and let $\{Z'_1, \ldots, Z'_N\}$ be the B-dual basis, so

$$B(Z_i, Z'_i) = 1.$$

Then the B_1-dual basis is $\{c^{-1}Z'_1, \ldots, c^{-1}Z'_N\}$. Hence the Casimir operator becomes

Sc 1. $\omega_{B_1} = c^{-1}\omega_B$.

Let us write $H_{B,\lambda}$ for the vector in $\mathfrak{a}$ representing a functional $\lambda \in \mathfrak{a}^\vee$ with respect to B. Then

Sc 2. $H_{B_1,\lambda} = c^{-1}H_{B,\lambda}$,

Sc 3. $B_1(\lambda, \lambda') = B_1(H_{B_1,\lambda}, H_{B_1,\lambda'}) = c^{-1}B(\lambda, \lambda')$.

Both formulas follow at once from the definition. Therefore:

Proposition 3.6. *The formula of Theorem 3.4 is invariant under scaling $B \mapsto B_1 = cB$, in other words we also have*

$$\mathbf{h}(\omega_{B_1}) = \omega_{B_1,\mathfrak{a}} - B_1(\rho, \rho).$$

As already pointed out, the formula is valid for a wide class of groups, especially complex groups viewed as real groups, when there are even more choices for the form B which can and have been made.

VII, §4. THE POLAR DIRECT IMAGE

We shall carry out the similar computation as in §3 for the polar direct image, a prescription for which is given by Theorems 5.2 and 5.5 of Chapter VI. We use the notation of Chapter VI, §4 and §5, for instance

$$\mathcal{J}(\mathfrak{k}^b, \mathfrak{k}) = \tilde{\mathfrak{k}}^b \,\mathrm{IDO}(G) + \mathrm{IDO}(G)\tilde{\mathfrak{k}}.$$

The polar direct image $(\mathrm{Pol}_{A^+})_*(D)$ is shown to be characterized by the property $(\mathrm{Pol}_{A^+})_*(D) = E$ such that

$$D \equiv E(b) \mod \mathcal{J}(\mathfrak{k}^b, \mathfrak{k}) \qquad \text{for all } b \in A',$$

and $E \in \mathfrak{R}\mathrm{IDO}(A)$, where $\mathfrak{R}$ is the algebra generated by the functions g_α $(\alpha \in \mathcal{R}(\mathfrak{n}))$, whose definitions we also recall.

As in calculus, we deal with the functions on A defined by

$$C_\alpha(a) = \tfrac{1}{2}(a^\alpha + a^{-\alpha}), \qquad S_\alpha(a) = \tfrac{1}{2}(a^\alpha - a^{-\alpha}), \qquad \chi_\alpha(a) = a^\alpha.$$

As in Chapter VI, we then have the real analytic function g_α defined on A' by

$$g_\alpha(b) = \frac{b^{-\alpha}}{S_\alpha(b)}.$$

By abuse of notation, some people write

$$1 + g_\alpha = \coth \alpha,$$

because

$$\frac{C_\alpha}{S_\alpha} = \frac{a^\alpha + a^{-\alpha}}{a^\alpha - a^{-\alpha}} = 1 + \frac{2a^{-\alpha}}{a^\alpha - a^{-\alpha}} = 1 + g_\alpha.$$

The correct notation for us would be $\coth \alpha \circ \log$, defined on A'.

For simplicity, we write ω instead of ω_B, but the scaling comment in §3 applies.

Theorem 4.1. *The polar direct image of Casimir on A^+ is given by*

$$(\mathrm{Pol}_{A^+})_*(\omega) = \omega_{B,\mathfrak{a}} + \sum_{\alpha \in \mathcal{R}(\mathfrak{n})} (1 + g_\alpha)\tilde{H}_\alpha.$$

Proof. Starting with Proposition 3.1, we get

$$\begin{aligned}\omega &= \omega_{B,\mathfrak{a}} + \sum(\tilde{E}_\alpha \tilde{E}_{-\alpha} + \tilde{E}_{-\alpha}\tilde{E}_\alpha) \\ &= \omega_{B,\mathfrak{a}} + \sum[\tilde{E}_\alpha, \tilde{E}_{-\alpha}] + 2\sum \tilde{E}_{-\alpha}\tilde{E}_\alpha \quad \text{[by Lemma 3.2]} \\ &\equiv \omega_{B,\mathfrak{a}} + \sum_{\alpha\in\mathcal{R}(\mathfrak{n})} \tilde{H}_\alpha + \sum g_\alpha(b)\tilde{Z}_\alpha\tilde{E}_\alpha \quad \operatorname{mod} \mathcal{J}(\mathfrak{k}^b, \mathfrak{k})\end{aligned}$$

by (6) of Chapter VI, §4 (note that $z_\alpha = 2w_\alpha$). We then write

$$\begin{aligned}\tilde{Z}_\alpha\tilde{E}_\alpha = [\tilde{Z}_\alpha, \tilde{E}_\alpha] + \tilde{E}_\alpha\tilde{Z}_\alpha &\equiv [\tilde{Z}_\alpha, \tilde{E}_\alpha] \quad \operatorname{mod} \mathrm{IDO}(G)\tilde{\mathfrak{k}} \\ &\equiv \tilde{H}_\alpha \quad \operatorname{mod} \mathrm{IDO}(G)\tilde{\mathfrak{k}}\end{aligned}$$

by Lemma 3.2. Applying Theorem 5.2 of Chapter VI completes the proof.

Warning. Our normalization of g_α is slightly different from [Har 58a], p. 266, et seq., and [GaV 88] following Harish-Chandra in their formula 4.2.2 and Proposition 4.2.1. Indeed, Harish's function g_α is one-half times our function g_α. Thus the formulas in [Har 58a] and [GaV 88] have a factor of 2 appearing frequently. This factor disappears if one uses the normalization we have adopted. Of course, we could have used a different letter, but there is a shortage of letters, and we chose to use the same letter, with a warning.

Because of Corollary 3.5, all the theorems of this section are theorems about the Laplace operator, which thus have their source in Harish-Chandra. For instance, Theorem 4.1 corresponds to [Hel 84], Chapter II, Proposition 3.9.

Corollary 4.2. $(\mathrm{Pol}_{A^+})_*(\omega) = \mathbf{h}(\omega) + \sum(1+g_\alpha)\tilde{H}_\alpha + \langle\rho,\rho\rangle.$

Proof. Apply Theorem 3.4.

Recall explicitly that $\mathbf{h}(\omega) = \delta^{-1/2}(\mathrm{Iw}_A)_*(\omega)\circ\delta^{1/2}$. We shall get a formula in the polar case analogous to Theorem 3.4, in other words, a result related to Theorem 4.1 in the same way that Theorem 3.4 is related to Theorem 3.3. As in §3, we let $\{H_i\}$ $(i = 1, \dots, r)$ be a B-orthonormal basis of $\mathfrak{a}$.

Theorem 4.3. *For the $\delta^{1/2}$-conjugation of the polar direct image, we get*

$$\begin{aligned}\delta^{1/2}(\mathrm{Pol}_{A^+})_*(\omega)\circ\delta^{-1/2} &= \sum_{i=1}^r \tilde{H}_i^2 - \langle\rho,\rho\rangle + \sum_{\alpha\in\mathcal{R}(\mathfrak{n})} g_\alpha(\tilde{H}_\alpha - \rho(H_\alpha)) \\ &= \mathbf{h}(\omega) + \sum_{\alpha\in\mathcal{R}(\mathfrak{n})} g_\alpha(\tilde{H}_\alpha - \rho(H_\alpha)).\end{aligned}$$

Proof. By Theorem 2.1 of Chapter III, and Theorem 4.1, we obtain

$$\delta^{1/2}(\mathrm{Pol}_{A^+})_*(\omega)\circ\delta^{-1/2} = \sum_{i=1}^{r}(\tilde{H}_i-\rho(H_i))^2+2(\tilde{H}_\rho-\rho(H_\rho)) + \sum_{\alpha\in\mathcal{R}(\mathfrak{n})} g_\alpha(\tilde{H}_\alpha-\rho(H_\alpha)),$$

We expand the square term and use the formula (2) of §3 to get a cancellation of a term $2\tilde{H}_\rho$. We then use formula (3) of §3 and Theorem 3.4 to conclude the proof.

Remark. There is an internal check for the formulas of Theorem 4.3, namely we can compare them with Proposition 5.5 of Chapter VI. This proposition gave a relation between the polar direct image and the Harish-Chandra map **h** by a conjugation, and some correcting term in $\mathfrak{R}_+\mathrm{IDO}(A)$ which is here of degree 1. Among other things, one can view Theorem 4.3 as giving a computation of this term for the Casimir operator.

Scaling. The same comment made at the end of §3 applies to this section. The theorems are valid when B is replaced by cB with $c > 0$. *For this purpose, one must not omit the subscript B.* For example, Theorem 4.1 should read without abbreviation

$$(\mathrm{Pol}_{A^+})_*(\omega_B) = \omega_{B,\mathfrak{a}} + \sum_{\alpha\in\mathcal{R}(\mathfrak{n})}(1+g_\alpha)\tilde{H}_{B,\alpha}.$$

Then one can say precisely that this formula is valid when B is replaced by cB with $c > 0$.

Next we shall conjugate with the function $J^{1/2}$ (see below) instead of $\delta^{1/2}$. We thereby get rid of the terms of degree 1 in the polar direct image of Casimir. This is also due to Harish-Chandra ([Har 58a], Corollary 1 of Lemma 27, p. 270). We found the exposition in [GaV 88], 4.2, useful. To carry out the conjugation, we recall the v-**logarithmic derivative** (with v in the Lie algebra of an arbitrary Lie group) from Chapter III, §2, namely for a positive function φ on an open set,

$$\tilde{v}(\log\varphi) = \varphi^{-1}\tilde{v}(\varphi).$$

We apply the logarithmic derivative on A with $v = H \in \mathfrak{a}$. We recall the function

$$J(a) = \prod_{\alpha\in\mathcal{R}(\mathfrak{n})}(a^\alpha - a^{-\alpha}).$$

For our present purposes, we don't care that this function is the Jacobian of the polar map, and we may take the product on the right as the definition of $J(a)$ on A^+, so J is positive on A^+. Its H-logarithmic derivative is given by

$$\tilde{H}(\log J) = \sum_{\alpha \in \mathcal{R}(n)} (1 + g_\alpha)\alpha(H). \tag{1}$$

Proof. This comes from using the homomorphic property of the logarithmic derivative, and the special case with the calculus functions, for which we have

(1a) $\tilde{H}(\chi_\alpha) = \alpha(H)\chi_\alpha$ and $\tilde{H}(\chi_{-\alpha}) = -\alpha(H)\chi_{-\alpha}$,

(1b) $\tilde{H}(S_\alpha) = \alpha(H)C_\alpha$,

(1c) $\tilde{H}(\log S_\alpha) = \alpha(H)C_\alpha/S_\alpha = \alpha(H)(1 + g_\alpha)$.

Formula (1) then drops out.

As a consequence, we find:

$$\sum_{i=1}^{r} \tilde{H}_i(\log J)\tilde{H}_i = \sum_{\alpha \in \mathcal{R}(n)} (1 + g_\alpha)\tilde{H}_\alpha. \tag{2}$$

Proof. This is immediate from (1) and the general formula for any $\lambda \in \mathfrak{a}^\vee$,

$$\sum_{i=1}^{r} \lambda(H_i)\tilde{H}_i = \tilde{H}_\lambda.$$

Combining Theorem 4.1 and (2) yields:

Theorem 4.4. *The polar direct image of Casimir also has the expression*

$$(\mathrm{Pol}_{A^+})_*(\omega) = \sum_{i=1}^{r} (\tilde{H}_i^2 + \tilde{H}_i(\log J)\tilde{H}_i).$$

We want a conjugation to get rid of first-order terms. For this, we need more formulas on an arbitrary Lie group G. As before, let $v \in \mathrm{Lie}(G)$. We are interested in differential operators of the form

$$\tilde{v}^2 + \tilde{v}(\log \varphi)\tilde{v}, \tag{3}$$

which appear in Theorem 4.4. To get at Theorem 4.7, we develop a systematic formalism concerning the logarithmic derivative. We recall that, since $\tilde{v}$ is a derivation on the ring of functions,

$$\tilde{v} \circ \varphi = \varphi \tilde{v} + \tilde{v}(\varphi), \tag{4}$$

$$\varphi^{-1}\tilde{v} \circ \varphi = \tilde{v} + \tilde{v}(\log \varphi) \quad \text{and} \quad \tilde{v} \circ \varphi^{-1} = \tilde{v} - \tilde{v}(\log \varphi). \tag{5}$$

Cf. Chapter III, §2. From these two formulas, we get conjugation of $\tilde{v}^2$, namely

$$\begin{aligned} \varphi^{-1}\tilde{v}^2 \circ \varphi &= \varphi^{-1}\tilde{v}(\varphi)\tilde{v} + \tilde{v}^2 + \varphi^{-1}\tilde{v} \circ (\tilde{v}(\varphi)) \\ &= 2\tilde{v}(\log \varphi)\tilde{v} + \tilde{v}^2 + \varphi^{-1}\tilde{v}^2(\varphi) \end{aligned} \tag{6}$$

so a factor 2 appears, and we can rewrite this formula as

$$\tilde{v}^2 + 2\tilde{v}(\log \varphi)\tilde{v} = \varphi^{-1}\tilde{v}^2 \circ \varphi - \varphi^{-1}\tilde{v}^2(\varphi). \tag{7}$$

The homomorphic property of the logarithmic derivative gives

$$\tfrac{1}{2}\tilde{v}(\log \varphi) = \varphi^{-1/2}\tilde{v}(\varphi^{1/2}). \tag{8}$$

Then from (7) with $\varphi^{1/2}$ instead of φ, we get

$$\tilde{v}^2 + \tilde{v}(\log \varphi)\tilde{v} = \varphi^{-1/2}\tilde{v}^2 \circ \varphi^{1/2} - \varphi^{-1/2}\tilde{v}^2(\varphi^{1/2}). \tag{9}$$

Applying this to $\varphi = J$ and using Theorem 4.4 yields:

Theorem 4.5. *Reminding the reader that* $\omega_{B,\mathfrak{a}} = \sum_{i=1}^{r} \tilde{H}_i^2$, *we have*

$$(\mathrm{Pol}_{A^+})_*(\omega) = J^{-1/2}\omega_{B,\mathfrak{a}} \circ J^{1/2} - J^{-1/2}\omega_{B,\mathfrak{a}}(J^{1/2}).$$

Hence

$$J^{1/2}(\mathrm{Pol}_{A^+})_*(\omega) \circ J^{-1/2} = \omega_{B,\mathfrak{a}} - J^{-1/2}\omega_{B,\mathfrak{a}}(J^{1/2}).$$

We shall derive an alternate expression for the function on the right, which is the degree 0 term of the differential operator. The term of degree 0 in (7) has the alternate expression

$$\varphi^{-1}\tilde{v}^2(\varphi) = \tilde{v}^2(\log \varphi) + (\tilde{v}(\log \varphi))^2. \tag{10}$$

Proof. The routine is:

$$\begin{aligned} \varphi^{-1}\tilde{v}^2(\varphi) = \varphi^{-1}\tilde{v}(\tilde{v}\varphi) &= \varphi^{-1}\tilde{v}(\varphi\varphi^{-1}\tilde{v}(\varphi)) \\ &= \varphi^{-1}\tilde{v}(\varphi\tilde{v}(\log \varphi)) \\ &= \tilde{v}^2(\log \varphi) + (\varphi^{-1}\tilde{v}(\varphi))\tilde{v}(\log \varphi) \end{aligned}$$

as claimed.

Summarizing (9) and (10), we get:

Lemma 4.6. *Let G be a Lie group, $v \in \mathrm{Lie}(G)$ and φ a positive function on an open subset of G. Let*

$$D = \tilde{v}^2 + \tilde{v}(\log\varphi)\tilde{v} \qquad \textit{and} \qquad q = \tfrac{1}{2}\tilde{v}(\log\varphi).$$

Then

$$\begin{aligned} \varphi^{1/2} D \circ \varphi^{-1/2} &= \tilde{v}^2 - (\tilde{v}(q) + q^2), \\ \varphi^{-1/2}\tilde{v}^2(\varphi^{1/2}) &= \tilde{v}(q) + q^2. \end{aligned}$$

We apply (10) to the term of degree 0 in Theorem 4.5, and get the special case relevant to Casimir.

Theorem 4.7. *Let $q_i = \frac{1}{2}\tilde{H}_i(\log J)$. Then*

$$J^{-1/2}\omega_{B,\mathfrak{a}}(J^{1/2}) = \sum_{i=1}^{r}(\tilde{H}_i(q_i) + q_i^2).$$

The function occurring in Theorems 4.5 and 4.7 will recur, and we give it a special notation, namely **we define**

$$\gamma_J = J^{-1/2}\omega_{B,\mathfrak{a}}(J^{1/2}) = \sum_{i=1}^{r}(\tilde{H}_i(q_i) + q_i^2).$$

Remark. It is of intrinsic interest for its own sake to consider the conjugation

$$J^{-1}\omega_{B,\mathfrak{a}} \circ J$$

without taking the square root. The expression on the right in Theorem 4.7 then simplifies to a constant. Interested readers can now read Chapter XII, §1, to see how the identity comes out.

CHAPTER VIII

The Harish-Chandra Series and Spherical Inversion

This chapter is fundamentally based on [Har 58a]. We define the Harish-Chandra series for eigenfunctions of Casimir, prove its basic properties, and show how the spherical functions can be expressed in terms of this series. We incorporate from the start a crucial estimate by Gangolli to insure the possibility of term by term differentiation [Gan 71]. The need for such an estimate had arisen in Helgason's approach to getting the inversion theorem on the Paley–Wiener space [Hel 66]. The applications to the inversion problem will come in the next chapter.

VIII, §0. LINEAR INDEPENDENCE OF CHARACTERS REVISITED

The exponential function may be viewed as a character (continuous homomorphism) from the additive group (of the complex numbers) into the multiplicative group. It is a simple fact of algebra that distinct characters are linearly independent (Artin's theorem). What will be needed is a series analogue of this theorem for infinite linear combinations which involve convergence questions. The material involves self-contained lemmas about exponential series, and we basically copy Harish-Chandra who placed this material in an appendix to [Har 58a]. Readers can skip this section until it is needed in §3. Wherever one places it interrupts the flow of spherical functions and inversion. The lemmas will again be used in Chapter X. They connect with Chapter III, §8, which gave information on eigenfunctions of W-invariant differential operators. The Harish-Chandra series will construct eigenfunctions. To know that they generate all eigenfunctions, one needs criteria for linear independence. The Harish-Chandra lemmas provide a necessary technical background, at the level of advanced calculus.

Lemma 0.1. *Let $y_1, \ldots, y_m$ be distinct real numbers, and t a positive real variable. Let $c_1, \ldots, c_m \in \mathbf{C}$. Then*

$$\limsup_{t\to\infty} \left| \sum_{j=1}^{m} c_j e^{iy_j t} \right| \geqq \left(\sum_{j=1}^{m} |c_j|^2 \right)^{1/2}.$$

Proof. Let

$$f(t) = \sum_{j=1}^{m} c_j e^{iy_j t}.$$

A direct computation shows that

$$\lim_{T\to\infty} \frac{1}{T} \int_0^T |f(t)|^2 \, dt = \sum_{j=1}^{m} |c_j|^2.$$

On the other hand, f is bounded, and it is obvious that

$$\lim_{T\to\infty} \frac{1}{T} \int_0^T |f(t)|^2 \, dt \leqq \left(\limsup_{t\to\infty} |f(t)| \right)^2.$$

This proves the lemma.

Lemma 0.2. *Let $z_1, \ldots, z_m$ be distinct non-zero complex numbers, and $p_0, \ldots, p_m$ polynomials in t with complex coefficients. Let $a \geqq 0$ be such that*

$$\limsup_{t\to\infty} \left| p_0(t) + \sum_{j=1}^{m} p_j(t) e^{z_j t} \right| \leqq a.$$

Then $\mathrm{Re}(z_j) \leqq 0$ *for all j; the polynomials p_0 and p_j with* $\mathrm{Re}(z_j) = 0$ *are constant; and $|p_0|^2 \leqq a^2$. More generally (but it won't be needed),*

$$|p_0|^2 + \sum_{\mathrm{Re}(z_j)=0} |p_j|^2 \leqq a^2.$$

Proof. Write $z_j = x_j + iy_j$ with real x_j, y_j. If $x_j < 0$ for some j, then $|p_j(t)e^{z_j t}| \to 0$ as $t \to \infty$, so without loss of generality, we can assume that $x_j \geqq 0$ for all j. We first prove that in this case, $\mathrm{Re}(z_j) = 0$

for all j. Let $x = \max x_j$ and suppose $x > 0$. Let d be the highest degree among the polynomials $p_j(t)$ such that $x_j = x$, and let

$$c_j = \lim_{t\to\infty} p_j(t)/t^d \qquad \text{for such } j.$$

Then not all c_j are equal to 0, and

$$\lim_{t\to\infty} t^{-d}e^{-xt}\left| p_0(t) + \sum_{j=1}^{m} p_j(t)e^{z_jt} \right| = 0.$$

Therefore, taking the sum just over those j such that $x_j = x$, we get

$$\lim_{t\to\infty} \left| \sum_{x_j = x} c_j e^{iy_jt} \right| = 0.$$

By assumption, the corresponding imaginary parts y_j are distinct, and Lemma 0.1 implies that $c_j = 0$, a contradiction proving that $\mathrm{Re}(z_j) = 0$ for all $j = 1, \ldots, m$.

Next we prove that p_j is constant for all j. Let e be the maximum of the degrees of $p_0, \ldots, p_m$. We claim $e = 0$. Otherwise, since $x_j = 0$ for all j, it follows that

$$\lim_{t\to\infty} t^{-e}\left| p_0(t) + \sum_{j=1}^{m} p_j(t)e^{iy_jt} \right| = 0.$$

Therefore

$$\lim_{t\to\infty} \left| c_0 + \sum_{j=1}^{m} c_j e^{iy_jt} \right| = 0.$$

Again Lemma 0.1 implies $c_j = 0$, a contradiction which proves p_j constant for $j \geqq 0$. Then finally Lemma 0.1 implies Lemma 0.2, thus concluding the proof.

Proposition 0.3. *Let E be a finite dimensional vector space over* $\mathbf{R}$. *Let V be an open subset of E such that $tV \subset V$ for all $t \geqq 1$. Let $\{\zeta_j\}$ be a sequence of* $\mathbf{R}$*-linear complex valued functions on E, and $\{p_j\}$ a sequence of polynomial functions on E. Suppose that the following two conditions hold:*

(1) *For every* $\mathbf{R}$*-linear complex valued function β on E, the series*

$$\sum_j p_j(H)e^{\beta(H)+\zeta_j(H)}$$

converges uniformly for $H \in V$.

(2) *We have*

$$\sum_j P_j(H)e^{\zeta_j(H)} = 0 \qquad \textit{for all } H \in V.$$

Then $P_j = 0$ *for all* j.

Proof. Let k be a given fixed index. We shall prove that $P_k = 0$. By (1), given ϵ, there exists $N > k$ such that

$$\left| \sum_{j>N} P_j(H)e^{(\zeta_j - \zeta_k)(H)} \right| \leqq \epsilon \qquad \text{for all } H \in V.$$

By (2), it follows that

$$\left| \sum_{j=1}^{N} P_j(H)e^{(\zeta_j - \zeta_k)(H)} \right| \leqq \epsilon \qquad \text{for all } H \in V.$$

Let $H_0 \in V$ be such that $(\zeta_j - \zeta_k)(H_0) \neq 0$ for $1 \leqq j \leqq N$ and $j \neq k$. Let

$$p_j(t) = P_j(tH_0).$$

Then for $t \geqq 1$,

$$\left| p_k(t) + \sum_{\substack{j=1 \\ j\neq k}}^{N} p_j(t)e^{t(\zeta_j - \zeta_k)(H_0)} \right| \leqq \epsilon.$$

By Lemma 0.2, each p_k is constant and $|p_k| \leqq \epsilon$. Hence $p_k = 0$, and so $P_k(H_0) = 0$. We can choose H_0 arbitrarily in some non-empty open set in V, so $P_k = 0$. This proves the proposition.

VIII, §1. EIGENFUNCTIONS OF CASIMIR

In pulling back functions of (χ, a) to $\mathfrak{a}^\vee \times \mathfrak{a}$, as we described in Chapter VI, §3, one can really go through two stages, pulling back first the first variable, but leaving the second variable on the multiplicative group A. To deal with the action of Casimir on spherical functions, we find it advantageous to use this intermediate pull-back. Therefore we shall now use the notation φ_ζ for the function on A previously denoted φ_{χ_ζ}, where

$$\chi_\zeta(a) = a^\zeta.$$

We pick $\zeta \in \mathfrak{a}_{\mathbf{C}}^{\vee}$, so ζ is complex valued. We let B be a Cartan type form so $B_{\mathfrak{a}}$ is positive definite on $\mathfrak{a}$. As in the preceding Chapter VII, we use an orthonormal basis $\{H_1, \ldots, H_r\}$, which makes $\mathfrak{a}$ into a euclidean space. A variable element can be written

$$H = h_1 H_1 + \cdots + h_r H_r,$$

and $(h_1, \ldots, h_r)$ are the euclidean coordinates. We have the change of variables

$$a_i = \exp H_i \qquad \text{and} \qquad a_i \frac{\partial}{\partial a_i} = \frac{\partial}{\partial h_i} = \tilde{H}_i.$$

The Casimir (negative Laplacian) operator on A is then given as

$$\omega_{B,\mathfrak{a}} = \sum_{i=1}^{r} \tilde{H}_i^2.$$

Under the above change of variables, it may be viewed as the ordinary Laplacian on $\mathfrak{a}$.

Let E be any invariant differential operator on A. Then E is in the polynomial algebra

$$E \in \mathbf{C}[\tilde{H}_1, \ldots, \tilde{H}_r] \qquad \text{so} \qquad E = P_E(\tilde{H}_1, \ldots, \tilde{H}_r)$$

with some polynomial P_E. For $\zeta \in \mathfrak{a}_{\mathbf{C}}^{\vee}$ we have by ordinary calculus

$$E\chi_\zeta = P_E(\zeta)\chi_\zeta$$

with this same polynomial P_E. For Casimir, we get

(1) $$\omega_{B,\mathfrak{a}} e^{\zeta} = \langle \zeta, \zeta \rangle e^{\zeta} \qquad \text{or} \qquad \omega_{B,\mathfrak{a}} \chi_\zeta = \langle \zeta, \zeta \rangle \chi_\zeta,$$

where $\langle \zeta, \zeta \rangle = B(\zeta, \zeta)$ is the value of the Cartan type form B (the trace form). Note that we are using here the symmetric form, so for ζ complex, $\langle \zeta, \zeta \rangle$ is not necessarily real, let alone positive.

From Chapter III, Proposition 1.5, we know that for $D \in \mathrm{IDO}(G)^K$,

(2) $$D\varphi_\zeta = \mathrm{ev}(D, \varphi_\zeta)\varphi_\zeta = \mathrm{ev}(\mathbf{h}(D), \chi_\zeta)\chi_\zeta.$$

By Chapter VII, Theorem 3.4, we know that $\mathbf{h}(\omega) = \omega_{B,\mathfrak{a}} - \langle \rho, \rho \rangle$. Hence from (1), we get

(3) $$\boxed{\mathrm{ev}(\omega, \varphi_\zeta) = \langle \zeta, \zeta \rangle - \langle \rho, \rho \rangle.}$$

In summary:

Theorem 1.1. *On spherical functions* φ_ζ *with* $\zeta \in \mathfrak{a}_{\mathbf{C}}^\vee$ *we have*

$$\omega(\varphi_\zeta) = \mathrm{ev}(\omega, \varphi_\zeta)\varphi_\zeta$$

with the eigenvalue

$$\mathrm{ev}(\omega, \varphi_\zeta) = \langle \zeta, \zeta \rangle - \langle \rho, \rho \rangle = \mathrm{ev}(\mathbf{h}(\omega), \chi_\zeta).$$

Remark. In the main spectral application, one deals with $\zeta = \mathbf{i}\lambda$ and $\lambda \in \mathfrak{a}^\vee$, in which case $\langle \mathbf{i}\lambda, \mathbf{i}\lambda \rangle = -\langle \lambda, \lambda \rangle$, and so $\mathrm{ev}(\omega, \varphi_{\mathbf{i}\lambda})$ is negative.

Since we deal only with the polar direct image for the present, we shall abbreviate the notation, and write

$$\pi_* = (\mathrm{Pol}_{A^+})_*.$$

Note that φ_ζ is K-bi-invariant, and can be viewed as a function on the group or a function on A. Although we sometimes use the same letter for both, in laying down the basic formulas we make the distinction, so we let

$$\varphi_\zeta^A = \text{restriction of } \varphi_\zeta \text{ to } A.$$

Directly from the definition of the direct image, φ_ζ^A is an eigenfunction of $\pi_*(\omega)$ and on A^+,

$$\pi_*(\omega)\varphi_\zeta^A = \mathrm{ev}(\omega, \varphi_\zeta)\varphi_\zeta^A. \tag{4}$$

We shall also consider a multiple of φ_ζ, so we make general remarks about multiples of eigenfunctions.

Let F, G be a pair of functions on a manifold, and suppose $G = \eta F$ with some positive function η. Let D be a differential operator. Then F is an eigenfunction of D, namely

$$DF = \mathrm{ev}(D, F)F,$$

if and only if G is an eigenfunction of $\eta D \circ \eta^{-1}$, and in this case the eigenvalue is the same, that is

(5) $\eta D \circ \eta^{-1} G = \mathrm{ev}(D, F)G$, or $\mathrm{ev}(\eta D \circ \eta^{-1}, G) = \mathrm{ev}(D, F)$.

In the application, we also translate a differential operator by a function γ. Then we have trivially

$DF = \mathrm{ev}(D, F)F$ if and only if $(D+\gamma)F = (\mathrm{ev}(D, F)+\gamma)F$, so

$$\mathrm{ev}(D+\gamma, F) = \mathrm{ev}(D, F) + \gamma.$$

We return to the special case at hand. Simultaneously with φ_ζ^A we shall consider the **function** ψ_ζ on A^+ **defined by** the $J^{1/2}$ multiple of φ_ζ^A, that is

$$\psi_\zeta = J^{1/2}\varphi_\zeta^A.$$

For J, see Chapter VII, §4. By Chapter VII, Theorem 4.5, we know that

(6)
$$\begin{aligned} \pi_*(\omega) &= J^{-1/2}\omega_{B,\mathfrak{a}} \circ J^{1/2} - \gamma_J \\ \text{or} \qquad \omega_{B,\mathfrak{a}} &= J^{1/2}\pi_*(\omega) \circ J^{-1/2} + \gamma_J, \end{aligned}$$

with the function γ_J defined following Theorem 4.7 of Chapter VII, namely

$$\gamma_J = J^{-1/2}\omega_{B,\mathfrak{a}}(J^{1/2}).$$

We can then apply the preceding remarks, and conclude that ψ_ζ is an eigenfunction of $\omega_{B,\mathfrak{a}}$ with the eigenvalue $\mathrm{ev}(\omega, \varphi_\zeta) + \gamma_J$, namely

(7)
$$\omega_{B,\mathfrak{a}}(\psi_\zeta) = \omega_{B,\mathfrak{a}}(J^{1/2}\varphi_\zeta^A) = (\mathrm{ev}(\omega, \varphi_\zeta) + \gamma_J)\psi_\zeta.$$

Thus we see that the pair of differential equations (4) and (7) go together. We shall determine more general solutions of this pair of equations, namely we shall determine pairs of functions on A^+,

$$(F_\zeta, G_\zeta) \qquad \text{with} \quad G_\zeta = J^{1/2}F_\zeta$$

satisfying the pair of differential equations which we call the **Harish-Chandra Casimir equations**:

EIGEN 1. $\pi_*(\omega)F_\zeta = (\langle\zeta,\zeta\rangle - \langle\rho,\rho\rangle)F_\zeta.$

EIGEN 2. $\omega_{B,\mathfrak{a}}G_\zeta = (\langle\zeta,\zeta\rangle - \langle\rho,\rho\rangle + \gamma_J)G_\zeta.$

We have quite generally:

Lemma 1.2. *Let (F_ζ, G_ζ) be a pair of functions on A^+, with $G_\zeta = J^{1/2}F_\zeta$. Then F_ζ satisfies* **EIGEN 1** *if and only if G_ζ satisfies* **EIGEN 2.**

Proof. This is just a more general application of the remarks concerning conjugation of a differential operator and its eigenfunctions, i.e. formula (5).

In the next section, we shall determine general solutions of the two differential equations, generating all solutions in a suitable sense. Note that φ_ζ is invariant under the group W, but our general solution F_ζ will not be. The function φ_ζ will be obtained by taking a linear combination of translates of F_ζ by elements of W (the W-trace).

VIII, §2. THE HARISH-CHANDRA SERIES AND GANGOLLI ESTIMATE

We start with a lemma on relevant characters. We define:

L = semigroup of linear combinations

$$\sum_{\alpha \in \mathcal{R}(\mathfrak{n})} n_\alpha \alpha \qquad \text{with } n_\alpha \in \mathbf{Z}, n_\alpha \geqq 0.$$

It turns out that this semigroup has a basis over $\mathbf{Z}_{\geqq 0}$.

Lemma 2.1. *As usual, let α_{ij} be the character of $\mathfrak{a}$ given by $\alpha_{ij}(\mathrm{diag}(x_1, \ldots, x_n)) = x_i - x_j$ $(i < j)$. Then*

$$\alpha_{12}, \alpha_{23}, \ldots, \alpha_{n-1,n}$$

are linearly independent over **R**, *and generate* **L** *over* $\mathbf{Z}_{\geqq 0}$.

Proof. Both the **R**-linear independence and the fact that these elements generate **L** over $\mathbf{Z}_{\geqq 0}$ are immediate. Note that if $i < j$ and $j \neq i + 1$,

$$\alpha_{ij} = \alpha_{i,i+1} + \alpha_{i+1,i+2} + \cdots + \alpha_{j-1,j}.$$

The elements forming the basis of Lemma 2.1 are called the **simple** $(\mathfrak{a}, \mathfrak{n})$ characters, and the set of these characters $\alpha_{i,i+1}$ is denoted by $\mathcal{S}(\mathfrak{n})$.

Remark. The general definition of a **simple character** α is that it cannot be written as a sum $\alpha = \beta + \gamma$ with $\beta, \gamma \in \mathcal{R}(\mathfrak{n})$. The general case of Lemma 2.1 is not difficult to prove, cf. [Hel 78], Chapter III, Theorem 5.7.

The simple $(\mathfrak{a}, \mathfrak{n})$-characters are usually denoted by $\alpha_1, \ldots, \alpha_r$. We let:

$\mathbf{L}_\mathbf{Z}$ = free abelian group generated by the simple $(\mathfrak{a}, \mathfrak{n})$-characters.

$\mathbf{L}^+$ = set of non-zero elements of **L**.

$$\begin{aligned}\mathbf{L}_{\mathbf{R}} &= \mathbf{R} \otimes \mathbf{L}_{\mathbf{Z}} \\ &= \mathbf{R}\text{-vector space with basis the simple } (\mathfrak{a}, \mathfrak{n})\text{-characters.}\end{aligned}$$

Then $\mathbf{L}_{\mathbf{Z}}$ is a lattice in $\mathbf{L}_{\mathbf{R}}$, namely the lattice of integral linear combinations

$$\mu = \sum_{\alpha \in \mathcal{S}(\mathfrak{n})} m_\alpha \alpha, \qquad \text{with } m_\alpha \in \mathbf{Z}.$$

We can also say that $\mathbf{L}$ itself is the semi lattice of elements whose coefficients m_α are $\geqq 0$. For $\mathbf{L}^+$ the condition is that $m_\alpha > 0$ for some α.

On the vector space $\mathbf{L}_{\mathbf{R}}$ we have the norm given by the sum of the absolute value of the coefficients. We shall be concerned with the restriction of this norm to $\mathbf{L}$, and we define the **degree** of an element $\mu \in \mathbf{L}$ to be

$$\deg(\mu) = |\mu| = \sum_{\alpha \in \mathcal{S}(\mathfrak{n})} m_\alpha.$$

We shall be interested in series of the form

$$f(H) = f_c(H) = \sum_{\mu \in \mathbf{L}} c_\mu e^{-\mu(H)} \tag{1}$$

with coefficients $c_\mu \in \mathbf{C}$. Thus f is determined by its sequence of coefficients $\{c_\mu\}_{\mu \in \mathbf{L}}$. We say that $\{c_\mu\}$ has **at most polynomial growth** if there exists a constant M such that

$$|c_\mu| = O(|\mu|^M) \qquad \text{for } |\mu| \to \infty.$$

It is immediate that if $\{c_\mu\}$ has at most polynomial growth, then the series (1) above converges absolutely for $H \in \mathfrak{a}^+$, i.e. the open positive cone of elements such that $\alpha(H) > 0$ for all $\alpha \in \mathcal{S}(\mathfrak{n})$ (or equivalently, all $\alpha \in \mathcal{R}(\mathfrak{n})$). The convergence is uniform on the domain defined by the inequality

$$\alpha(H) \geqq \epsilon > 0 \qquad \text{for every } \alpha \in \mathcal{S}(\mathfrak{n}).$$

On this region, the series is dominated by the series

$$\sum |\mu|^M e^{-\epsilon|\mu|},$$

which converges because the number of $\mu \in \mathbf{L}$ such that $|\mu| = n$ is $\leqq n^{r-1}$. Note that the function f_c can then be extended to an analytic function on the complex domain of $H \in \mathfrak{a}_{\mathbf{C}}$ such that $\mathrm{Re}(H) \in \mathfrak{a}^+$, because the series (1) for f_c converges absolutely uniformly on the region defined by

$$\alpha(\mathrm{Re}(H)) \geqq \epsilon > 0 \qquad \text{for every simple } \alpha \in \mathcal{R}(\mathfrak{n}), \text{ i.e. } \alpha \in \mathcal{S}(\mathfrak{n}).$$

We denote the vector space of series (1) whose coefficients have at most polynomial growth by Gan(**L**), and call it the **Gangolli algebra.** It is indeed an algebra, because it is immediately verified that if $f, g \in \text{Gan}(\mathbf{L})$, then also $fg \in \text{Gan}(\mathbf{L})$. Note that if

$$g = \sum_{\mu \in \mathbf{L}} b_\mu e^{-\mu} \quad \text{and} \quad fg = \sum_{\mu \in \mathbf{L}} d_\mu e^{-\mu},$$

then d_μ is the convolution

$$d_\mu = (c * b)_\mu = \sum_{\nu+\sigma=\mu} c_\nu b_\sigma.$$

The above sum is taken for $\nu, \sigma \in \mathbf{L}$, with $\nu + \mu = \sigma$.

Examples. Given a function f on A, we let $f^{\mathfrak{a}}$ or $f_{\mathfrak{a}}$ be the corresponding function on $\mathfrak{a}$ via the exponential map, that is

$$f_{\mathfrak{a}}(H) = f^{\mathfrak{a}}(H) = f(\exp H).$$

In the reverse direction, given a function f on $\mathfrak{a}$, we define f^A or f_A on A to be the function

$$f_A(a) = f^A(a) = f(\log a).$$

We let $\text{Gan}(\mathbf{L})^A$ be the image of Gan(**L**) under the map $f \mapsto f^A$. We have $g_\alpha, S_\alpha^{-1} \in \text{Gan}(\mathbf{L})$, and we have the explicit series

$$\tag{2} \tfrac{1}{2} g_\alpha^{\mathfrak{a}} = \sum_{k=1}^{\infty} e^{-2k\alpha},$$

$$\tag{3} \tfrac{1}{2} S_{\alpha,\mathfrak{a}}^{-1} = \sum_{k=1}^{\infty} e^{-(2k-1)\alpha}.$$

We won't need such explicit series; we shall only need to know that some functions are in Gan(**L**), as well as the terms of lowest degree in (2), (3). In any case, we see that

$$\mathfrak{R} = \mathbf{R}[\ldots, g_\alpha, \ldots] \subset \text{Gan}(\mathbf{L}).$$

Define $\text{Gan}^+(\mathbf{L})$ to be the ideal of Harish-Chandra series in Gan(**L**) whose constant term is 0, in other words, $c_0 = 0$. Then we also get the inclusion of ideals

$$\mathfrak{R}_+ \subset \text{Gan}^+(\mathbf{L}).$$

The function J is defined by

$$J_{\mathfrak{a}} = \prod_{\alpha \in \mathcal{R}(\mathfrak{n})} (e^\alpha - e^{-\alpha}) = \prod_{\alpha \in \mathcal{R}(\mathfrak{n})} e^\alpha (1 - e^{-2\alpha}).$$

Therefore the following functions are in the Harish-Chandra algebra, and we give their constant term:

(4) $J_{\mathfrak{a}}^{-1/2} = e^{-\rho}(1 + \text{higher terms}),$

(5) $e^{-\rho} J_{\mathfrak{a}}^{1/2} = 1 + \text{higher terms},$

(6) $$\gamma_{J,\mathfrak{a}} = \sum_{\mu \in \mathbf{L}} h_\mu e^{-\mu} = h_0 + \text{higher terms, and } h_0 = \langle \rho, \rho \rangle,$$

where the higher terms are in $\text{Gan}^+(\mathbf{L})$. The function γ_J originally comes from Theorems 4.5 and 4.7 of Chapter X, and appeared in (6) of the preceding section. The constant term $\langle \rho, \rho \rangle$ comes from a direct application of the ordinary euclidean Laplacian to $e^{\rho(H)}$, in terms of ordinary coordinates, so by (3) of Chapter VII, §3, the constant term of (6) is

$$\sum_{i=1}^{r} \rho(H_i)^2 = \langle \rho, \rho \rangle.$$

In his construction of series for spherical functions, Harish-Chandra used exponential growth for the coefficients [Har 58a], Lemma 28. Gangolli eventually proved the polynomial growth estimates [Gan 71], see [GaV 88], Chapter 4, Lemmas 4.3.1 and 4.3.5, used to prove Corollary 4.3.6. A stronger estimate than Harish-Chandra's had been needed by Helgason [Hel 66] to get the Paley–Wiener surjection. In his book [Hel 84], Helgason uses an estimate implied by Gangolli's, but sufficient to carry out Paley–Wiener, and rapidly proved. Gangolli's estimates take a couple of pages, whereas the intermediate estimate of small exponential growth comes out faster, so we use it.

We say that a sequence $\{c_\mu\}$ has **arbitrarily small exponential growth** if given $H \in \mathfrak{a}^+$ there exists a constant K_H such that for all μ, we have

(7) $$|c_\mu| \leqq K_H e^{\mu(H)}.$$

It is then clear that under this condition, the series

$$\sum_{\mu \in \mathbf{L}} c_\mu e^{-\mu(H)}$$

converges absolutely, because for every $\epsilon > 0$ it is dominated by

$$\sum_{\mu \in \mathbf{L}} e^{\mu(\epsilon H) - \mu(H)} = \sum_{\mu \in \mathbf{L}} e^{-\mu((1-\epsilon)H)}.$$

For simplicity, we omit the word "arbitrarily".

Lemma 2.2. *If $\{b_\mu\}$ has polynomial growth and $\{c_\mu\}$ has small exponential growth, then $\{(b * c)_\mu\}$ has small exponential growth.*

Proof. Immediate.

A Harish-Chandra series $\sum c_\mu e^{-\mu}$ will be said to be of **small exponential growth type** if the sequence of its coefficients has small exponential growth.

A change of variables makes a Harish-Chandra series into a power series, even of complex variables, as follows. Take $H \in \mathfrak{a}_{\mathbf{C}}$,

$$H = \mathrm{Re}(H) + \sqrt{-1}\,\mathrm{Im}(H).$$

Let $\alpha_1, \dots, \alpha_r$ be the simple $(\mathfrak{a}, \mathfrak{n})$-characters, and let

$$z_i = e^{-\alpha_i(H)} = a^{-\alpha_i}.$$

Let $\epsilon > 0$. Then a region defined by the inequalities $\alpha_i(\mathrm{Re}(H)) \geqq \epsilon > 0$ (so $\mathrm{Re}(H) \in \mathfrak{a}^+$) is mapped to discs defined by $|z_i| \leqq e^{-\epsilon}$ for all $i = 1, \dots, r$. Writing

$$\mu = \sum m_i \alpha_i \quad \text{and} \quad c_\mu = c_{(m)},$$

we have

$$\sum c_\mu e^{-\mu(H)} = \sum c_{(m)} z^{(m)} \quad \text{where} \quad z^{(m)} = z_1^{m_1} \cdots z_r^{m_r}.$$

The small exponential growth of the coefficients then implies that the power series converges absolutely and uniformly on the product of the above discs. In particular, the coefficients are uniquely determined by the function defined by the series, as follows from Cauchy's theorem applied by taking the repeated integral over circles around the center of each disc.

We shall now define for each $\zeta \in \mathfrak{a}_{\mathbf{C}}^\vee$ a pair of functions $F_\zeta, G_\zeta = J^{1/2} F_\zeta$ satisfying the Casimir eigenfunction differential equations coming from the theory of spherical functions. Cf. [Har 58a], §8. We do this by series of small exponential growth type. Gangolli [Gan 71] showed that they are actually of polynomial growth type in the above mentioned references. From Lemma 2.2 it is clear that we may define either one first, and then use Lemma 1.2 to get the other. It turns out that it is easier to deal with G_ζ first. We shall use the method of undetermined coefficients. We consider the series

$$\text{(8a)} \qquad F_\zeta^{\mathfrak{a}} = e^{\zeta - \rho} \sum_{\mu \in \mathbf{L}} f_\mu(\zeta) e^{-\mu},$$

$$\text{(8b)} \qquad G_\zeta^{\mathfrak{a}} = J_{\mathfrak{a}}^{1/2} F_\zeta^{\mathfrak{a}} = e^{\zeta} \sum_{\mu \in \mathbf{L}} g_\mu(\zeta) e^{-\mu}.$$

After a while, as in freshman calculus, one sometimes omits the index $\mathfrak{a}$, leaving to the reader to interpret a formula on A in terms of a variable $a = \exp H$, or on $\mathfrak{a}$ in terms of a variable $H = \log a$. It's relatively harmless.

In any case, we note from (4) and (5) that

$$f_0(\zeta) = g_0(\zeta). \tag{8c}$$

In this section, we essentially deal with a fixed ζ, and determine solutions of the eigenfunction equations, although we do give some additional information about some possible uniformities for ζ in certain domains. A later section will provide a more detailed analysis on the analytic and meromorphic dependence on ζ. We view first $f_\mu(\zeta)$ and $g_\mu(\zeta)$ as unknowns for $\mu \neq 0$. The Casimir eigenfunction equation

$$\omega_{B,\mathfrak{a}} G_\zeta = (\langle \zeta, \zeta \rangle - \langle \rho, \rho \rangle + \gamma_J) G_\zeta \tag{9}$$

is equivalent to the **recursion relations**

$$(\langle \mu, \mu \rangle - 2\langle \mu, \zeta \rangle) g_\mu(\zeta) = \sum_{\substack{\nu \neq 0 \\ \nu, \mu - \nu \in \mathbf{L}}} h_\nu g_{\mu - \nu}(\zeta). \tag{10}$$

This is immediate since we can differentiate term by term, and

$$\omega_{B,\mathfrak{a}} e^\zeta = \langle \zeta, \zeta \rangle e^\zeta$$

for $\zeta \in \mathfrak{a}_{\mathbf{C}}^\vee$. We also used the fact that the constant term in the series expansion of $\gamma_{J,\mathfrak{a}}$ cancels the $-\langle \rho, \rho \rangle$ in the eigenvalue for the differential operator. This feature was exploited first in [Gan 71].

It is now clear that we can solve recursively for the coefficients $g_\mu(\zeta)$ in terms of lower level coefficients, starting with $g_0 = 1$, provided that the factor $\langle \mu, \mu \rangle - 2\langle \mu, \zeta \rangle$ is not 0. Thus we define the **hyperplane** (for $\mu \in \mathbf{L}^+$):

hyp(μ) = hyperplane of all ζ such that $\langle \mu, \mu \rangle - 2\langle \mu, \zeta \rangle = 0$.

Theorem 2.3. *Let $g_0 = 1$, and let g_μ be determined recursively by (10). Then g_μ is a rational function whose poles are among the hyperplanes* hyp$(\mu - \nu)$ *such that $\mu - \nu \in \mathbf{L}$ and $\nu \neq 0$. Let $\zeta \in \mathfrak{a}_{\mathbf{C}}^\vee$ but $\zeta \notin$* hyp(μ) *for all $\mu \in \mathbf{L}$, $\mu \neq 0$. Then the coefficients $\{g_\mu(\lambda)\}$ have small exponential growth. Let $G_\zeta^{\mathfrak{a}}$ be the function having series (8b), and let $F_\zeta^{\mathfrak{a}} = J_{\mathfrak{a}}^{-1/2} G_\zeta^{\mathfrak{a}}$, with series (8a). Then the coefficients $\{f_\mu(\zeta)\}$ also have small exponential growth. In other words, for each $H \in \mathfrak{a}^+$ there exists $K_{\zeta,H}$ such that for all μ we have*

$$|f_\mu(\zeta)|, |g_\mu(\zeta)| \leqq K_{\zeta,H} e^{\mu(H)}.$$

The constant $K_{\zeta,H}$ can be taken uniformly for ζ in a compact set not intersecting the above hyperplanes, and for H in a compact subset of $\mathfrak{a}^+$.

Proof. The quadratic term $\langle\mu, \mu\rangle$ dominates the linear term $2\langle\zeta, \mu\rangle$, so there exists a constant $C > 0$ such that for all $\mu \neq 0$ we have

$$(*) \qquad |g_\mu(\zeta)| \leqq C|\mu|^{-2} \sum_{\substack{\nu\neq 0\\ \nu,\mu-\nu\in \mathbf{L}}} |h_\nu||g_{\mu-\nu}(\zeta)|.$$

The next estimate applies to any sequence $\{g_\mu\}$ satisfying $(*)$ for all μ, with any sequence $\{h_\nu\}$ having at most polynomial growth, or even such that the series $\sum |h_\nu|e^{-\nu(H)}$ converges for all $H \in \mathfrak{a}^+$. We leave ζ out of the notation for this general estimate. For given $H \in \mathfrak{a}^+$, let

$$N_0^2 \geqq C \sum_{\nu\in\mathbf{L}} |h_\nu|e^{-\nu(H)}.$$

Let K be a constant such that

$$(**) \qquad |g_\sigma| \leqq Ke^{\sigma(H)} \qquad \text{for all } \sigma \text{ with } |\sigma| \leqq N_0.$$

Let $N \in \mathbf{Z}^+$ and $N > N_0$. Assume $(**)$ inductively for g_σ with $|\sigma| < N$. Let $|\mu| = N$. Then

$$\begin{aligned} |g_\mu| &\leqq C|\mu|^{-2} \sum |h_\nu|Ke^{\mu(H)-\nu(H)} \\ &\leqq Ke^{\mu(H)} \end{aligned}$$

thereby proving the desired inequality. The uniformity statements are clear, so the theorem is proved.

The uniformity statement of Theorem 2.3 came from the very first inequality $(*)$, with the constant C. Any region in $\mathfrak{a}^\vee$ for which we can take such a uniform constant will then give rise to the conclusion of Theorem 2.4 uniformly for λ in that region. For $\lambda \in \mathfrak{a}^\vee$ **we define**:

$$\begin{aligned} \lambda \geqq 0 \quad &\iff \quad \langle\alpha, \lambda\rangle \geqq 0 \quad \text{for all } (\mathfrak{a}, \mathfrak{n})\text{-characters } \alpha \\ &\iff \quad \langle\alpha_i, \lambda\rangle \geqq 0 \quad \text{for all simple characters } \alpha_i, \\ &\qquad\qquad\qquad\qquad\qquad i = 1, \ldots, r. \end{aligned}$$

We define $\lambda \leqq 0$ similarly, using the inequality $\leqq 0$ instead of $\geqq 0$. We call the sets of such λ **semipositive** or **seminegative** respectively. We define $\lambda > 0$ and $\lambda < 0$ similarly, with the inequalities > 0 and < 0 respectively. Cf. Chapter I, §4. For $\lambda \leqq 0$ we have

$$(11) \qquad |\langle\mu, \mu\rangle - 2\langle\mu, \lambda\rangle| \geqq \langle\mu, \mu\rangle.$$

This means that for ζ such that $\mathrm{Re}(\zeta) \leqq 0$ we can take the constant C uniformly, in fact we can take $C = 1$. Thus we obtain the small exponential estimate, which suffices but is weaker than Gangolli's polynomial estimate [Gan 71].

Theorem 2.4. *In Theorem 2.3, the constant* $K_{\zeta,H}$ *is uniform for* $\text{Re}(\zeta) \leqq 0$. *In other words, given* $H \in \mathfrak{a}^+$, *there exists* K_H *such that for all* $\mu \in \mathbf{L}$,

$$|f_\mu(\zeta)|, |g_\mu(\zeta)| \leqq K_H e^{\mu(H)} \qquad \text{for all } \zeta \text{ with } \text{Re}(\zeta) \leqq 0.$$

Remark. Both in [Hel 84] and [GaV 88], Theorem 2.3 is proved first for the function F_ζ instead of G_ζ. Since the recursion conditions for F_ζ are more complicated, it simplifies the presentation to do both Theorems 2.3 and 2.4 for G_ζ directly.

We strengthen (11) in the next lemma. For a real number ϵ, it is convenient to extend the definition of positivity to an inequality, that is:

$\lambda < \epsilon$ is **defined** to mean that $\lambda \in \mathfrak{a}^\vee$ and $\langle \alpha_i, \lambda \rangle < \epsilon$ for $i = 1, \ldots, r$.

The imaginary axis will play an important role later, and the next lemma provides information in a half plane slightly to the right of this axis.

Lemma 2.5.

(i) *A compact subset of* $\mathfrak{a}_{\mathbf{C}}^\vee$ *meets only a finite number of hyperplanes* $\text{hyp}(\mu)$ *with* $\mu \in \mathbf{L}^+$.

(ii) *There exists* $\epsilon > 0$ *such that if* $\text{Re}(\zeta) < \epsilon$ *then* $\zeta \notin \text{hyp}(\mu)$ *for all* $\mu \neq 0$.

Proof. The first assertion is immediate from the quadratic growth of the term $\langle \mu, \mu \rangle$, and the linear growth of the term $\langle \mu, \zeta \rangle$. The second assertion also comes from the fact that the quadratic form on $\mathfrak{a}$ is positive definite, so there exists a positive number b such that for all $\mu \in \mathbf{L}^+$ we have

$$\langle \mu, \mu \rangle \geqq b m(\mu)^2,$$

where $m(\mu)$ is the degree, that is, if $\mu = \sum_{i=1}^r m_i \alpha_i$ then $m(\mu) = \sum_{i=1}^r m_i$. Under the hypothesis of (ii), we then get

$$\text{(12)} \qquad \begin{aligned} |\langle \mu, \mu \rangle - 2\langle \mu, \zeta \rangle| &\geqq \langle \mu, \mu \rangle - \sum_{i=1}^{r} 2m_i \langle \alpha_i, \text{Re}(\zeta) \rangle \\ &\geqq b m(\mu)^2 - 2m(\mu)\epsilon. \end{aligned}$$

We can choose ϵ so that $bm^2 - 2m\epsilon > 0$ for all positive integers m, thus concluding the proof.

We stated Lemma 2.5 right away to continue easily the trend of thoughts presented by (11). However, we won't use (ii) until somewhat later. We turn to the eigenfunction property.

As observed in Lemma 1.2, the function F_ζ satisfies the Casimir equation **EIGEN 1** of §1. Theorem 2.3 and Lemma 2.2 show that it has a Harish-Chandra series of small exponential growth type. Using the explicit determination of the lowest degree term for $J^{1/2}$ and G_ζ, we get:

Proposition 2.6. *Let $\zeta \in \mathfrak{a}_{\mathbf{C}}^\vee$ but $\zeta \notin \mathrm{hyp}(\mu)$ for all $\mu \in \mathbf{L}^+$. Then $F_\zeta^{\mathfrak{a}}$ is the unique solution of equation* **EIGEN 1** *having a series expression of the form*

$$e^{\zeta-\rho}\sum_{\mu\in\mathbf{L}} f_\mu(\zeta)e^{-\mu},$$

with small exponential growth of its coefficients, and such that $f_0(\zeta)=1$. Given an arbitrary solution of **EIGEN 1**,

$$F_\zeta^{\#} = e^{\zeta-\rho}\sum_{\mu\in\mathbf{L}} f_\mu^{\#}(\zeta)e^{-\mu}$$

where the series has small exponential growth type, we have

$$F_\zeta^{\#} = f_0^{\#}(\zeta)F_\zeta.$$

Proof. The first statement comes from using the explicit determination of the lowest degree term for $J^{1/2}$ and G_ζ. The second statement comes from transferring to the same assertion concerning an arbitrary solution $G_\zeta^{\#}$ of equation **EIGEN 2**, §1, and using the recursion relations (10), which show that

$$G_\zeta^{\#} = g_0^{\#}(\zeta)G_\zeta.$$

This proves the proposition.

The next theorem is due to Harish-Chandra [Har 58a]. The proof we give is a substantial simplification by Helgason [Hel 72], [Hel 84], Chapter IV, Proposition 5.4.

Theorem 2.7. *Let $\pi: G^+ \to A^+$ be the polar projection. Then F_ζ is an eigenfunction of the direct image $\pi_*\mathrm{IDO}(G)^K$. In fact, for $D \in \mathrm{IDO}(G)^K$, F_ζ has the same eigencharacter as φ_ζ^A (cf.* §1, (2)) *namely*

$$(\pi_*D)F_\zeta = \mathrm{ev}(D,\varphi_\zeta)F_\zeta = \mathrm{ev}(\mathbf{h}(D),\chi_\zeta)F_\zeta.$$

Proof. By Proposition 5.5 of Chapter VI, $(\pi_* D)F_\zeta$ can be expressed by a Harish-Chandra series on A^+ (the series being given on $\mathfrak{a}^+$)

$$(13) \qquad (\pi_* D)F_\zeta = e^{\zeta-\rho} \sum_{\mu \in \mathbf{L}} f_\mu^D(\zeta) e^{-\mu}$$

of small exponential growth type. Here we are allowing the abuse of notation since the left side is a function on A whereas the right side is a function on $\mathfrak{a}$. By the commutativity of invariant differential operators on A, it follows that $\pi_* D$ and $\pi_* \omega$ commute, so $(\pi_* D)F_\zeta$ is an eigenfunction of $\pi_* \omega$ with the same eigenvalue as F_ζ itself. Differentiating with respect to the variable coordinates of H introduces only factors with polynomial growth in the coefficients, so $(\pi_* D)F_\zeta$ has an expression as a Harish-Chandra series of small exponential growth type. We apply Proposition 2.6, and get

$$(\pi_* D)F_\zeta = f_0^D(\zeta)F_\zeta.$$

We have to prove that $f_0^D(\zeta) = \mathrm{ev}(\mathbf{h}(D), \chi_\zeta)$. To do this, we combine $f_0(\zeta) = 1$ and the fact that the series for F_ζ in Proposition 2.6 starts with $e^{\zeta(H)}e^{-\rho(H)}$, with the more precise information from Proposition 5.5 of Chapter VI, which tells us that

$$\pi_* D = e^{-\rho}\mathbf{h}(D) \circ e^\rho + \text{terms in } \mathfrak{R}_+ \,\mathrm{IDO}(A).$$

Only the first term on the right of the equality sign matters to compute $f_0^D(\zeta)$. But then

$$\begin{aligned} e^{-\rho}\mathbf{h}(D)(e^\rho e^{\zeta-\rho}) &= e^{-\rho}\,\mathrm{ev}(\mathbf{h}(D), \chi_\zeta)e^\zeta \\ &= \mathrm{ev}(\mathbf{h}(D), \chi_\zeta)e^{\zeta-\rho}. \end{aligned}$$

This shows that $f_0^D(\zeta) = \mathrm{ev}(\mathbf{h}(D), \chi_\zeta)$, and concludes the proof of the theorem.

VIII, §3. THE c-FUNCTION AND THE W-TRACE

The last theorem of §2 shows that F_ζ satisfies one of the three properties characterizing spherical functions φ_ζ^A. The other two are the Weyl group invariance and the normalization (taking value 1 at the unit element). To obtain Weyl group invariance, it is natural to take the trace with respect to W, but an additional linear combination is necessary, as we now show following [Har 58a]. We found [Hel 84] and [GaV 88] useful.

It will be necessary to stay away from more hyperplanes than already encountered in §2. Thus we define the **hyperplane** $\mathrm{hyp}_\nu(w, w')$ for

$\nu \in \mathbf{L_Z}$ and $w \neq w' \in W$ by the condition:

$$\text{hyp}_\nu(w, w') = \text{subset of } \zeta \in \mathfrak{a}_\mathbf{C}^\vee \text{ such that } w\zeta - w'\zeta = \nu.$$

The action of w on $\mathbf{L}$ is the functorial action on functions, so $(w\zeta)(H) = \zeta(w^{-1}H)$ for $H \in \mathfrak{a}_\mathbf{C}$.

We define the **exceptional subset** of $\mathfrak{a}_\mathbf{C}^\vee$ by

$$\text{Exc}_W(\mathfrak{a}_\mathbf{C}^\vee) = \bigcup_{\substack{w\in W\\ \mu\in\mathbf{L}^+}} w\,\text{hyp}(\mu) \cup \bigcup_{\substack{w\neq w'\in W\\ \nu\in\mathbf{L_Z}}} \text{hyp}_\nu(w, w').$$

As was the case for the hyperplanes $\text{hyp}(\mu)$, a compact subset of $\mathfrak{a}_\mathbf{C}^\vee$ meets only a finite number of hyperplanes $\text{hyp}_\nu(w, w')$ with $\nu \in \mathbf{L_Z}$, $w \neq w' \in W$. Note that the hyperplanes whose union is the exceptional set are complex hyperplanes, and hence the complement of this union is open, connected, and dense in $\mathfrak{a}_\mathbf{C}^\vee$. It is also W-invariant, directly from its definition. We call this complement the **generic subset of** $\mathfrak{a}_\mathbf{C}^\vee$, and give it a symbol,

$$\text{Gen}_W(\mathfrak{a}_\mathbf{C}^\vee) = \mathfrak{a}_\mathbf{C}^\vee - \text{Exc}_W(\mathfrak{a}_\mathbf{C}^\vee).$$

Theorem 3.1. *For $\zeta \in \text{Gen}_W(\mathfrak{a}_\mathbf{C}^\vee)$, the functions $\{F_{w\zeta}\}_{w\in W}$ on A^+ are linearly independent.*

Proof. Each function $e^\rho F_{w\zeta}^\mathfrak{a}$ has its Harish-Chandra series expression

$$e^\rho F_{w\zeta}^\mathfrak{a} = \sum_{\mu\in\mathbf{L}} f_\mu(w\zeta)e^{w\zeta-\mu}.$$

By the assumption on ζ, we have

$$w\zeta - \mu \neq w'\zeta - \nu \qquad \text{for all } w \neq w' \in W \text{ and } \mu, \nu \in \mathbf{L}.$$

There exists an element $H \in \mathfrak{a}$ such that

$$(1) \qquad w\zeta(H) - \mu(H) \neq w'\zeta(H) - \nu(H)$$

for all $w \neq w' \in W$ and $\mu, \nu \in \mathbf{L}$. A non-trivial linear relation among the functions $e^\rho F_{w\zeta}^\mathfrak{a}$ $(w \in W)$ would then contradict Proposition 0.3, thus concluding the proof.

Remark. Having picked ζ generic we are in the "unramified" case, cf. Chapter III, Theorem 8.6. Although the ramified case was used by Harish-Chandra, the simplifications by Helgason, Gangolli, Rosenberg, and Anker have made its use unnecessary for the spherical inversion theory.

Corollary 3.2. *For $\zeta \in \mathrm{Gen}_W(\mathfrak{a}_{\mathbf{C}}^\vee)$, the functions $\{F_{w\zeta}\}_{w\in W}$ form a basis of the space of (real) analytic functions on A^+ which are eigenfunctions of $(\mathrm{Pol}_{A^+})_*(\mathrm{IDO}(G)^K)$ with the same eigencharacter as φ_ζ.*

Proof. Theorem 6.2 of Chapter VI giving the upper bound $|W|$ for the dimension of this eigenspace, combines with Theorem 3.1 which gives exactly $|W|$ linearly independent elements in it, to conclude the proof.

For simplicity of notation, we now work on the Lie algebra, so for $\zeta \in \mathfrak{a}_{\mathbf{C}}^\vee$,

$$\boxed{\varphi_\zeta(\exp H) = \varphi_\zeta^{\mathfrak{a}}(H) = \varphi^{\mathfrak{a}}(\zeta, H).}$$

By Corollary 3.2, for each $w \in W$ there is a constant $\mathbf{c}_w(\zeta)$ such that

$$\varphi_\zeta^{\mathfrak{a}} = \sum_{w\in W} \boldsymbol{c}_w(\zeta) F_{w\zeta}^{\mathfrak{a}} \quad \text{on } \mathfrak{a}^+,$$

and similarly in the multiplicative context with φ_ζ^A and $F_{w\zeta}^A$ on A^+. We define

$$\boldsymbol{c}(\zeta) = \boldsymbol{c}_{\mathrm{id}}(\zeta)$$

where id is the unit element of W. It is of course not accidental that we are using the same letter $\boldsymbol{c}$ that we used for $\boldsymbol{c}_{\mathrm{Har}}$, because it will be proved in a moment that $\boldsymbol{c} = \boldsymbol{c}_{\mathrm{Har}}$, and then we can omit the subscript Har or put it in, at will, in either context. Replacing ζ by $w'\zeta$ with any $w' \in W$ we conclude that

(2) $$\boldsymbol{c}_w(\zeta) = \boldsymbol{c}(w\zeta) \qquad \text{for all } w \in W.$$

Thus we obtain one of Harish-Chandra's main theorems.

Theorem 3.3. *Suppose $\zeta \in \mathfrak{a}_{\mathbf{C}}^\vee$ is generic, i.e. not in the exceptional set. Then $\varphi_\zeta^{\mathfrak{a}}$ can be expressed as a series on $\mathfrak{a}^+$, and more precisely as a linear combination*

$$\varphi_\zeta = \sum_{w\in W} \boldsymbol{c}(w\zeta) F_{w\zeta}^{\mathfrak{a}}$$

so that

$$e^\rho \varphi_\zeta = \sum_{w\in W} \boldsymbol{c}(w\zeta) e^{w\zeta} \sum_{\mu\in\mathrm{L}} f_\mu(w\zeta) e^{-\mu}$$

converging absolutely. Given ϵ, *the convergence is uniform on the set of elements* $H \in \mathfrak{a}^+$ *such that*

$$\alpha_i(H) \geqq \epsilon > 0 \qquad \textit{for every } i = 1, \ldots, r.$$

Note the structure of the expansion for φ_ζ, namely a degree 0 term, plus an infinite series over $\mu \neq 0$. More precisely, since $f_0(\zeta) = 1$ (all ζ),

$$e^\rho \varphi_\zeta = \sum_{w \in W} c(w\zeta) e^{w\zeta} + \sum_{\mu \neq 0} \sum_{w \in W} c(w\zeta) e^{w\zeta} f_\mu(w\zeta) e^{-\mu}.$$

Note how $e^\rho \varphi_\zeta$ is more natural than φ_ζ itself. In Chapter I, §4, and Chapter V, §7, we defined for $\lambda \in \mathfrak{a}^\vee$,

$$\lambda > 0 \quad \text{if and only if} \quad \langle \alpha, \lambda \rangle > 0 \quad \text{for all } \alpha \in \mathcal{R}(\mathfrak{n}).$$

Corollary 3.4. *Let* ζ *be generic in* $\mathfrak{a}_{\mathbf{C}}^\vee$ *and such that* $\mathrm{Re}(\zeta) > 0$. *Define* $H \to \infty$ *to mean* $\alpha_i(H) \to \infty$ *for all* $i = 1, \ldots, r$. *Similarly,* $a \to \infty$ *if* $a = \log H$ *and* $H \to \infty$. *Then*

$$\begin{aligned} c(\zeta) &= \lim_{H \to \infty} e^{-\zeta(H)} e^{\rho(H)} \varphi_\zeta^{\mathfrak{a}}(H) \\ &= \lim_{a \to \infty} a^{-\zeta} a^\rho \varphi_\zeta(a). \end{aligned}$$

Proof. What is happening is that $c(\zeta)$ appears as part of the constant term in the series on the right, and all the other terms go to 0 when $H \to \infty$. But we have to make this precise. We can write

$$\begin{aligned} e^{-(\zeta-\rho)(H)} \varphi_\zeta^{\mathfrak{a}}(H) &= c(\zeta) + c(\zeta) \sum_{\mu \neq 0} f_\mu(\zeta) e^{-\mu(H)} \\ &\quad + \sum_{w \neq \mathrm{id}} c(w\lambda) e^{(w\zeta - \zeta)(H)} \sum_{\mu \in \mathrm{L}} f_\mu(w\zeta) e^{-\mu(H)}. \end{aligned}$$

As $H \to \infty$, the sum over $\mu \neq 0$ on the right goes to 0, so it remains only to prove that the sum over $w \neq \mathrm{id}$ goes to 0. Here the sum has a non-zero first term, so the limit 0 comes from the other factor, taken care of by the following lemma, putting $\xi = \mathrm{Re}(\zeta)$.

Lemma 3.5. *Let* $w \in W$ *and* $w \neq \mathrm{id}$, *and* $\xi > 0$. *Then*

$$\lim_{H \to \infty} e^{(w\xi - \xi)(H)} = 0.$$

In other words, $(w\xi - \xi)(H) \to -\infty$ *as* $H \to \infty$.

Proof. In accordance with Chapter I, §4, and Chapter V, Proposition 7.1, write

$$\xi = x_1\lambda_1 + \cdots + x_r\lambda_r \qquad \text{with } x_i \in \mathbf{R}, x_i > 0.$$

Let $H = \operatorname{diag}(h_1, \ldots, h_n)$ with $n = r + 1$, so

$$w^{-1}H = \operatorname{diag}(h_{w(1)}, \ldots, h_{w(n)}),$$

viewing the element of the Weyl group as a permutation of the indices. Then

$$\begin{aligned}(w\xi)(H) - \xi(H) = x_1 h_{w(1)} + \cdots + x_r(h_{w(1)} + \cdots + h_{w(r)}) \\ - x_1 h_1 - \cdots - x_r(h_1 + \cdots + h_r).\end{aligned}$$

By hypothesis $h_{w(1)} - h_1 \leqq 0$ and tends to $-\infty$ as $H \to \infty$ if $w(1) \neq 1$. If $w(1) = 1$, we proceed inductively. Note first that for all i we have

$$h_{w(1)} + \cdots + h_{w(i)} - (h_1 + \cdots + h_i) \leqq 0,$$

and since $w \neq \mathrm{id}$ we must have strict inequality for some i, with the difference tending to $-\infty$. This concludes the proof of the lemma, and also of the corollary.

In Chapter V, we had defined $c_{\mathrm{Har}}(\zeta)$ and obtained a meromorphic continuation in terms of the gamma function, actually a holomorphic continuation on the set $\operatorname{Re}(\zeta) > 0$. We now put the two c-functions together.

Theorem 3.6. *For ζ generic we have $c(\zeta) = c_{\mathrm{Har}}(\zeta)$.*

Proof. First assume that ζ is not only generic but $\operatorname{Re}(\zeta) > 0$. By Corollary 3.4 and Chapter V, Theorem 6.5, or its reformulation Theorem 7.2, we conclude that $c(\zeta) = c_{\mathrm{Har}}(\zeta)$ (both functions satisfy the same asymptotic relation vis-à-vis the spherical function). To get equality on the whole open generic set, we have to give a further argument. For each $H \in \mathfrak{a}^+$ the function $\zeta \mapsto \varphi^{\mathfrak{a}}(\zeta, H)$ is analytic in ζ. Furthermore, the coefficients $f_\mu(w\zeta)$ are rational functions in ζ, and the series

$$\sum_{\mu \in \mathbf{L}} f_\mu(w\zeta)e^{-\mu(H)}$$

converges absolutely, with uniformity as stated in Theorem 3.3. Hence $\zeta \mapsto F^{\mathfrak{a}}_{w\zeta}(H)$ is analytic. The equality

$$\varphi^{\mathfrak{a}}(\zeta, H) = \sum_{w \in W} c_{\mathrm{Har}}(w\zeta)F^{\mathfrak{a}}_{w}(H)$$

is valid for $\mathrm{Re}(\zeta) > 0$, as we have just seen. By analyticity, it is valid on the generic set (which is connected). On the generic set, the coefficients $c(w\zeta)$ of $F^{\mathfrak{a}}_{w\zeta}$ are uniquely determined since the functions $\{F_{w\zeta}\}_{w\in W}$ are linearly independent. Hence $c(w\zeta) = c_{\mathrm{Har}}(w\zeta)$ for ζ in the generic set. This concludes the proof.

Remark. Harish-Chandra originally introduced the c-function as the coefficient in taking the linear combination to get the series expansion for the spherical function. He gave the other approach via the integral representation only afterwards. We have carried the approaches separately, and reversed the order. One advantage of this reversal is that we can use the single characterizing asymptotic property, and then plug in at once the properties following from the integral, e.g. that the c-function is holomorphic on the appropriate domain. This avoids giving an extra argument, as is done in expositions following Harish-Chandra (who didn't have the Bhanu-Murty Gindikin-Karpelevic formula in 1958).

VIII, §4. THE HELGASON AND ANKER SUPPORT THEOREMS

In Chapter IV, §7, we recalled the classical Paley–Wiener theorem which describes the Mellin transform of $C_c^\infty(A)$. This is part of the Mellin inversion theory and spherical inversion. Harish-Chandra gave a complete spherical inversion theorem on a space corresponding to the Schwartz space in ordinary Fourier analysis. Helgason proved the corresponding result for the Paley–Wiener space [Hel 66], up to a step requiring some estimates for the Harish-Chandra series for spherical functions in order to justify integrating such a series term by term. Gangolli [Gan 71] completed this step. The essential part consists in describing the support of what will turn out to be the inverse spherical transform. The present section carries out this step. Actually, Anker [Ank 91] pointed out that the support theorem could be extended to more general convex sets than balls, and we carry out right away this generalization, based on Chapter IV, §8.

We write the **Harish-Chandra series** as a sum of terms

$$(1)\quad \varphi^{\mathfrak{a}}(\zeta, H) = \sum_{\mu\in\mathrm{L}} \Phi^{(\mu)}(\zeta, H) \qquad \text{for } \zeta \text{ generic in } \mathfrak{a}^\vee_{\mathbf{C}},\ H \in \mathfrak{a}^+,$$

with the μ-term

$$(2)\qquad \Phi^{(\mu)}(\zeta, H) = \sum_{w\in W} c(w\zeta) f_\mu(w\zeta) e^{(w\zeta-\rho-\mu)(H)}$$

defined for ζ generic in $\mathfrak{a}_{\mathbf{C}}^{\vee}$ and $H \in \mathfrak{a}$. Functions of two variables give rise to integral operators. We consider the integral operators associated with the functions $\varphi^{\mathfrak{a}}(\zeta, H)$ and $\Phi^{(\mu)}(\zeta, H)$. Referring to Chapter V, Lemma 8.1, we define

$$\eta(\lambda) = |c^{-1}(\mathbf{i}w\lambda)|^2 \qquad \text{for } \lambda \in \mathfrak{a}^{\vee},$$

so η is real semipositive continuous on $\mathfrak{a}^{\vee}$. We define the **Harish-Chandra inverse spherical operator** ${}^t\Phi_\eta$ on a test function F by the integral

$$(3) \qquad ({}^t\Phi_\eta F)(H) = \int_{\mathfrak{a}^{\vee}} F(\mathbf{i}\lambda)\varphi^{\mathfrak{a}}(\mathbf{i}\lambda, H)\eta(\lambda)\,d\lambda.$$

We shall use various spaces of test functions. We first deal with the absolute convergence of the integral and possible derivatives. There is nothing delicate about this. We say that a bounded measurable function F on the imaginary space $\mathbf{i}\mathfrak{a}^{\vee}$ has **super polynomial decay** if for each positive integer N,

$$F(\mathbf{i}\lambda) = O(|\lambda|^{-N}) \qquad \text{for } |\lambda| \to \infty.$$

Lemma 4.1. *If F has super polynomial decay on $\mathbf{i}\mathfrak{a}^{\vee}$, then the integral for ${}^t\Phi_\eta F$ is absolutely convergent, and defines a linear map*

$${}^t\Phi_\eta\colon \textit{super polynomial decay}(\mathbf{i}\mathfrak{a}^{\vee}) \to C^{\infty}(\mathfrak{a}).$$

Proof. The function η has at most polynomial growth, so $F\eta$ has super polynomial decay. The function $\lambda \mapsto \varphi^{\mathfrak{a}}(\mathbf{i}\lambda, H)$ is bounded, uniformly for H in a compact set. Picking a basis for $\mathfrak{a}$, one can differentiate under the integral sign because such partial derivatives amount to invariant differential operators on the multiplicative group A applied to the lifted function $\varphi_{\mathbf{i}\lambda}(\exp H)$. This function is defined from characters by the Harish-Chandra integral of Chapter III, §1, taken over a compact set, so no convergence problem arises. The spherical functions are eigenfunctions of such differential operators, and their eigencharacters are polynomials in λ, so have polynomial growth. This justifies differentiating under the integral sign, and shows that ${}^t\Phi_\eta F$ is C^{∞}, thus concluding the proof of the lemma.

Our main concern in the present section and chapter is the Paley–Wiener space which we already discussed in Chapter IV, §8, where we defined the subspaces

$$\mathrm{PW}_C(\mathfrak{a}_{\mathbf{C}}^{\vee}) \qquad \text{and} \qquad \mathrm{PW}_C(\mathfrak{a}_{\mathbf{C}}^{\vee})^W$$

with respect to a convex set C which is W-stable.

Theorem 4.2. *Let $F \in PW_C(\mathfrak{a}_C^\vee)^W$ be a Paley–Wiener function of Paley–Wiener order $\leqq q^C$, invariant under W. Then ${}^t\Phi_\eta F$ has support in C.*

Proof. By the W-invariance of F, without loss of generality, it suffices to show that $({}^t\Phi_\eta F)(H) = 0$ for $H \in \mathfrak{a}^+$, $H \notin C$. We shall reduce the vanishing of the integral to the vanishing of the corresponding integral for each term of the double sum arising from (1) and (2). Then Corollary 8.4 of Chapter IV is applicable. By definition, we have

$$\begin{aligned}({}^t\Phi_\eta F)(H) &= \int_{\mathfrak{a}^\vee} F(\mathbf{i}\lambda)\varphi^{\mathfrak{a}}(\mathbf{i}\lambda, H)\eta(\lambda)\,d\lambda \\ &= \int_{\mathfrak{a}^\vee} F(\mathbf{i}\lambda)\sum_{\mu\in \mathrm{L}} \Phi^{(\mu)}(\mathbf{i}\lambda, H)\eta(\lambda)\,d\lambda \\ &= \sum_{\mu\in \mathrm{L}}\int_{\mathfrak{a}^\vee} F(\mathbf{i}\lambda)\Phi^{(\mu)}(\mathbf{i}\lambda, H)\eta(\lambda)\,d\lambda\end{aligned}$$

by Theorem 2.3 (Gangolli).

Each function $\Phi^{(\mu)}$ defines an **integral operator** ${}^t\Phi_\eta^{(\mu)}$ by

$$({}^t\Phi_\eta^{(\mu)} F)(H) = \int_{\mathfrak{a}} F(\mathbf{i}\lambda)\Phi^{(\mu)}(\mathbf{i}\lambda, H)\eta(\lambda)\,d\lambda.$$

It will therefore suffice to prove:

Theorem 4.3 ([Hel 66], [Ank 91]). *Let C be a W-stable compact convex set containing 0. Let $F \in PW_C(\mathfrak{a}_C^\vee)^W$. If $H \notin C$ then $({}^t\Phi_\eta^{(\mu)} F)(H) = 0$, i.e. ${}^t\Phi_\eta^{(\mu)} F$ has support in C.*

Proof. By definition, after bringing a λ-constant factor to the left, we get

$$\begin{aligned}&e^{(\rho+\mu)(H)}({}^t\Phi_\eta^{(\mu)} F)(H) \\ &\quad = \sum_{w\in W}\int_{\mathfrak{a}^\vee} F(\mathbf{i}\lambda)\,\mathbf{c}(\mathbf{i}w\lambda) f_\mu(\mathbf{i}w\lambda)e^{\mathbf{i}w\lambda(H)}\,\mathbf{c}(\mathbf{i}w\lambda)^{-1}\overline{\mathbf{c}(\mathbf{i}w\lambda)}^{-1}\,d\lambda\end{aligned}$$

by using Lemma 8.1(iii) of Chapter V (invariance of absolute value of $\mathbf{c}$ under the Weyl group, on the imaginary axis). The term $\mathbf{c}(\mathbf{i}w\lambda)$ cancels $\mathbf{c}(\mathbf{i}w\lambda)^{-1}$. By the above reference, we replace the conjugation by a minus sign inside the parentheses. We then make the change of variables $\lambda \mapsto w^{-1}\lambda$. We are then reduced to proving

$$\int_{\mathfrak{a}^\vee} F(\mathbf{i}\lambda) f_\mu(\mathbf{i}\lambda)e^{\mathbf{i}\lambda(H)}\,\mathbf{c}(-\mathbf{i}\lambda)^{-1}\,d\lambda = 0.$$

The function F is in the Paley–Wiener space. The function f_μ is a rational function. It is obtained as a linear combination of functions g_ν coming from the relation

$$F_\zeta = J^{-1/2} G_\zeta$$

of §2, so f_μ is a linear combination of rational functions g_ν whose poles ζ satisfy

$$\langle \mu, \mu \rangle = 2\langle \mu, \zeta \rangle \qquad \text{with } \mu \neq 0.$$

Thus the poles occur only on the positive $\mathfrak{a}^\vee$-axis, uniformly away from the imaginary axis. Hence Ff_μ satisfies the hypotheses of Corollary 8.4 of Chapter IV, whose application concludes the proof.

VIII, §5. AN L^2-ESTIMATE AND LIMIT

The Harish-Chandra series splits in two parts: a constant term which is an exponential (trigonometric) polynomial, and a series, summed over $\mu \neq 0$ in the lattice generated by the $(\mathfrak{a}, \mathfrak{n})$-characters. We view this second part as a perturbation of the first, that is a perturbation from the ordinary case of Fourier expansions. We want to estimate the effect of the perturbation term in some contexts, especially with a view to an application to spherical inversion in Chapter IX. The present section provides a technical estimate taken from Rosenberg's "quick proof" of Harish-Chandra inversion [Ros 77], p. 148, and is meant to be read in conjunction with Chapter IX, Axiom **W 3** and Lemma 3.3.

Referring back to the definition of $J_{\mathfrak{a}}^{1/2} F = G$ in §2, we abbreviate the notation and write

$$(1) \qquad \theta_\zeta = \sum_{\mu \in \mathbf{L}} g_\mu(\zeta) e^{-\mu} = 1 + \sum_{\mu \neq 0} g_\mu(\zeta) e^{-\mu},$$

because the series on the right amounts to a theta type series (in several variables). The g_μ are rational functions satisfying the Gangolli estimate as in Theorem 2.3. In the notation of §2, we then have

$$(2) \qquad G_{\mathbf{i}\lambda}^{\mathfrak{a}} = e^{\mathbf{i}\lambda} \theta_{\mathbf{i}\lambda} \qquad \text{for } \lambda \in \mathfrak{a}^\vee.$$

As in Chapter V, Lemma 8.3, we shall divide by $\boldsymbol{c}(\mathbf{i}\lambda)$ and use the function

$$\gamma(w, \lambda) = \frac{\boldsymbol{c}(\mathbf{i}w\lambda)}{\boldsymbol{c}(\mathbf{i}\lambda)} \qquad \text{with } |\gamma(w, \lambda)| = 1,$$

to regularize the coefficients on the imaginary axis (actually at 0). From §2, (8b), and from Theorem 3.3, we obtain for $H \in \mathfrak{a}^+$,

$$(3) \qquad c^{-1}(\mathbf{i}\lambda)J_{\mathfrak{a}}^{1/2}(H)\varphi^{\mathfrak{a}}(\mathbf{i}\lambda, H) = \sum_{w\in W}\gamma(w,\lambda)e^{\mathbf{i}w\lambda}\theta_{\mathbf{i}w\lambda}(H)$$
$$= \sum_{w\in W}\gamma(w,\lambda)e^{\mathbf{i}w\lambda} + E(\mathbf{i}\lambda, H)$$

where $E(\mathbf{i}\lambda, H)$ is an error (perturbation) term given by

$$(4) \qquad E(\mathbf{i}\lambda, H) = \sum_{w\in W} E_w(\mathbf{i}\lambda, H)$$

with

$$E_w(\mathbf{i}\lambda, H) = \gamma(w,\lambda)e^{\mathbf{i}w\lambda}[\theta(\mathbf{i}w\lambda, H) - 1]$$
$$= \gamma(w,\lambda)e^{\mathbf{i}w\lambda}\sum_{\mu\neq 0} g_\mu(\mathbf{i}w\lambda)e^{-\mu(H)}.$$

The next section and Chapter IX will formulate analogues of Fourier inversion in the spherical case. The error term measures the extent to which spherical inversion differs from what is essentially a Fourier term, namely the character sum. In Rosenberg's proof determining the constant in the inversion formula, one needs to know that a certain error term depending on a parameter ϵ has an L^2-limit equal to 0 as $\epsilon \to 0$. See Axiom **W 3** of Chapter IX, §3, used in the proof of Chapter IX, Theorem 3.1. We are now going to prove that the axiom is satisfied in the concrete case under consideration.

Let $f \in C_c^\infty(\mathfrak{a}^+)$. for $\epsilon > 0$ and $\lambda \in \mathfrak{a}^\vee$, define

$$E_{f,\epsilon}(\lambda) = \epsilon^{-r/2}\int_{\mathfrak{a}^+} f(H)E(\mathbf{i}\lambda, H/\epsilon)\,dH.$$

Lemma 5.1. *For $f \in C_c^\infty(\mathfrak{a}^+)$ we have for the $L^2(d\lambda)$ norm:*

$$\lim_{\epsilon\to 0}\|E_{f,\epsilon}\|_2 = 0.$$

Proof. The expression $E(\mathbf{i}\lambda, H/\epsilon)$ is a sum of terms $E_w(\mathbf{i}\lambda, H/\epsilon)$ so $E_{f,\epsilon}$ is a sum of terms

$$E_{f,\epsilon} = \sum_{w\in W} E_{w,f,\epsilon}.$$

Also $\|E_{f,\epsilon}\|_2 \leqq \sum_w \|E_{w,f,\epsilon}\|_2$ so it suffices to prove that for such $w \in W$,

$$\lim_{\epsilon\to 0}\|E_{w,f,\epsilon}\|_2^2 = 0.$$

Here goes. We have

$$\|E_{w,f,\epsilon}\|_2^2$$

$$= \int_{\mathfrak{a}^\vee} \left| \epsilon^{-r/2}\gamma(w,\lambda) \int_{\mathfrak{a}^+} f(H)e^{\mathbf{i}w\lambda(H)/\epsilon}[\theta(\mathbf{i}w\lambda, H/\epsilon) - 1]\,dH \right|^2 d\lambda$$

$$= \epsilon^{-r} \int_{\mathfrak{a}^\vee} \left| \int_{\mathfrak{a}^+} f(H)e^{-\mathbf{i}\lambda(H)/\epsilon}[\theta(-\mathbf{i}\lambda, H/\epsilon) - 1]\,dH \right|^2 d\lambda$$

[because $|\gamma(w,\lambda)| = 1$, and we make the change of variables $\lambda \mapsto w^{-1}\lambda$]

$$= \int_{\mathfrak{a}^\vee} \left| \int_{\mathfrak{a}^+} f(H)e^{-\mathbf{i}\lambda(H)}[\theta(-\mathbf{i}\epsilon\lambda, H/\epsilon) - 1]\,dH \right|^2 d\lambda$$

[letting $\lambda \mapsto \epsilon\lambda$]

$$= \int_{\mathfrak{a}^\vee} \left| \sum_{\mu\neq 0} g_\mu(-\mathbf{i}\epsilon\lambda) \int_{\mathfrak{a}^+} f(H)e^{-(\mathbf{i}\lambda+\mu/\epsilon)(H)}\,dH \right|^2 d\lambda.$$

But the $\mathfrak{a}^+$-integral is an ordinary Fourier transform $f^\wedge$, the function f being set equal to 0 outside $\mathfrak{a}^+$. Therefore

$$\|E_{w,f,\epsilon}\|_2^2 = \int_{\mathfrak{a}^\vee} \sum_{\mu,\nu\neq 0} g_\mu(-\mathbf{i}\epsilon\lambda)\overline{g_\nu(-\mathbf{i}\epsilon\lambda)}\, f^\wedge(\lambda - \mathbf{i}\mu/\epsilon)\overline{f^\wedge(\lambda - \mathbf{i}\nu/\epsilon)}\,d\lambda.$$

For $\mu \in \mathbf{L}$, let $m(\mu)$ be the degree as defined in §1. Let $b > 0$ be such that f has support in the set of $H \in \mathfrak{a}^+$ such that $\alpha_i(H) \geqq b$, for $i = 1, \ldots, r$. Then $\mu(H) \geqq m(\mu)b$ for all $\mu \neq 0$, so

$$e^{-(\mu/\epsilon)(H)} \leqq e^{-m(\mu)b/\epsilon} \quad \text{for all } H \in \operatorname{supp}(f).$$

The usual integration by parts applies to show that given a positive integer N there exists a constant C_N such that (see below)

$$|f^\wedge(\lambda - \mathbf{i}\mu/\epsilon)| \leqq C_N(1 + |\lambda|)^{-N} e^{-m(\mu)b/\epsilon},$$

and similarly with ν instead of μ. On the other hand, the estimate of Theorem 2.4 applies with any given $H \in \mathfrak{a}^+$, so with a multiple δH instead of H, and thus the coefficients have the fixed exponential growth

$$|g_\mu(-\mathbf{i}\epsilon\lambda)\overline{g_\nu(-\mathbf{i}\epsilon\lambda)}| \ll e^{m(\mu)b+m(\nu)b}$$

if we pick δ small enough. Then the integrand is dominated by

$$(1 + |\lambda|)^{-N} \sum_{\mu,\nu\neq 0} e^{(m(\mu)+m(\nu))b} e^{-(m(\mu)+m(\nu))b/\epsilon}.$$

For N large enough, the function $(1+|\lambda|)^{-N}$ is in L^1. As $\epsilon \to 0$, the series not only converges, but approaches 0, which proves Lemma 5.1.

Remark 1. The above estimate is routine. Just look at one variable. We are in a situation where f has support in an interval (b, B) with $0 < b < B$. We want to estimate

$$\int_b^B f(x)e^{-(\mathbf{i}\lambda+c)x}\,dx.$$

The boundary terms vanish under integration by parts, which we do N times. A factor $(\mathbf{i}\lambda + c)^{-N}$ comes out in front of an integral which is estimated absolutely by $(B-b)$ times the absolute value of the N-th derivative $|f^{(N)}|$, times e^{-cb}. In the application, $c = m(\mu)/\epsilon$.

Remark 2. In the application to Axiom **W 3** in Chapter IX, §3, one is given a function f, and then the lemma is applied to the function $J_{\mathfrak{a}}^{1/2} f$, which is also in $C_c^\infty(\mathfrak{a}^+)$. There was no need to carry the extra factor throughout the proof of the lemma.

VIII, §6. SPHERICAL INVERSION

This section finally puts everything that precedes together. We let φ be the spherical kernel, so

$$\varphi = \varphi(\zeta, x) \qquad \text{with } \zeta \in \mathfrak{a}_{\mathbf{C}}^\vee \text{ and } x \in G.$$

Thus φ is pulled back to the Lie algebra only in its first variable. Then φ defines convolution integral operators going from left to right and right to left respectively, namely for suitable test functions f and F, with f being K-bi-invariant and F invariant under the Weyl group W:

$$(\mathbf{S}f)(\zeta) = \int_G \varphi(\zeta, x) f(x)\,dx.$$

Actually we view the image $\mathbf{S}f$ restricted to the imaginary axis $\mathbf{i}\mathfrak{a}^\vee$, so the argument is written $\mathbf{i}\lambda$ with $\lambda \in \mathfrak{a}^\vee$. The problem considered and solved by Harish-Chandra was to determine the measure on $\mathbf{i}\mathfrak{a}^\vee$ which inverts the spherical transform. Harish-Chandra found that this measure is given by $\eta(\lambda)\,d\lambda$ where

$$\eta(\lambda) = |c^{-1}(\mathbf{i}\lambda)|^2.$$

In other words, let us define

$$({}^t\mathbf{S}_\eta F)(x) = \int_{\mathfrak{a}^\vee} F(\mathbf{i}\lambda)\varphi(\mathbf{i}\lambda, x)\eta(\lambda)\,d\lambda.$$

The Haar measures have to be normalized as follows.

The function $J_{\mathfrak{a}}$ is the W-invariant function such that

$$J_{\mathfrak{a}} = \prod_{\alpha \in \mathcal{R}(\mathfrak{n})} (e^{\alpha} - e^{-\alpha}) \quad \text{on } \mathfrak{a}^{+}.$$

The Haar measure dx on G is normalized to be the **polar Haar measure**, which we define to be the unique Haar measure giving K total measure 1, and such that

$$\int_G f(x)\,dx = \int_{\mathfrak{a}^+} f_{\mathfrak{a}}(H) J_{\mathfrak{a}}(H)\,dH \qquad \text{for } f \in C_c^{\infty}(K\backslash G/K).$$

Such a measure exists by Chapter VI, Theorem 1.5.

The Haar measure $d\lambda$ on $\mathfrak{a}^{\vee}$ will be said to be **Fourier normalized** if the ordinary Fourier inversion formula (for functions on $\mathfrak{a}$, $\mathfrak{a}^{\vee}$ respectively) holds with constant factor equal to 1 for the measures $dH, d\lambda$. Starting with any one of Haar measures $dx, dH, d\lambda$ on $G, \mathfrak{a}, \mathfrak{a}^{\vee}$ respectively, one can determine the other two uniquely by the polar and Fourier conditions. One then says that the triple of measures is **polar Fourier normalized.** We assume such normalization in the sequel.

Of course the spherical transform $\mathbf{S}f$ is defined only for functions with rapid decay. We shall make precise each time the spaces on which this transform is defined. Similarly, we define the **transpose spherical transform** with respect to the measure $\eta(\lambda)\,d\lambda$ on functions F with rapid decay on $\mathbf{i}\mathfrak{a}^{\vee}$ by

$$({}^{t}\mathbf{S}_{\eta}F)(x) = \int_{\mathfrak{a}^{\vee}} F(\mathbf{i}\lambda)\varphi(\mathbf{i}\lambda, x)\eta(\lambda)\,d\lambda.$$

As usual in Fourier analysis, there is a minus sign coming in somewhere, so we use the notation

$$(\mathbf{S}^{-}f)(\zeta) = (\mathbf{S}f)(-\zeta) = (\mathbf{S}f^{-})(\zeta).$$

Recall Harish-Chandra's result that $\varphi(-\zeta, x) = \varphi(\zeta, x^{-1})$ (Proposition 7.3 of Chapter III), so the minus sign can be put on either variable.

Theorem 6.1. *Assume the measures $dx, dH, d\lambda$ are polar Fourier normalized. Then*

$$\mathbf{S}^{-}: C_c^{\infty}(K\backslash G/K) \to \mathrm{PW}(\mathbf{i}\mathfrak{a}^{\vee})^{W}$$

is a linear isomorphism, and also an L^2-isometry extending to an isometry

$$\mathbf{S}^{-}: L^2(K\backslash G/K, dx) \to L^2(\mathbf{i}\mathfrak{a}^{\vee}, \eta(\lambda)\,d\lambda)^{W}.$$

The inverse is given by the Harish-Chandra transpose $|W|^{-1}\,{}^{t}\mathbf{S}_{\eta}$.

The inverse relation may also be written out in full explicitly, in the form

$$'\mathbf{S}_\eta \circ \mathbf{S}^- = |W| \cdot \mathrm{id} \quad \text{on } C_c^\infty(K\backslash G/K).$$

In other words,

$$f(x) = \frac{1}{|W|}\int_{\mathfrak{a}^\vee} (\mathbf{S}^- f)(\mathbf{i}\lambda)\varphi(\mathbf{i}\lambda, x)\eta(\lambda)\,d\lambda.$$

Harish-Chandra originally defined a Schwartz space of K-bi-invariant functions, and proved that the two integral operators $\mathbf{S}^-$ and $|W|^{-1}\,'\mathbf{S}_\eta$ are inverse to each other on the Schwartz spaces, extending to L^2-isometries [Har 58], [Har 66]. In particular, the Harish-Chandra inversion in one direction applies to $C_c^\infty(K\backslash G/K)$. This inversion and the L^2-isometry will be proved in Chapter IX. In the meantime, we give a consequence which pursues Chapter IV, Corollaries 8.4 and 8.5, as well as Chapter V, Proposition 8.5. For a W-stable convex set C, we define:

$C_C^\infty(K\backslash G/K) =$ subspace of functions $f \in C^\infty(K\backslash G/K)$ such that $f_\mathfrak{a}$ has support in C, or alternatively, such that $\mathrm{supp}(f) \subset G_C$, where

$$G_C = K(\exp C)K.$$

Theorem 6.2. *Let C be compact, convex, W-stable and U-polar expending. Then the spherical transform induces a linear isomorphism*

$$\mathbf{S}\colon C_C^\infty(K\backslash G/K) \to \mathrm{PW}_C(\mathfrak{a}_C^\vee)^W.$$

Corollary 6.3. *With C as in Theorem* 6.2, *the Harish transform induces a linear isomorphism*

$$\mathbf{H}\colon C_C^\infty(K\backslash G/K) \to C_C^\infty(A)^W.$$

Proof. Combine Theorem 6.2 with Chapter IV, Corollary 8.4, and the W-invariant version using Chapter IV, Proposition 8.6.

In the above we have followed [Hel 66], [Ros 77], p. 146, indicated by Helgason, [Gan 71], and [Hel 84], Chapter IV, Theorem 7.1. Injectivity will be proved in Chapter IX, Theorem 1.6.

The theorems can be formulated by pulling back to $\mathfrak{a}$ and $\mathfrak{a}^\vee$. We let $\mathbf{S}_{\mathrm{Im}}$ (Im for imaginary) be defined by

$$(\mathbf{S}_{\mathrm{Im}} f)(\lambda) = (\mathbf{S}f)(\mathbf{i}\lambda) \qquad \text{for } \lambda \in \mathfrak{a}^\vee.$$

Similarly we define $\mathrm{PW}(\mathfrak{a}^\vee)$ (the imaginary restriction of $\mathrm{PW}(\mathfrak{a}_{\mathbf{C}}^\vee)$). We have a commutative diagram

(1)

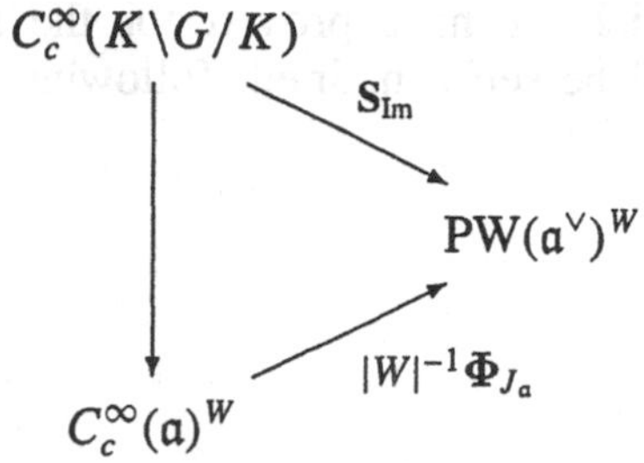

where the transform $\Phi_{J_\mathfrak{a}}$ is defined for $\lambda \in \mathfrak{a}^\vee$ by

$$(\Phi_{J_\mathfrak{a}} f_\mathfrak{a})(\lambda) = \int_\mathfrak{a} \varphi^\mathfrak{a}(\mathbf{i}\lambda, H) f_\mathfrak{a}(H) J_\mathfrak{a}(H)\, dH$$

Similarly, we have the diagram going in the opposite direction

(2)

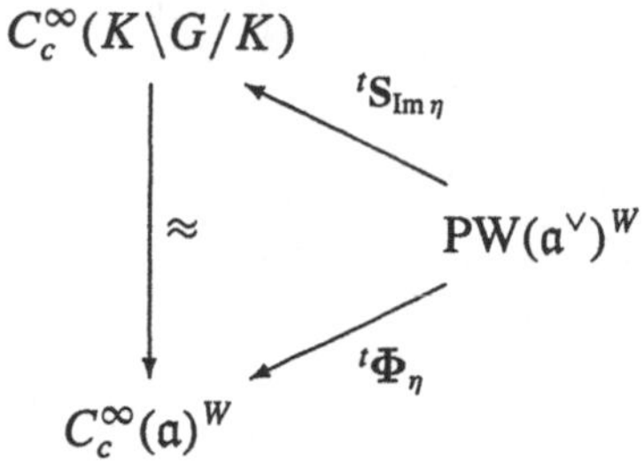

where ${}^t\Phi_\eta$ is defined by

$$({}^t\Phi_\eta F)(H) = \int_{\mathfrak{a}^\vee} F(\mathbf{i}\lambda)\varphi^\mathfrak{a}(\mathbf{i}\lambda, H)\eta(\lambda)\, d\lambda.$$

Then one sees that Theorem 6.1 is equivalent with:

Theorem 6.4. *Assume the measures $dx, dH, d\lambda$ are polar Fourier normalized. Then*

$$|W|^{-1}\Phi_{J_\mathfrak{a}}^- : C_c^\infty(\mathfrak{a})^W \to \mathrm{PW}(\mathfrak{a}^\vee)^W$$

is a linear isomorphism, and also an L^2-isometry extending to an isometry

$$|W|^{-1}\Phi_{J_\mathfrak{a}}^- : L^2(\mathfrak{a}, dH)^W \to L^2(\mathfrak{a}^\vee, \eta(\lambda)\, d\lambda)^W.$$

The inverse is given by the Harish-Chandra transpose $|W|^{-1}\,{}^t\Phi_\eta$.

In this form, Theorem 6.4 is a theorem on euclidean spaces, where G has disappeared. To emphasize the euclidean nature of the theorem, we shall give a proof in the next chapter, depending on an axiomatization of the properties which we have proved for the spherical transform. The next chapter will be self-contained, following Rosenberg's "quick proof" [Ros 77].

CHAPTER IX

General Inversion Theorems

The Fourier inversion formula is a standard fact of elementary analysis. Harish-Chandra developed an inversion for K-bi-invariant functions on a semisimple Lie group G, in other words he developed the theory of a spherical transform [Har 58a], [Har 58b], which is an integral transform, with a kernel called the spherical kernel. There are variations to the setting of this transform, involving various factors. Harish-Chandra's general Plancherel inversion in the non-K-bi-invariant case is a lot more complicated, and highly non-abelian. The K-bi-invariant case turns out to be abelian. However, in this case, one can look at the inversion on various function spaces, ranging over C_c^∞, C_c, Schwartz space, L^1, L^2, ad lib. Harish-Chandra dealt fundamentally with the Schwartz space, which he defined in the context of semisimple Lie groups, and L^2. Helgason pointed to the correspondence between C_c^∞ and the Paley–Wiener space [Hel 66], complemented by Gangolli [Gan 71]. Then Rosenberg [Ros 77] gave a much simpler version of some parts of Harish inversion, using some lemmas from Helgason and Gangolli. Here we give this Rosenberg material, suitably axiomatized in a way which gives rise to a setting for pure inversion theorems in ordinary euclidean space, essentially along the lines of the similar elementary and standard theory of Fourier inversion. Roughly speaking, the situation is as follows.

Let $\mathfrak{a}$ be a finite dimensional vector space over $\mathbf{R}$, with dual space $\mathfrak{a}^\vee$. Let the variables be λ, H in $\mathfrak{a}^\vee$, $\mathfrak{a}$ respectively. Let $dH, d\lambda$ be given Haar–Lebesgue measures. Let $\Phi = \Phi(\lambda, H)$ be a function on $\mathfrak{a}^\vee \times \mathfrak{a}$, and let ν, η be positive functions on $\mathfrak{a}, \mathfrak{a}^\vee$ respectively. We obtain two integral operators going respectively from left to right and right to left, the first of which being Φ_ν, defined on suitable test functions f by

$$(\Phi_\nu f)(\lambda) = \int_{\mathfrak{a}} \Phi(\lambda, H) f(H) \nu(H)\, dH,$$

and the other operator ${}^t\Phi_\eta$ being defined similarly in the opposite direction. The problem is to describe all triples (Φ, ν, η) giving rise to the inversion formulas

$${}^t\Phi_\eta \circ \Phi_\nu^- = \mathrm{id} \qquad \text{and} \qquad \Phi_\nu^- \circ {}^t\Phi_\eta = \mathrm{id}.$$

There are variations to the problem, for instance given (Φ, ν) when does there exist η satisfying inversion? And of course, on which function spaces.

In the application to Chapter VIII, §6, Φ is the function pulled back from the Harish-Chandra spherical kernel, that is for this chapter,

$$\boxed{\Phi(\lambda, H) = \varphi^{\mathfrak{a}}(\mathbf{i}\lambda, H) = \varphi(\mathbf{i}\lambda, \exp H),}$$

using a notation which emphasizes the analogy with the classical Fourier inversion. The present chapter was designed for a wide audience, and can be used as hand-out for analysts who might provide other situations fitting the axioms.

The function ν of this chapter is the function $J_{\mathfrak{a}}$ in Chapter VIII, §6, in other words, it is the Jacobian in the change of variable formula from the non-abelian group to the euclidean space which is the tangent space of the A-Iwasawa component at the origin.

The function η is the pull-back of the Harish-Chandra $|\mathbf{c}^{-2}(\mathbf{i}\lambda)|$ with $\lambda \in \mathfrak{a}^\vee$.

However, no knowledge whatsoever of anything about semisimple Lie groups or Iwasawa decompositions is required for this chapter, which has been kept self-contained and can be read independently of all others, by axiomatization.

The original motivation for Harish-Chandra to define his $\mathbf{c}$-function to get the "Plancherel" measure, i.e. in our set up, the function η which satisfies the inversion theorem, came from linear differential equations. It would be useful to have an exposition how inversion arises from the linear differential equations, showing explicitly how the axioms are satisfied in this case. For second-order differential equations (rank 1), this is likely contained in [FlJ 72].

Finally, we note that before Harish-Chandra's work, there had been an earlier paper of Gelfand [Gel 50] on spherical functions, and Godement had proved a general theorem about the existence of a measure on the space of spherical functions such that L^2–Plancherel inversion holds [God 52b]. Readers will find an account of this and other general matters in Chapter I of [GaV 88].

IX, §1. THE ROSENBERG ARGUMENTS

Lemma 1.1. *Let X be a manifold, and T a functional on $C_c^\infty(X)$, continuous for the sup norm. Suppose the support of T is a point x_0, i.e. if $f \in C_c^\infty(X)$ and $x_0 \notin \operatorname{supp}(f)$, then $Tf = 0$. Then there exists a constant C such that for all $f \in C_c^\infty(X)$ we have $Tf = Cf(x_0)$.*

Proof. First we note that $C_c^\infty(X)$ is dense in $C_c(X)$ for the sup norm. By using a partition of unity, it suffices to prove this denseness for an open subset of euclidean space, and then one can use a C^∞ Dirac sequence whose support shrinks to the origin to approximate a continuous function with compact support uniformly, by convolution with the elements of this sequence. Thus a sup norm continuous linear map on $C_c^\infty(X)$ extends to a sup norm continuous linear map on $C_c(X)$, with the same norm (as linear map). If $f \in C_c(X)$ has support not containing x_0, then by the above procedure, sufficiently closely approximating C_c^∞ functions also have support not containing x_0, so $Tf = 0$. In other words, the extension of T to $C_c(X)$ also has support at x_0. By the Riesz theorem of measure theory, there exists a complex valued measure μ such that for all $f \in C_c(X)$ we have

$$Tf = \int f\, d\mu.$$

Cf. most books on standard measure theory and integration, e.g. [Lan 93], Chapter IX, Theorem 4.2. Let $\{h_n\}$ be a sequence in $C_c^\infty(X)$ with values in $[0, 1]$, $h_n = 1$ on some neighborhood of x_0, and the support of h_n shrinks to x_0 as $n \to \infty$. Then $f - h_n f$ has support not containing x_0, so $T(f - h_n f) = 0$, and

$$\begin{aligned} Tf = T(h_n f) = \lim_{n\to\infty} \int h_n f\, d\mu &= \int \lim_{n\to\infty} h_n f\, d\mu \\ &= \int f(x_0) 1_{x_0}\, d\mu \end{aligned}$$

where 1_{x_0} is the function with value 1 at x_0 and 0 otherwise, which proves the lemma with the constant C equal to the integral of this function.

Remark. Larry Guth observed that one doesn't even need the measure theory for the proof of the lemma, by arguing as follows. We claim that if $f_1, f_2 \in C_c(X)$ have the same value at x_0 then $Tf_1 = Tf_2$. Indeed, for a sequence $\{h_n\}$ as above, we have $Tf_1 = T(h_n f_1)$ and $Tf_2 = T(h_n f_2)$. But $\{h_n(f_1 - f_2)\}$ converges to 0 in sup norm, whence $T(h_n(f_1 - f_2))$ converges to 0, whence $Tf_1 = Tf_2$. Hence Tf depends only on $f(x_0)$ for $f \in C_c(X)$. Given $u \in \mathbf{R}$, define Lu to be Tf

for any $f \in C_c(X)$ such that $f(x_0) = u$. Then L is a linear map of $\mathbf{R}$ into itself, so of the form $Lu = Cu$ for some constant C. Hence $Tf = Cf(x_0)$, thus concluding the proof.

Next we describe a general situation about inverting an integral transform.

We let $\mathfrak{a}$ be a finite dimensional vector space of dimension r over $\mathbf{R}$, and we let $\mathfrak{a}^\vee$ be its dual space. We denote the variables in $\mathfrak{a}$ and $\mathfrak{a}^\vee$ by H and λ respectively. We fix Haar–Lebesgue measures dH and $d\lambda$.

Let ν and η be semipositive continuous functions on $\mathfrak{a}$ and $\mathfrak{a}^\vee$ respectively. We assume η even, that is, $\eta(-\lambda) = \eta(\lambda)$ for all $\lambda \in \mathfrak{a}^\vee$.

Then we have the corresponding functionals $\nu(H)\,dH$ and $\eta(\lambda)\,d\lambda$ on $C_c(\mathfrak{a})$ and $C_c(\mathfrak{a}^\vee)$ respectively, and the corresponding measures, with corresponding spaces $L^1(\nu)$, $L^2(\nu)$, etc.

Let Φ be a continuous function on $\mathfrak{a}^\vee \times \mathfrak{a}$. In practice we shall have Φ even, that is, $\Phi(-\lambda, -H) = \Phi(\lambda, H)$ (Proposition 7.3 of Chapter III), but we don't need to assume this now. We let Φ_ν be the convolution operator defined on a function f for which the integral is absolutely convergent by

$$(\Phi_\nu f)(\lambda) = (\Phi_\nu * f)(\lambda) = \int_{\mathfrak{a}} \Phi(\lambda, H) f(H) \nu(H)\, dH.$$

Similarly, on a space of functions on $\mathfrak{a}^\vee$ for which the integral is absolutely convergent, we define the **transpose** ${}^t\Phi_\eta$ by

$$({}^t\Phi_\eta f)(H) = (F * \Phi_\eta)(H) = \int_{\mathfrak{a}^\vee} F(\lambda)\Phi(\lambda, H)\eta(\lambda)\, d\lambda.$$

Example. The most elementary is that of the ordinary **Fourier transform**, with the integral kernel

$$\Phi_{\mathrm{euc}}(\lambda, H) = e^{i\lambda(H)}.$$

Thus the **Fourier transform** with respect to dH is defined by

$$(\Phi_{\mathrm{euc}}^- f)(\lambda) = f^\wedge(\lambda) = \int_{\mathfrak{a}} f(H) e^{-i\lambda(H)}\, dH,$$

and similarly for the transpose transform, without the minus sign. In this case, using a positive definite scalar product on $\mathfrak{a}$ identifying $\mathfrak{a}$ with its dual $\mathfrak{a}^\vee$, one usually takes $d\lambda = dH$. Then there is some constant factor C_{euc} for which Fourier inversion holds, that is

$$f^{\wedge\wedge} = C_{\mathrm{euc}} f^-.$$

With the general notation, we say that **ordinary Fourier inversion** holds if $C_{\text{euc}} = 1$, that is

$$'\Phi_{\text{euc}} \circ \Phi^{-}_{\text{euc}} = \text{id}.$$

We want to describe the analogue of Fourier inversion in a situation applicable to Harish inversion. Thus defining

$$\Phi^{-}(\lambda, H) = \Phi(-\lambda, H),$$

we want conditions under which there exists a constant C such that

$$'\Phi_{\eta} \circ \Phi^{-}_{\nu} = C \cdot \text{id} \qquad \text{and} \qquad \Phi^{-}_{\nu} \circ {'\Phi_{\eta}} = C \cdot \text{id}.$$

We shall first deal with inversion at the origin, in the sense made precise in Theorem 1.2. In practice, we have $\Phi(\lambda, 0) = 1$, so we expect the $\Phi(\lambda, 0)$ factor to be invisible in computing $('\Phi_{\eta} * F)(0)$. We assume the following boundedness and growth conditions:

INV 1. The function Φ is bounded on $\mathfrak{a}^{\vee} \times S$ for every compact S in $\mathfrak{a}$.

INV 2. There exist integers $M, N \geqq 0$ such that:

(a) $\nu(H) = O(|H|^N)$ for $|H| \to 0$;

(b) $\eta(\lambda) = O(|\lambda|^M)$ for $|\lambda| \to \infty$, and $M \leqq N$.

The absolute value signs refer to some fixed norms on $\mathfrak{a}$ and $\mathfrak{a}^{\vee}$. We recall that all norms on a finite dimensional vector space are equivalent, so the choice of norm is irrelevant for such Big Oh growth conditions. Thirdly, we have:

INV 3. The Φ_{ν}-transform maps $C_c^{\infty}(\mathfrak{a})$ into the space of super polynomially decreasing functions on $\mathfrak{a}^{\vee}$.

By **super polynomially decreasing**, we mean functions F such that for every polynomial P, the function PF is bounded on $\mathfrak{a}^{\vee}$. This is of course a condition like that of the Schwartz space, but we don't need estimates on the derivatives for what follows. We only need the rapid decay of the function itself.

Example. The spherical transform is the composite of the Harish transform, whose image is contained in $C_c^{\infty}(A)$; and the Mellin transform, which is super polynomially decreasing. Cf. Chapter IV, §7. We shall come back to a more detailed interpretation later.

We denote by $f^\wedge$ or $F^\wedge$ the ordinary Fourier transform of a function on $\mathfrak{a}$ or $\mathfrak{a}^\vee$ respectively with respect to the measures $dH, d\lambda$ respectively. We then have a fourth condition, concerning ${}^t\Phi_\eta$.

INV 4. Let $\psi \in C_c^\infty(\mathfrak{a})^\wedge$ be the Fourier transform of a function in $C_c^\infty(\mathfrak{a})$. For $\epsilon \to 0$ define $\psi_\epsilon(\lambda) = \psi(\epsilon\lambda)$. Then

$$\operatorname{supp} {}^t\Phi_\eta \psi_\epsilon \subset \mathbf{B}_{O(\epsilon)} \qquad \text{for } \epsilon \to 0.$$

As usual, $\mathbf{B}_R$ denotes the ball of radius R centered at 0.

Remark 1. The function ψ and all its derivations are actually super polynomially decreasing, so is ψ_ϵ, and by **INV 1**, ${}^t\Phi_\eta\psi_\epsilon$ is automatically defined.

Remark 2. Let $\psi = \beta^\wedge$. Then directly from a change of variables in the integral defining the Fourier transform, one sees that up to a constant factor,

$$\psi_\epsilon = \text{Fourier transform of } \epsilon^{-r}\beta \circ \epsilon^{-1},$$

where $r = \dim \mathfrak{a}$. Thus $\psi \circ \epsilon$ is the Fourier transform of a function with support in $\mathbf{B}_{O(\epsilon)}$ for $\epsilon \to 0$. Roughly speaking, the convolution operator ${}^t\Phi_\eta$ behaves not so differently from the inverse Fourier transform as far as the order of magnitude of the support of its image is concerned.

Remark 3. In the standard development, in the context of semisimple Lie groups, a stronger version of **INV 4** giving the exact support is provided by Helgason's support theorem [Hel 66], see Theorem 4.2 of Chapter VIII.

To deal with inversion at the origin, for $f \in C_c^\infty(\mathfrak{a})$, we **define**

$$Tf = \int_{\mathfrak{a}^\vee} (\Phi_\nu f)(\lambda)\eta(\lambda)\, d\lambda = \int_{\mathfrak{a}^\vee} (\Phi_\nu f)(-\lambda)\eta(\lambda)\, d\lambda,$$

the second equality because η is assumed even.

Theorem 1.2. *Assume that the four conditions* **INV 1–INV 4** *are satisfied. Then there exists a constant C such that for all $f \in C_c^\infty(\mathfrak{a})$ we have*

$$Tf = Cf(0), \qquad \textit{that is,} \quad T = C\delta_0,$$

where δ_0 is the Dirac functional at the origin.

Proof. We shall prove that the conditions of Lemma 1.1 are satisfied. Essentially, we copy [Ros 77]. Let ψ and ψ_ϵ be as **INV 4**, even, and such that $\psi \geqq 0$, $\psi(0) = 1$. For instance, one can take $\psi = (\beta * \beta)^\wedge$ with a real even function β having compact support, and multiplied by a suitable constant to get $\psi(0) = 1$. By **INV 2** and **INV 3**,

$$\int (\Phi_\nu f)(\lambda)\eta(\lambda)\,d\lambda$$

is absolutely convergent, and

$$\begin{aligned}
(1)\quad Tf &= \lim_{\epsilon\to 0}\int_{\mathfrak{a}^\vee} (\Phi_\nu f)(\lambda)\psi(\epsilon\lambda)\eta(\lambda)\,d\lambda \quad \text{(by dominated convergence)}\\
&= \lim_{\epsilon\to 0}\int_{\mathfrak{a}^\vee}\left[\int_{\mathfrak{a}} \Phi(\lambda, H) f(H)\nu(H)\,dH\right]\psi(\epsilon\lambda)\eta(\lambda)\,d\lambda\\
&= \lim_{\epsilon\to 0}\int_{\mathfrak{a}} f(H) g_\epsilon(H)\nu(H)\,dH \qquad \text{by Fubini,}
\end{aligned}$$

where, using that η, ψ are even,

$$(2)\qquad g_\epsilon(H) = \int_{\mathfrak{a}^\vee} \Phi(\lambda, H)\psi(\epsilon\lambda)\eta(\lambda)\,d\lambda = ({}^t\Phi_\eta \psi_\epsilon)(H).$$

By **INV 4**, g_ϵ has support in $\mathbf{B}_{O(\epsilon)}$ for $\epsilon \to 0$, so $\operatorname{supp} T = \{0\}$, which proves the first condition.

As for the second, we have to show that T is continuous for the sup norm. By (1) we have

$$|Tf| \leqq \|f\|_\infty \|g_\epsilon\|_{1,\nu}.$$

Hence it will suffice to prove that the $L^1(\nu)$-norm $\|g\|_{1,\nu}$ is bounded for $\epsilon \to 0$. Since $\operatorname{supp} g_\epsilon \subset \mathbf{B}_{O(\epsilon)}$, we have

$$(3)\quad \|g_\epsilon\|_{1,\nu} \ll \|g_\epsilon\|_\infty \int_{\mathbf{B}_{O(\epsilon)}} |\nu(H)|\,dH \ll \|g_\epsilon\|_\infty \epsilon^{r+N} \quad \text{by } \mathbf{INV\ 2}\text{(a).}$$

On the other hand, by (2), by the support condition, and by **INV 1** using balls of small radius, we have

$$\begin{aligned}
(4)\qquad \|g_\epsilon\|_\infty &\ll \int_{\mathfrak{a}^\vee} \psi(\epsilon\lambda)\eta(\lambda)\,d\lambda = \epsilon^{-r}\int_{\mathfrak{a}^\vee}\psi(\lambda)\eta(\lambda/\epsilon)\,d\lambda\\
&\ll \epsilon^{-r}\int_{\mathfrak{a}^\vee}\psi(\lambda)(1 + |\lambda|/\epsilon)^M\,d\lambda \quad \text{by } \mathbf{INV\ 2}\text{(b)}\\
&= O(\epsilon^{-r-M}) \quad \text{for } \epsilon \to 0.
\end{aligned}$$

From (3) and (4) we conclude that $\|g_\epsilon\|_{1,\nu}$ is bounded. This concludes the proof.

Remark 1. Actually, one does not need the full condition of rapid decrease for $\Phi_\nu^- f$ in the above proof. One only needs that $(\Phi_\nu^- f)\eta$ is in L^1 (of Haar–Lebesgue measure).

Remark 2. Since Haar measures are determined only up to a constant factor, the question arises how to normalize the measures to that the constant is 1. In the applications we have in mind, the answer is that Fourier inversion should hold exactly (with constant 1) for the standard kernel e^{ixy}. See below.

Remark 3. There are other choices which can be made because of an important additional structure which we shall introduce next, namely a certain finite group W of automorphisms acting on $\mathfrak{a}$, whence on $\mathfrak{a}^\vee$. We shall consider functions invariant under this group. It is then reasonable that the transforms Φ_ν and $'\Phi_\eta$ be defined on $\mathfrak{a}/W$ and $\mathfrak{a}^\vee/W$, thus introducing a factor $|W|^{-2}$ in the inversion constant. With such a normalization, the constant C turns out to be equal to 1, cf. the remark at the end of §3.

The above arguments gave a punctual inversion at the origin. To translate this result to a global inversion, we need another axiom showing how the transform Φ_ν^- behaves vis-à-vis translations. We denote

$$\tau_H \colon \mathfrak{a} \to \mathfrak{a}$$

translation by H, that is, $\tau_H(v) = v + H$. To make the axiom applicable to the cases we have in mind, we must introduce another structure.

We suppose given a finite group W of linear automorphisms of $\mathfrak{a}$.

Then W also induces a group of linear automorphisms of $\mathfrak{a}^\vee$, or for that matter, of other spaces functorially associated with $\mathfrak{a}$, e.g. certain function spaces. As usual, we let $C_c^\infty(\mathfrak{a})^W$ be the subspace of $C_c^\infty(\mathfrak{a})$ consisting of those functions fixed under the action of W. We shall use an additional condition, which in the applications comes from Theorem 5.6 of Chapter IV and Theorem 3.3 of Chapter VI.

INV 5. For $f \in C_c^\infty(\mathfrak{a})^W$, and every $H \in \mathfrak{a}$ we have

$$(\Phi_\nu^-(f \circ \tau_H))(\lambda) = \Phi(\lambda, H)(\Phi_\nu^- f)(\lambda).$$

Theorem 1.3. *Assume condition* **INV 5** *is satisfied in addition to* **INV 1 – INV 4**. *Let C be the constant of Theorem* 1.2. *For*

$f \in C_c^\infty(\mathfrak{a})^W$ and all $H \in \mathfrak{a}$, we have

$$\int_{\mathfrak{a}^\vee} (\Phi_\nu^- f)(\lambda)\Phi(\lambda, H)\eta(\lambda)\,d\lambda = Cf(H).$$

That is,

$${}^t\Phi_\eta \circ \Phi_\nu^- = C \cdot \mathrm{id} \quad \textit{on } C_c^\infty(\mathfrak{a})^W.$$

Proof. We apply Theorem 1.2 to the function $f \circ \tau_H$. Since $(f \circ \tau_H)(0) = f(H)$, the theorem falls out of **INV 5**.

Thus we have one-sided inversion, corresponding to Rosenberg's proof of Harish-Chandra inversion on one side in the context of semisimple Lie groups. Of course, the constant C may be 0, because we have made no assumption of non-degeneracy on Φ, ν, η which would guarantee $C \neq 0$. We now look into this possibility.

INV 6. (a) The function ν is > 0 except possibly for a set of measure 0 in $\mathfrak{a}$, and it is W-invariant outside a set of measure 0.

(b) The function η is > 0 except possibly for a set of measure 0 in $\mathfrak{a}^\vee$, and is W-invariant outside a set of measure 0.

Each part of **INV 6** will serve a different purpose. Here we are first concerned with (a), and its relevance for the non-degeneracy of Φ_ν. We shall also need:

INV 7. (a) The functions $\Phi(\lambda, \cdot)$ and $\Phi(\mu, \cdot)$ on $\mathfrak{a}$ are equal if and only if there exists $w \in W$ such that $\mu = w\lambda$.

(b) We also have $\Phi(\lambda, wH) = \Phi(\lambda, H)$ for all $\lambda \in \mathfrak{a}^\vee$, $H \in \mathfrak{a}$, $w \in W$.

Note that in the applications, **INV 7** corresponds to Theorem 5.6 of Chapter III.

Our space of test functions is the usual one $C_c^\infty(\mathfrak{a})$. For our purposes here, we take as definition of the **Paley–Wiener space**

$$\mathrm{PW}(\mathfrak{a}^\vee) = C_c^\infty(\mathfrak{a})^\wedge = \text{Fourier transform of } C_c^\infty(\mathfrak{a}).$$

Then directly from the definitions, for any linear automorphism w of $\mathfrak{a}$, and $\beta \in C_c^\infty(\mathfrak{a})$, we have

$$(5) \qquad \beta^\wedge(w\lambda) = (\beta \circ w)^\wedge(\lambda).$$

If $\beta \in C_c^\infty(\mathfrak{a})^W$, it follows that $\beta^\wedge \in \mathrm{PW}(\mathfrak{a}^\vee)^W$, namely the Fourier transform of β is also fixed under the action of W. In symbols,

$$\left(C_c^\infty(\mathfrak{a})^\wedge\right)^W = \left(C_c^\infty(\mathfrak{a})^W\right)^\wedge.$$

Lemma 1.4. *Assume* **INV 3, INV 6(a)** *and* **INV 7**. *Then the family of functions* $\{\Phi_\nu^- g\}$ *with* $g \in C_c^\infty(\mathfrak{a})^W$ *separates points on* $\mathfrak{a}^\vee/W$ *and vanishes at infinity.*

Proof. **INV 3** guarantees that each function $\Phi_\nu g$ vanishes at infinity. As to separating points, suppose $\lambda, \mu \in \mathfrak{a}^\vee$ are such that

$$\int_\mathfrak{a} \Phi(-\lambda, H)g(H)\nu(H)\,dH = \int_\mathfrak{a} \Phi(-\mu, H)g(H)\nu(H)\,dH$$

for all g. Let $F(H) = \Phi(-\lambda, H) - \Phi(-\mu, H)$. Then

$$\int F(H)g(H)\nu(H)\,dH = 0$$

for all $g \in C_c^\infty(\mathfrak{a})^W$. But then because of the invariance under W in **INV 6(a)** and **INV 7**, for all $g \in C_c^\infty(\mathfrak{a})$ and $w \in W$ we have by $H \mapsto wH$,

$$\int F(H)g(H)\nu(H)\,dH = \int F(H)g(wH)\nu(H)\,dH.$$

Taking the sum over $w \in W$ (i.e. taking the W-trace) the right side becomes equal to 0, so is the left side, whence it follows that $F(H)\nu(H) = 0$ for almost all H. By **INV 6(a)**, we conclude that $F(H) = 0$ for almost all H, so $F = 0$ by continuity. Then **INV 7(a)** concludes the proof.

We need more conditions to get enough functions of the form $\Phi_\nu^- g$, with $g \in C_c^\infty(\mathfrak{a})^W$. The conditions say that Φ_ν^- is a homomorphism for some operations. See Theorem 3.2 of Chapter VI.

INV 8. (a) There is a bilinear product $(f, g) \mapsto f \# g$ of $C_c^\infty(\mathfrak{a})^W$ into itself such that Φ_ν^- transforms this product into ordinary multiplication, that is

$$\Phi_\nu^-(f \# g) = (\Phi_\nu^- f)(\Phi_\nu^- g).$$

(b) Define $f^*(H) = \overline{f(-H)}$ for $H \in \mathfrak{a}$. Then

$$\Phi_\nu^-(f^*) = \overline{(\Phi_\nu^- f)}.$$

Lemma 1.5. *Under* **INV 8**, *the set of functions* $\{\Phi_\nu^- g\}$ *with* $g \in C_c^\infty(\mathfrak{a})^W$ *is an algebra, closed under complex conjugation.*

Proof. Immediate from (a) and (b).

We note that the properties proved in Lemmas 1.4 and 1.5 constitute the hypotheses in the Stone–Weierstrass theorem on $(\mathfrak{a}^\vee/W, \infty)$, the one-point compactification of $\mathfrak{a}^\vee/W$. The conclusion of the Stone–Weierstrass theorem then implies that the family of functions $\{\Phi_\nu^- g\}$ is dense in the function space $C(\mathfrak{a}^\vee/W, \infty)$, consisting of continuous functions on $\mathfrak{a}^\vee/W$ vanishing at infinity.

Theorem 1.6. *Under* **INV 1–INV 8** *the constant C of Theorem* 1.2 *is* $\neq 0$. *In particular,* Φ_ν^- *is injective on* $C_c^\infty(\mathfrak{a})^W$.

Proof. Suppose $C = 0$. By Theorem 1.2 we have

$$\int_{\mathfrak{a}^\vee} (\Phi_\nu^-)(\lambda)\eta(\lambda)\, d\lambda = 0 \qquad \text{for all } g \in C_c^\infty(\mathfrak{a})^W.$$

By the Stone–Weierstrass theorem, it follows that $\eta(\lambda) = 0$ for almost all λ, contradicting **INV 6(b)**, and proving the theorem.

The conditions in **INV 8** are still qualitative. To get more precise information on the inversion constant C, we have to pin down the #-product, for instance as in the next condition giving its quadratic form at the origin. Cf. Theorem 3.4 of Chapter VI.

INV 9. The product of **INV 8** satisfies

$$(f \# f^*)(0) = \|f\|_{2,\nu}^2.$$

Theorem 1.7. *Under* **INV 1–INV 4, INV 8** *and* **INV 9**, *for*

$$f \in C_c^\infty(\mathfrak{a})^W,$$

we have

$$\|\Phi_\nu f\|_{2,\eta}^2 = \|\Phi_\nu^- f\|_{2,\eta}^2 = C\|f\|_{2,\nu}^2.$$

Proof. The first equality is trivial because η is assumed even. As to the second equality, we have

$$\begin{aligned}
\|\Phi_\nu^- f\|_{2,\eta}^2 &= \int_{\mathfrak{a}^\vee} (\Phi_\nu^- f)\overline{(\Phi_\nu^- f)}(\lambda)\eta(\lambda)\, d\lambda \\
&= \int_{\mathfrak{a}^\vee} \Phi_\nu^-(f \# f^*)(\lambda)\eta(\lambda)\, d\lambda && \text{by } \textbf{INV 8} \\
&= C(f \# f^*)(0) && \text{by Theorem 1.2} \\
&= C\|f\|_{2,\nu}^2 && \text{by } \textbf{INV 9},
\end{aligned}$$

as was to be shown.

IX, §2. HELGASON INVERSION ON PALEY–WIENER AND THE L^2-ISOMETRY

Harish-Chandra proved that the map Φ_ν (in the context of semisimple Lie groups) preserves the L^2-metric up to the constant C, which he determined. He also proved inversion on the other side for his Schwartz space. In this section, which is a direct continuation of the preceding one, we give Helgason's proof for the inversion on the other side of $C_c^\infty(\mathfrak{a})$, which amounts to the surjectivity on the Paley–Wiener space [Hel 66], [Ros 77], p. 146, and [Hel 84], p. 453.

Theorem 2.1. *Assume conditions* **INV 1–INV 8**. *Then on* $\mathrm{PW}(\mathfrak{a}^\vee)^W$ *with the constant* C *of Theorems* 1.2 *and* 1.3, *we have inversion*

$$\Phi_\nu^- \circ {}^t\Phi_\eta = C \cdot \mathrm{id}.$$

In particular, the image of Φ_ν^- *contains* $\mathrm{PW}(\mathfrak{a}^\vee)^W$. *If* Φ_ν^- *maps* $C_c^\infty(\mathfrak{a})^W$ *into* $\mathrm{PW}(\mathfrak{a}^\vee)^W$ *then* $\Phi_\nu^-: C_c^\infty(\mathfrak{a})^W \to \mathrm{PW}(\mathfrak{a}^\vee)^W$ *is a linear isomorphism.*

Proof. Let $\psi \in \mathrm{PW}(\mathfrak{a}^\vee)^W$ be the Fourier transform of a function in $C_c^\infty(\mathfrak{a})$, invariant under W. Let $F = \Phi_\nu^- \circ {}^t\Phi_\eta\psi - C\psi$. We want to show $F = 0$. We have

$$\begin{aligned} {}^t\Phi_\eta F &= {}^t\Phi_\eta\Phi_\nu^- \circ {}^t\Phi_\eta\psi - C\,{}^t\Phi_\eta\psi \\ &= 0 \quad \text{by Theorem 1.3.} \end{aligned}$$

Then from the definition of ${}^t\Phi_\eta$ as integral transform, we obtain

$$\int_{\mathfrak{a}^\vee} F(\lambda)\Phi(\lambda, H)\eta(\lambda)\,d\lambda = {}^t\Phi_\eta F(H) = 0.$$

Hence for every $g \in C_c^\infty(\mathfrak{a})^W$ we get

$$\begin{aligned} 0 &= \int_{\mathfrak{a}}\int_{\mathfrak{a}^\vee} F(\lambda)\Phi(\lambda, H)g(H)\eta(\lambda)\nu(H)\,d\lambda\,dH \\ &= \int_{\mathfrak{a}^\vee} F(\lambda)(\Phi_\nu g)(\lambda)\eta(\lambda)\,d\lambda \quad \text{by Fubini} \\ &= \int_{\mathfrak{a}^\vee} F(-\lambda)(\Phi_\nu^- g)(\lambda)\eta(-\lambda)\,d\lambda. \end{aligned}$$

By Lemmas 1.4 and 1.5, and the Stone–Weierstrass theorem, it follows that

$$F(-\lambda)\eta(-\lambda) = 0$$

for almost all λ. By the assumption **INV 6(b)** on η, it follows that $F(\lambda) = 0$ for almost all λ. Since F is continuous, it follows that $F = 0$, thus proving the inversion formula. Since $C \neq 0$ by Theorem 1.5, we get the surjectivity and isomorphism property by using Theorem 1.3. This concludes the proof.

Example. In the application to the semisimple Lie groups, the spherical transform is composed of the Harish transform and the Mellin transform, and hence one knows that its image is contained in the Paley–Wiener space. We have not specified a condition in our general axiomatization which would guarantee this a priori.

IX, §3. THE CONSTANT IN THE INVERSION FORMULA

In this section, we follow Rosenberg's argument to show that under certain conditions, arising from a proof analysis of [Ros 77], the constant C of Theorems 1.2 and 1.5 is equal to 1. Not surprisingly, more data must be made part of the structure. This section is a continuation of §1, but is independent of §2. We consider the following W-invariance conditions. We suppose that there is an open positive cone $\mathfrak{a}^+$ which is a fundamental domain for W on $\mathfrak{a}$ from which a set of measure 0 has been deleted. We assume:

W 1. The function ν is W-invariant on $\bigcup_{w \in W} w\mathfrak{a}^+$, $\nu > 0$ on $\mathfrak{a}^+$, and ν is C^∞ on $\mathfrak{a}^+$.

W 2. The function $\eta(\lambda)$ has the form:

(a) $\eta(\lambda) = |\boldsymbol{c}^{-2}(\mathbf{i}\lambda)|$ with a continuous function $\boldsymbol{c}^{-1}$ on $\mathbf{i}\mathfrak{a}^\vee$.

(b) $\boldsymbol{c}(-\mathbf{i}\lambda) = \overline{\boldsymbol{c}(\mathbf{i}\lambda)}$ (wherever $\boldsymbol{c}^{-1}(\mathbf{i}\lambda) \neq 0$).

(c) For each $w \in W$, let

$$\gamma(w, \lambda) = \frac{\boldsymbol{c}(\mathbf{i}w\lambda)}{\boldsymbol{c}(\mathbf{i}\lambda)}.$$

Then the limit $\lim_{\epsilon \to 0} \gamma(w, \epsilon\lambda)$ exists for every $w \in W$ and $\lambda \in \mathfrak{a}^\vee$ outside a set of measure 0, and is independent of λ.

We denote the above limit by $\gamma(w, 0)$. The W-invariance of η shows the W-invariance of the absolute value $|\boldsymbol{c}(\mathbf{i}\lambda)|$. In particular, the limit in (c) is taken over terms of absolute value 1, that is

$$|\gamma(w, \lambda)| = \frac{|\boldsymbol{c}(w\mathbf{i}\lambda)|}{|\boldsymbol{c}(\mathbf{i}\lambda)|} = 1 \quad \text{so} \quad |\gamma(w, 0)| = 1.$$

W 3. The kernel function Φ is a perturbation of the W-trace of the multiple $c(i\lambda)e^{i\lambda(H)}$ of the Fourier kernel on $\mathfrak{a}^+$, that is for $H \in \mathfrak{a}^+$,

$$c^{-1}(i\lambda)\nu^{1/2}(H)\Phi(\lambda, H) = \sum_{w\in W} \gamma(w, \lambda)e^{iw\lambda(H)} + E(\lambda, H)$$

where $E(\lambda, H)$ is an error term satisfying the following L^2-condition (so it doesn't matter that $c(i\lambda)$ is not defined on a set of measure 0). For each $f \in C_c^\infty(\mathfrak{a}^+)$, define the perturbation term

$$R_{f,\epsilon}(\lambda) = \epsilon^{-r/2} \int_{\mathfrak{a}^+} f(H)\nu^{1/2}(H)E(\lambda, H/\epsilon)\, dH.$$

Then

$$\lim_{\epsilon\to 0} \|R_{f,\epsilon}\|_2 = 0.$$

Note that for any function ψ on $\mathfrak{a}^\vee$, we have $\|\psi\|_2^2 = \|\psi^-\|_2^2$, so $\psi(\lambda) \mapsto \psi(-\lambda)$ does not affect the L^2-norm. The axiom **W 3** was verified in Chapter VIII, Lemma 5.1, with the function $fJ_{\mathfrak{a}}^{1/2}$ which is also in $C_c^\infty(\mathfrak{a}^+)$, so $\nu = J_{\mathfrak{a}}$ on $\mathfrak{a}^+$.

Theorem 3.1. *Assume conditions* **INV 1–INV 9** *and* **W 1, W 2, W 3**. *Suppose the Lebesgue measures on $\mathfrak{a}$ and $\mathfrak{a}^\vee$ are such that ordinary Fourier inversion holds for them. Let C be the inversion constant of* §1 *and* §2. *Then*

$$C = |W|^2.$$

Proof. We follow Rosenberg's proof [Ros 77]. Let $f \in C_c^\infty(\mathfrak{a})^W$ have compact support in the union of the open sets $w\mathfrak{a}^+$, $w \in W$. For $\epsilon > 0$, let f_ϵ be the function on $\mathfrak{a}/W$ defined for $H \in \mathfrak{a}^+$ by the formula

$$f_\epsilon(H) = \epsilon^{r/2} f(\epsilon H)\nu^{1/2}(\epsilon H)\nu^{-1/2}(H).$$

Lemma 3.2. *We have*

$$\|f_\epsilon\|_{2,\nu}^2 = \|f\|_{2,\nu}^2.$$

Proof. This relation is a mere change of variables in the integral defining the $L^2(\nu\, dH)$-norm square. We start with

$$\|f_\epsilon\|_{2,\nu}^2 = \epsilon^r \int_{\mathfrak{a}} |f(\epsilon H)\nu^{1/2}(\epsilon H)|^2 \nu^{-1}(H)\nu(H)\, dH,$$

and let $H \mapsto \epsilon^{-1}H$ to get the desired equality.

Note that by Theorem 1.7, **W 2(a)**, and Lemma 3.2, we get

$$(1) \qquad \|c^{-1}\Phi_\nu f_\epsilon\|_2^2 = \|\Phi_\nu f_\epsilon\|_{2,\eta}^2 = C\|f_\epsilon\|_{2,\nu}^2 = C\|f\|_{2,\nu}^2.$$

Next, we use **W 3** to decompose $\Phi_\nu^- f_\epsilon$. We have

$$\begin{aligned}\frac{1}{|W|}(\Phi_\nu f_\epsilon)(\lambda) &= \int_{\mathfrak{a}^+} f_\epsilon(H)\Phi(\lambda, H)\nu(H)\,dH \\ &= \epsilon^{r/2}\int_{\mathfrak{a}^+} f(\epsilon H)\nu^{1/2}(\epsilon H)\nu^{1/2}(H)\Phi(\lambda, H)\,dH \\ &\qquad\qquad\qquad\qquad\qquad\qquad [\text{by definition of } f_\epsilon] \\ &= \epsilon^{-r/2}\int_{\mathfrak{a}^+} f(H)\nu^{1/2}(H)\nu^{1/2}(H/\epsilon)\Phi(\lambda, H/\epsilon)\,dH \\ &\qquad\qquad\qquad\qquad\qquad\qquad [\text{by } H \mapsto H/\epsilon].\end{aligned}$$

Therefore, normalizing with c^{-1}, we get

$$\begin{aligned}(2)\quad &|W|^{-1}c^{-1}(\mathbf{i}\lambda)(\Phi_\nu f_\epsilon)(\lambda) \\ &= \epsilon^{-r/2}\sum_{w\in W}\int_{\mathfrak{a}^+} f(H)\nu^{1/2}(H)\gamma(w,\lambda)e^{\mathbf{i}w\lambda(H)/\epsilon}\,dH \\ &\quad + \epsilon^{-r/2}\int_{\mathfrak{a}^+} f(H)\nu^{1/2}(H)E(\lambda, H/\epsilon)\,dH \quad [\text{by } \mathbf{W\ 3}].\end{aligned}$$

Note that the second term on the right is precisely $R_{f,\epsilon}(\lambda)$, cf. **W 3**. We define

$$F_\epsilon(\lambda) = \epsilon^{-r/2}\sum_{w\in W}\gamma(w,\lambda)\int_{\mathfrak{a}^+} f(H)\nu^{1/2}(H)e^{\mathbf{i}w\lambda(H)/\epsilon}\,dH.$$

Then we obtain

$$(3) \qquad |W|^{-1}c^{-1}(\mathbf{i}\lambda)(\Phi_\nu f_\epsilon)(\lambda) = F_\epsilon(\lambda) + R_{f,\epsilon}(\lambda).$$

Lemma 3.3. *We have*

$$\lim_{\epsilon\to 0}\|F_\epsilon\|_2^2 = \|f\|_{2,\nu}^2.$$

Putting (1), (3), Lemma 3.3 and the limit in **W 3** together, we see immediately that $C = |W|^2$, thereby concluding the proof of Theorem 3.1.

There remains to prove the lemma.

Proof of Lemma 3.3. Starting with the definitions, we get

$$(4)\quad \|F_\epsilon\|_2^2 = \epsilon^{-r}\int_{\mathfrak{a}^\vee}\left|\sum_{w\in W}\gamma(w,\lambda)\int_{\mathfrak{a}^+} f(H)\nu^{1/2}(H)e^{iw\lambda(H)/\epsilon}\,dH\right|^2 d\lambda$$

$$= \int_{\mathfrak{a}^\vee}\left|\sum_{w\in W}\gamma(w,\epsilon\lambda)\int_{\mathfrak{a}^+} f(H)\nu^{1/2}(H)e^{iw\lambda(H)}\,dH\right|^2 d\lambda$$

[by $\lambda \mapsto \epsilon\lambda$]

$$= \int_{\mathfrak{a}^\vee}\left|\sum_{w\in W}\gamma(w,\epsilon\lambda)|W|^{-1}(f\nu^{1/2})^\wedge(-\lambda)\right|^2 d\lambda.$$

By **W 1**, the function $f\nu^{1/2}$ is in $C_c^\infty(\mathfrak{a})$, so its Fourier transform is in the Schwartz space, rapidly decreasing. Since $|\gamma(w,\epsilon\lambda)| = 1$, we can apply the dominated convergence theorem and take the limit as $\epsilon \to 0$ under the integral sign. From **W 2(c)**, we then get

$$(5)\qquad \lim_{\epsilon\to 0}\|F_\epsilon\|_2^2 = \|(f\nu^{1/2})^\wedge\|_2^2 = \|f\nu^{1/2}\|_2^2 = \|f\|_{2,\nu}^2$$

by ordinary L^2-isometry of the Fourier transform and the assumption that the measures $dH, d\lambda$ are so normalized. This proves the lemma, and Theorem 3.1.

Remarks. Is there a stronger estimate to the effect that

$$E(\lambda, H/\epsilon) = o(\epsilon^{r/2})E_0(\lambda, H) \qquad \text{for } \epsilon \to 0,$$

where E_0 is locally in L^2, because the compactly supported f cuts down the support to a compact set?

In the applications to complex groups, the error term is 0, and hence the whole machinery used to estimate the error term including the Harish-Chandra series is unnecessary. The Flensted-Jensen reduction to complex groups may give something which is not given by the Harish-Chandra series and Gangolli estimates for this series. In any case, it gives an alternative approach to the whole inversion, and may give a simpler approach for the limit of the negligible term in **W 3**. We would find it valuable to have an axiomatization of the Flensted-Jensen method and an investigation whether it applies to other situations.

CHAPTER X

The Harish-Chandra Schwartz Space (HCS) and Anker's Proof of Inversion

As we have already mentioned, Harish-Chandra first set up his inversion theorem on a space which he defined, as an analogue of the Schwartz space in ordinary Fourier analysis. Classically, such analysis takes place on the imaginary axis, with the Fourier kernel e^{ixy}. The symmetry of this kernel function makes it so that one views the kernel as a function of two real variables. One can do Fourier inversion on many spaces, the most important of which are C_c^∞, the Schwartz space, and L^2. Under the Fourier transform, C_c^∞ goes to the Paley–Wiener space. As one considers functions which are less restricted than C_c^∞, the image space changes accordingly, and the Fourier transforms are not necessarily entire. At a pivotal stage, the Schwartz space is self-dual, with functions on the given euclidean space or its corresponding imaginary axis $\mathbf{i}\mathfrak{a}^\vee$. This leads naturally into the L^2-duality. The Fourier transform of functions in spaces between the Schwartz space and C_c^∞ although not entire may have analytic continuation to a domain larger than the imaginary axis, namely tube domains, which provide an added structure to the situation.

Having the most restricted case of duality C_c^∞–Paley–Wiener, one hopes that the other cases of duality can be obtained by continuity with respect to their respective topologies: Schwartz topology, L^2, L^p, etc. However, for decades after Harish-Chandra's general Schwartz space and L^2-spherical inversion, the impression was that a proof of spherical inversion for the Schwartz space required the elaboration of substantial machinery beyond the C_c^∞–Paley–Wiener inversion, namely machinery elaborated by Harish-Chandra. See for instance comments to this effect in [Hel 84], p. 493, and [GaV 88], p. 299. In 1991, Anker [Ank 91] showed how to carry out the continuity argument directly and briefly. We shall follow his paper. Anker also deals with L^p–Schwartz spaces

($0 < p \leqq 2$) introduced by Trombi and Varadarajan. In this chapter, we stay with $p = 2$. The other values of p will be dealt with in the next chapter. Anker's proof is based on a variation of the Paley–Wiener theorem, making greater use of positivity and convexity, so we start with a continuation of Chapter I, §4, namely Harish-Chandra positivity results. These are first used to prove estimates for the spherical functions, using both the Harish-Chandra integral and the Harish-Chandra series. We then give Anker's proof of Harish-Chandra inversion on the Harish-Chandra Schwartz space, using his variation of Paley–Wiener and a continuity argument.

The L^p–Schwartz spaces for $p \neq 2$ are needed for various applications, and will be treated in the next chapter. Already in §7 of the present chapter, we see the need for the case $p = 1$, which is the most important. More comments will be made in the introduction to Chapter XI. We hope that having presented the simplest case $p = 2$ separately makes the exposition more accessible. The general case has practically the same proofs, except for the introduction of tube domains.

X, §1. MORE HARISH-CHANDRA CONVEXITY INEQUALITIES

The present section may be viewed as a continuation of Chapter I, §4 and §5. It will be used to estimate spherical functions in §2 and subsequently. Readers might want to look at the beginning of §2 right away for motivation.

We start by tabulating together more Harish-Chandra inequalities having to do with positivity. See [Har 58a], Lemma 35, and its corollaries.

As in Chapter I, §4, we have the standard positive definite scalar product on $\mathfrak{a}$ (the trace form), whose associated norm is denoted by $|H|$, so $\langle H, H\rangle = |H|^2$.

We use systematically the notation already indicated in Chapter I. For $x \in G$, with Iwasawa and polar decompositions

$$x = uak = k_1 b k_2,$$

we let $x^+ = b^+$ be the unique representative in $\mathrm{Cl}(A^+)$ for the orbit b^W. Thus x^+ and b^+ depend on the polar decomposition. We write

$$a = \mathrm{Iw}_A(x), \qquad b = \mathrm{pol}_A(x), \qquad b^+ = \mathrm{pol}_{A^+}(x).$$

Of course, a is uniquely determined, but b is determined only up to the action of W. Only b^+ is uniquely determined by x. We define the **(polar) height**

$$\sigma(x) = |\log b| = |\log \mathrm{pol}_A(x)| = \sigma(\mathrm{pol}_A(x)),$$

which is independent of b in b^W. Trivially, the height is K-bi-invariant, and

$$\sigma(x) = \sigma(x^{-1}).$$

Proposition 1.1. *The height satisfies the* **Harish-Chandra "triangle inequality"**

$$\sigma(xy) \leqq \sigma(x) + \sigma(y).$$

Proof. We shall give Harish-Chandra's proof [Har 66], §7, Lemma 10.

First we make some remarks showing how the polar height can be expressed in terms of the Cartan decomposition. We have the Cartan Lie decomposition

$$\mathfrak{g} = \mathfrak{k} + \mathfrak{p} \qquad \text{with } \mathfrak{p} = \mathbf{c}(K)\mathfrak{a},$$

and putting $\mathbf{P} = \exp \mathfrak{p}$, the global Cartan decomposition

$$G = K\mathbf{P}.$$

The norm on $\mathfrak{a}$ is of course the restriction of the norm associated with the positive definite scalar product B on $\mathfrak{p}$, which in our concrete case is again the trace form. This extended norm is also invariant under K-conjugation, that is, for $H \in \mathfrak{a}$, $k \in K$,

$$|\mathbf{c}(k)H| = |H|.$$

We may use the Cartan decomposition to express the polar height, namely if

$$x = k \cdot \exp X \qquad \text{with } X \in \mathfrak{p}, k \in K$$

then

$$\sigma(x) = |X| = |\log \mathrm{Ca}_{\mathbf{P}}(x)|$$

where $\mathrm{Ca}_{\mathbf{P}}(x) \in \mathbf{P}$ is the $\mathbf{P}$-projection of x in the Cartan decomposition, and

$$\log\colon \mathbf{P} \to \mathfrak{p} \quad \text{is the inverse of} \quad \exp\colon \mathfrak{p} \to \mathbf{P}.$$

Now let

$$x = \exp X, \qquad y = \exp Y, \qquad xy = k \cdot \exp Z,$$

with $X, Y, Z \in \mathfrak{p}$ and $k \in K$. Considering ${}^t(xy)(xy)$ one immediately verifies that

$$\exp(2Z) = \exp(Z)^2 = (\exp Y)(\exp 2X)(\exp Y).$$

We may define a curve $t \mapsto Z(t)$ in $\mathfrak{p}$ by the formula

$$\text{(1)} \qquad \exp(2Z(t)) = \exp(tY)\exp(2X)\exp(tY).$$

The right side being positive definite symmetric, we can take its log to define $2Z(t)$. We differentiate with respect to t. We need a computation, which we express as a lemma.

Lemma 1.2. *Recalling the definition* $\langle A, B\rangle = \operatorname{tr}(AB)$, *we have*

$$\langle Z(t), Z'(t)\rangle = \langle Z(t), Y\rangle.$$

Proof. The derivative of the exponential map is expressed by a power series f (cf. [Mos 53] on Cartan's work, and [Lan 99], Chapter XII, Theorem 2.4, where the computation is reproduced), namely letting $[v]$ denote the regular Lie algebra representation of an element v in $\mathfrak{p}$, we have

$$\text{(2)} \qquad \exp'(v)w = e^{v/2} \cdot f([v])w \cdot e^{v/2}.$$

Furthermore, the constant term of the power series is 1. An element $u \in \mathfrak{g}$ which commutes with v also commutes with $\exp(v)$ or $\exp(v/2)$. Since $\exp(tY) = \sum t^n Y^n/n!$, differentiating (1) yields

$$\text{(3)} \qquad \exp'(2Z(t))2Z'(t) = Y\exp(2Z(t)) + \exp(2Z(t))Y.$$

We put $v = 2Z(t)$, $w = 2Z'(t)$. We multiply (3) on the right and left by $e^{-v/2}$. From (2) we then obtain

$$\text{(4)} \qquad f([v])w = e^{-v/2}Ye^{v/2} + e^{v/2}Ye^{-v/2}, \qquad \text{with } v/2 = Z(t).$$

We write $f([v]) = 1 + f_1([v])$ where f_1 is a power series starting with a term of degree $\geqq 1$. For an integer $m \geqq 1$, an element $[v]^m w$ is of the form $vw_m - w_m v$ for some w_m. For $m \geqq 1$ we get

$$\operatorname{tr}(v[v]^m w) = \operatorname{tr}(v(vw_m - w_m v)) = 0.$$

Hence in the series for $\operatorname{tr}(vf([v])w)$ all the terms are 0 except the first term which is $\operatorname{tr}(vw)$. Thus we get

$$\begin{aligned} \langle 2Z(t), 2Z'(t)\rangle = \operatorname{tr}(vw) &= \operatorname{tr}(vf([v])w) \\ &= 2\operatorname{tr}(vY) \qquad \text{by the right side of (4)} \\ &= \langle 2Z(t), 2Y\rangle \quad \text{by definition.} \end{aligned}$$

This concludes the proof of the lemma.

Now we come to the essential part of the proof of Theorem 1.1. We distinguish two cases. Suppose first $Z(t_0) = 0$ for some $t_0 \in \mathbf{R}$. Then

$$I = \exp(t_0 Y)\exp(2X)\exp(t_0 Y),$$

whence $X = -t_0 Y$, so X, Y commute. From its definition $Z = X + Y$, and we are done in this case. Next suppose $Z(t) \neq 0$ for all $t \in \mathbf{R}$. Then

$$\frac{1}{2}\frac{d}{dt}|Z(t)|^2 = \langle Z(t), Z'(t)\rangle = |Z(t)|\frac{d}{dt}|Z(t)|.$$

Let $E(t) = Z(t)/|Z(t)|$ be the unit vector in the direction of $Z(t)$. Then by Lemma 1.2,

$$\left|\frac{d}{dt}|Z(t)|\right| = |\langle E(t), Y\rangle| \leqq |Y| \qquad \text{for all } t \in \mathbf{R}.$$

Hence

$$|Z(1)| - |Z(0)| \leqq |Y|.$$

But directly from the definitions, $Z(1) = Z$ and $Z(0) = X$, so Proposition 1.1 is proved.

Remark. By using the differential geometry of G/K (or SPos_n), Proposition 1.1 is immediately seen to be equivalent to the ordinary triangle inequality on this coset space. This proof is given in [GaV 88], Proposition 4.6.11. Gangolli told us that the result is already in Cartan.

We now recall Theorems 6.1 and 6.2 of Chapter I, for the reader's convenience. For the notation see Chapter I, §4.

Theorem 1.3. *Let $a, b \in \mathrm{Cl}(A^+)$.*

(i) *If $u \in U$, $k_1, k_2 \in K$ and $au = k_1 b k_2$ then*

$$\log a \leqq_{A'} \log b.$$

(ii) *For all $k \in K$,*

$$-\log b \leqq_{A'} \log(kb^{-1})_A.$$

(iii) *For all $x \in G$, $\sigma(\mathrm{Iw}_A(x)) \leqq \sigma(\mathrm{pol}_A(x)) = \sigma(x)$.*

Proof. Assertion (i) comes from Theorem 6.1 of Chapter I by taking the log. For (ii), we note that $u^{-1}a^{-1} = k_2^{-1}b^{-1}k_1^{-1}$ so $u^{-1}a^{-1}k_1 = k_2^{-1}b^{-1}$. But

$$(k_2^{-1}b^{-1})_A = a^{-1}$$

by definition of the A–Iwasawa projection, so

$$-\log a = \log(k_2^{-1}b^{-1})_A,$$

which shows that (i) implies (ii). Finally (iii) is a rephrasing of Theorem 6.2 of Chapter I, namely in (i), $\log a$ is in the convex closure of $W \log b$.

We use still another symmetry. Let $w_0 \in W$ be such that

$$w_0\mathfrak{a}^+ = -\mathfrak{a}^+.$$

This is the "reversal" element, which on $\mathrm{SL}_n(\mathbf{R})$ is represented by conjugation with the anti-diagonal matrix having 1 on the anti-diagonal, except for a possible minus sign, say in the top row entry to make the matrix have determinant 1. We let this matrix be denoted by k_0, so $w_0 = \mathbf{c}(k_0)$ (conjugation by k_0). *Define*

$$H^* = -w_0 H \quad \text{for } H \in \mathfrak{a} \qquad \text{and} \qquad \lambda^* = -w_0\lambda \quad \text{for } \lambda \in \mathfrak{a}^\vee.$$

Then this star operator has order 2,

$$H \mapsto H^* \text{ is a permutation of } \mathfrak{a}^+, \text{ or also } \mathrm{Cl}(\mathfrak{a}^+),$$

and of course the permutation is positive-linear, i.e. it commutes with positive linear combinations. The star operation also extends to A, namely if $a = \exp H$ then *we define* $a^* = \exp H^*$ so that

$$\log a^* = (\log a)^* \qquad \text{and} \qquad a^* = k_0 a^{-1} k_0^{-1}$$

with the above notation.

From the action of W on $\mathfrak{a}^\vee$, it is easily verified that $\alpha \mapsto \alpha^*$ is a permutation of $\mathcal{R}(\mathfrak{n})$. Since $2\rho = \sum_{\alpha \in \mathcal{R}(\mathfrak{n})} \alpha$, it follows that

(5) $$\rho = \rho^* \qquad \text{and} \qquad \rho(H^*) = \rho(H) \quad \text{for all } H \in \mathfrak{a}.$$

Note in passing that this implies $-\rho = w_0\rho$, so $-\rho$ is in the orbit $W\rho$.

Corollary 1.4. *Let $a \in \mathrm{Cl}(A^+)$ and $k \in K$. Then*

(i) $$-\log a \leqq_{A'} \log(ka^{-1})_A \leqq_{A'} \log a^*.$$

Thus for $\lambda \in \mathfrak{a}^\vee$, $\lambda \geqq 0$ (cf. Chapter I, **Pos 3**),

(ii) $$-\lambda(\log a) \leqq \lambda(\log(ka^{-1})_A) \leqq \lambda(\log a^*).$$

Furthermore, for any $\lambda \in \mathfrak{a}^\vee$, $a = \exp H$,

(iii) $$|\lambda(\log(ka^{-1})_A| \leqq \max(|\lambda(H)|, |\lambda(H^*)|) \leqq \lambda^+(H^+).$$

Proof. Let $b = a^* = k_0 a^{-1} k_0^{-1}$. Then $a^{-1} = k_0^{-1} b k_0$, so for $k' = kk_0^{-1}$,

$$\begin{aligned}\log(a^*) - \log(ka^{-1})_A &= \log b - \log(k'bk_0)_A \\ &\geqq_{\mathcal{A}'} 0 \quad \text{by Theorem 1.3(i).}\end{aligned}$$

This proves the right inequality of (i). The left inequality is a repetition of Theorem 1.3(ii). The second inequality with $\lambda \geqq 0$ comes from the fact that λ is a semipositive linear combination of the elements α_i' in $\mathcal{A}'$ (**Pos 3**). The third inequality comes from (ii) and Chapter I, Proposition 4.4(ii). This concludes the proof.

Next we apply the above inequalities in a context when we take the norm, which is invariant under the Weyl group so allows a reduction to positive elements.

Proposition 1.5. *For $a \in A$, we have*

$$|\log a| = |\log a^*|.$$

There exists a constant $C > 0$ such that for all $a \in A$ and $k \in K$,

$$|\log(ka)_A| \text{ and } |\log(ka^{-1})_A| \leqq C|\log a|.$$

Proof. The first equality comes from the fact that the norm on $\mathfrak{a}$ is invariant under the Weyl group. For the inequality, it does not matter whether we use a or a^{-1} on the left side, because $|\log a| = |-\log a| = |\log a^{-1}|$. Using $\alpha_i' \in \mathcal{A}'$ in Corollary 1.4(ii), and using the right equality, gives us

$$|\alpha_i'(\log(ka^{-1})_A)| \leqq C_i |\log a| \quad \text{for some constant } C_i.$$

Since $\{\alpha_1', \ldots, \alpha_r'\}$ is a basis of $\mathfrak{a}^\vee$, it follows that

$$|\log(ka^{-1})_A| \leqq C|\log a| \quad \text{for some constant } C.$$

This concludes the proof.

Proposition 1.6. *For $a \in \mathrm{Cl}(A^+)$ and $k \in K$,*

$$(ka^{-1})_A^{\pm\rho} \leqq a^\rho \quad \text{or equivalently} \quad |\rho(\log((ka^{-1})_A)| \leqq \rho(\log a).$$

Proof. Immediate from Corollary 1.4(iii) and (5), using the fact that

$$\rho = \lambda_1 + \cdots + \lambda_r \geqq 0.$$

As usual, we use the notation $\lambda_i = \alpha_i'$ for the dual basis of the simple $(\mathfrak{a}, \mathfrak{n})$-characters.

The next lemma gives a coarser estimate for arbitrary multiplicative translations.

Lemma 1.7. *Let S be a compact subset of G. For each i there exists a constant $C_i > 0$ such that for $y \in S$ and $x \in G$,*

$$x_A^{\lambda_i} \leqq C_i (xy)_A^{\lambda_i} \quad \text{so} \quad x_A^{\rho} \leqq C(xy)_A^{\rho} \quad \text{with } C = \prod C_i.$$

Proof. Let C_i be a bound for the operator $\bigwedge^i(y^{-1})$. Then

$$\begin{aligned}(xy)_A^{-\lambda_i} &= \left\| \bigwedge^i (y^{-1}x^{-1})(e_1 \wedge \cdots \wedge e_i) \right\| \\ &\leqq C_i \left\| \bigwedge^i (x^{-1})(e_1 \wedge \cdots \wedge e_i) \right\| \\ &= C_i x_A^{-\lambda_i}\end{aligned}$$

which proves the lemma.

Remark. The above lemma provides another approach to prove Proposition 2.7 of the next section, following the general pattern of first proving inequalities for the values of characters, then integrating to get similar inequalities for the spherical functions.

X, §2. MORE HARISH-CHANDRA INEQUALITIES FOR SPHERICAL FUNCTIONS

The present section deals mostly with inequalities for spherical functions, and culminates with what Harish-Chandra called "an important inequality" [Har 58a], §9, especially Theorem 3. We recall a basic definition of the spherical functions and show how one special value $\zeta = 0$ controls estimates for other values. We use the **Iwasawa decomposition** $G = UAK$, and for $\zeta \in \mathfrak{a}_{\mathbf{C}}^{\vee}$, the Harish-Chandra integral for the spherical function is given for $x \in G$ by

$$(1) \qquad \varphi_\zeta(x) = \int_K \mathrm{Iw}_A(kx)^{\zeta+\rho}\, dk = \int_K (kx)_A^{\zeta+\rho}\, dk.$$

Trivially, $\varphi_\lambda(x) > 0$ for all $\lambda \in \mathfrak{a}^{\vee}$, $x \in G$. By Proposition 7.3 of Chapter III,

$$(2) \qquad \varphi_{-\zeta}(x) = \varphi_\zeta(x^{-1}).$$

So for $\zeta = 0$ the spherical function φ_0 is even, namely

(3) $$\varphi_0(x^{-1}) = \varphi_0(x) \qquad \text{with} \qquad \varphi_0(x) = \int_K (kx)_A^\rho \, dk.$$

We shall prove that for $\log a \geqq 0$,

$$a^{-\rho} \leqq \varphi_0(a) << a^{-\rho}(1 + |\log a|)^{|\mathcal{R}(\mathfrak{n})|}.$$

The left inequality depends only on the inequalities of §1 and will be given immediately. The right inequality is more substantial and will be given in Theorem 2.8. In between, we show how to reduce estimates for φ_ζ to estimates for φ_0. For a refinement of the above inequalities, see [Ank 87].

Proposition 2.1. *For $a \in \mathrm{Cl}(A^+)$ and $H \in \mathrm{Cl}(\mathfrak{a}^+)$, we have*

$$a^{-\rho} \leqq \varphi_0(a) \qquad \text{and} \qquad e^{-\rho(H)} \leqq \varphi_0(\exp H).$$

Proof. Theorem 1.3(ii) gives for $k \in K$,

$$-\lambda_i(\log a) \leqq \lambda_i(\log(ka^{-1})_A) \qquad \text{for } i = 1, \ldots, r.$$

Since $\rho = \lambda_1 + \cdots + \lambda_r$, we see that applying ρ instead of λ_i preserves the inequality. We then exponentiate and integrate over K, to obtain

$$a^{-\rho} \leqq \int_K (ka^{-1})_A^\rho \, dk = \varphi_0(a^{-1}) = \varphi_0(a).$$

This proves the inequality, and concludes the proof of the proposition.

We note in passing an estimate of φ_λ in terms of φ_0, given in [GaV 88], Proposition 4.6.1.

Proposition 2.2. *Let $\lambda \in \mathfrak{a}^\vee$ and $H \in \mathfrak{a}$, with $a = \exp H$. Let λ^+, H^+ be the $\mathcal{A}$-semipositive elements in the orbits $W\lambda$ and WH of λ and H respectively under the Weyl group (cf. Chapter* I, *Propositions* 4.3 *and* 4.3′*). Then*

$$\varphi_\lambda(a) \leqq e^{\lambda^+(H^+)}\varphi_0(a).$$

For $\zeta \in \mathfrak{a}_{\mathbf{C}}^\vee$ we have the estimate in terms of its real part

$$|\varphi_\zeta(x)| \leqq \varphi_{\mathrm{Re}(\zeta)}(x).$$

Proof. Recall that $\varphi_\lambda(a)$ is invariant under the action of the Weyl group on each variable. For the a-variable this was seen directly from the definition, and for the λ-variable, it is Theorem 5.2 of Chapter III. In particular,

$$\varphi_\lambda(a) = \varphi_{\lambda^+}(a^+) \qquad \text{where } a^+ = \exp(H^+).$$

By Corollary 1.4(iii) we know that

$$|\lambda(\log(ka^{-1})_A)| \leqq \lambda^+(H^+).$$

Using $\varphi_{-\lambda}(a^{-1}) = \varphi_\lambda(a)$ (cf. (2)) we get

$$\varphi_\lambda(a) = \int_K (ka^{-1})_A^{-\lambda}(ka^{-1})_A^\rho \, dk \leqq e^{\lambda^+(H^+)} \int_K (ka^{-1})_A^\rho \, dk,$$

which concludes the proof of the first inequality. The second is trivial.

We return to getting estimates for the basic spherical function φ_0. The next proposition can be used to shift estimates away from the boundary of the positive fundamental domain, for instance as in the proof of Theorem 2.8.

Proposition 2.3. *For $a \in A$, $b \in \mathrm{Cl}(A^+)$, we have*

$$b^{-\rho}\varphi_0(ab) \leqq \varphi_0(a) \leqq b^\rho \varphi_0(ab).$$

Sometimes the notation $\psi(a) = a^\rho \varphi_0(a)$ is used, so the inequality reads

$$\psi(a) \leqq \psi(ab).$$

For the proof, it is convenient to extract a general lemma in which we change variables. For this, it will be convenient to use the notation

$$\rho_G(uak) = a^\rho \qquad \text{so} \qquad \rho_G(x) = x_A^\rho.$$

Lemma 2.4. *For $x, y \in G$, we have*

$$\varphi_0(x^{-1}y) = \int_K \rho_G(kx)\rho_G(ky)\, dk.$$

Proof. Write $G = PK$ with $P = UA$. Make the change of variables

$$kx^{-1}p_k = k' \qquad \text{so that} \qquad k = p_k k' x.$$

Then $dk/dk' = \rho_G^{-2}(p_k)$ by Chapter III, Lemma 7.2, and since $\rho_G(k) = 1$,

$$\begin{aligned}\int_K \rho_G(kx^{-1}y)\,dk &= \int_K \rho_G(kx^{-1}y)\rho_G(k)\,dk \\ &= \int_K \rho_G(p_k)\rho_G(k'y)\rho_G(p_kk'x)\frac{dk}{dk'}\,dk' \\ &= \int_K \rho_G^2(p_k)\rho_G(k'y)\rho_G(k'x)\frac{dk}{dk'}\,dk'\end{aligned}$$

which proves the lemma because dk/dk' cancels $\rho_G^2(p_k)$.

We apply the lemma first to finish the proof of Proposition 2.3. We have

$$\begin{aligned}\varphi_0(a) = \varphi_0(a^{-1}) &= \varphi_0(bb^{-1}a^{-1}) \\ &= \int_K \rho_G(kb^{-1})\rho_G(kb^{-1}a^{-1})\,dk \quad \text{by Lemma 2.4} \\ &\leqq b^\rho \int_K (ka^{-1}b^{-1})_A^\rho\,dk \quad \text{by Proposition 1.6} \\ &= b^\rho \varphi_0(ab) \quad \text{by (3),}\end{aligned}$$

thus proving the right inequality. The left then follows from the right because

$$\varphi_0(ab) = \varphi_0(a^{-1}b^{-1}) \leqq b^\rho \varphi_0(a^{-1}) = b^\rho \varphi_0(a).$$

The left inequality was first noted by Anker [Ank 91], as part of a general program to get optimal inequalities for spherical functions.

Next we give another application of Lemma 2.4.

Theorem 2.5. *The function φ_0 is* **positive definite**, *in the following sense. For all $x_i, x_j \in G$ and c_i, c_j real, we have*

$$\sum \varphi_0(x_1^{-1}x_j)c_ic_j \geqq 0.$$

Proof. By Lemma 2.4,

$$\begin{aligned}\sum_{i,j} \varphi_0(x_i^{-1}x_j)c_ic_j &= \int_K \sum \rho_G(kx_i)\rho_G(kx_j)c_ic_j\,dk \\ &= \int_K \left(\sum \rho_G(kx_i)c_i\right)^2 dk,\end{aligned}$$

which proves the theorem.

Corollary 2.6. *We have* $0 < \varphi_0 \leqq 1 = \varphi_0(e)$.

Proof. Put $c_1 = \varphi_0(x)$, $c_2 = -\varphi_0(x)$, $x_1 = e$ and $x_2 = x$.

From the corollary and Proposition 2.2, we obtain the inequality (bound)

$$|\varphi_{i\lambda}(x)| \leqq 1 \qquad \text{for all } \lambda \in \mathfrak{a}^\vee \text{ and } x \in G.$$

Furthermore the positive definiteness of Theorem 2.5 also applies to the spherical functions $\varphi_{i\lambda}$ $(\lambda \in \mathfrak{a}^\vee)$ in the hermitian case (c_i, c_j complex, and a bar over c_j). The broader context for this positive definiteness, which stems from Cartan's definition of spherical functions as coefficient functions in a Hilbert space representation, is treated very generally in self-contained details in $SL_2(\mathbf{R})$ [Lan 75/85], Chapter IV, §5, see especially formula (3). We thought it useful for the reader to see the direct argument freed from that context, giving the sharp structural upper bound for the spherical function.

Proposition 2.7. *Uniformly for y_1, y_2 in a compact subset of G, we have*

$$\varphi_0(y_1xy_2) \gg\ll \varphi_0(x) \qquad \textit{for } x \in G.$$

In other words, $\varphi_0(y_1xy_2)/\varphi_0(x)$ as a function of $x \in G$ is bounded away from 0 and ∞ for y_1, y_2 in a compact set.

Proof. This is an immediate consequence of Lemma 2.4, since we can estimate $\rho_G(ky)$ by a constant when y ranges over a compact set. We then use the identity $\varphi_0(x) = \varphi_0(x^{-1})$ to conclude the proof.

Propositions 2.3 and 2.7 have to do with estimates of multiplicative translations of the spherical function. In particular, Proposition 2.3 is useful among other things because it shows how one can reduce certain estimates of the spherical function to points which are away from the boundary of A^+, by using such translations. We shall use Proposition 2.3 in the proof of the "important inequality", which depends on more elaborate arguments, one way or another. We use the Harish-Chandra series expression for the spherical functions. The inequality implies that φ_0 decreases exponentially at infinity.

Theorem 2.8 (The "Important Inequality"). *There is a constant c_1 such that for $a \in \mathrm{Cl}(A^+)$*

$$\varphi_0(a) \leqq c_1 a^{-\rho}(1 + |\log a|)^{|\mathcal{R}(\mathfrak{n})|};$$

or in additive notation, for $H \in \mathrm{Cl}(\mathfrak{a}^+)$, letting $\Phi_\lambda(H) = \varphi_\lambda(\exp H)$,

$$\Phi_0(H) \leqq c_1 e^{-\rho(H)}(1 + |H|)^{|\mathcal{R}(\mathfrak{n})|}.$$

Proof. We recall the product

$$\Pi_+^\vee(\lambda) = \prod_{\alpha \in \mathcal{R}(\mathfrak{n})} \langle \alpha, \lambda \rangle$$

from Chapter V, Lemma 8.2, which tells us that this product defines the polar divisor of the $\boldsymbol{c}$-function $\boldsymbol{c}(\mathbf{i}\lambda)$ in a neighborhood of 0. We note that

$$(4) \qquad \mathcal{D}(\Pi_+^\vee)(\Pi_+^\vee(\lambda)\Phi_\lambda(H))|_{\lambda=0} = c_0\Phi_0(H)$$

with some constant c_0, actually $c_0 = \mathcal{D}(\Pi_+^\vee)\Pi_+^\vee$ because if some partial derivatives are applied to Φ_λ, then there remains some factor $\langle \lambda, \alpha \rangle$ in the product which is not differentiated, and hence yields 0 when we make $\lambda = 0$.

Now we have the Harish-Chandra series

$$(5) \qquad \Phi(\zeta, H) = \sum_{w \in W} \boldsymbol{c}(w\zeta)e^{(w\zeta - \rho)(H)} \sum_{\mu} f_\mu(w\zeta)e^{-\mu(H)},$$

with $f_0 = 1$, f_μ being a rational function with arbitrarily small exponential growth in μ, uniformly for ζ in a neighborhood of 0. Cauchy's theorem implies a similar small exponential growth for partial derivatives of given order. Thus we have

$$(6) \qquad e^{\rho(H)}\Pi_+^\vee(\zeta)\Phi(\zeta, H) = \sum_{\mu}\sum_{w \in W} s(w)\mathbf{b}(w\zeta)f_\mu(w\zeta)e^{w\zeta(H)}e^{-\mu(H)},$$

where $s(w)$ is a sign. Furthermore, the series can be differentiated term by term. From Theorem 2.3 of Chapter VIII, we have the Gangolli estimate for the coefficients f_μ, uniformly for ζ in a neighborhood of the origin. By Cauchy's theorem, a similar estimate applies to derivatives. More precisely, let $P \in S(\mathfrak{a}^\vee)$ be a polynomial of degree d. Given a small neighborhood of the origin, and $H_0 \in \mathfrak{a}$, there exists a constant C (depending on the neighborhood, P, H_0) such that for all μ, and ζ in this neighborhood,

$$(7) \qquad |\mathcal{D}(P)f_\mu(\zeta)| \leqq Ce^{\mu(H_0)}.$$

We select H_0 such that $\alpha_i(H_0) = 1$ for $i = 1, \ldots, r$. Let

$$\begin{aligned} F(t) &= e^{t\rho(H_0)}c_0\Phi(0, tH_0) \quad \text{for } t > 0 \\ &= e^{t\rho(H_0)}\mathcal{D}(\Pi_+^\vee)(\Pi_+^\vee(\zeta)\Phi(\zeta, tH_0))\big|_{\zeta=0}. \end{aligned}$$

We put $H = tH_0$ in (6) and differentiate term by term. Let $d = |\mathcal{R}(\mathfrak{n})|$, and note that $\Pi_+^\vee$ is a polynomial of degree d. The derivatives of the exponential introduce powers of t of at most degree d, so we obtain

$$(8) \qquad F(t) = \sum_{\mu \in \mathbf{L}} p_\mu(t)e^{-t\mu(H_0)} \qquad \text{with} \quad \deg p \leqq d.$$

By Gangolli's estimates, we know that the series converges uniformly for $t \geqq t_1$ for some t_1. Hence there exists N such that

$$\left| F(t) - \sum_{m(\mu) \leqq N} p_\mu(t) e^{-t\mu(H_0)} \right| \leqq 1 \quad \text{for all } t \geqq t_1.$$

Since $\mu(H_0) = m(\mu)$ by the way we chose H_0, we obtain

$$\lim_{t \to \infty} \sum_{0 < m(\mu) \leqq N} p_\mu(t) e^{-t\mu(H_0)} = 0.$$

In particular, for $t \geqq t_2$ we get $|F(t)| \leqq 2 + |p_0(t)|$, whence

$$|F(t)| \ll (1+t)^d \qquad \text{for } t \to \infty.$$

For $H \in \mathfrak{a}$, let

$$\beta(H) = \max \alpha_i(H) \qquad \text{for } i = 1, \ldots, r.$$

Recalling that $\mathrm{Cl}(\mathfrak{a}^+) = \mathfrak{a}_{A \geqq 0}$, and $\alpha_i(H_0) = 1$ for all i, it is immediate that

$$\beta(H)H_0 - H \in \mathrm{Cl}(\mathfrak{a}^+).$$

Hence by Proposition 2.3 we get with $a = \exp(H)$,

$$\begin{aligned} e^{\rho(H)}\varphi_0(\exp H) &\ll F(\beta(H)) \\ &\ll (1+\beta(H))^d \quad \text{for all } H \in \mathrm{Cl}(\mathfrak{a}^+) \\ &\ll (1+|H|)^d \quad \text{because } \beta(H) \ll |H| \text{ for } H \in \mathrm{Cl}(\mathfrak{a}^+), \end{aligned}$$

thus proving the theorem.

Remark 1. The above proof of Harish-Chandra's inequality is essentially the one he gave in [Har 58a], §9, with a simplification due to Helgason [Hel 84], Solutions to Exercises, Chapter IV, **B1**. To start the proof, i.e. to get to (8), Harish-Chandra uses somewhat more structure about his series at the totally W-ramified point $\zeta = 0$. See Chapter III, Theorem 8.8. Helgason bypasses this structure (which we have not considered), by using the differential operator $\mathcal{D}(\Pi_+^\vee)$. This variation was used by Anker [Ank 87] to get an optimal Big Oh estimate for the spherical function φ_0, carefully expanding the above arguments.

Knapp gave a proof using only the Harish-Chandra integral for spherical functions in [Kna 86] Chapter VII, §8, Proposition 7.15(c).

Remark 2. In Chapter XII, Theorem 4.3, we shall give an explicit formula for the spherical function on $\mathrm{SL}_n(\mathbf{C})$, which not only makes the Important Inequality obvious, but gives an exact asymptotic estimate. The present proof in the real case depends on the Harish-Chandra series. It remains to be seen whether one can get the real case directly from the complex case by the Flensted-Jensen reduction, and get as well an asymptotic expansion refining Anker's estimate.

Remark 3. From Theorem 2.8, we see that the basic spherical function $\varphi_0(x)$ has exponential decay for $\sigma(x) \to \infty$, up to a factor with polynomial growth. This function will be used in an essential way to define the Harish-Chandra Schwartz space in the next section. In preparation, we give a corollary of the Important Inequality.

Corollary 2.9. *We have*

$$-\log \varphi_0(x) \gg\ll \sigma(x) \qquad \text{for } x \in G, \sigma(x) \to \infty.$$

Proof. The "important inequality" and Proposition 2.1 show that for $a \in \mathrm{Cl}(A^+)$, $a = \exp(H)$, $H \in \mathfrak{a}^+$, and $N = |\mathcal{R}(\mathfrak{n})|$ we have

$$\rho(H) - N\log(1 + |H|) - O(1) \leqq -\log \varphi_0(a) \leqq \rho(H).$$

By Chapter I, §4, (1), we know that $\rho(H) \gg\ll |H| = \sigma(a)$, so the corollary follows.

X, §3. THE HARISH-CHANDRA SCHWARTZ SPACE

The topology of seminorms

Before going into the Harish-Chandra version of the Schwartz space involving the group G, we recall briefly the way one uses a family of seminorms on a space to define a topology.

Let $\mathbf{E}$ be a vector space over $\mathbf{R}$ or $\mathbf{C}$, and let $\mathcal{S}$ be a family of seminorms on $\mathbf{E}$. In what follows, one of the seminorms is a norm. Let $\mathcal{S}'$ be another family. We say that $\mathcal{S}$ is **equivalent** to $\mathcal{S}'$ if given a seminorm $p \in \mathcal{S}$ there exists a positive linear combination of seminorms

$$c_1 p_1' + \cdots + c_N p_N' \qquad (p_i' \in \mathcal{S}')$$

which is $\geqq p$ on $\mathbf{E}$, and vice versa, interchanging the role of the two families. We define the $\mathcal{S}$-topology on $\mathbf{E}$ by defining a fundamental system of open neighborhoods of 0 to consist of the sets $U(p_1, \ldots, p_N; s)$ where $p_1, \ldots, p_N$ ranges over finite subsets of $\mathcal{S}$, s ranges over positive reals (which could be taken to be a sequence tending to 0), and $f \in U(\boldsymbol{p}_1, \ldots, \boldsymbol{p}_N; s)$ if and only if $\boldsymbol{p}_i(f) < s$ for $i = 1, \ldots, N$, or equivalently $\sum_{i=1}^N \boldsymbol{p}_i(f) < s$. This means that f is close to 0 in the $\mathcal{S}$-topology if and only if it is close to 0 for a large finite subset of $\mathcal{S}$. Two equivalent families of seminorms define the same topology. The sets $U(\boldsymbol{p}_1, \ldots, \boldsymbol{p}_N; s)$ are analogous to balls of radius s, defined by a single norm, but using a finite number of seminorms. A neighborhood of an arbitrary element is obtained by translation of a neighborhood of 0. One can take limits as usual, for neighborhoods shrinking around a point. So for $f, g \in \mathbf{E}$, f, g are close together if and only if $f - g$ is close to 0, that is, $\boldsymbol{p}_i(f-g)$ is close to 0 for a large set of seminorms $\boldsymbol{p}_i$.

The ordinary Schwartz space

Next we recall the definition of the ordinary **Schwartz space** on an ordinary finite dimensional vector space over the reals, say V. We denote this space by **Sch(V)**. It consists of all complex valued C^∞ functions h on V such that, for every $P \in S(V)$ and every integer $m \geqq 0$, there is a constant $c > 0$ such that

$$|\mathcal{D}(P)h(v)| \leqq c(1 + |v|)^{-m} \qquad \text{for all } v \in V.$$

In this definition, we use any norm on V. Note that $\mathcal{D}(P)$ is just an invariant differential operator on V, so an ordinary polynomial differential operator (with constant coefficients). We can then define the seminorm

$$p_{m,D}(h) = \sup_{v \in V} (1 + |v|)^m |(Dh)(v)|, \qquad \text{with } D \in \mathrm{IDO}(V).$$

The Schwartz topology is the one defined by the family of seminorms $\{p_{m,D}\}$. Note that it can be defined by a denumerable sequence, because for the family $\{D\}$ we can pick the monomial differential operators, which are denumerable.

If instead of $|v|$ in the definition of $p_{m,D}$ we write $|v|^2$, we obtain an equivalent family of seminorms. Even more importantly, let us take the norm to be definitely the one associated with a positive definite scalar product. Then the norm squared is C^∞. Let $v^2 = \langle v, v \rangle$. Define

$$p'_{m,D}(h) = \sup_{v \in V} |\mathcal{D}(P)\{(1 + |v|^2)^m h(v)\}|.$$

It is easily proved by induction on the degree of D that the family $\{p'_{m,D}\}$ is equivalent to the family $\{p_{m,D}\}$.

Finally we note that instead of the constant 1 in the expression $(1+|v|^2)^m$ above, we could take any constant $c > 0$, i.e. use $(c+|v|^2)^m$, and get an equivalent family. In applications dealing with spherical functions it will be convenient to take $c = \rho^2$. See Theorem 4.2.

We shall deal with the two spaces $V = \mathfrak{a}$ and $V = \mathbf{i}\mathfrak{a}^\vee$, where $\mathfrak{a}$ is finite dimensional over $\mathbf{R}$.

The Paley–Wiener space is related to the Schwartz space as we shall now describe.

Let $F \in \mathrm{PW}(\mathfrak{a}_\mathbf{C}^\vee)$ be in the Paley–Wiener space. We restrict F to the imaginary space $\mathbf{i}\mathfrak{a}^\vee$, and thus obtain a natural injection

$$\mathrm{PW}(\mathfrak{a}_\mathbf{C}^\vee) \hookrightarrow C^\infty(\mathbf{i}\mathfrak{a}^\vee).$$

We let $F_{\mathbf{i}\mathfrak{a}^\vee}$ be the image of F under this injection, which we call the **imaginary restriction**. Since an entire function is determined by its imaginary part, the restriction is indeed an injection. Using the Fourier transform and the fact that the Fourier transform interchanges multiplication by a polynomial and applying the differential operator determined by this polynomial, one sees at once that the imaginary restriction maps $\mathrm{PW}(\mathfrak{a}_\mathbf{C}^\vee)$ into the Schwartz space $\mathrm{Sch}(\mathbf{i}\mathfrak{a}^\vee)$. The argument will be given in detail in a more general situation later, proof of Theorem 4.3. The image of $\mathrm{PW}(\mathfrak{a}_\mathbf{C}^\vee)$ in $\mathrm{Sch}(\mathbf{i}\mathfrak{a}^\vee)$ will be denoted by $\mathrm{PW}(\mathbf{i}\mathfrak{a}^\vee)$. Thus we have a natural embedding

$$\mathrm{PW}(\mathbf{i}\mathfrak{a}^\vee) \hookrightarrow \mathrm{Sch}(\mathbf{i}\mathfrak{a}^\vee).$$

For a seminorm p on $\mathrm{Sch}(\mathbf{i}\mathfrak{a}^\vee)$ and $F \in \mathrm{PW}(\mathfrak{a}_\mathbf{C}^\vee)$ we **define**

$$p(F) = p(F_{\mathbf{i}\mathfrak{a}^\vee}).$$

By definition, the Schwartz topology on $\mathrm{PW}(\mathfrak{a}_\mathbf{C}^\vee)$ is that induced by the above embedding.

The five properties **M 1** through **M 5** of Chapter IV, §7, hold for the Mellin transform (Fourier transform) on the ordinary Schwartz space

$$\mathbf{M}_\mathfrak{a}\colon \mathrm{Sch}(\mathfrak{a}) \to \mathrm{Sch}(\mathbf{i}\mathfrak{a}^\vee).$$

All but one state in a precise way that differentiation by a polynomial differential operator corresponds to multiplication by the polynomial (with the sign alternation) on the other side. The proofs are the same, and all that was needed was the absolute convergence of the integral at each step. This convergence is guaranteed by the defining properties of the Schwartz space. Hence these properties will be quoted by referring back to Chapter IV, §7. Cf. also [Lan 93], Chapter XV.

The following theorem summarizes the main properties of the Schwartz space. The main purpose of this chapter will be to transpose these properties to a context with the group G.

Theorem 3.1. *Let* $\mathrm{Sch}(\mathfrak{a})$ *be the ordinary Schwartz space of a euclidean space* $\mathfrak{a}$.

(i) $\mathrm{Sch}(\mathfrak{a})$ *is complete for the Schwartz topology (defined by the seminorms).*

(ii) *The Mellin transform*

$$\mathbf{M}_{\mathfrak{a}}\colon \mathrm{Sch}(\mathfrak{a}) \to \mathrm{Sch}(\mathbf{i}\mathfrak{a}^{\vee})$$

$$\textit{given by } \mathbf{M}_{\mathfrak{a}} f(\mathbf{i}\lambda) = \int_{\mathfrak{a}} f(H) e^{\mathbf{i}\lambda(H)}\, dH$$

is a linear isomorphism, and is bi-continuous for the Schwartz topologies. Its inverse is given by

$$({}^{t}\mathbf{M}_{\mathfrak{a}}^{-} g)(H) = \int_{\mathfrak{a}^{\vee}} g(\mathbf{i}\lambda) e^{-\mathbf{i}\lambda(H)}\, d\lambda.$$

The Lebesgue measures are assumed normalized so that no extra constant factor appears in the inversion.

(iii) *The space* $C_c^{\infty}(\mathfrak{a})$ *is Schwartz dense in* $\mathrm{Sch}(\mathfrak{a})$.

(iv) *The Paley–Wiener space* $\mathrm{PW}(\mathbf{i}\mathfrak{a}^{\vee})$ *is Schwartz dense in* $\mathrm{Sch}(\mathbf{i}\mathfrak{a}^{\vee})$.

(v) *The Schwartz space is closed under the convolution product, and this product is continuous. Similarly for the ordinary product. Also* $\mathbf{M}_{\mathfrak{a}}$ *transforms the convolution product to the ordinary product.*

The proofs are at the level of advanced calculus, starting with what is standard in textbooks, e.g. [Lan 93], Chapter XV. We shall prove more complicated versions of these properties in the following sections involving the Harish-Chandra Schwartz space in connection with the non-commutative group G. In the classical case, the complications do not occur.

(i) The completeness of $\mathrm{Sch}(\mathfrak{a})$ just comes from differentiating a sequence term by term, and using the fact that derivatives and limits commute under conditions of rapid uniform convergence.

(ii) That $\mathbf{M}_{\mathfrak{a}}$ is a linear isomorphism can be found in standard books, e.g. the above reference. It is no more than Fourier inversion, once one knows that $\mathbf{M}_{\mathfrak{a}}$ maps the Schwartz space into the Schwartz space. As for this, the proof that $\mathbf{M}_{\mathfrak{a}}$ maps $\mathrm{Sch}(\mathfrak{a})$ into $\mathrm{Sch}(\mathbf{i}\mathfrak{a}^{\vee})$ is a direct application of the way $\mathbf{M}_{\mathfrak{a}}$ transforms multiplication by a polynomial into a polynomial differential operator, just as it was for the Paley–Wiener space. The situation is actually easier here for the Schwartz space because it concerns only the super polynomial decay, and one does not have to provide the additional argument moving the vertical space of integration which we had to do in the Paley–Wiener theory.

To show that the Mellin transform is bicontinuous, it suffices to show that it is continuous, because up to the minus sign, the transform and its inverse are defined by "the same" kernel function. (Dealing with situations when the inverse transform measure is different from the direct transform measure in cases to be considered later will give rise to an additional complication.) The proof of continuity uses the alternative family of seminorms mentioned at the beginning of this section. Indeed, let Q be the polynomial function on $\mathbf{i}\mathfrak{a}^\vee$ defined by

$$Q(\mathbf{i}\lambda) = (1 - (\mathbf{i}\lambda)^2)^m = (1 + \langle\lambda, \lambda\rangle)^m.$$

So $Q \in S(\mathfrak{a})$. Let $P \in S(\mathbf{i}\mathfrak{a}^\vee)$, and let $D = \mathcal{D}(P)$. We sometimes write $D = \mathcal{D}_{\mathbf{i}\lambda}(P)$ to denote the variable. Then we use the seminorms

$$p'_{m,D}(h) = \sup_{\lambda \in \mathfrak{a}^\vee} |\mathcal{D}_{\mathbf{i}\lambda}(P)(Qh(\mathbf{i}\lambda))|.$$

For the continuity of $\mathbf{M}_\mathfrak{a}$, it suffices to prove that given such a seminorm $p'_{m,D}$ there exists a seminorm $q_{s,E}$ such that for all $f \in \mathrm{Sch}(\mathfrak{a})$ we have

$$p'_{m,D}(\mathbf{M}_\mathfrak{a} f) \leqq c_{m,D} q_{s,E}(f)$$

with a suitable constant $c_{m,D}$. We now do this. We have

$$\begin{aligned} Q(\mathbf{i}\lambda)\mathbf{M}_\mathfrak{a} f(\mathbf{i}\lambda) &= \int_\mathfrak{a} Q(\mathbf{i}\lambda) e^{\mathbf{i}\lambda(H)} f(H)\, dH \\ &= \int_\mathfrak{a} \mathcal{D}_H(Q) e^{\mathbf{i}\lambda(H)} f(H)\, dH \\ (*) \qquad &= \int_\mathfrak{a} e^{\mathbf{i}\lambda(H)} \mathcal{D}_H(Q) f(H)\, dH. \end{aligned}$$

The integration by parts is justified by the rapid decay of f and its derivatives, so the boundary terms vanish at infinity in the process. Also $Q^- = Q$ trivially. Now we apply $\mathcal{D}(P)$ viewed as a differential operator on $\mathbf{i}\mathfrak{a}^\vee$, for some polynomial P on $\mathbf{i}\mathfrak{a}^\vee$. Using $(*)$ we get

$$\begin{aligned} \mathcal{D}(P)(Q\mathbf{M}_\mathfrak{a} f)(\mathbf{i}\lambda) &= \int_\mathfrak{a} \mathcal{D}_{\mathbf{i}\lambda}(P) e^{\mathbf{i}\lambda(H)} \mathcal{D}(Q) f(H)\, dH \\ &= \int_\mathfrak{a} P(H) e^{\mathbf{i}\lambda(H)} \mathcal{D}(Q) f(H)\, dH \\ &= \int_\mathfrak{a} \frac{P(H) e^{i\lambda(H)}}{(1 + |H|^2)^N} (1 + |H|^2)^N \mathcal{D}(Q) f(H)\, dH. \end{aligned}$$

We pick N so large that the integral of the first factor in the integrand converges absolutely, thereby giving rise to the constant $c_{m,D}$. Then

$$p'_{m,D}(\mathbf{M}_\mathfrak{a} f) \leqq c_{m,D} q_{2N,\mathcal{D}(Q)}(f),$$

This concludes the proof that $\mathbf{M}_\mathfrak{a}$ is continuous.

Remark. In the analogous situation on the group G, instead of using the eigenvalue property with respect to the "other" variable, we shall use a lemma giving the desired estimate. See Theorem 4.3 and Lemma 4.1. Readers acquainted with the above proof will thus be in a better position to appreciate the structures of both proofs.

(iii) To prove that $C_c^\infty(\mathfrak{a})$ is dense in $\mathrm{Sch}(\mathfrak{a})$ we use a bump function g_R for real $R \geqq 1$, say, whose graph is shown on the figure with respect to a positive definite scalar product on $\mathfrak{a}$.

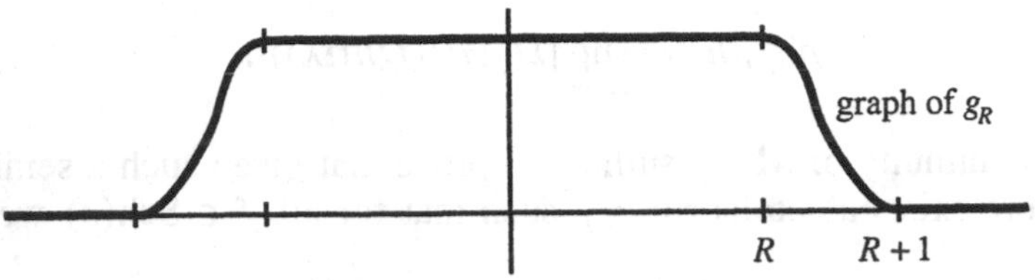

Thus g_R smoothes out the characteristic function of the closed ball $\bar{\mathbf{B}}_R$ at the boundary, and it is clear that its derivatives are bounded independently of R, because the pieces where g_R declines from 1 to 0 are obtained by translation from one fixed piece, and derivatives are invariant under translation.

Given $f \in \mathrm{Sch}(\mathfrak{a})$, we claim that the product $g_R f$ is Schwartz-convergent to f when $R \to \infty$. Given a polynomial differential operator D, there exist such operators D_i, E_i such that

$$D(g_R f) = \sum (D_i g_R)(E_i f),$$

just by using Leibniz's rule for the derivative of a product. Note that $g_R f - f = 0$ on $\mathbf{B}_R$. Then for each $m \geqq 0$,

$$\begin{aligned} p_{m,D}(g_R f - f) &\leqq \sup_{|H| \geqq R} (1 + |H|)^m D(g_R f - f)(H) \\ &\leqq (1+R)^{-1}(p_{m+1,D}(g_R f) + p_{m+1,D}(f)) \quad \text{by definition of } p_{m+1,D} \\ &\leqq \text{constant} \cdot (1+R)^{-1} \left(\sum_i p_{m+1,E_i}(f) + p_{m+1,D}(f) \right). \end{aligned}$$

This last expression tends to 0 as $R \to \infty$, so the proof of (iii) is concluded.

(iv) That $\mathrm{PW}(\mathbf{i}\mathfrak{a}^\vee)$ is dense in the Schwartz space is proved with the Mellin transform using its Schwartz continuity, namely given

$g \in \mathrm{Sch}(i\mathfrak{a}^\vee)$, then $\mathbf{M}_\mathfrak{a}^{-1}g$ is in $\mathrm{Sch}(\mathfrak{a})$, and there exists $f \in C_c^\infty(\mathfrak{a})$ such that f is Schwartz close to $\mathbf{M}_\mathfrak{a}^{-1}g$. Then applying $\mathbf{M}_\mathfrak{a}$, the functions $g, \mathbf{M}_\mathfrak{a}f$ (on $i\mathfrak{a}^\vee$) are Schwartz close by the Schwartz continuity of $\mathbf{M}_\mathfrak{a}$, and $\mathbf{M}_\mathfrak{a}f$ is in the Paley–Wiener space, as desired.

(v) As to products, let $f, g \in \mathrm{Sch}(\mathfrak{a})$. Using the rule for the derivative of a product (as we did in (iii) above), it is immediately verified that the ordinary product fg is also in the Schwartz space. As for convolution, we have

$$\mathbf{M}_\mathfrak{a}(f * g) = (\mathbf{M}_\mathfrak{a}f)(\mathbf{M}_\mathfrak{a}g)$$

by a straight application of Fubini's theorem (see, for instance, [Lan 93], Chapter XV, Theorem 4.2). Actually, the formal argument of Chapter IV, Theorem 6.2 is valid here. Furthermore, for a polynomial differential operator D, we have

$$D(f * g) = Df * g,$$

because we can differentiate under the integral sign of the convolution since f is in the Schwartz space. Finally, let P be a polynomial. There are polynomials P_j, Q_j such that for all $x, y \in \mathfrak{a}$ we have

$$P(x) = P(x - y + y) = \sum P_j(x - y)Q_j(y).$$

Then

$$\begin{aligned} P(x)(f * g)(x) &= P(x)\int_\mathfrak{a} f(x-y)g(y)\,dy \\ &= \sum \int_\mathfrak{a} P_j f(x-y)Q_j g(y)\,dy. \end{aligned}$$

Since $P_j f$ and $Q_j g \in \mathrm{Sch}(\mathfrak{a})$, the right side is bounded, so $f * g \in \mathrm{Sch}(\mathfrak{a})$. That the convolution product is continuous is shown equally easily.

Remarks. From (iv) it is immediate that in the standard situation when we have a finite group W acting on $\mathfrak{a}$, then $\mathrm{PW}(i\mathfrak{a}^\vee)^W$ is Schwartz dense in $\mathrm{Sch}(i\mathfrak{a}^\vee)^W$. Indeed, we just apply the result without the W, and then take the W-trace of the approximating function, normalized by dividing with the order of W.

The Harish-Chandra Schwartz space

Harish-Chandra defined the analogue of the Schwartz space of ordinary Fourier analysis in the context of functions on the non-commutative G. The ordinary case is characterized by the decay of a function and all its partial derivatives, faster than any polynomial decay. One uses the same idea on G, except that the Jacobian factor in the change of variables formula when pulling back to the Lie algebra of A has to be taken into account. Following Harish-Chandra [Har 66], §9, §12, we use the spherical function φ_0 as a factor, because φ_0^2 has the right order of decay at infinity and φ_0 is C^∞, strictly positive, and K-bi-invariant, which is convenient.

As in §1, we use the notation

$$\sigma(x) = |\log \mathrm{pol}_A(x)| = |H|$$

if $\mathrm{pol}_A(x) = \exp H$. The norm $|H|$ is the B-norm, associated with the trace form on $\mathfrak{a}$. Let f be a measurable function on G. We say that f has **polynomial growth** if there exists a positive integer m such that

$$|f| \ll (1+\sigma)^m \qquad \text{that is} \qquad |f(x)| \leqq c_{f,m}(1+\sigma(x))^m \qquad \text{for } x \in G.$$

We then say that the growth is of **order** m. We say that f has polynomial **decay of order** $\geqq m$ if

$$|f| \ll (1+\sigma)^{-m} \qquad \text{that is} \qquad |f(x)| \leqq c'_{f,m}(1+\sigma(x))^{-m} \qquad \text{for } x \in G.$$

On $\mathfrak{a}^+$, as before let

$$J_\mathfrak{a} = \prod_{\alpha \in \mathcal{R}(\mathfrak{n})} (e^\alpha - e^{-\alpha}).$$

Then trivially, for $H \in \mathrm{Cl}(\mathfrak{a}^+)$, we have the estimate with some constant c_J,

$$0 \leqq J_\mathfrak{a}(H) \leqq c_J e^{2\rho(H)}.$$

So $J_\mathfrak{a}$ grows exponentially. From Theorem 2.8 we know that $\varphi_0^2(a)$ has order of decay $O(a^{-2\rho})$ times polynomial growth in H. Hence the product $\varphi_0^2 J$ has only polynomial growth in H.

We define the **Harish-Chandra Schwartz space** $\mathrm{HCS}(K\backslash G/K)$ to be the subspace of functions $f \in C^\infty(K\backslash G/K)$ having the following property. For each $D \in \mathrm{IDO}(G)$ and positive integer m, there exists a constant $C_{m,D}(f)$ such that for all $x \in G$,

$$(1) \qquad |(Df)(x)| \leqq C_{m,D}\varphi_0(x)(1+\sigma(x))^{-m}.$$

For each m, D we then get a **seminorm** $\mathbf{q}_{m,D}$ on the Harish-Chandra Schwartz space, defined by

$$q_{m,D}(f) = \sup_{x \in G} (1+\sigma(x))^m \varphi_0(x)^{-1} |(Df)(x)|.$$

Thus we can take $C_{m,D}(f) = q_{m,D}(f)$. The family of such seminorms defines a topology on $\mathrm{HCS}(K\backslash G/K)$ called the **HCS topology**. Thus an element of HCS$(K\backslash G/K)$ has arbitrary large polynomial decay beyond the exponential decay of φ_0. One can get other topologies by taking powers of φ_0. The most important case is when one uses φ_0^2 instead of φ_0, that is, one replaces (1) by

$$(2) \qquad |(Df)(x)| \leqq C_{m,D}\varphi_0^2(x)(1+\sigma(x))^{-m}.$$

For the corresponding seminorm, one replaces φ_0^{-1} by φ_0^{-2}. This seminorm will be denoted by $q_{m,D}^{(1)}$. We might have written $q_{m,D}^{(2)}$ instead of $q_{m,D}$ because of Proposition 3.4 below. The space of functions satisfying (2) instead of (1) will be denoted by HCS$^{(1)}(K\backslash G/K)$. One might write HCS$^{(2)}(K\backslash G/K)$ instead of HCS$(K\backslash G/K)$. For the general system behind this, see Chapter XI.

Proposition 3.2. *The spaces* HCS$(K\backslash G/K)$ *and* HCS$^{(1)}(K\backslash G/K)$ *are complete for their* HCS (resp. HCS$^{(1)}$) *topologies.*

Proof. This is just standard calculus, that the limit of a derivative is the derivative of the limit under a uniformity assumption. One has to insert the extra factors $(1+\sigma(x))^m$ and $\varphi_0(x)^{-1}$ while taking the limit of a sequence of functions $\{f_k\}$. We leave the details to the reader.

We give various inclusions among some standard spaces.

Lemma 3.3. *Let $f \in C(K\backslash G/K)$. Assume that for every positive integer m, we have*

$$|f| \ll_m \varphi_0(1+\sigma)^{-m} \quad \text{resp.} \quad \ll_m \varphi_0^2(1+\sigma)^{-m},$$

or equivalently, for some constant $c_{m,f}$ and all $a \in A^+$,

$$|f(a)| \leqq c_{m,f}a^{-\rho}(1+|\log a|)^{-m} \quad \text{resp.} \quad \leqq c_{m,f}a^{-2\rho}(1+|\log a|)^{-m}.$$

Then $f \in L^2(G)$ resp. $f \in L^1(G)$.

Proof. Say for the L^2 case, by the polar decomposition of Haar measure and Theorem 2.9, the L^2-integral of f is dominated by

$$\int_A [\varphi_0(a)(1+|\log a|)^{-m}]^2 J(a)\, da$$

$$\ll \int_{\mathfrak{a}^+} e^{-2\rho(H)}(1+|H|)^{-2m+2N} e^{2\rho(H)}\, dH,$$

with N fixed ($= |\mathcal{R}(\mathfrak{n})|$). The factor $e^{2\rho(H)}$ cancels, and the integral converges for m sufficiently large, thus proving the L^2 case. The L^1 case is done similarly.

The next results come also from [Har 66], §10 to §16.

Proposition 3.4. *We have the inclusions*

$$\mathrm{HCS}(K\backslash G/K) \subset L^2(G) \text{ and } \mathrm{HCS}^{(1)}(K\backslash G/K) \subset L^1(G).$$

Proof. Special case of the lemma and the definitions.

Proposition 3.5. *We have the inclusions*

$$C_c^\infty(K\backslash G/K) \subset \mathrm{HCS}^{(1)}(K\backslash G/K) \subset \mathrm{HCS}(K\backslash G/K),$$

and $C_c^\infty(K\backslash G/K)$ is dense in this inclusion.

Proof. The inclusion is obvious, because for a function

$$f \in C_c^\infty(K\backslash G/K),$$

all its derivatives have at most the support of f, and so vanish outside a fixed compact set. So the inequality defining the HCS spaces also applies to Df for any invariant differential operator D. As for the denseness, it is due to Harish-Chandra, [Har 66], Lemma 20. We found useful the exposition in [GaV 88], Lemma 6.1.7 and Corollary 6.1.8. First we need a bump function. For $R > 0$ we let

$$G_R = K(\exp \bar{\mathbf{B}}_R)K = \{x \in G \text{ such that } \sigma(x) \leqq R\}.$$

One sometimes calls G_R the **R-ball in G**. It is obviously compact.

From the definitions and Proposition 1.1, we get

(3) $$G_R = G_R^{-1}, \qquad G_R G_{R'} \subset G_{R+R'} \qquad \text{for } R, R' > 0.$$

Lemma 3.6. *Given $s > 0$, there exists a function*

$$g_R = g_{R,s} \in C_c^\infty(K\backslash G/K)$$

such that:

(i) $0 \leqq g_R \leqq 1$;

(ii) $g_R = 1$ *on* G_R *and* $= 0$ *outside* G_{R+s};

(iii) *Given a differential operator $D \in \mathrm{IDO}(G)$, there exists a constant c_D such that for all $R > 0$,*

$$\|Dg_R\|_\infty \leqq c_D.$$

Proof. Let h_R be the characteristic function of $\mathbf{B}_{R+s/2}$ and let g be an element of $C_c^\infty(K\backslash G/K)$ such that

$$g(x) = g(x^{-1}) \geqq 0 \quad \text{for all } x \in G, \qquad \int_G g(x)\,dx = 1,$$
$$\text{and } \operatorname{supp}(g) \subset G_{s/4}.$$

Thus g is a standard bump function around the origin, of the type used to make up Dirac families. Let

$$g_R = h_R * g.$$

Then we get at once that g_R is in $C_c^\infty(K\backslash G/K)$ and is bounded by 1 so (i) is satisfied. Property (ii) follows from (3) (the triangle inequality for the height function) and the formula

$$g_R(x) = \int_{G_{s/4}} h_R(xy^{-1})g(y)\,dy.$$

As for (iii), it follows at once from the standard formula (DUTIS)

$$D(h_R * g) = h_R * Dg \qquad \text{using} \qquad g_R(x) = \int_G h_R(y^{-1})g(yx)\,dy.$$

We can then take $c_D = \|Dg\|_1$.

The next lemma concludes the proof of Proposition 3.5.

Lemma 3.7. *Let* $f \in \mathrm{HCS}(K\backslash G/K)$ resp. $\mathrm{HCS}^{(1)}(K\backslash G/K)$. *Then* $g_R f \in C_c^\infty(K\backslash G/K)$, *and*

$$\lim_{R\to\infty} g_R f = f,$$

the limit being taken for the HCS resp. $\mathrm{HCS}^{(1)}$*-topology.*

Proof. Given $D \in \mathrm{IDO}(G)$, there exist $D_i, D_i' \in \mathrm{IDO}(G)$ such that for all f we have

$$D(g_R f) = \sum (D_i g_R)(D_i' f).$$

Note that $g_R f - f = 0$ on G_R. Then for each $m \geqq 0$, and, say, the HCS case,

$$\begin{aligned} q_{m,D}(g_R f - f) &\leqq \sup_{\sigma(x)\geqq R} \varphi_0(x)^{-1}(1+\sigma(x))^m |D(g_R f - f)(x)| \\ &\leqq (1+R)^{-1}(q_{m+1,D}(g_R f) + q_{m+1,D}(f)) \quad \text{by definition of } q_{m+1,D} \end{aligned}$$

$$\leqq \text{constant} \cdot (1+R)^{-1} \left(\sum_i q_{m+1,D_i'}(f) + q_{m+1,D}(f) \right).$$

This last expression tends to 0 as $R \to \infty$, and this concludes the proof of Lemma 3.7 as well as Proposition 3.5.

Remark 1. In the standard references (Harish-Chandra, [GaV 88]) they define the Harish-Chandra Schwartz space on G, not necessarily for K-bi-invariant functions. Then they have to use seminorms involving both right and left invariant differential operators. For K-bi-invariant functions, they prove that the seminorms involving only the left invariant differential operators suffice to give the same topology, cf. [GaV 88], Corollary 6.1.12. For our purposes in the present book, we are only concerned with the K-bi-invariant case, so we took the restricted definition at once.

Remark 2. The cases of HCS and $\text{HCS}^{(1)}$ are only special cases of general $\text{HCS}^{(p)}$ spaces $0 < p \leqq 2$, investigated by Trombi [Tro 70] and Trombi–Varadarajan [TrV 71], cf. [GaV 88], Chapter 7. In the present section, we considered them together because there is essentially no difference in the arguments. For subsequent considerations, there is a greater difference, so the $\text{HCS}^{(1)}$ space will be considered separately in Chapter XI.

The next proposition extends a C_c^∞ formula of Chapter II to the HCS space. We take it from [GaV 88], Proposition 6.1.18.

Proposition 3.8. *Let φ be a K-bi-invariant measurable function on G. Assume that $\varphi_0^{-1}\varphi$ has polynomial growth. Then for $f \in \text{HCS}(K\backslash G/K)$, the integral*

$$T_\varphi(f) = \int_G f(y)\varphi(y)\,dy = [f, \varphi]_G$$

converges absolutely, and T_φ defines an HCS *continuous functional on the space* $\text{HCS}(K\backslash G/K)$. *If φ is C^∞ in addition, and $\varphi_0^{-1}D\varphi$ has polynomial growth for all $D \in \text{IDO}(G)$, then letting ${}^tD$ be the transpose from Chapter* II, ${}^t\mathcal{D}(P) = \mathcal{D}(P^-)$, *the transpose formula is valid,*

$$[{}^tDf, \varphi]_G = [f, D\varphi]_G,$$

Proof. For the transpose formula, one may verify directly that the argument used to prove Proposition 6.3 of Chapter II goes through in the present case. Or one may show first that T_φ is HCS continuous, as follows. Let m' be a positive integer, and suppose $\varphi_0^{-1}\varphi$ has polynomial growth or order $\leqq m$. We have the bound

$$|f\varphi| \leqq c_m(1+\sigma)^{m'+m}\varphi_0^{-1}f\frac{\varphi_0^2}{(1+\sigma)^{m'}}.$$

We select m' so that $\varphi_0^2/(1+\sigma)^{m'}$ is absolutely integrable. This yields the inequality

$$|T_\varphi(f)| \leqq c'_{m,m'}q_{m'+m,1}(f).$$

Replacing f by Df (with $D \in \mathrm{IDO}(G)$) ends up with $q_{m'+m,D}$ instead of $q_{m'+m,1}$ on the right side, and thus proves the HCS continuity of T_φ. This concludes the proof.

X, §4. SCHWARTZ CONTINUITY OF THE SPHERICAL TRANSFORM

For this section and the next, we follow Anker closely [Ank 91]. The general pattern is the same as for Theorem 3.1(ii) in the euclidean case, but is more complicated because the lack of symmetry in the spherical kernel screws up the integration by parts. There will still be some integration by parts, but it will be combined with another type of estimate for the derivatives of the spherical integrals.

We start with a lemma of Harish-Chandra estimating derivatives of the spherical functions [Har 58a], Lemma 46. Only Lemma 4.1(i) will be used in this section. Lemma 4.1(ii) will be used in corresponding fashion in §5, Lemma 5.3. We use the proof given in [GaV 88], Proposition 4.6.2. The anti-Iwasawa decomposition $G = KAU$ occurs, rather than $G = UAK$. If $x = k'a'u'$ with $k' \in K$, $a' \in A$, $u' \in U$, then we write the anti-Iwasawa projection

$$a' = \mathrm{Iw}'_A(x).$$

Lemma 4.1.

(i) *Let* $P \in S(\mathfrak{a}_{\mathbf{C}}^\vee)$. *There is a constant* c_2 *such that for* $x \in G$, $\zeta \in \mathfrak{a}_{\mathbf{C}}^\vee$,

$$|\mathcal{D}_\zeta(P)\varphi(\zeta,x)| \leqq c_2(1+\sigma(x))^{\deg P}\varphi_{\mathrm{Re}(\zeta)}(x).$$

(ii) *Let* $D \in \mathrm{IDO}(G)$. *There is a constant* c_3 *such that for* $x \in G$, $\zeta \in \mathfrak{a}_{\mathbf{C}}^\vee$,

$$|(D\varphi_\zeta)(x)| \leqq c_3(1+\|\zeta\|)^{\deg D}\varphi_{\mathrm{Re}(\zeta)}(x).$$

Proof. For (i), we differentiate the integral expression for the spherical function under the integral sign, so for the polar A-projection a of x,

$$\begin{aligned}\mathcal{D}_\zeta(P)\int_K (ka)_A^{\zeta+\rho}\,dk &= \int_K \mathcal{D}_\zeta(P)(ka)_A^{\zeta+\rho}dk\\ &= \int_K P(\log(ka)_A)(ka)_A^{\zeta+\rho}dk.\end{aligned}$$

The value of the polynomial is estimated using Proposition 1.5, and the remaining integral is estimated using the real part of ζ. This concludes the proof of (i).

As for (ii), let $g_\zeta(x) = \mathrm{Iw}'_A(x)^{\zeta-\rho}$, so that by Proposition 7.5 of Chapter III,

$$\varphi_\zeta(x) = \int_K g_\zeta(xk^{-1})\,dk = \int_K g_\zeta(\mathbf{c}(k)x)\,dk,$$

where $\mathbf{c}(k)x = kxk^{-1}$ (conjugation). Given $D \in \mathrm{IDO}(G)$, we can differentiate under the integral sign,

$$(1)\qquad (D\varphi_\zeta)(x) = \int_K D_x g_\zeta(\mathbf{c}(k)x)\,dk.$$

Lemma 4.2. *Let $G = UAK$ be an Iwasawa decomposition of a Lie group G. Let $D \in \mathrm{IDO}(G)$. There exist functions*

$$\psi_1, \ldots, \psi_N \in C^\infty(K)$$

and elements $D_1, \ldots, D_N \in \mathrm{IDO}(G)$ such that for all $k \in K$,

$$\mathbf{c}(k)_* D = \sum \psi_i(k) D_i \quad \textit{with} \ \deg D_i \leqq \deg D.$$

Proof. By Chapter I, Proposition 1.5, and the fact that $\mathfrak{g}$ (and so $\tilde{\mathfrak{g}}$) is stable under conjugation, given a basis of $\mathfrak{g}$ and $v \in \mathfrak{g}$, for $k \in K$ we can write v^k as a linear combination of the basis elements with coefficients which depend on k, and the dependence is C^∞. By Chapter II, Corollary 1.3, we see that the desired relation holds for monomials $\tilde{v}_1 \ldots \tilde{v}_M$, hence for a linear combination of such monomials, hence for any $D \in \mathrm{IDO}(G)$. This proves the lemma.

For any function g, using the formalism of Chapter II, **FUNCT 4**, with $\theta = \mathbf{c}(k)$,

$$D(g \circ \mathbf{c}(k)) = (\mathbf{c}(k)_* D)g) \circ \mathbf{c}(k).$$

Hence by Lemma 4.2,

$$(*)\quad D(g \circ \mathbf{c}(k)) = \sum \psi_i(k) D_i(g \circ \mathbf{c}(k)) = \sum \psi_i(k)(D_i g) \circ \mathbf{c}(k).$$

Putting x back into the notation yields

$$(**) \qquad D_x g_\zeta(\mathbf{c}(k)x) = \sum \psi_i(k)(D_i g_\zeta)(\mathbf{c}(k)x).$$

Thus (1) becomes a sum of integrals. By Chapter II, Corollary 1.3, applied to the anti-Iwasawa decomposition, an element D_i of $\mathbb{IDO}(G)$ has a decomposition into a sum of elements in $\mathbb{IDO}(G)\tilde{\mathfrak{n}}$, $S(\mathfrak{k})^{\sim}S(\mathfrak{a})^{\sim}$, and $S(\mathfrak{a})^{\sim}$. By the definition of g_ζ with the anti-Iwasawa projection, elements of $\mathbb{IDO}(G)\tilde{\mathfrak{n}}$ annihilate g_ζ. The corresponding term in the sum expressing (1) is 0.

Next consider an element of $S(\mathfrak{a})^{\sim}$. Let E_i be the pure $\mathfrak{a}$-component of D_i. Then $\deg E_i \leqq \deg D$, and $E_i = \mathcal{D}_G(P_i)$ for some $P_i \in S(\mathfrak{a})$. We apply Proposition 7.6 of Chapter III to conclude that the eigenvalue $P_i(\zeta - \rho)$ gives rise to the desired estimate $(1 + \|\zeta\|)^{\deg D}$. Furthermore, $|g_\zeta|$ is estimated by $g_{\mathrm{Re}(\zeta)}$. The remaining integral is then estimated by $\varphi_{\mathrm{Re}(\zeta)}(x)$, as desired.

Thus we are reduced to proving the estimate of Lemma 4.1 when

$$D = \tilde{w}_1 \cdots \tilde{w}_m \tilde{H}_1 \cdots \tilde{H}_n$$

with $w_1, \ldots, w_m \in \mathfrak{k}$ and $H_1, \ldots, H_n \in \mathfrak{a}$. Applying the $\mathfrak{a}$-component and using the eigenfunction property of g_ζ reduces the estimate to the case when $D = \tilde{w}_1 \ldots \tilde{w}_m$ with $w_j \in \mathfrak{k}$ all j. We do this estimate using integration by parts. Let $w \in \mathfrak{k}$. For a given C^∞ function ψ on K, we have

$$\frac{d}{dt}\int_K \psi(k) g_\zeta(xk^{-1}\exp(tw))\,dk = \frac{d}{dt}\int_K \psi(\exp(tw)k) g_\zeta(xk^{-1})\,dk.$$

The derivative d/dt can be moved in and out of the integral, and we may then evaluate the derived expression at $t = 0$. Thus we have determined a (right invariant) differential operator E_w such that

$$\int_K \psi(k)(\tilde{w} g_\zeta)(xk^{-1})\,dk = \int_K (E_w \psi)(k) g_\zeta(xk^{-1})\,dk.$$

We may continue by induction, finding a differential operator E such that

$$\int_K \psi(k)(D g_\zeta)(xk^{-1})\,dk = \int_K (E\psi)(k) g_\zeta(xk^{-1})\,dk.$$

Note that E depends only on the original D. We then estimate the integral by the sup norm of the derivatives of the function ψ, which depend only on D, and the integral

$$\int_K g_{\mathrm{Re}(\zeta)}(xk^{-1})\,dk,$$

which is the desired factor $\varphi_{\mathrm{Re}(\zeta)}(x)$. This concludes the proof of Lemma 4.1(ii).

Remark. The two groups U and K do not play a symmetric role here because K is compact and U is not. Thus agreeing from the start that we shall deal with left invariant differential operators, and wanting $\tilde{\mathfrak{n}}$ to annihilate the Harish-Chandra integral for the spherical function, the anti-Iwasawa decomposition is forced if one is going to use the present method.

When taking the spherical transform of C^∞ functions with compact support, we got entire functions of a variable $\zeta \in \mathfrak{a}_{\mathbf{C}}^\vee$. To deal with functions in the HCS-space, we restrict the spherical transform to the imaginary axis. For a K-bi-invariant function f and $\lambda \in \mathfrak{a}^\vee$ we write

$$\begin{aligned}(\mathbf{S}f)(\mathbf{i}\lambda) &= \int_G \varphi(\mathbf{i}\lambda, x) f(x)\, dx \\ &= \int_{\mathfrak{a}^+} \varphi(\mathbf{i}\lambda, \exp H) f_{\mathfrak{a}}(H) J(H)\, dH.\end{aligned}$$

For $f \in \mathrm{HCS}(K\backslash G/K)$, and every positive integer m by Theorem 2.8 we have

(2a) $$|\varphi(\mathbf{i}\lambda, \exp H) f_{\mathfrak{a}}(H)| \ll e^{-2\rho(H)}(1 + |H|)^{-m} \qquad \text{for } |H| \to \infty.$$

Note that both φ and $f_{\mathfrak{a}}$ contribute one factor $e^{-\rho(H)}$. As noted in §3,

(2b) $$0 \leqq J(H) \ll e^{2\rho(H)} \qquad \text{for } |H| \to \infty.$$

Thus the integral is absolutely convergent, uniformly for $\lambda \in \mathfrak{a}^\vee$. Actually:

Theorem 4.3. *For $f \in \mathrm{HCS}(K\backslash G/K)$, its spherical transform $\mathbf{S}f$ lies in the ordinary Schwartz space $\mathrm{Sch}(\mathbf{i}\mathfrak{a}^\vee)$, and*

$$\mathbf{S}\colon \mathrm{HCS}(K\backslash G/K) \to \mathrm{Sch}(\mathbf{i}\mathfrak{a}^\vee)$$

is continuous.

Proof. By Lemma 4.1(i), differentiating under the integral sign with respect to $\mathbf{i}\lambda$ introduces only a factor of polynomial growth in $|H|$, which is nullified by the assumed arbitrarily rapid polynomial decay of f. We shall prove that $\mathbf{S}f$ is in the Schwartz space and $\mathbf{S}$ is continuous simultaneously, by showing that given a seminorm $p'_{m,D}$ as at the beginning of §3, its value on the image of $\mathbf{S}$ is finite, and there exists a seminorm $q_{s,E}$ on HCS, as well as a constant $c_{m,D}$ such that for all $f \in \mathrm{HCS}(K\backslash G/K)$ we have

$$p'_{m,D}(\mathbf{S}f) \leqq c_{m,D}q_{s,E}(f).$$

By Chapter III, Proposition 1.5, and Chapter VII, Theorem 3.4, the eigenvalue of Casimir on the spherical function is a polynomial Q, specifically

$$\mathrm{ev}(\omega, \varphi_{\mathbf{i}}\lambda) = (\mathbf{i}\lambda)^2 - \rho^2 = Q(\mathbf{i}\lambda).$$

A polynomial $P \in S(\mathbf{i}\mathfrak{a}^\vee)$ gives rise to the differential operator $\mathcal{D}(P) = \mathcal{D}_{\mathbf{i}\lambda}(P)$. For any positive integer m, we get

$$\begin{aligned}\mathcal{D}_{\mathbf{i}\lambda}(P)(Q^m\mathbf{S}f)(\mathbf{i}\lambda) &= \mathcal{D}_{\mathbf{i}\lambda}(P)\int_G Q^m(\mathbf{i}\lambda)\varphi(\mathbf{i}\lambda, x)f(x)\,dx \\ &= \mathcal{D}_{\mathbf{i}\lambda}(P)\int_G (\omega_x^m\varphi)(\mathbf{i}\lambda, x)f(x)\,dx \\ &= \int_G \mathcal{D}_{\mathbf{i}\lambda}(P)\varphi(\mathbf{i}\lambda, x)(\omega^m f)(x)\,dx \qquad (3)\end{aligned}$$

by Proposition 3.8, and the fact that ω is trivially equal to its transpose. So far we have reached the same point as (*) in the proof of continuity of Theorem 3.1. Instead of working with an exact formula, we now estimate the two factors under the integral. Let $D = \mathcal{D}(P)$. By Lemma 4.1(i) and Theorem 2.8, we get

$$(4) \qquad |\mathcal{D}_{\mathbf{i}\lambda}(P)\varphi(\mathbf{i}\lambda, a)| \leqq c_1(D)(1+|H|)^{\deg P + |\mathcal{R}(\mathfrak{n})|}e^{-\rho(H)}$$

for $a = \exp H$, $H \in \mathrm{Cl}(\mathfrak{a}^+)$. Let $E = \omega^m$ From Theorem 2.8 and the definition of the HCS-topology, there is a constant $c_2(m)$ and s large, say $s = \deg P + |\mathcal{R}(\mathfrak{n})| + r + 1$, such that for all $H \in \mathrm{Cl}(\mathfrak{a}^+)$,

$$(5) \qquad |(\omega^m f(\exp H)| \leqq c_2(m)q_{s,E}(f)(1+|H|)^{-s}.$$

Then estimating the integral (3) in polar coordinates and using (4), (5) yields

$$p'_{m,D}(\mathbf{S}f) \leqq c_3(m, D)q_{s,E}(f).$$

Thus the Schwartz space estimate for $\mathbf{S}f$ came out simultaneously with the continuity, thus proving the theorem.

X, §5. CONTINUITY OF THE INVERSE TRANSFORM AND SPHERICAL INVERSION ON HCS($K\backslash G/K$)

We start with a simple lemma giving a criterion for the inverse spherical transform to be defined on a fairly large space of functions.

Let h be a function on $\mathbf{i}\mathfrak{a}^\vee$. We define $\mathbf{p}_m(h)$ by

$$\mathbf{p}_m(h) = \sup_{\lambda \in \mathfrak{a}^\vee} (1 + |\lambda|)^m |h(\mathbf{i}\lambda)|.$$

Thus $\mathbf{p}_m$ is a norm on the space of functions for which the right side is finite. We call such functions **$\mathbf{p}_m$-bounded.**

We let $\eta(\lambda)\, d\lambda$ be the measure used to define the inverse spherical transform in Chapter VIII, §4. By Chapter V, Lemma 8.4, the function $\eta(\lambda)$ has polynomial growth.

Lemma 5.1. *Let $\{h_k\}$ be a sequence of $\mathbf{p}_m$-bounded continuous functions, $\mathbf{p}_m$-convergent to a continuous function h. Then for m sufficiently large,*

$$\int_{\mathfrak{a}^\vee} h_k(\mathbf{i}\lambda)\varphi(\mathbf{i}\lambda, x)\eta(\lambda)\, d\lambda \quad \textit{converges to} \quad \int_{\mathfrak{a}^\vee} h(\mathbf{i}\lambda)\varphi(\mathbf{i}\lambda, x)\eta(\lambda)\, d\lambda,$$

uniformly in x.

Proof. By proposition 2.2, $|\varphi(\mathbf{i}\lambda, x)|$ is bounded by $\varphi_0(x)$, which is bounded by 1 according to Corollary 2.6. Hence

$$\left| \int_{\mathfrak{a}^\vee} (h_k - h)\varphi(\mathbf{i}\lambda, x)\eta(\lambda)\, d\lambda \right|$$
$$\ll \int_{\mathfrak{a}^\vee} |h_k - h|(\mathbf{i}\lambda)(1 + |\lambda|)^{M+1+\dim \mathfrak{a}} \frac{\eta(\lambda)}{(1 + |\lambda|)^M} \frac{1}{1 + |\lambda|)^{1+\dim \mathfrak{a}}}\, d\lambda.$$

We take M to be the polynomial order of growth of η. The exponent $1+\dim \mathfrak{a}$ is just what is needed to make the integral absolutely convergent, so we take

$$m = M + 1 + \dim \mathfrak{a}$$

to conclude the proof.

Remark. No derivatives occur in estimates such as those used in Lemma 5.1, which are basically routine estimates involving only absolute convergence of the integral, and sup norm convergence of the functions.

Next, we shall use Anker's refined version of the support theorem (Chapter IV, Theorems 8.2 and 8.7, and Chapter VIII, §6) to prove first:

Theorem 5.2. *The inverse spherical transform*

$$\mathbf{S}^{-1} \colon \mathrm{PW}(\mathfrak{a}_{\mathbf{C}}^\vee)^W \to C_c^\infty(K\backslash G/K)$$

is continuous for the Schwartz topology on the left, and HCS *topology on the right.*

We recall from Chapter VIII, §6, that $\mathbf{S}^{-1}$ is defined by the integral

$$(1)\qquad (\mathbf{S}^{-1}h)(x) = \frac{1}{|W|}\int_{\mathfrak{a}^\vee} h(\mathbf{i}\lambda)\varphi(-\mathbf{i}\lambda, x)\eta(\lambda)\,d\lambda$$

As Anker himself says, the proof of the above theorem is the heart of his paper. The proof will be carried out in several steps. Made explicit, the theorem asserts:

Given a seminorm $q_{m,D}$ on $C_c^\infty(K\backslash G/K)$, there exists a Schwartz-continuous seminorm $\mathbf{p}$ on $\mathrm{Sch}(\mathbf{i}\mathfrak{a}^\vee)$ *such that for all $f \in C_c^\infty(K\backslash G/K)$ or $h \in \mathrm{PW}(\mathbf{i}\mathfrak{a}^\vee)$ we have*

$$q_{m,D}(f) \leqq \mathbf{p}(\mathbf{S}f) \qquad \text{that is} \qquad q_{m,D}(\mathbf{S}^{-1}h) \leqq \mathbf{p}(h).$$

For formal purposes here, let $\mathbf{H}_\mathfrak{a} = \mathbf{M}_\mathfrak{a}^{-1}\mathbf{S}$ so $\mathbf{S} = \mathbf{M}_\mathfrak{a}\mathbf{H}_\mathfrak{a}$. Below we shall use a non-formal property, to be made explicit when the time comes. For now, we note only that $\mathbf{M}_\mathfrak{a}$ is Schwartz bicontinuous by Theorem 3.1(ii). Therefore it suffices to prove:

There exists a Schwartz-continuous seminorm $\mathbf{p}''$ such that

$$q_{m,D}(f) \leqq c(m, D)\mathbf{p}''(\mathbf{H}_\mathfrak{a} f) \qquad \text{for all } f \in C_c^\infty(K\backslash G/K).$$

This will be achieved by the next three lemmas.

For definiteness, by a **continuous seminorm** on a vector space whose topology is defined by a family of seminorms, the reader may understand a finite linear combination of seminorms in the given family, with positive real coefficients.

Anker's convex sets were defined in Chapter IV, §8, see also Chapter VIII, Theorem 6.2 and Corollary 6.3. We recall the definition. Let Λ be the unit vector in the direction of ρ, that is

$$\Lambda = \rho/|\rho|.$$

For $j = 1, 2, \ldots$ we let:

$C_j = C_{\Lambda,j}$ = set of $H \in \mathfrak{a}$ such that $\Lambda(wH) \leqq j$ for all $w \in W$ or equivalently, $|\Lambda(wH)| \leqq j$ for all $w \in W$ (cf. §1(5));

$G_j = G_{\Lambda,j} = K(\exp C_j)K$,

thus obtaining a filtration of $\mathfrak{a}$ and G respectively. Note that each C_j and G_j are compact. Furthermore, by Chapter I, §4(1), we have

$$\rho(H) \gg\ll |H| \quad \text{on } \mathrm{Cl}(\mathfrak{a}^+).$$

In other words, on $\mathfrak{a}^+$ the functions ρ and the norm are each bounded by a positive constant times the other.

For a function F on a set S, we shall use the notation for the **sup norm**

$$\|F\|_S = \sup_{x \in S} |F(x)|.$$

Lemma 5.3. *Given $D \in \mathrm{IDO}(G)$ there exist a constant c_D and a positive integer d, depending only on D, such that for all $f \in C_c^\infty(K\backslash G/K)$ we have*

$$\|\varphi_0^{-1} D f\|_G \leqq c_D \boldsymbol{p}_{d,1}(\mathbf{S}f).$$

Let

$$f_{m,D} = (1+\sigma)^m \varphi_0^{-1} Df.$$

Then

$$\|f_{m,D}\| G_j \leqq c_D (1+j)^m \boldsymbol{p}_{d,1}(\mathbf{S}f).$$

Proof. From the definition (1), we have for some constant c_1,

$$(Df)(x) = c_1 \int_{\mathfrak{a}^\vee} (\mathbf{S}f)(\mathbf{i}\lambda) D_x \varphi(-\mathbf{i}\lambda, x) \eta(\lambda)\, d\lambda.$$

We know that $\eta(\lambda)$ has polynomial growth (going back to Chapter V, Lemma 8.4), so by Lemma 4.1(ii), for some integer $s > 0$,

$$\begin{aligned} |(Df)(x)| &\leqq c_2 \varphi_0(x) \int_{\mathfrak{a}^\vee} |(\mathbf{S}f)(\mathbf{i}\lambda)| (1+|\lambda|)^s \, d\lambda \\ &\leqq c_2 \varphi_0(x) \int_{\mathfrak{a}^\vee} |\mathbf{S}f(\mathbf{i}\lambda)| (1+|\lambda|)^{s+r+1} (1+|\lambda|)^{-(r+1)} \, d\lambda \end{aligned}$$

the exponent $r+1$ being designed to make $\int (1+|\lambda|)^{-(r+1)}\, d\lambda < \infty$. Then we get the further inequality

$$\leqq c_D \varphi_0(x) p_{d,1}(\mathbf{S}f) \qquad \text{with } d = s + r + 1.$$

This proves the first inequality. The second follows trivially with $\sigma(x) \leqq j$. This proves Lemma 5.3.

Now putting $j = 2$, we get with some constant $c_3 = c_3(m, D)$:

(2) $\quad \|f_{m,D}\|G_2 \leqq c_3 p_{d,1}(\mathbf{S}f) \qquad$ for some $d > 0$ (as above).

Next we deal with $j \geqq 3$, and estimate on $G_j - G_{j-1}$. We need a lemma.

Lemma 5.4. *There exists* $\gamma_j \in C_c^\infty(\mathfrak{a})^W$ *such that* $\gamma_j = 1$ *on* C_{j-1}, $\gamma_j = 0$ *outside* C_j, *and the derivatives of* γ_j $(j = 2, 3, \ldots)$ *are bounded uniformly in* j.

Proof. Let $\gamma \in C^\infty(\mathbf{R})$ be such that:

$\gamma = 0$ on $(-\infty, 0]$;

$\gamma = 1$ on $[1, \infty)$;

γ is increasing on $[0, 1]$.

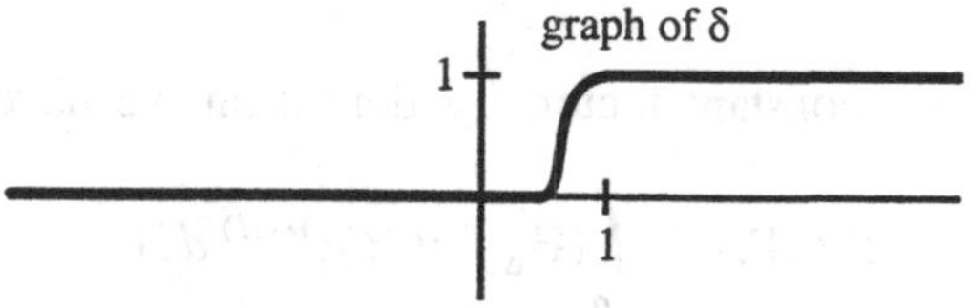

Define γ_j on $\mathfrak{a}$ by

$$\gamma_j(H) = \prod_{w \in W} \gamma(j - \Lambda(wH)).$$

Then γ_j satisfies the required conditions.

We let $\mathbf{H}_\mathfrak{a}$ be the Harish transform pulled back to $\mathfrak{a}$, that is, composed with the logarithm so the a-variable gets pulled back to the H-variable. For $f \in C_c^\infty(K\backslash G/K)$ we write

$$\mathbf{H}_\mathfrak{a} f = \gamma_j \mathbf{H}_\mathfrak{a} f + (1 - \gamma_j)\mathbf{H}_\mathfrak{a} f.$$

By the Harish-Chandra inversion theorem, there exists $f_j \in C_c^\infty(K\backslash G/K)$ such that

$$(1 - \gamma_j)\mathbf{H}_\mathfrak{a} f = \mathbf{H}_\mathfrak{a} f_j \qquad \text{and so} \qquad \mathbf{H}_\mathfrak{a}(f - f_j) = \gamma_j \mathbf{H}_\mathfrak{a} f.$$

Since $\gamma_j \mathbf{H}_\mathfrak{a} f$ has support in C_j, it follows from Chapter VIII, Corollary 6.3, that

$$\operatorname{supp}(f - f_j) \subset G_j.$$

This is precisely where we make use of Anker's variation of the Helgason support theorem.

We can write

$$f_{m,D} = (f - f_j)_{m,D} + (f_j)_{m,D}.$$

Since $f - f_j$ has support in G_j, we get the sup norm estimate

$$\|f_{m,D}\|_{G_{j+1}-G_j} = \|(f_j)_{m,D}\|_{G_{j+1}-G_j}.$$

By Lemma 5.3 applied to $(f_j)_{m,D}$ we have with $c_4 = c_4(m, D)$,

$$\|f_{m,D}\|_{G_{j+1}-G_j} \leqq c_4 j^m \mathbf{p}_{d,1}(\mathbf{S}f_j). \tag{3}$$

Lemma 5.5. *Given positive integers m, d there exists a Schwartz continuous seminorm $\mathbf{p}''$ on* $\mathrm{PW}(\mathbf{i}\mathfrak{a}^\vee)$ *and a constant $c_{m,d}$ such that for all j and all $f \in C_c^\infty(K\backslash G/K)$,*

$$j^m \mathbf{p}_{d,1}(\mathbf{S}f_j) \leqq c_{m,d}\mathbf{p}''(\mathbf{H}_\mathfrak{a} f).$$

Proof. Up to a constant factor, by definition we have

$$\mathbf{S}f_j(\mathbf{i}\lambda) = \int_\mathfrak{a} (\mathbf{H}_\mathfrak{a} f_j)(H)e^{\mathbf{i}\lambda(H)}\,dH.$$

For a polynomial Q, which we take to be $((\mathbf{i}\lambda)^2 - 1)^d$, we get

$$\begin{aligned} Q(\mathbf{i}\lambda)(\mathbf{S}f_j)(\mathbf{i}\lambda) &= \int_\mathfrak{a} (H_\mathfrak{a} f_j)(H)Q(\mathbf{i}\lambda)e^{\mathbf{i}\lambda(H)}\,dH \\ &= \int_\mathfrak{a} (\mathbf{H}_\mathfrak{a} f_j)(H)\mathcal{D}_H(Q)e^{\mathbf{i}\lambda(H)}\,dH \\ &= \int_\mathfrak{a} D(\mathbf{H}_\mathfrak{a} f_j)(H)e^{\mathbf{i}\lambda(H)}dH, \end{aligned}$$

where D is the transpose of $\mathcal{D}(Q)$. We multiply inside the integral sign by

$$(1 + H^2)^{1+r} \cdot (1 + H^2)^{-(1+r)} \qquad \text{with } r = \dim \mathfrak{a}.$$

Then

$$\mathbf{p}_{d,1}(\mathbf{S}f_j) \leqq c_5(m, d) \sup_{H\in\mathfrak{a}} (1 + H^2)^{1+r}|D(\mathbf{H}_\mathfrak{a} f_j)(H)|. \tag{4}$$

We estimate $D(\mathbf{H}_\mathfrak{a} f_j) = D((1 - \gamma_j)\mathbf{H}_\mathfrak{a} f)$ using the Leibniz rule for the derivative of a product as in Chapter X, Theorem 3.1(iii). Since $1 - \gamma_j = 0$ on $\mathcal{C}_{j-1}$ and is bounded together with any given finite number of derivatives uniformly in j, it follows that there exist a finite number of polynomial differential operators D_k depending only on D such that

$$p_{d,1}(\mathbf{S}f_j) \leqq c_6(m,d) \sum_k \sup_{H \in \mathfrak{a} - C_{j-1}} (1+H^2)^{1+r} |D_k \mathbf{H}_{\mathfrak{a}} f(H)|.$$

We multiply by j^m and put $m' = 1 + r + m$ to get

$$(5)\quad j^m p_{d,1}(\mathbf{S}f_j) \leqq c_7(m,d) \sum_k \sup_{H \in \mathfrak{a} - C_{j-1}} (1+H^2)^{m'} |D_k \mathbf{H}_{\mathfrak{a}} f(H)|$$

$$(6)\quad \leqq c_7(m,d) \sum_k \sup_{H \in \mathfrak{a}} (1+H^2)^{m'} |D_k \mathbf{H}_{\mathfrak{a}} f(H)|.$$

This gives the desired estimate in terms of the sum of seminorms p_{2m', D_k}, and concludes the proof of Lemma 5.5.

As already noted, we have also concluded the proof of Theorem 5.2. We are now in a position to prove the analogue of Theorem 3.1 for the Harish-Chandra Schwartz space $\mathrm{HCS}(K \backslash G/K)$. All the work has been done, so this final step is short.

Theorem 5.6. *Let* $\mathbf{S}$ *be the spherical transform. Then* $\mathbf{S}$ *induces a linear isomorphism*

$$\mathbf{S}: \mathrm{HCS}(K \backslash G/K) \to \mathrm{Sch}(i\mathfrak{a}^\vee)^W,$$

which is bi-continuous for the Schwartz topologies. The inverse transform is given by the same integral formula as between $\mathrm{PW}(i\mathfrak{a}^\vee)^W$ *and* $C_c^\infty(K \backslash G/K)$.

Proof. The injectivity of $\mathbf{S}$ comes, for instance, from the fact that the spherical transform is L^2-isometric, and Proposition 3.4.

For the surjectivity, given $h \in \mathrm{Sch}(i\mathfrak{a}^\vee)^W$, by Theorem 3.1(iv) there is a sequence $\{h_k\}$ in $\mathrm{PW}(i\mathfrak{a}^\vee)^W$ such that $\{h_k\}$ is Schwartz convergent to h. By Theorem 5.2, $\{\mathbf{S}^{-1}h_k\}$ is HCS Cauchy (in an obvious sense) in $C_c^\infty(K \backslash G/K)$, and so is HCS convergent to some $f \in \mathrm{HCS}(K \backslash G/K)^W$ by Proposition 3.2, Applying $\mathbf{S}$ and using Theorem 4.3 shows that $h = \mathbf{S}f$ as desired. Note that the justification for taking the limit comes form Lemma 5.1, which is also needed to prove the last assertion of Theorem 5.6, about the integral formula. This concludes the proof of Theorem 5.6.

X, §6. EXTENSION OF FORMULAS BY HCS CONTINUITY

A number of formulas concerning convolutions were listed in Chapter IV, §5 and §6, with test functions which have compact support. These formulas are valid for a much wider class of test functions, essentially those which make all the steps valid, with absolutely convergent integrals. We organized the material so that we would strive as fast as possible to the inversion theorem on $C_c^\infty(K\backslash G/K)$, without cluttering up the proofs and exposition with the extension to other spaces of test functions. Now that we are putting emphasis on the Schwartz space and its Harish-Chandra extension to take G into account, we return to these formulas and make the necessary comment about their extended validity. Some of the arguments needed are much simpler than those involving the Schwartz or HCS-spaces, because they do not involve taking derivatives. They need only rapid decay of the test functions to insure the desired convergence. We tabulate some relevant past formulas in the specific case of spherical inversion on the HCS-space.

This section has its source in Harish-Chandra [Har 66], with some simplifications from [Var 73], [GaV 88], Chapter 6, §1, and [Ank 91].

In treating the HCS space, we dealt with functions f such that $\varphi_0^{-1} f$ has polynomial decay of arbitrarily high order, as well as all derivatives Df with $D \in \mathrm{IDO}(G)$. In taking the integral scalar product, one of the functions may have polynomial growth instead of decay.

We shall use Proposition 3.8 and its notation, so that

$$T_\varphi(f) = \int_G f(y)\varphi(y)\,dy.$$

Proposition 6.1. *Let $\varphi \in C(K\backslash G/K)$ be C-spherical and such that $\varphi_0^{-1}\varphi$ has polynomial growth. Let $f \in \mathrm{HCS}(K\backslash G/K)$. Then*

$$T_\varphi(f^{xK}) = \varphi(x^{-1})T_\varphi(f).$$

Proof. The arguments used to prove the result when f has compact support are now valid for the HCS-space in light of Proposition 3.8.

Proposition 6.2. *For $f \in \mathrm{HCS}(K\backslash G/K)$ and $\lambda \in \mathfrak{a}^\vee$, we have the following formulas:*

(1) $\mathbf{S}(f^*)(\mathbf{i}\lambda) = \overline{\mathbf{S}f(\mathbf{i}\lambda)},$ [Ch. III, Cor. 7.4],

(2) $\mathbf{S}(f^{xk})(\mathbf{i}\lambda) = \varphi_{\mathbf{i}\lambda}(x^{-1})(\mathbf{S}f)(\mathbf{i}\lambda).$ [Ch. IV, Cor. 5.6].

For $D \in \mathrm{IDO}(G)^K$,

(3) $(\mathbf{S}Df)(\mathbf{i}\lambda) = \mathrm{ev}({}^t\mathbf{h}(D), \chi_{\mathbf{i}\lambda})\mathbf{S}f(\mathbf{i}\lambda)$ [Ch. III, Props. 1.5, 1.7]

Proof. The arguments used to prove these formulas when

$$f \in C_c^\infty(K\backslash G/K)$$

as in the references to the right are valid for the more general f. The convergence of the integrals is immediate from the definitions (no worse than in Lemma 5.1) and does not involve derivatives for (1) and (2), but (3) does. However, the formula extends from $C_c^\infty(K\backslash G/K)$ to $\mathrm{HCS}(K\backslash G/K)$ because of the HCS continuity of the spherical transform S proved in Theorem 4.3, and that of the integral in Proposition 3.8.

The convolution product

The next item of business is the continuity of the convolution product, and its extension to the HCS-space.

First we extend the commutativity of Chapter IV, Theorem 4.6.

Lemma 6.3. *Let f, g, be measurable K-bi-invariant functions on G. If for each $x \in G$ the function*

$$y \mapsto f(xy^{-1})g(y)$$

is in L^1, then the convolution integrals for both $f * g$ and $g * f$ are absolutely convergent, and they are equal.

Proof. Nothing else was used in the proof of the above mentioned reference.

Proposition 6.4. *The space* $\mathrm{HCS}(K\backslash G/K)$ *is stable under the convolution product, which is commutative and continuous.*

Proof. We give a proof going back to Varadarajan, see [GaV 88], Lemma 6.1.9 and Theorem 6.1.10. We start with a lemma. For a real number s, we define the K-bi-invariant function

$$\psi_s(x) = (1+\sigma(x))^{-s}\varphi_0(x) \quad \text{that is} \quad \psi_s = (1+\sigma)^{-s}\varphi_0.$$

Lemma 6.5. *Let s_0 be a real number $\geqq 0$ such that $\psi_{s_0} \in L^2(G)$. Let $s \geqq 0$. If $s' \geqq s + 2s_0$ and $c_0 = \|\psi_{s_0}\|_2^2$, then*

$$\psi_{s'} * \psi_s(x) \leqq c_0\psi_s(x) \quad \text{for all } x \in G.$$

The convolution integral on the left is absolutely uniformly convergent for x ranging over a compact subset of G, and we have commutativity,

$$\psi_{s'} * \psi_s = \psi_s * \psi_{s'}.$$

Proof. We estimate everything from above as follows. From the Harish-Chandra triangle inequality of Proposition 1.1, we have for all

$y \in G$,

$$1+\sigma(x) \leqq (1+\sigma(xky^{-1}))(1+\sigma(y)). \tag{4}$$

Then from the fact that K has total measure 1, we get

$$\begin{aligned}
\psi_s * \psi_{s'}(x) &= \int_G \psi_s(xy^{-1})\psi_{s'}(y)\,dy \\
&= \int_G\int_K \psi_s(xky^{-1})\psi_{s'}(y)\,dy \\
&\qquad\qquad [\text{by } y \mapsto yk^{-1} \text{ and Fubini}] \\
&\leqq \int_G\int_K \frac{(1+\sigma(y))^s}{(1+\sigma(x))^s}\varphi_0(xky^{-1})(1+\sigma(y))^{-s'}\varphi_0(y)\,dk\,dy \\
&\qquad\qquad [\text{by (4)}] \\
&= \psi_s(x)\int_G (1+\sigma(y))^{s-s'}\varphi_0(y^{-1})\varphi_0(y)\,dy \\
&\qquad\qquad [\text{by Ch. IV, Th. 5.2}].
\end{aligned}$$

Since $\varphi_0(y^{-1}) = \varphi_0(y)$, and $s'-s \geqq 2s_0$, the integral has a finite value, which determines the constant. The commutativity comes from Lemma 6.3. This concludes the proof.

We are now ready to prove Proposition 6.4. Let $f, g \in \mathrm{HCS}(K\backslash G/K)$. The integrals

$$\int_G |f(xy^{-1})||g(y)|\,dy = \int_G |f(y^{-1})||g(yx)|\,dy$$

are absolutely convergent, uniformly for x in a compact set, and similarly if we replace f by Df or g by Dg for $D \in \mathrm{IDO}(G)$. Then we can differentiate under the integral sign to get

$$D(f*g)(x) = D_x\int_G f(y^{-1})g(yx)\,dy = \int_G f(y)(Dg)(y^{-1}x)\,dy.$$

So for s, s' and s_0 as in Lemma 6.5, we get

$$\begin{aligned}
|D(f*g)(x)| &\leqq \int_G q_{s',1}(f)\psi_{s'}(y)q_{s,D}(g)\psi_s(y^{-1}x)\,dy \\
&\leqq q_{s',1}(f)q_{s,D}(g)\psi_s(x) \quad [\text{by Lemma 6.5}].
\end{aligned}$$

Hence with the constant c_0 of Lemma 6.5, we find from the definitions

$$q_{s,D}(f*g) \leqq c_0 q_{s',1}(f)q_{s,D}(g). \tag{5}$$

Putting $s = m$, $s' = m + s_0$ proves the HCS continuity of the convolution product, and also proves that $f * g \in \mathrm{HCS}(K\backslash G/K)$, thereby finishing the proof of Proposition 6.4.

Corollary 6.6. *On* $\mathrm{HCS}(K\backslash G/K)$ *the spherical transform is a multiplicative homomorphism, that is*

$$\mathbf{S}(f * g) = \mathbf{S}f \cdot \mathbf{S}g \qquad \textit{for } f, g \in \mathrm{HCS}(K\backslash G/K).$$

Proof. Immediate by HCS continuity from the corresponding Proposition 6.3 of Chapter IV.

X, §7. AN EXAMPLE: THE HEAT KERNEL

This section is taken from Gangolli [Gan 68].

Let $\mathfrak{a}$ as usual be the space of diagonal $n \times n$ real matrices with trace 0, with the trace form defining a positive definite scalar product. We denote by λ elements of $\mathfrak{a}^\vee$. For $\zeta \in \mathfrak{a}^\vee_{\mathbf{C}}$ we write

$$\zeta^2 = \langle \zeta, \zeta \rangle \quad \text{(the } \mathbf{C}\text{-bilinear scalar product).}$$

We let

$$\text{(1)} \qquad E_t(\zeta) = e^{(\zeta^2 - \rho^2)t} \text{ and } E_t^\#(\lambda) = e^{-(\lambda^2+\rho^2)t} \text{ so } E_t(\mathbf{i}\lambda) = E_t^\#(\lambda),$$

so $E_t^\#$ is the classical Gauss function in the Schwartz space $\mathrm{Sch}(\mathfrak{a}^\vee)$. We take its inverse spherical transform, i.e. we define the **Gauss function** on G to be

$$g_t(x) = |W|^{-1} \int_{\mathfrak{a}^\vee} e^{-(\lambda^2+\rho^2)t} \varphi(-\mathbf{i}\lambda, x)\eta(\lambda)\, d\lambda,$$

so

$$g_t = \mathbf{S}^{-1} E_t.$$

Since both E_t and η are even functions, it follows that we can omit the minus sign on $\mathbf{i}\lambda$, so we also get

$$g_t(x) = |W|^{-1} \int_{\mathfrak{a}^\vee} e^{-(\lambda^2+\rho^2)t} \varphi(\mathbf{i}\lambda, x)\eta(\lambda)\, d\lambda.$$

Let ω be the Casimir operator. By Chapter VII, Theorem 1.1, we know the eigenvalue of Casimir on the spherical function $\varphi_{\mathbf{i}\lambda}$ (or $\varphi_{-\mathbf{i}\lambda}$), namely

$$\omega\varphi_{\mathbf{i}\lambda} = -(\lambda^2 + \rho^2)\varphi_{\mathbf{i}\lambda} = ((\mathbf{i}\lambda)^2 - \rho^2)\varphi_{\mathbf{i}\lambda}.$$

We can differentiate under the integral sign, so we get at once:

Theorem 7.1. *The function $(t, x) \mapsto g_t(x)$ on $\mathbf{R}^+ \times G$ (or $\mathbf{R}^+ \times G/K$) satisfies the heat equation, namely*

$$(-\omega_x + \partial_t) g_t(x) = 0.$$

By Theorem 5.6, the function g_t is in $\mathrm{HCS}(K\backslash G/K)$.

Theorem 7.2. *For $s, t > 0$ we have $g_t * g_s = g_{t+s}$.*

Proof. The following formal steps are justified by spherical inversion on the HCS space:

$$\mathbf{S}(g_t * g_s) = \mathbf{S}g_t \cdot \mathbf{S}g_s = E_t E_s = E_{t+s} = \mathbf{S}g_{t+s},$$

whence the theorem follows since $\mathbf{S}$ is injective.

Theorem 7.3. *Let $f \in \mathrm{HCS}(K\backslash G/K)$. Then*

$$L^2\text{-}\lim_{t\to 0} g_t * f = f.$$

Proof. We have

$$\begin{aligned} \|g_t * f - f\|_2^2 &= \int_G |g_t * f(x) - f(x)|^2 \, dx \\ &= \frac{1}{|W|} \int_{\mathfrak{a}^\vee} |(\mathbf{S}g_t - 1)\mathbf{S}f(\mathbf{i}\lambda)|^2 \eta(\lambda) \, d\lambda \end{aligned}$$

by the L^2-isometric property of Chapter IX, Theorem 1.7. Since

$$(\mathbf{S}g_t)(\mathbf{i}\lambda) = e^{-(\lambda^2+\rho^2)t}$$

approaches 1 as $t \to 0$, we can apply the dominated convergence theorem to conclude the proof.

Gangolli at this point invokes without reference certain results of analysis and functional analysis to conclude that the function h_t defined by

$$h_t(x, y) = g_t(y^{-1}x)$$

is the heat kernel. What one is after are the three properties:

DIR 1. We have $g_t \geqq 0$ for all $t > 0$.

DIR 2. $\int_G g_t(x)\,dx = 1$.

DIR 3. Given $\delta > 0$, $\displaystyle\lim_{t\to 0}\int_{\sigma(x)\geqq\delta} g_t(x)\,dx = 0$.

Each one of these properties depends on further analysis which has not yet been covered in previous sections. That g_t is real is of course obvious since g_t is invariant under complex conjugation ($i\lambda \mapsto -i\lambda$). The two conditions **DIR 2** and **DIR 3** are L^1-conditions, and so far we have only discussed L^2 properties. In Chapter XI, Theorem 2.3, we shall prove that g_t is in L^1; and in Theorem 4.1 we show that $\varphi(\zeta, x)$ is bounded for $x \in G$ and ζ in a tube containing $-\rho$. Then as Gangolli pointed out, the argument showing **DIR 2** runs as follows. By the inversion of Chapter XI, Theorem 2.3, we get

$$E_t(\zeta) = (\mathbf{S}g_t)(\zeta) = \int_G \varphi(\zeta, x) g_t(x)\,dx$$

for ζ in the tube. We can substitute $\zeta = -\rho$ and use the value $\varphi(-\rho, x) = 1$ for all $x \in G$ because of the Harish-Chandra integral

$$\varphi(\zeta, x) = \int_K (kx)_A^{\zeta+\rho}\,dk.$$

This gives **DIR 2**. This important application may be used as motivation for the L^1 theory.

X, §8. THE HARISH TRANSFORM

In Chapter II we defined the **Harish transform** by the integral

$$(\mathbf{H}f)(a) = a^\rho \int_U f(au)\,du \quad \text{(whenever absolutely convergent).}$$

In particular, $\mathbf{H}$ is defined for $f \in C_c^\infty(K\backslash G/K)$, and we have the Harish-Chandra commutative diagram expressed by the formula

$$\mathbf{S} = \mathbf{MH},$$

relating the spherical, Mellin, and Harish transforms as in Chapter III, Proposition 5.1. Since both **S** and **M** have been proved to be bicontinuous isomorphisms for the HCS and Schwartz space topologies respectively in §4 and §5, it follows that **H** is also bicontinuous. Hence **H** extends to a bicontinuous linear isomorphism

$$\mathrm{HCS}(K\backslash G/K) \xrightarrow{\approx} \mathrm{Sch}(A)^W,$$

where $\mathrm{Sch}(A)^W$ may be defined as the multiplicative version under exp of the Schwartz space $\mathrm{Sch}(\mathfrak{a})^W$. However, there remains to show that this extension is expressible in terms of the above integral, which still needs to be proved absolutely convergent for $f \in \mathrm{HCS}(K\backslash G/K)$. Actually, only the continuity of the spherical transform $\mathbf{S}$ is involved here in addition to the inverse Mellin transform, because $\mathbf{H} = \mathbf{M}^{-1}\mathbf{S}$. Hence this section is independent of §5 (and of course §6).

Two methods are given in [GaV 88]. One method goes through an argument of Harish-Chandra, which describes the implication of the "important inequality" for the Gelfand–Naimark decomposition. See especially [Har 66], §43, Lemma 89. The other also has its source in Harish-Chandra [Har 66], put in a different form by Varadarajan [Var 76], §9.6, part II. See [GaV 88], Theorems 6.1.20, 6.2.3 and 6.2.4, which we now follow dealing exclusively with K-bi-invariant functions.

Theorem 8.1. *Let $\mathcal{M} = \{\mu\}$ be a set of semi positive Borel measures on G. Suppose that there is an* HCS*-continuous seminorm q such that for all $f \in C_c^\infty(K\backslash G/K)$ and all $\mu \in \mathcal{M}$ we have*

$$\left|\int_G f\,d\mu\right| \leqq q(f).$$

Then there exists a positive integer m_0 and $C > 0$ such that for all $\mu \in \mathcal{M}$,

$$\int_G (1+\sigma)^{-m_0}\varphi_0\,d\mu \leqq C,$$

and the integrals on the left converge uniformly for $\mu \in \mathcal{M}$.

The proof will be based on a separate lemma. Let:

$A_R = \{x \in G$ such that $\varphi_0(x) \geqq e^{-R}\}$;

$\gamma_R =$ characteristic function of A_R.

By Proposition 2.1 and Corollary 2.6, we know that $a^{-\rho} \leqq \varphi_0(a) \leqq 1$ for $a \in A^+$, and by Chapter I, §4, (1), $\rho(H) \gg\ll |H|$ for $H \in \mathrm{Cl}(\mathfrak{a}^+)$, so A_R is compact.

Lemma 8.2. *There is a constant c_1 and a positive integer m' such that for all $\mu \in \mathcal{M}$,*

$$\mu(A_R) \leqq c_1(1+R)^{m'}e^R.$$

Proof. Let $g \in C_c^\infty(K\backslash G/K)$ be such that

$$0 \leqq g(x) = g(x^{-1}) \quad \text{for all } x \in G \qquad \text{and} \qquad \int_G g(x)\,dx = 1.$$

By Proposition 2.7, we can select $s > 0$ such that

$$e^{-s}\varphi_0(x) \leqq \varphi_0(y_1 x y_2) \leqq e^{s}\varphi_0(x)$$

for all $x \in G$ and $y_1, y_2 \in \mathrm{supp}(g)$. Then for $R > s$ and $y_1, y_2 \in \mathrm{supp}(g)$, we have

(1) $$A_{R-s} \subset \{x \in G \mid \varphi_0(y_1 x y_2) \geqq e^{-R}\} \subset A_{R+s},$$

(2) $$g * \gamma_R = \gamma_R * g = 1 \quad \text{on } A_{R-s}.$$

Furthermore, if $f \in C_c(K\backslash G/K)$ and $\mathrm{supp}(f) \subset \mathrm{supp}(g)$, then

(3) $$\mathrm{supp}(f * \gamma_R) \subset A_{R+s}.$$

Both (2) and (3) are immediate from the definition of the convolution integral.

Now without loss of generality, we may assume that q is a finite sum

$$q = \sum_i q_{m',D_i}.$$

Then for all $\mu \in \mathcal{M}$, and $R > s$,

$$\begin{aligned} \mu(A_R) = \int_G \gamma_R \, d\mu &\leqq \int_G (g * \gamma_{R+s}) d\mu \quad \text{by (2)} \\ &\leqq \sum_i q_{m',D_i}(g * \gamma_{R+s}) \\ &= \sum_i q_{m'}((D_i g) * \gamma_{R+s}) \\ &\leqq \sup\nolimits_{A_{R+2s}} (1+\sigma)^{m'} \varphi_\circ^{-1} \cdot \sup\nolimits_G |D_i g|. \end{aligned}$$

The term $\sup_G |D_i g|$ contributes to the constant factor. On the set $A_{R+2s'}$ we have the bound $\varphi_0^{-1} \leqq e^{R+2s}$, which is $\leqq e^R$ up to a constant factor. As to the term $(1+\sigma)^{m'}$, we consider values $\sigma(a)$ with

$$a \in A_{R+2s} \cap \mathrm{Cl}(A^+),$$

and use Corollary 2.9 to conclude the proof of the lemma.

We now apply the lemma to prove Theorem 8.1. For $j = 1, 2, \ldots$ we let:

$$E_j = \{x \in G | e^{-(j+1)} \leqq \varphi_0(x) \leqq e^{-j}\}.$$

So E_j is like an annulus in G and is K-bi-invariant. Again using Corollary 2.9, there exists a constant c_2 such that for all μ and j, letting

$m_0 = m' + 2$, we get

$$\int_{E_j} (1+\sigma)^{-m_0}\varphi_0 \, d\mu \leqq c_2^{m_0}(1+j)^{-m_0}e^{-j}\mu(E_j)$$
$$\leqq c_1 c_2^{m_0}(1+j)^{-2} \quad \text{[by Lemma 8.2]}.$$

Hence there is a constant c_3 such that

$$\sum_{j=1}^{\infty} \sup_{\mu} \int_{E_j} (1+\sigma)^{-m_0}\varphi_0 \, d\mu \leqq c_3.$$

Since the sets $E_1 \cup \cdots \cup E_j$ are compact and their union is G, we get

$$\int_G (1+\sigma)^{-m_0}\varphi_0 \, d\mu \leqq c_3 \qquad \text{for all } \mu \in \mathcal{M},$$

and the convergence of the above series also shows that the integrals over G converge uniformly for $\mu \in \mathcal{M}$. This proves Theorem 8.1.

We apply Theorem 8.1 as follows. For each $a \in A$ we define the positive functional $d\mu_a$ on $C_c(G)$ by the integral

$$\int_G f \, d\mu_a = a^{\rho} \int_U f(au) \, du \qquad \text{for } f \in C_c(G).$$

Restricted to $C_c^{\infty}(K\backslash G/K)$, the right side is what we called the Harish-Chandra transform $(\mathbf{H}f)(a)$. As we remarked at the beginning of this section, it is continuous for the Schwartz topology. In particular, the hypothesis of Theorem 8.1 is satisfied. Therefore, in the present case which amounts to an example as well as a result in its own right, we obtain:

Corollary 8.3. *There exists $m_0 \geqq 0$ such that the integral*

$$a^{\rho} \int_U (1+\sigma(au))^{-m_0}\varphi_0(au) \, du$$

is convergent, uniformly for $a \in A$. Omitting the factor a^{ρ}, the convergence is still uniform in any region for which a^{ρ} is bounded away from 0.

We also obtain the estimate:

Corollary 8.4. *Given an integer $m \geqq 0$ there exists $m_1 \geqq m_0$ and a constant $c_4 > 0$ such that for all $a \in A$, we have*

$$a^{\rho} \int_U (1+\sigma(au))^{-m_1}\varphi_0(au) \, du \leqq c_4(1+\sigma(a))^{-m}.$$

Proof. For this inequality, we use the family of measures μ_a defined by

$$\int_G f\,d\mu_a = (1+\sigma(a))^m a^\rho \int_U f(au)\,du \qquad \text{for } f \in C_c(G),$$

and again apply Theorem 8.1.

Theorem 8.5. *Let $f \in \mathrm{HCS}(K\backslash G/K)$. Then the Harish-Chandra integral*

$$a^\rho \int_U f(au)\,du$$

converges absolutely, and uniformly for a in any subset such that a^ρ is bounded away from 0 (or equivalently, $\rho(\log a)$ bounded away from $-\infty$). This integral is equal to the continuous extension of the Harish-Chandra transform to $\mathrm{HCS}(K\backslash G/K)$, so we may write

$$(\mathbf{H}f)(a) = a^\rho \int_U f(au)\,du \qquad \textit{for } f \in \mathrm{HCS}(K\backslash G/K).$$

Proof. Immediate from Corollary 8.3, by definition of the Schwartz space whereby for all positive integers m,

$$|f(au)| \ll_m (1+\sigma(au))^{-m}\varphi_0(au).$$

We may now tabulate additional formulas, extended from the case of compact support to the Schwartz space.

Corollary 8.6. *Let $D \in \mathrm{IDO}(G)^K$. On $\mathrm{HCS}(K\backslash G/K)$ we have*

$$\mathbf{H} \circ D = \mathbf{h}(D) \circ \mathbf{H}.$$

Proof. This is just Chapter II, Proposition 5.1, extended to the HCS space by continuity.

CHAPTER XI

Tube Domains and the L^1 (Even L^p) HCS Spaces

The HCS space of Chapter X corresponds exactly to the classical Schwartz space on $\mathbf{i}\mathfrak{a}^\vee$ under the spherical transform. When one restricts the spherical transform to various spaces intermediate to $C_c^\infty(K\backslash G/K)$ and $\mathrm{HCS}(K\backslash G/K)$ but $\neq \mathrm{HCS}(K\backslash G/K)$, then their spherical transforms extend beyond the imaginary axis $\mathbf{i}\mathfrak{a}^\vee$ of $\mathfrak{a}_{\mathbf{C}}^\vee$, and the faster the functions on G decay, the larger is the domain of analytic continuation. The intermediate spaces can be defined by a very broad range of conditions, depending on applications. In this chapter we treat the L^p–Schwartz spaces $(0 < p < 2)$ originally considered by Trombi–Varadarajan [TrV 71], whose treatment is essentially reproduced in [GaV 88], Chapter 7. Special cases in rank 1 had occurred before, e.g. in Trombi's thesis (1970) and others, cf. the references given in [Ank 91]. Here we again follow Anker's very simple and direct proofs by continuity from his version of the Paley–Wiener case [Ank 91]. Except for one item about tube domains, the proofs are the same as in Chapter X. The L^1 case is especially important for the usual reasons that one needs L^1 analysis, and also for our purposes in particular to deal with the heat kernel as in Chapter X, §7. Indeed, the conditions **DIR 2** and **DIR 3** are L^1 conditions defining Dirac families.

There are other spaces of test functions between C_c^∞ and the standard Harish-Chandra Schwartz space besides the L^p–Schwartz spaces, for instance the spaces of exponential decay as in Delorme [Del 86] who proves the inversion theorem for them. His space is the intersection of all L^p spaces with $0 < p \leqq 2$. There are also spaces of exponential square decay, especially relevant for the heat kernel.

XI, §1. THE SCHWARTZ SPACE ON TUBES

Instead of the finite dimensional euclidean space $i\mathfrak{a}^\vee$ on which ordinary Fourier analysis is carried out, we consider some tube domains of the form

$$C^\vee + i\mathfrak{a}^\vee$$

where $C^\vee$ is a convex set. The relevant convex sets for the application will be of the form

$$\mathrm{Co}(RW\rho) = R\,\mathrm{Co}(W\rho), \qquad \text{with } R > 0,$$

where $\mathrm{Co}(W\rho)$ is the convex closure of $W\rho$ in $\mathfrak{a}^\vee$. The case $R = 0$ was treated in Chapter X, so we now are interested in $R > 0$. The case $R = 1$ is in some sense the most important, because it corresponds under the spherical transform to the L^1–Schwartz space $\mathrm{HCS}^{(1)}$, as we shall see.

We summarize some properties of the convex set $\mathrm{Co}(WR\rho)$.

$WR\rho > 0$, $-WR\rho \subset WR\rho$ and so the convex set $\mathrm{Co}(WR\rho)$ *is symmetric. Furthermore* $\mathrm{Co}(WR\rho)$ *is Minkowskian (cf. Chapter* IV, *Proposition* 8.9), *that is, stable under multiplication by* $[0, 1]$, *compact, and containing a neighborhood of* 0. *Also* $\mathrm{Co}(WR\rho)$ *is W-stable.*

See Chapter X, §1, (5), for the symmetry. We then let

$$\mathrm{Tub}_{R\rho} = \mathrm{Co}(WR\rho) + i\mathfrak{a}^\vee,$$

and call $\mathrm{Tub}_{R\rho}$ the **tube domain determined by** $R\rho$ or with **base** $\mathrm{Co}(WR\rho)$.

We repeat here Corollary 5.2 of Chapter I for the convenience of the reader.

Let $\xi \in \mathfrak{a}^\vee$. Then $\xi \in \mathrm{Co}(WR\rho)$ if and only if

$$|w\xi(H)| \leqq R\rho(G) \qquad \textit{for all } w \in W \textit{ and } H \in \mathrm{Cl}(\mathfrak{a}^+).$$

We define the **Schwartz space** $\mathrm{Sch}(\mathrm{Tub}_{R\rho})$ to be the vector space of complex valued functions h such that:

(i) h is holomorphic on the interior of $\mathrm{Tub}_{R\rho}$.

(ii) For every differential operator D on $\mathfrak{a}^\vee_{\mathbf{C}}$ with constant coefficients, i.e., $D \in \mathrm{IDO}(\mathfrak{a}^\vee_{\mathbf{C}})$, the function Dh has arbitrary large polynomial decay in the interior of $\mathrm{Tub}_{R\rho}$.

Readers concerned at once with boundary behavior are referred to Proposition 1.2. On the Schwartz space of the tube domain we can then define the seminorm $p_{m,D}^{(R\rho)}$ by

$$p_{m,d}^{(R\rho)}(h) = \sup_{\zeta \in \operatorname{Int}\operatorname{Tub}_{R\rho}} (1 + |\zeta|)^m |Dh(\zeta)|.$$

The family of these seminorms then defines what we shall call the **Schwartz $R\rho$-tube topology**, or **$R\rho$-tube Schwartz topology** (it's commutative), or **$\mathrm{Sch}^{(R\rho)}$-topology.**

For the norm, we take the euclidean norm on $\mathfrak{a}^\vee + \mathbf{i}\mathfrak{a}^\vee$, such that for $\xi, \eta \in \mathfrak{a}^\vee$,

$$|\zeta|^2 = |\xi + \mathbf{i}\lambda|^2 = |\xi|^2 + |\lambda|^2 = \langle \xi, \xi \rangle + \langle \lambda, \lambda \rangle,$$

and $\langle \xi, \xi \rangle$ is the usual scalar product on $\mathfrak{a}^\vee$, namely the trace scalar product.

The trace scalar product has a symmetric bilinear extension to $\mathfrak{a}_{\mathbf{C}}^\vee$, but of course this symmetric extension is not real, let alone positive definite. We use the notation ζ^2 for the scalar product of ζ with itself.

Proposition 1.1.

(i) *The space* $\mathrm{Sch}(\mathrm{Tub}_{R\rho})$ *is complete for the Schwartz $R\rho$-*topology, *and so is the subspace* $\mathrm{Sch}(\mathrm{Tub}_{R\rho})^W$ *of W-invariant functions.*

(ii) $\mathrm{Sch}(\mathrm{Tub}_{R\rho})$ *is an algebra under ordinary multiplication of functions, and this multiplication is continuous.*

(iii) *Let $D \in \mathrm{IDO}(\mathfrak{a}_{\mathbf{C}}^\vee)$. Then the map $h \mapsto Dh$ is continuous from* $\mathrm{Sch}(\mathrm{Tub}_{R\rho})$ *into itself.*

Proof. This is just advanced calculus, analogous to that of Chapter X, Theorem 3.1.

Proposition 1.2. *A function $h \in$* $\mathrm{Sch}(\mathrm{Tub}_{R\rho})$ *(and hence Dh for all $D \in \mathrm{IDO}(\mathfrak{a}_{\mathbf{C}}^\vee)$) is uniformly continuous on* $\operatorname{Int}\operatorname{Tub}_{R\rho}$, *and hence extends by continuity to the boundary, so to the whole tube* $\mathrm{Tub}_{R\rho}$.

Proof ([GaV 88], p. 332). A function in $\mathrm{Sch}(\mathrm{Tub}_{R\rho})$ and its derivatives Dh for $D \in \mathrm{IDO}(\mathfrak{a}_{\mathbf{C}}^\vee)$ are bounded, directly from the definition property (ii). Since $\mathrm{Tub}_{R\rho}$ is convex, the mean value theorem shows that the function is uniformly continuous, whence it extends by continuity to the boundary, thus proving the proposition.

In light of Proposition 1.2, for the definition of the semiform $p_{m,D}^{(R\rho)}$ we could have taken the sup over the whole tube $\mathrm{Tub}_{R\rho}$ instead of the interior.

Just as in the case of Chapter X, we note that the $R\rho$-tube Schwartz topology can be defined by the equivalent family of seminorms

$$p'^{(R\rho)}_{m,D}(h) = \sup_{\zeta \in \operatorname{Int} \operatorname{Tub}_{R\rho}} |D_\zeta\{(1+\zeta^2)^m h(\zeta)\}|,$$

in other words, this family is equivalent to the family of seminorms $p^{(R\rho)}_{m,D}$. This family will be used in the next section. Of course, instead of $(1+\zeta^2)^m$ we could have used $(c+\zeta^2)^m$ with any constant $c > 0$. In particular, the constant $c = \rho^2$ is the one which will be useful in the application similar to Chapter X, §4.

Proposition 1.3. *The Paley–Wiener space* $\mathrm{PW}(\mathfrak{a}^\vee_{\mathbf{C}})^W$ *restricted to the tube* $\operatorname{Tub}_{R\rho}$ *is dense in* $\operatorname{Sch}(\operatorname{Tub}_{R\rho})^W$.

Proof. We give Anker's sketch of the proof, [Ank 91], Lemma 7. Let $A_W(e^{R\rho})$ be the function on $\mathfrak{a}$ given by

$$A_W(e^{R\rho})(H) = |W|^{-1} \sum_{w \in W} e^{(Rw\rho)(H)}.$$

So it's the W-average of the $R\rho$-exponential map. Let:

$\operatorname{Sch}_{R\rho}(\mathfrak{a})$ = space of all functions $g \in C^\infty(\mathfrak{a})$ such that

$$\sup_{H \in \mathfrak{a}} (1+|H|^2)^m A_W(e^{R\rho})(H) |\mathcal{D}(P) g(H)| < \infty$$

for all positive integers m and all $P \in S(\mathfrak{a})$. Then $\operatorname{Sch}_{R\rho}(\mathfrak{a})$ has the topology defined by the corresponding seminorms, and is complete. The corresponding seminorms are given by

$$s^{(R\rho)}_{m,D}(g) = \sup_{H \in \mathfrak{a}} (1+|H|^2)^m A_W(e^{R\rho})(H) |Dg(H)|.$$

Lemma 1.4. *We have two topological W-isomorphisms:*

(i) *The map $g \mapsto A_W(e^{R\rho})g$ is such an isomorphism*

$$\operatorname{Sch}_{R\rho}(\mathfrak{a}) \xrightarrow{\approx} \operatorname{Sch}(\mathfrak{a}).$$

(ii) *The Fourier–Mellin$_{\mathfrak{a}}$ transform is such an isomorphism*

$$\mathbf{M}_{\mathfrak{a}} \colon \operatorname{Sch}_{R\rho}(\mathfrak{a}) \to \operatorname{Sch}(\operatorname{Tub}_{R\rho}).$$

Proof. The first isomorphism (i) is immediate from the definitions. The second (ii) follows by repeated application of the interchange between polynomial multiplication and differentiation under the Mellin–Fourier transform, i.e. by **M 2, M 3, M 4** of Chapter IV. We leave the details to the reader.

By (i) it follows that $C_c^\infty(\mathfrak{a})^W$ is dense in $\mathrm{Sch}_{R\rho}(\mathfrak{a})^W$, and by (ii) it follows that $\mathrm{PW}(\mathfrak{a}_\mathbf{C})^W$ is dense in $\mathrm{Sch}(\mathrm{Tub}_{R\rho})^W$, thus proving Proposition 1.3.

Example 1. Let E_t be the function on Tub_ρ defined by

$$E_t(\zeta) = e^{(\zeta^2-\rho^2)t}.$$

Writing $\zeta = \xi + \mathbf{i}\lambda$ with $\lambda,\ \xi \in \mathfrak{a}^\vee$ we see that $E_t \in \mathrm{Sch}(\mathrm{Tub}_\rho)$, so $R = p = 1$ in this case. Thus Theorem 2.3 below will apply to this function, whose inverse spherical transform is the function g_t of Chapter X, §7.

XI, §2. THE FILTRATION $\mathrm{HCS}^{(p)}(K\backslash G/K)$ WITH $0 < p \leqq 2$

For p in the above interval, we define the subspace $\mathrm{HCS}^{(p)}(K\backslash G/K)$ to consist of functions $f \in C^\infty(K\backslash G/K)$ such that for all positive integers m and all $D \in \mathrm{IDO}(G)$, the function

$$(1+\sigma)^m \varphi_0^{-2/p}|Df|$$

is bounded. We may then define the corresponding seminorm $q_{m,D}^{(p)}$ on $\mathrm{HCS}^{(p)}(K\backslash G/K)$ by

$$q_{m,D}^{(p)}(f) = \|(1+\sigma)^m \varphi_0^{-2/p} Df\|_G.$$

The expression on the right is the sup norm of the function on G as in Chapter X. The topology defined by the family of these seminorms for fixed p is called the **$\mathrm{HCS}^{(p)}$-topology.**

We note that for $p' < p$ we have the inclusion

$$\mathrm{HCS}^{(p')}(K\backslash G/K) \subset \mathrm{HCS}^{(p)}(K\backslash G/K)$$

and

$$q_{m,D}^{(p')}(f) \leqq q_{m,D}^{(p)}(f) \qquad \text{for } f \in \mathrm{HCS}^{(p')}(K\backslash G/K).$$

Thus the spaces $\mathrm{HCS}^{(p)}$ give a descending filtration of HCS. Spherical inversion duality for these spaces was considered by Trombi and Varadarajan [TrV 71]. We shall follow Anker's simple arguments just as we did for $p = 2$ in Chapter X.

Proposition 2.1.

(i) *The space* $\mathrm{HCS}^{(p)}(K\backslash G/K)$ *is complete for the* $\mathrm{HCS}^{(p)}$ *topology.*

(ii) *We have* $\mathrm{HCS}^{(p)}(K\backslash G/K) \subset L^p(K\backslash G/K)$.

Proof. The two parts correspond to Propositions 3.2 and 3.4 of Chapter X. There are essentially no changes in the proofs. The first part is the same advanced calculus, justifying the limit of a derivative being the derivative of the limit. The second part comes from the same immediate writing down of the integrand, using $(\varphi_0(x)^{-2/p})^p = \varphi_0(x)^{-2}$ to cancel the growth of the Jacobian factor, thus making the L^p-integral absolutely convergent.

Proposition 2.2. *The space* $C_c^\infty(K\backslash G/K)$ *is* $\mathrm{HCS}^{(p)}$*-dense in* $\mathrm{HCS}^{(p)}(K\backslash G/K)$.

Proof. Essentially no change in the proof of Proposition 3.5 in Chapter X. The key argument is in Lemma 3.7 of Chapter X, the seminorm $q_{m,D}$ is replaced by $q^{(p)}_{m,D}$, and $\varphi_0(x)^{-1}$ is replaced by $\varphi_0(x)^{-2/p}$. That's all.

Given $0 < p < 2$, we let

$$R = \frac{2}{p} - 1 \quad \text{so} \quad p = \frac{2}{R+1}.$$

As p goes from 2 to 0, R goes from 0 to ∞. Note that

$$R = 1 \quad \text{corresponds to} \quad p = 1.$$

Thus functions in $\mathrm{HCS}^{(1)}(K\backslash G/K)$ are in $L^1(G)$. Thus $p = 1$ is the most important value of p in the range under consideration. In light of the inclusion

$$\mathrm{HCS}^{(p)}(K\backslash G/K) \subset \mathrm{HCS}^{(2)}(K\backslash G/K) = \mathrm{HCS}(K\backslash G/K),$$

and Chapter X, the spherical transform is defined by the usual integral

$$\mathbf{S}f(\zeta) = \int_G \varphi(\zeta, x) f(x)\, dx$$

for $f \in \mathrm{HCS}^{(p)}(K\backslash G/K)$, $0 < p < 2$.

The problem is to determine a range for ζ to guarantee absolute convergence. The inverse transform is defined on the image of $\mathbf{S}$ for each value of p. We are concerned with the characterization of this image. The next theorem is the first main goal of this chapter.

Theorem 2.3. *The spherical transform integral*

$$(\mathbf{S}f)(\zeta) = \int_G \varphi(\zeta, x) f(x)\, dx$$

is absolutely convergent for $f \in \mathrm{HCS}^{(p)}(K\backslash G/K)$ and $\zeta \in \mathrm{Tub}_{R\rho}$, with $R = 2/p - 1$. It defines a topological isomorphism

$$\mathbf{S}\colon \mathrm{HCS}^{(p)}(K\backslash G/K) \xrightarrow{\approx} \mathrm{Sch}(\mathrm{Tub}_{R\rho})^W.$$

The inverse is given as before by the integral operator ${}^t\mathbf{S}^-_\eta$, namely

$$(\mathbf{S}^{-1}h)(x) = |W|^{-1} \int_{\mathfrak{a}^\vee} h(\mathbf{i}\lambda)\varphi(-\mathbf{i}\lambda, x)\eta(\lambda)d\lambda.$$

For $p = 2$, this is Harish-Chandra's theorem. For other values of p, it is Trombi–Varadarajan's extension [TrV 71].

Proof. We follow Anker's simple proof [Ank 91] as before. We start with the first statement and the continuity of the spherical transform, which is the (p, R)-analogue of Chapter X, Theorem 4.3, The proof can be repeated essentially verbatim, and actually Anker gives the proof immediately in the more general case. We have split the cases only for expository reasons, to keep separate the added feature of the tube domain. Thus we now show the continuity of $\mathbf{S}$, namely given a seminorm $p'^{(R\rho)}_{m,D}$ as in §1, its value on the image $\mathbf{S}(\mathrm{HCS}^{(p)})$ is finite, and there exists a seminorm $q^{(p)}_{s,E}$ on $\mathrm{HCS}^{(p)}$ such that for all $f \in \mathrm{HCS}^{(p)}(K\backslash G/K)$ we have

$$p'^{(R\rho)}_{m,D}(\mathbf{S}f) \leqq c_{m,D}\, q^{(p)}_{s,E}(f) \tag{1}$$

with a suitable constant $c_{m,D}$.

By Chapter VII, Theorem 1.1, we know that the eigenvalue of Casimir on the spherical function is given by

$$\mathrm{ev}(\omega, \varphi_\zeta) = \zeta^2 - \rho^2.$$

Then for $P \in S(\mathfrak{a}^\vee_{\mathbf{C}})$ and an integer $m \geqq 0$ we get (provided the integrals are absolutely convergent)

$$\begin{aligned} \mathcal{D}_\zeta(P)\{(\zeta^2 - \rho^2)^m \mathbf{S}f(\zeta)\} &= \mathcal{D}_\zeta(P) \int_G (\zeta^2 - \rho^2)^m \varphi(\zeta, x) f(x)\, dx \\ &= \mathcal{D}_\zeta(P) \int_G \omega^m \varphi_\zeta(x) f(x)\, dx \\ &= \int_G \mathcal{D}_\zeta(P)\varphi(\zeta, x)\omega^m f(x)\, dx. \end{aligned} \tag{2}$$

We estimate $\mathcal{D}_\zeta(P)\varphi(\zeta, x)$ by Lemma 4.1(i), Theorem 2.8, Proposition 2.2 of Chapter X, and Proposition 1.1. (So we require more information at this point than for the proof of Chapter X, Theorem 4.3, because of the tube.) We get for $H \in \mathrm{Cl}(\mathfrak{a}^+)$,

$$|\mathcal{D}_\zeta(P)\varphi(\zeta, \exp H| \leqq c_1(D)(1 + |H|)^{m'} e^{(R-1)\rho(H)} \tag{3}$$

with some exponent m'. Let $E = \omega^m$. From Theorem 2.8 and the definition of $f \in \mathrm{HCS}^{(p)}$, we get for $H \in \mathrm{Cl}(\mathfrak{a}^+)$,

$$|(\omega^m f)(\exp H)| \leqq c_2(m) q^{(p)}_{m'',E}(f)(1 + |H|)^{-m''} e^{-(2/p)\rho(H)} \tag{4}$$

with sufficiently large m''. The product of the two estimates in (2) and (3) is bounded by

$$c_3 q^{(p)}_{m'',E}(f)(1 + |H|)^{-m''+m'} e^{-2\rho(H)}.$$

The factor $e^{-2\rho(H)}$ cancels the Jacobian estimate $e^{2\rho(H)}$ in polar coordinates. For m'' sufficiently large, the remaining integrand is in L^1. Thus finally estimating (2) using (3) and (4) gives the desired inequality

$$p'^{(R\rho)}_{m,D}(\mathbf{S}f) \leqq c_4 q^{(p)}_{m'',E}(f). \tag{5}$$

This inequality proves that $\mathbf{S}f$ is $\mathrm{Sch}^{(R\rho)}$ bounded on the tube $\mathrm{Tub}_{R\rho}$, and shows that $\mathbf{S}$ is continuous. It also allows differentiation under the integral sign, and then by the fact that $\varphi(\zeta, x)$ is entire in ζ, we conclude by the Cauchy–Riemann equations that $\mathbf{S}f$ is holomorphic in the interior of the tube, so $\mathbf{S}f$ is in $\mathrm{Sch}(\mathrm{Tub}_{R\rho})$. This concludes the first part of the proof of Theorem 2.3.

While we are still on the transform itself, we give the usual complement about differentiation and its transpose. Cf. [GaV 88], Lemma 7.8.5.

Proposition 2.4. *For $D \in \mathrm{IDO}(G)$, $f \in \mathrm{HCS}^{(p)}(K\backslash G/K)$, and $\zeta \in \mathrm{Tub}_{R,\rho}$, we have*

$$\int_G \varphi(\zeta, x) Df(x)\, dx = \int_G {}^t D\varphi_\zeta(x) f(x)\, dx.$$

Proof. This is immediate from Proposition 2.2, the known formula when $f \in C_c^\infty(K\backslash G/K)$, and the continuity of the spherical transform which we have proved in Theorem 2.3(i). One can also use Proposition 3.8 of Chapter X.

XI, §3. THE INVERSE TRANSFORM

Next we consider the inverse map and surjectivity. First note that the restriction to $\mathbf{i}\mathfrak{a}^\vee$ gives a continuous embedding

$$\mathrm{Sch}(\mathrm{Tub}_{R\rho}) \hookrightarrow \mathrm{Sch}(\mathbf{i}\mathfrak{a}^\vee).$$

In particular, by Chapter X, §5, the inverse spherical transform $\mathbf{S}^{-1} = {}^t\mathbf{S}_\eta^-$ is defined as before by the prescribed integral operator, giving one-sided inversion

$${}^t\mathbf{S}_\eta^- \circ \mathbf{S} = \mathrm{id}\,.$$

However, we need Chapter X, Theorem 5.2, formulated more generally, as follows.

Theorem 3.1. *Given $R = 2/p - 1$, $0 < p < 2$, the map*

$${}^t\mathbf{S}_\eta^- : \mathrm{PW}(\mathfrak{a}_\mathbf{C}^\vee)^W \to C_c^\infty(K\backslash G/K)$$

is continuous for the $\mathrm{Sch}^{R\rho}$ *and* $\mathrm{Sch}^{(p)}$ *topologies respectively.*

Proof. The proof given previously for the case when $p = 2$ goes over verbatim except that one writes ζ instead of $\mathbf{i}\lambda$, and one pays attention to the needed occurrence of R and p respectively. We run through Anker's proof again. Note that

$$-R - 1 = -\frac{2}{p}$$

As before, we use the notation for the **sup norm** of a function F on a set S

$$\|F\|_S = \sup_{x \in S} |F(x)|.$$

Lemma 3.2. *Given $D \in \mathbb{IDO}(G)$ there exists a constant c_D and a positive integer d, depending on D, such that for all $f \in C_c^\infty(K\backslash G/K)$ we have*

$$\sup_{x \in G} \varphi_0(x)^{-1} |(Df)(x)| \leqq c_D p_{d,1}(\mathbf{S}f).$$

For $R > 0$ we let

$$f_{m,D}^{(R)} = (1+\sigma)^m \varphi_0^{-R-1} Df.$$

then

$$\|f_{m,D}^{(R)}\|G_j \leqq c_D(1+j)^m e^{R|\rho|j} p_{d,1}(Sf).$$

Proof. This is the R-version of Lemma 5.3 of Chapter X. The first inequality does not have R and so nothing is changed from this reference. We thus get the first inequality

$$|(Df)(x)| \leqq c_D \varphi_0(x) p_{d,1}(\mathbf{S}f).$$

We multiply both sides by $\varphi_0(x)^{-R-1}$ and use Proposition 2.1 (namely $\varphi_0^{-1} \leqq e^{\rho}$) as well as the definition of G_j to get the second inequality estimating over G_j. Indeed, C_j is defined by $(\rho/|\rho|)(H) \leqq j$, that is, $\rho(H) \leqq |\rho| j$. This concludes the proof.

With R fixed and $j = 2$, say, we can absorb the exponential term of the second inequality into the constant factor, and so we get the inequality

(1) $$\|f_{m,D}^{(R)}\|_{G_2} \leqq c_5 p_{d,1}(\mathbf{S}f),$$

just as in the case when $R = 0$. We use the auxiliary function γ_j in Lemma 5.4 of Chapter X, C^∞ decreasing from 1 on the convex set C_{j-1} to 0 outside C_j, and with derivatives bounded uniformly in j. We let $\mathbf{H}_{\mathfrak{a}}$ be the Harish transform pulled back to $\mathfrak{a}$, that is composed with the logarithm. For $f \in C_c^\infty(K\backslash G/K)$ we write

$$\mathbf{H}_{\mathfrak{a}} f = \gamma_j H_{\mathfrak{a}} f + (1-\gamma_j)\mathbf{H}_{\mathfrak{a}} f,$$

and we let $f_j \in C_c^\infty(K\backslash G/K)$ be such that

$$(1-\gamma_j)\mathbf{H}_{\mathfrak{a}} f = \mathbf{H}_{\mathfrak{a}} f_j.$$

Since $\gamma_j \mathbf{H}_{\mathfrak{a}} f$ has support in C_j, it follows from Chapter VIII, Corollary 6.3, that $\operatorname{supp}(f - f_j) \subset G_j$. By Lemma 3.2 we get the inequality

(2) $$\|f_{m,D}^{(R)}\|_{G_{j+1}-G_j} \leqq c_6 j^m e^{R|\rho|j} p_{d,1}(\mathbf{S}f_j),$$

which has an extra exponential factor. So we need an R-version of Chapter X, Lemma 5.5.

Lemma 3.3. *Given positive integers m, d there exists an* $\mathrm{Sch}_{R\rho}$-*continuous seminorm* s *and a constant* $c_{m,d}$ *such that for all* j, *and all* $f \in C_c^\infty(K\backslash G/K)$,

$$j^m e^{R|\rho|j} p_{d,1}(\mathbf{S}f_j) \leqq c_{m,d}\, s(\mathbf{H}_{\mathfrak{a}} f).$$

Proof. We start as in Lemma 5.5 of Chapter X, namely the spherical transform

$$\mathbf{S}f_j(\mathbf{i}\lambda) = \int_{\mathfrak{a}} \mathbf{H}_{\mathfrak{a}} f_j(H) e^{\mathbf{i}\lambda(H)} dH.$$

We reach the inequality

$$p_{d,1}(\mathbf{S}f_j) \leqq c_7 \sum_{k=0}^{d} \sup_{H \in \mathfrak{a}^+ - C_{j-1}} (1+H^2)^{1+r} |D_k \mathbf{H}_{\mathfrak{a}} f(H)|.$$

We use $m' = 1 + r + m$ as before, and now find the R-inequality

$$j^m e^{R|\rho|j} p_{d,1}(\mathbf{S}f_j)$$

$$\text{(3)} \qquad \leqq c_8 \sum_k \sup_{H \in \mathfrak{a}^+ - C_{j-1}} (1+H^2)^{m'} e^{R\rho(H)} |D_k \mathbf{H}_{\mathfrak{a}} f(H)|$$

$$\text{(4)} \qquad \leqq c_8 \sum_k \sup_{H \in \mathfrak{a}^+} (1+H^2)^{m'} e^{R\rho(H)} |D_k \mathbf{H}_{\mathfrak{a}} f(H)|$$

This concludes the proof of Lemma 3.3.

Lemma 3.4. *Given $D \in \mathrm{IDO}(\mathfrak{a})$ there exists a constant c'_D and a continuous seminorm $p''^{(R\rho)}$ on* $\mathrm{Sch}(\mathrm{Tub}_{R\rho})$ *such that for all $g \in C_c^\infty(\mathfrak{a})$ and $H \in \mathfrak{a}$,*

$$e^{R\rho(h)} |Dg(H)| \leqq c'_D p''^{(R\rho)}(\mathbf{M}_{\mathfrak{a}} g).$$

Proof. Let $h = \mathbf{M}_{\mathfrak{a}} g$. By Mellin–Fourier inversion, we have

$$g(H) = \int_{\mathfrak{a}} h(\mathbf{i}\lambda) e^{-\mathbf{i}\lambda(H)}\, dH.$$

We apply formulas **M 2, M 3, M 4** of Chapter IV, §7, to get for any $P \in S(\mathbf{i}\mathfrak{a}^\vee)$ and $Q \in S(\mathfrak{a})$,

$$P(H) e^{R\rho(H)} \mathcal{D}(Q) g(H) = \int_{\mathfrak{a}^\vee} \mathcal{D}(P)(Qh)(\mathbf{i}\lambda + R\rho) e^{-\mathbf{i}\lambda(H)} d\lambda.$$

Hence we get an estimate

$$\text{(5)} \quad |e^{R\rho(H)} Dg(H)| \leqq c_9 \sum_{k'} \int_{\mathfrak{a}^\vee} (1+|\lambda|)^{m''} D_{k'} h(\mathbf{i}\lambda + R\rho)\, d\lambda$$

$$\leqq c_{10} p''^{(R\rho)}(h) \quad \text{with some seminorm } p''^{(R\rho)}.$$

This concludes the proof of Lemma 3.4.

Theorem 3.1 then follows from (4), Lemma 3.3, and (2). Hence Theorem 2.3 is also proved.

XI, §4. BOUNDED SPHERICAL FUNCTIONS

Helgason–Johnson [HelJ 69] determined all bounded spherical functions as follows, giving further insight in the L^1 case.

Theorem 4.1. *A spherical function φ_ζ on G is bounded if and only if*

$$\zeta \in \mathrm{Tub}_\rho = \mathrm{Co}(W\rho) + \mathbf{i}\mathfrak{a}^\vee.$$

The proof is reproduced in [Hel 84]. Here we shall give only the argument for half the theorem, i.e. if ζ is in the tube, then φ_ζ is bounded. We write

$$\zeta = \xi + \mathbf{i}\lambda \text{ with } \xi \in \mathrm{Co}(W\rho) \text{ and } \lambda \in \mathfrak{a}^\vee.$$

By general measure theory, it suffices to prove that for every function $f \in L^1(K\backslash G/K)$, the integral is bounded:

$$\int_G |\varphi_\zeta(x) f(x)|\, dx < \infty,$$

and it also suffices to prove that

$$\int_G \varphi_\xi(x) |f(x)|\, dx < \infty,$$

because in the definition of φ_ζ, the factor $(kx)_A^{\mathbf{i}\lambda}$ has absolute value 1. Here goes.

By Fubini's theorem, we have by Chapter I, Propositions 2.1 and 2.3,

$$\begin{aligned}\int_G f(x)dx &= \int_A \int_U f(au)\, da\, du = \int_A (\mathbf{H}f)(a) a^{-\rho} da \\ &= \int_A (\mathbf{H}f)(a) a^{\rho}\, da\end{aligned}$$

where the Harish transform $(\mathbf{H}f)(a)$ is defined by the usual integral for almost all $a \in A$, because $\mathbf{H}f$ is invariant under W and there is an element $w' \in W$ such that $w'\rho = -\rho$. The integral over A is absolutely convergent, and without loss of generality, we may replace f by $|f|$, so assume $f \geqq 0$. Then

$$\int_G \varphi_\xi(x) f(x)\,dx = \int_G \int_K (kx)_A^{\xi+\rho} f(x)\,dk\,dx$$
$$= \int_G x_A^{\xi+\rho} f(x)\,dx$$

[by Fubini and the K-invariance of f]

$$= \int_A \int_U a^\xi a^\rho f(au)\,da\,du$$
$$= \int_A \mathbf{H}f(a) a^\xi\,da = \sum_{w\in W} \int_{A^+} (\mathbf{H}f)(a) a^{w\xi}\,da$$

[because $\mathbf{H}f$ is invariant under W]

$$\leqq |W| \int_{A^+} (\mathbf{H}f)(a) a^\rho\,da$$

by Chapter I, Corollary 5.2. We have already noted that this last integral is absolutely convergent, so this concludes the proof.

XI, §5. BACK TO THE HEAT KERNEL

We return to Chapter X, §7, and observe that the arguments there give some approximation result as Gangolli pointed out. We carry the argument out for $p = 1$, which is the most important case, but it goes over to arbitrary p with $0 < p < 2$ mutatis mutandis.

Theorem 5.1. *Let $f \in \mathrm{HCS}^{(1)}(K\backslash G/K)$. Let $g_t = \mathbf{S}^{-1}E_t$ as in Chapter X, §7. Then $g_t * f \to f$ uniformly on G, and the convergence is also in $L^1(G)$.*

Proof. We follow exactly the same steps as for L^2 approximation, but with a tighter structure. Both g_t and f satisfy the spherical inversion formula. By Corollary 6.7 of Chapter X, the spherical transform changes the convolution product into the ordinary product even on HCS, and $\mathrm{HCS}^{(1)} \subset \mathrm{HCS}$. Thus we obtain

$$(g_t * f)(x) - f(x) = |W|^{-1} \int_{\mathfrak{a}^\vee} (\mathbf{S}g_t(\mathbf{i}\lambda) - 1)(\mathbf{S}f(\mathbf{i}\lambda))\varphi(-\mathbf{i}\lambda, x)\eta(\lambda)\,d\lambda.$$

Estimating and using Chapter X, Propositions 2.2 and 2.6, we find

$$|(g_t * f)(x) - f(x)| \leqq \varphi_0(x) \int |\mathbf{S}g_t(\mathbf{i}\lambda) - 1||\mathbf{S}f(\mathbf{i}\lambda)|\eta(\lambda)\,d\lambda$$
$$\leqq \int |\mathbf{S}g_t(\mathbf{i}\lambda) - 1||\mathbf{S}f(\mathbf{i}\lambda)|\eta(\lambda)\,d\lambda.$$

But $\mathbf{S}g_t = E_t$ and $E_t - 1 \to 0$ boundedly as $t \to 0$, so we get the first statement that $g_t * f \to f$ uniformly on G as $t \to 0$. Furthermore

$$\|g_t * f\|_1 \leqq \|g_t\|_1 \|f\|_1 = \|f\|_1$$

by **DIR 2** (justified by Theorem 2.3). Hence by dominated convergence, we conclude that the limit $g_t * f \to f$ is a limit in L^1 also for $f \in \mathrm{HCS}^{(1)}$. This concludes the proof.

Corollary 5.2. *If $f \in L^1(G)$, then $g_t * f \to f$ in L^1.*

Proof. First approximate f by a function f_1 in $\mathrm{HCS}^{(1)}$ and then approximate f_1 by $g_t * f_1$ by using the theorem.

Theorem 5.1 is only a middle ground to proving **DIR 3**. Gangolli would prove **DIR 3** by invoking a lemma of Garding [Gar 60], see [War 72], Vol. I, pp. 279–286, Lemma 4.4.5.14; or he would invoke the theory of stochastic integrals. Whichever way, those methods require more work. In addition, Gangolli also recognizes that the proof of **DIR 1** (positivity or even semipositivity) of g_t also requires more work, involving what the trade calls the maximum principle (for parabolic equations). So his 1969 paper leaves a lot for the reader to fill out. We shall go around these problems by considering the heat kernel in the complex case and dealing with the possibility of reducing the properties to the complex case following Flensted-Jensen's method.

CHAPTER XII

$SL_n(\mathbf{C})$

The fundamental formulas for spherical functions are much simpler on "complex groups," of which $SL_n(\mathbf{C})$ is the prototype. These formulas become essentially algebraic, and information can thus be read directly from them without additional complications of convergence and analytic estimates which can become elaborate. This chapter tabulates a number of such formulas. The first section gives a complement to the real theory on A or $\mathfrak{a}$. The identity proved in §1 is crucial for describing the collapse of formulas on the complex group, for spherical functions and the heat kernel.

Since we are leaving $SL_n(\mathbf{R})$ and what precedes was specifically directed at $SL_n(\mathbf{R})$, some compromises now have to be made in giving exhaustively detailed proofs. In some instances, we invoke the possibility of essentially repeating previous proofs, with minor adjustments to take into account a pure imaginary component and some multiplicities which were equal to 1, and are now equal to 2. On the other hand, the need is beginning to appear to have a general framework for real groups, subsuming complex groups by viewing them as real groups. Harish-Chandra of course functioned in such a framework, as do [Hel 84] and [GaV 88]. For our purposes, [GaV 88] is better adjusted because as in Harish, they deal with the Casimir operator rather than the Laplacian, and thus make knowledge of differential geometry unnecessary for the strict logical development of the Lie group formulas. We shall make more comments on the respective roles of the Lie theory and differential geometry at the end of §6. We shall also expand comments on the real case and its relationship with the complex case.

In any case, we hope to have achieved our main purpose, which was to provide a straight, narrow and concrete development on $SL_n(\mathbf{R})$, both for its own sake and its own use, accessible to someone with no knowledge of Lie theory; while at the same time, such a treatment makes it easier to get into the general theory which is somewhat more formidable in its technical details, but whose main features are already apparent on SL_n.

XII, §1. A FORMULA OF EXPONENTIAL POLYNOMIALS

This section follows Cartier's Bourbaki seminar talk [Car 54/55]. We let:

$$\mathfrak{a} = \text{real vector space of real diagonal matrices with trace } 0.$$

We define a function f on $\mathfrak{a}$ to be **skew symmetric** if $wf = (\det w)f$ for all $w \in W$. We shall be concerned with skew-symmetric polynomials and exponential polynomials. We let $\alpha_1, \ldots, \alpha_r$ be the simple $(\mathfrak{a}, \mathfrak{n})$-characters, so if $H = \operatorname{diag}(h_1, \ldots, h_n)$ then $\alpha_i(H) = h_i - h_{i+1}$. We let $\mathcal{R}(\mathfrak{n}) = \{\alpha\} = \{\alpha_{ij}\}$ $(i < j)$ be the set of all $(\mathfrak{a}, \mathfrak{n})$-characters, so in our concrete set up,

$$\alpha_{ij} = \alpha_i + \cdots + \alpha_{j-1}.$$

For each $i = 1, \ldots, r$ we define the **reflections** S_i and S_α by

$$S_i\lambda = \lambda - \langle\lambda, \alpha_i\rangle\alpha_i \quad \text{and} \quad S_\alpha\lambda = \lambda - \langle\lambda, \alpha\rangle\alpha.$$

We may view S_i as a linear map on $\mathfrak{a}^\vee$, or similarly on $\mathfrak{a}$. Taking squares shows that the map is real unitary. Its effect on H is to interchange h_i, h_{i+1} so $S_i \in W$ and the set of all S_i generates W. Also $\det S_i = -1$.

Lemma 1.1. *We have $S_i(\alpha_i) = -\alpha_i$, and otherwise, S_i permutes $\mathcal{R}(\mathfrak{n}) - \{\alpha_i\}$.*

Proof. It is trivial that $S_i(\alpha_i) = -\alpha_i$ since $\langle\alpha_i, \alpha_i\rangle = 2$. The permutation statement is immediate from the property that $S_i(H)$ interchanges h_i and h_{i+1}. Readers can find a general conceptual proof in [Car 54/55].

Corollary 1.2. *The polynomial function (on $\mathfrak{a}$)*

$$\Pi_+ = \prod_{\alpha \in \mathcal{R}(\mathfrak{n})} \alpha$$

is skew symmetric.

Proof. Immediate by expressing an element $w \in W$ as product of reflections.

In the sequel, we shall be dealing more with exponential polynomials than with polynomials.

As before, we let $\{\lambda_1, \ldots, \lambda_r\}$ be the dual basis of $\{\alpha_1, \ldots, \alpha_r\}$. We let:

$\mathbf{L}_{\mathbf{Z}}$ = lattice generated over $\mathbf{Z}$ by $\alpha_1, \ldots, \alpha_r$ so $\mathbf{L}_{\mathbf{Z}} = \sum \mathbf{Z}\alpha_i$;

$\mathbf{L}_{\mathbf{Z}}^{\vee}$ = dual lattice generated over $\mathbf{Z}$ by $\lambda_1, \ldots, \lambda_r$, so $L_{\mathbf{Z}}^{\vee} = \sum \mathbf{Z}\lambda_i$;

$\mathcal{E}$ = algebra generated over $\mathbf{R}$ by the exponentials e^{λ} with $\lambda \in \mathbf{L}_{\mathbf{Z}}^{\vee}$.

We may call $\mathcal{E}$ the algebra of **$\mathbf{L}_{\mathbf{Z}}^{\vee}$-exponential polynomials**. Trivially,

$$\mathcal{E} = \mathbf{R}[e^{\lambda_1}, e^{-\lambda_1}, \ldots, e^{\lambda_r}, e^{-\lambda_r}] = \mathbf{R}[\ldots, e^{\lambda}, \ldots]_{\lambda \in \mathbf{L}_{\mathbf{Z}}^{\vee}}.$$

The functions $e^{\lambda_1}, \ldots, e^{\lambda_r}$ are algebraically independent, and $\mathcal{E}$ is the quotient ring of the polynomial algebra $\mathbf{R}[e^{\lambda_1}, \ldots, e^{\lambda_r}]$ by the multiplicative group generated by the elements $e^{\lambda_1}, \ldots, e^{\lambda_r}$. Since the polynomial algebra has unique factorization, so does $\mathcal{E}$. The irreducible elements of $\mathcal{E}$ are the same as those of the polynomial algebra, except for the "variables" $X_i = e^{\lambda_i}$ $(i = 1, \ldots, r)$ which are turned into units in $\mathcal{E}$.

The Weyl group acts on $\mathcal{E}$ by defining

$$we^{\lambda} = e^{w\lambda}.$$

We let Sk be the **skew-symmetric operator** on $\mathcal{E}$ defined on an exponential polynomial P by the formula

$$\mathrm{Sk}(P) = \sum_{w \in W} (\det w) w P.$$

Note that $\det w = \pm 1$ for any number of reasons, e.g., w has finite order. Because multiplication by a given $w \in W$ permutes the elements of W, we get trivially for each $w \in W$,

$$w\,\mathrm{Sk}(P) = \sum_{w'} (\det w') ww' P = (\det w)^{-1}\,\mathrm{Sk}(P). \tag{1}$$

Thus Sk is a linear map of $\mathcal{E}$ into the space of skew-symmetric elements. Furthermore if P is **skew symmetric**, that is, $wP = (\det w)P$ for all $w \in W$, then $\mathrm{Sk}(P) = |W|P$. Hence $|W|^{-1}\,\mathrm{Sk}$ is a projection operator on the space of skew symmetric exponential polynomials.

Theorem 1.3. *Let $P \in \mathcal{E}$ be skew symmetric. Then P is a linear combination of the elements* $\mathrm{Sk}(e^{\lambda})$ *with λ in the positive semilattice $\mathbf{L}_{+}^{\vee}$, where*

$$\mathbf{L}_{+}^{\vee} = \{m_1\lambda_1 + \cdots + m_r\lambda_r \text{ with } m_i \in \mathbf{Z},\ m_i > 0\}.$$

Let $\lambda \in \mathbf{L}_{\mathbf{Z}}^{\vee}$, $\lambda = \sum m_i \lambda_i$ *with* $m_i \in \mathbf{Z}$. *If* $m_i = 0$ *for some* i *then* $\mathrm{Sk}(e^{\lambda}) = 0$.

Proof. Let $H_1, \dots, H_r$ be the diagonal matrices representing

$$\alpha_1, \dots, \alpha_r$$

respectively. The elements λ in $\mathbf{L}_{+}^{\vee}$ are precisely those such that $\lambda(H_i) > 0$ for all $i = 1, \dots, r$. By (1), we know that for all $w \in W$,

$$\mathrm{Sk}(e^{w\lambda}) = (\det w)\,\mathrm{Sk}(e^{\lambda}).$$

Since P is linear combination of elements $\mathrm{Sk}(e^{\lambda})$ with $\lambda \in \mathbf{L}_{\mathbf{Z}}^{\vee}$ by the projection operator property, it follows that it is also a linear combination of elements $\mathrm{Sk}(e^{\lambda)})$ with $\lambda \geqq 0$, by Proposition 4.3′ of Chapter I. Thus we have proved the weaker version of Theorem 1.3, with $\lambda = \sum m_i \lambda_i$ and $m_i \in \mathbf{Z}$, $m_i \geqq 0$. Suppose finally that $m_i = 0$ for some i, that is, $\lambda(H_i) = 0$. Then $S_i \lambda = \lambda$. Viewing S_i as an element of W, having order 2, write the coset decomposition

$$W = V \cup V S_i.$$

Then

$$\mathrm{Sk}(e^{\lambda}) = \sum_{w \in V} (\det w) e^{w\lambda} + \sum_{w \in V} \det(w S_i) e^{w S_i \lambda} = 0$$

because $S_i \lambda = \lambda$ and $\det S_i = -1$. This shows that e^{λ} with λ non-regular is annihilated by the skew-symmetric projection, and concludes the proof of the theorem.

Having to deal with $SL_n(\mathbf{C})$, we adjust past notation and now let

$$\rho_0 = \frac{1}{2} \sum_{\alpha \in \mathcal{R}(\mathrm{n})} \alpha \quad \text{while} \quad \rho = 2\rho_0.$$

Thus putting $G = SL_n(\mathbf{C})$ and $G_0 = SL_n(\mathbf{R})$ the notation is adjusted so that $\rho = \rho_G$ while $\rho_0 = \rho_{G_0}$. We consider the element

$$D = \prod_{\alpha \in \mathcal{R}(\mathrm{n})} (e^{\alpha/2} - e^{-\alpha/2}) = e^{\rho_0} \prod_{\alpha \in \mathcal{R}(\mathrm{n})} (1 - e^{-\alpha}).$$

Theorem 1.4. *We have the identity*

$$D = \sum_{w \in W} (\det w) e^{w\rho_0}.$$

Proof. By Lemma 1.1, each S_i permutes the elements of $\mathcal{R}(\mathrm{n})$ other than α_i, and maps α_i to $-\alpha_i$, so $S_i D = -D = (\det S_i) D$. Since

$S_1, \ldots, S_r$ generate W, it follows that D is skew symmetric. By Theorem 1.3, we can write D in the form

$$D = c_{\rho_0} \operatorname{Sk}(e^{\rho_0}) + \sum_{\lambda \neq \rho_0} c_\lambda \operatorname{Sk}(e^\lambda), \tag{2}$$

where the sum is taken for $\lambda \neq \rho_0$ in the semilattice of positive $\mathbf{Z}$-linear combinations of $\lambda_1, \ldots, \lambda_r$. We want to show that $c_\lambda = 0$ for $\lambda \neq \rho_0$. From the factorization $D = e^{\rho_0} \prod (1 - e^{-\alpha})$ we can also write D in the form

$$D = e^{\rho_0} + \sum_\mu c'_\mu e^{\rho_0 - \mu}, \tag{3}$$

where μ is a positive $\mathbf{Z}$-linear combination of $\alpha_1, \ldots, \alpha_r$. If $c_\lambda \neq 0$ for some $\lambda \neq \rho_0$, then $\lambda - \rho_0 \geqq 0$, and $\lambda = \rho_0 - \mu$ for some μ, whence $-\mu \geqq 0$, which contradicts $\mu > 0$. Hence $D = c_{\rho_0} \operatorname{Sk}(e^{\rho_0})$, and $c_{\rho_0} = 1$ by (3), proving the theorem.

Theorem 1.5. *The element D divides all skew elements in $\mathcal{E}$.*

Proof. By Theorem 1.3 it suffices to prove that D divides $\operatorname{Sk}(e^\lambda)$ when $\lambda \in \mathbf{L}_+^\vee$, so λ is regular. Let $\alpha \in \mathcal{R}(\mathfrak{n})$. Then

$$e^\lambda - S_\alpha e^\lambda = e^\lambda - e^{\lambda - \langle \lambda, \alpha \rangle \alpha} = e^\lambda (1 - e^{-\langle \lambda, \alpha \rangle \alpha}).$$

Putting $T = e^\alpha$ or $e^{-\alpha}$ and noting that $\langle \lambda, \alpha \rangle \in \mathbf{Z}$ we conclude that $1 - e^\alpha$ or $1 - e^{-\alpha}$ divides $e^\lambda - S_\alpha e^\lambda$, and hence $1 - e^{-\alpha}$ divides $e^\lambda - S_\alpha e^\lambda$ because

$$1 - e^\alpha = -e^\alpha (1 - e^{-\alpha}).$$

Hence $1 - e^{-\alpha}$ divides $\operatorname{Sk}(e^\lambda) - S_\alpha \operatorname{Sk}(e^\lambda) = 2 \operatorname{Sk}(e^\lambda)$. Since the elements $1 - e^{-\alpha}$ are relatively prime for $\alpha \in \mathcal{R}(\mathfrak{n})$, it follows that their product divides $\operatorname{Sk}(e^\lambda)$, which concludes the proof.

We shall now differentiate the relation of Theorem 1.4. We let B_0 be the trace form, positive definite on $\mathfrak{a}$, and we let $\omega_{B_0, \mathfrak{a}}$ be the Casimir operator which is the standard ordinary calculus Laplacian $\sum (\partial / \partial x_i)^2$ with respect to the coordinates $\{x_j\}$ of an orthonormal basis of $\mathfrak{a}$. Denote $B_0 = \langle \cdot, \cdot \rangle_0$. Then we have:

Theorem 1.6. *Let $\rho_0 = \frac{1}{2} \sum_{\alpha \in \mathcal{R}(\mathfrak{n})} \alpha$. Then D is an eigenfunction of $\omega_{B_0, \mathfrak{a}}$, namely*

$$\omega_{B_0, \mathfrak{a}} D = \langle \rho_0, \rho_0 \rangle_0 D.$$

Proof. Just differentiate the right side of the identity of Theorem 1.3, and use the fact that each $w \in W$ preserves the form B_0.

Corollary 1.7. *Let* $J_0(a) = \prod_{\alpha \in \mathcal{R}(n)} (a^\alpha - a^{-\alpha})$. *Then*

$$\omega_{B_0,\mathfrak{a}}(J_0) = \langle 2\rho_0, 2\rho_0 \rangle_0 J_0 \quad \text{or also} \quad J_0^{-1}\omega_{B_0,\mathfrak{a}}(J_0) = \langle 2\rho_0, 2\rho_0 \rangle_0.$$

Proof. Apply the theorem, or differentiate directly the identity of Theorem 1.4, applied to $2H$ instead of H.

XII, §2. CHARACTERS AND JACOBIANS

One would like to put **R** or **C** as index functorially to denote objects associated with $SL_n(\mathbf{R})$ and $SL_n(\mathbf{C})$ respectively. However, there are complications because for instance the group which plays the role of K for $SL_n(\mathbf{C})$ in the Iwasawa decomposition is the unitary group $\mathbf{U}_n(\mathbf{C})$, fixed point set of the involution θ of $SL_n(\mathbf{C})$ given by

$$\theta x = {}^t\bar{x}^{-1},$$

where the bar denotes complex conjugation. On the other hand, there is another natural complex group related to K, whose Lie algebra is

$$\mathfrak{k}_{\mathbf{R}} + \mathbf{i}\mathfrak{k}_{\mathbf{R}},$$

letting $\mathfrak{k}_{\mathbf{R}}$ denote what we have called $\mathfrak{k}$ up to now, i.e. the real Lie algebra of $\mathbf{U}_n(\mathbf{R}) = O(n)$. So there is a shortage of letters in the alphabet. The Lie industry has agreed to put the subscript zero to denote objects associated with the real group, and no subscript to denote the similar objects associated with the complex group. We shall go along with this for the most part.

Thus we let:

$G_0 = SL_n(\mathbf{R})$, and $G_0 = U_0A_0K_0$ is the Iwasawa decomposition;

$G = SL_n(\mathbf{C}) = G(\mathbf{C})$, and $G = UAK$ is the Iwasawa decomposition, with $K = \mathbf{SU}_n(\mathbf{C})$.

The group $U = U(\mathbf{C})$ is the group of unipotent complex matrices, so with complex coordinates above the diagonal. If $\mathfrak{n}_0$ denotes the real Lie algebra of U_0, and $\mathfrak{n}$ the complex Lie algebra of U, then $\mathfrak{n}$ is the complexification of $\mathfrak{n}_0$, that is

$$\mathfrak{n} = \mathbf{C}\mathfrak{n}_0 = \mathfrak{n}_0 + \mathbf{i}\mathfrak{n}_0.$$

It turns out fortunately that the group A is the same as A_0, so we don't have to put the index on A_0. We have actually both

$$G_0 = U_0AK_0 \quad \text{and} \quad G = UAK = UA_0K.$$

Indeed, the proof that $G = UA_0K$ is done in the same way in the complex case as in the real case. One maps $G = \mathrm{SL}_n(\mathbf{C})$ to the hermitian positive definite matrices by

$$x \mapsto x^t\bar{x}.$$

The same orthogonalization argument as in the real case then proves the Iwasawa decomposition in the complex case, and shows that the A-components are the same, because the eigenvalues of a complex hermitian positive definite matrix are positive.

We let $\mathfrak{g}_0, \mathfrak{g}$ be the Lie algebras of G_0 and G respectively. Then

$$\mathfrak{g}_0 = \mathfrak{n}_0 + \mathfrak{a} + \mathfrak{k}_0 \quad \text{and} \quad \mathfrak{g} = \mathfrak{n} + \mathfrak{a} + \mathfrak{k}$$

where $\mathfrak{k}$ is the real vector space of skew hermitian matrices. From the direct sum decomposition $\mathfrak{n} = \mathfrak{n}_0 + i\mathfrak{n}_0$, we see that $\mathfrak{n}$ decomposes under the Lie regular action of $\mathfrak{a}$ just as in the real case, with the same $(\mathfrak{a}, \mathfrak{n})$ characters as the $(\mathfrak{a}, \mathfrak{n}_0)$ characters encountered previously, except that in the complex case these characters occur with multiplicity 2. In the standard notation, one writes

$$m_\alpha = 2$$

for the multiplicity. The set of these characters is still denoted by $\mathcal{R}(\mathfrak{n})$.

Unlike $\mathfrak{n}$ which is the complexification of $\mathfrak{n}_0$, $\mathfrak{k}$ is not the complexification of $\mathfrak{k}_0$. We have

(1) $$\mathfrak{k} = \mathfrak{k}_0 + i\mathfrak{p}_0 = \mathrm{Lie}(\mathbf{U}_n(\mathbf{C})),$$

where $\mathfrak{p}_0$ is the space (not Lie algebra) of symmetric real matrices.

On the other hand, we denote by

$$\mathbf{C}\mathfrak{k}_0 = \mathfrak{k}_0 + i\mathfrak{k}_0$$

the complexified Lie algebra, and let

(2) $$K_0^{\mathbf{C}} = \text{Lie subgroup having Lie algebra } \mathbf{C}\mathfrak{k}_0.$$

A priori, $K_0^{\mathbf{C}}$ may be bigger than $\exp(\mathbf{C}\mathfrak{k}_0)$. Similarly, we let

$$\mathbf{C}\mathfrak{a}_0 = \mathfrak{a}_0 + \mathbf{i}_{\mathfrak{a}_0} \quad \text{and} \quad A_0^{\mathbf{C}} = \exp(\mathbf{C}\mathfrak{a}_0),$$

so $A_0^{\mathbf{C}}$ is the abelian subgroup corresponding to the Lie algebra $\mathbf{C}\mathfrak{a}_0$. Thus $\exp(i\mathfrak{a}_0)$ is a torus $\mathbf{T}$, and

$$A_0^{\mathbf{C}} = A_0\mathbf{T}.$$

We let $\mathfrak{p} = i\mathfrak{k}$ so that we have the two **Cartan Lie decompositions**

(3) $$\mathfrak{g}_0 = \mathfrak{k}_0 + \mathfrak{p}_0 \quad \text{and} \quad \mathfrak{g} = \mathfrak{k} + \mathfrak{p}.$$

Note that we can rewrite $\mathfrak{g} = \mathfrak{k} + \mathfrak{p}$ in the form

$$\mathfrak{g} = \mathfrak{g}_{\text{Skh}} + \mathfrak{g}_{\text{her}}$$

where Skh means **skew hermitian**, and **her** means **hermitian**.

The above decompositions correspond to the standard **global Cartan decompositions**

$$G_0 = \mathbf{P}_0 K_0 \quad \text{and} \quad G = \mathbf{P}K$$

where $\mathbf{P}$ is the set of special positive definite hermitian matrices, and K is the special unitary group. As usual, the word **special** means having determinant 1.

We let θ be the **involution** given by the formula

$$\theta Z = -{}^t\overline{Z} \qquad \text{for } Z \in \mathfrak{g}.$$

The invariant form B. The symmetric form B_0 on $\mathfrak{g}_0$ was taken to be the trace form, i.e.

$$B_0(X, X') = \operatorname{tr}(XX') \qquad \text{for } X, X' \in \mathfrak{g}_0.$$

On $\mathfrak{g}$, we let B be the **real trace form**, that is defined by

$$B(Z, Z') = \operatorname{Re}\operatorname{tr}(ZZ') \qquad \text{for } Z, Z' \in \mathfrak{g} = \mathfrak{sl}_n(\mathbf{C}).$$

Writing $Z = X + \mathbf{i}Y$ and $Z' = X' + \mathbf{i}Y'$ with X, X', Y, $Y' \in \mathfrak{sl}_n(\mathbf{R})$ we conclude at once that

(4) $$B(Z, Z') = B_0(X, X') - B(Y, Y').$$

Directly from the definition, we see that *B is G-invariant and θ-invariant*, that is for $g \in G$

$$B(gZg^{-1}, gZ'g^{-1}) = B(Z, Z') \quad \text{and} \quad B(\theta Z, \theta Z') = B(Z, Z').$$

Furthermore:

the restriction of B to $\mathfrak{g}_0$ is equal to B_0;

B is symmetric;

B is negative definite on $\mathfrak{k}$;

B is positive definite on $\mathfrak{p}$.

Thus the situation with the symmetric Cartan form is exactly analogous to the situation in the real case.

Remark. In most of the literature the form B is taken to be the Killing form with respect to the complex Lie algebra viewed as an **R**-algebra, and is equal to 2 times the real part, so its restriction to $\mathfrak{g}_0$ is $2B_0$. We prefer our normalization, dividing by 2, which corresponds to a standard absolute normalization of objects in extensions of the base, obtained by dividing certain invariants by the degree of the extension. This degree is here equal to $[\mathbf{C} : \mathbf{R}] = 2$.

The **Weyl group** is defined for G just as for G_0, and may be identified with the Weyl group for G_0, that is, $W = W_0$ can be interpreted again as the group of permutations of the diagonal coordinates of an element of A. Since the notion of positivity was defined entirely on $\mathfrak{a}$, it does not change for the group G, and $\mathrm{Cl}(A^+)$ is again a fundamental domain for the action of W on A. We have the **polar decompositions**

$$G_0 = K_0 A K_0 \qquad \text{and} \qquad G = KAK.$$

We also have isomorphisms

$$C^\infty(K\backslash G/K) \xrightarrow{\approx} C^\infty(A)^W \xleftarrow{\approx} C^\infty(K_0\backslash G_0/K_0),$$

so the K-bi-invariant C^∞ functions on G correspond exactly to the K_0-bi-invariant C^∞ functions on G_0, and similarly for those with compact support. The proof is the same as in the real case, Chapter VI, §2.

The $\mathbf{C}^*$-valued characters on A are the same in the real and complex cases $\mathrm{SL}_n(\mathbf{R})$ and $\mathrm{SL}_n(\mathbf{C})$ because $A = A_0$. However, the Haar modular function δ on

$$P_{\mathbf{C}} = AU = P$$

satisfies

$$\delta = \delta_0^2,$$

because $\mathfrak{n}_0$ occurs twice in the direct sum decomposition of $\mathfrak{n}$. Then

$$\delta^{1/2} = \delta_0 \qquad \text{and} \qquad \rho = 2\rho_0 = \sum_{\alpha \in \mathcal{R}(\mathfrak{n})} \alpha.$$

Casimir operator

Let us denote $\mathbf{i} = \sqrt{-1}$, and

$$\mathbf{i}E_\alpha = E_\alpha^{(\mathbf{i})}$$

because in certain contexts $\mathbf{i}$ occurs essentially as an indexing symbol. Let $\omega_{G,B} = \omega_B = \omega$ be the **Casimir operator on** G. Let $\{H_j\}$ $(j = 1, \ldots, r)$ be an orthonormal basis of $\mathfrak{a} = \mathfrak{a}_0$. Then a basis for $\mathfrak{g}$ is given by

$$\{H_j\}, \; \{E_\alpha, E_{-\alpha}, E_\alpha^{(\mathbf{i})}, E_{-\alpha}^{(\mathbf{i})}\}_{\alpha \in \mathcal{R}(\mathfrak{n})}.$$

Then the dual basis is

$$\{H_j\},\ \{E_{-\alpha}, E_\alpha, -E^{(i)}_{-\alpha}, -E^{(i)}_\alpha\}_{\alpha\in\mathcal{R}(n)}.$$

Note how $-E^{(i)}_\alpha$ is the dual element of $E^{(i)}_\alpha$, with the extra minus sign. Thus

$$(5)\quad \omega = \sum_{j=1}^{r} \tilde{H}_j^2 + \sum_{\alpha\in\mathcal{R}(n)} (\tilde{E}_\alpha \tilde{E}_{-\alpha} + \tilde{E}_{-\alpha}\tilde{E}_\alpha) - \sum_{\alpha\in\mathcal{R}(n)} (\tilde{E}^{(i)}_\alpha \tilde{E}^{(i)}_{-\alpha} + \tilde{E}^{(i)}_\alpha \tilde{E}^{(i)}_\alpha).$$

The **Harish-Chandra image** of $D \in \mathrm{IDO}(G)^K$ or $\mathrm{IDO}(G/K)^G$ is defined by the same formula as in the real case, namely

$$(6)\qquad \mathbf{h}(D) = \delta^{-1/2}(\mathrm{Iw}_A) * (D) \circ \delta^{1/2}.$$

Note that δ is the delta function on the complex group, so $\delta^{1/2} = \delta_0$ or $\delta = \delta_0^2$. The formula of Chapter VII, Theorem 3.4, holds with $\langle\lambda, \lambda'\rangle = B(\lambda, \lambda')$.

Proposition 2.1. $\mathbf{h}(\omega) = \omega_{B,\mathfrak{a}} - \langle\rho, \rho\rangle = \omega_{B,\mathfrak{a}} - B(\rho, \rho)$.

Of course, the symbols $\mathbf{h}(\omega)$ have the new interpretation on G, so might be indexed by G, for instance writing $\mathbf{h}_G$ instead of $\mathbf{h}$. The Harish-Chandra conjugation on G_0 could also be written $\mathbf{h}_0$. Since $\rho = 2\rho_0$ we obtain

$$\mathbf{h}_G(\omega_{G,B}) = \omega_{B,\mathfrak{a}} - 4\langle\rho_0, \rho_0\rangle.$$

The comments made after Theorem 3.4 of Chapter VII of course apply here, concerning the choice of the form B. The formula for the direct image looks the same no matter what positive scalar multiple of B is chosen.

The proofs can be done by repeating the proofs of the real case, or by dealing right away with groups more general than SL_n, as in Harish-Chandra and [GaV 88], Lemma 2.6.10. As already mentioned in the introduction of this chapter, something is now taking over with different boundary conditions than those to which we gave priority up to now.

XII, §3. THE POLAR DIRECT IMAGE

Although we have the polar decomposition $G = KAK$, there is now a non-trivial extra factor coming in. Let $M = M_G$ be the centralizer of A in K, so M is a closed subgroup of the unitary group K. In the real case, the corresponding group was just the group M_0 consisting of ± 1 on the diagonal. Now we leave to the reader the verification that:

$$M_G = M = \text{group of diagonal matrices with determinant 1, having elements of absolute value 1 on the diagonal.}$$

In particular, M is a product of circles (real torus), whose Lie algebra is

$$\text{Lie}(M) = i\mathfrak{a}_0.$$

Then the polar decomposition gives a differential isomorphism

$$K/M \times \text{Cl}(A^+) \times K \to G,$$

where K/M is the coset space, not factor group because M is not normal in K. Of course, we can also write the polar map as

$$K/M \times A \times K \to G,$$

with the action of W as in the real case, giving a covering over the regular elements, so restricting the polar map to the regular elements

$$K/M \times A' \times K \to G'.$$

We **define the function** J corresponding to what we now write J_0 by the product

$$J(a) = J_0(a)^2 = \prod_{\alpha \in \mathcal{R}(n)} (a^\alpha - a^{-\alpha})^2 \qquad \text{for } a \in A^+.$$

Proposition 3.1. *Up to a constant factor (depending on a normalization), J is the Jacobian of the polar map*

$$\mathbf{p}: K/M \times A \times K \to G.$$

This corresponds to Chapter VI, Theorem 1.5. We can then define a Haar measure dx on G to be **polar normalized** in a manner similar to Chapter VIII, §6, namely K has measure 1 and for $f \in C_c^\infty(K\backslash G/K)$,

$$\int_G f(x)\,dx = \int_{A^+} f(a)J(a)\,da = \int_{\mathfrak{a}^+} f_{\mathfrak{a}}(H)J_{\mathfrak{a}}(H)\,dH.$$

Theorem 1.4 was really stated for ρ_0, in the real case. *With our new convention*, view the functions of Theorem 1.4 on $\mathfrak{a}$, and replace H by $2H$ for $H \in \mathfrak{a}$. We now find the key relation

$$(1) \qquad \boxed{J_0(a) = J(a)^{1/2} = \prod_{\alpha \in \mathcal{R}(n)} (a^\alpha - a^{-\alpha}) = \sum_{w \in W} (\det w) a^{w\rho}.}$$

Thus $J^{1/2}$ for the complexified group is an exponential polynomial, whose effect is to make subsequent formulas essentially algebraic identities, as we shall see.

We use the same function g_α as for the real case, so that

$$\frac{e^\alpha + e^{-\alpha}}{e^\alpha - e^{-\alpha}} = \coth \alpha = 1 + g_\alpha.$$

The analogue of Theorems 4.1 and 4.4 of Chapter VII is:

Theorem 3.2. *The polar direct image of Casimir on A^+ is given by*

$$\begin{aligned}(\mathrm{Pol}^G_{A^+})_*(\omega) &= \omega_{B,\mathfrak{a}} + \sum_{\alpha \in \mathcal{R}(\mathfrak{n})} 2(1 + g_\alpha)\tilde{H}_\alpha \\ &= \omega_{B,\mathfrak{a}} + \sum_j \tilde{H}_j(\log J)\tilde{H}_j.\end{aligned}$$

As defined in §2, $\{H_j\}$ is a B-orthonormal basis of $\mathfrak{a} = \mathfrak{a}_0$.

Note that Chapter VIII, Theorem 4.4, holds without change in its notation, but the factor 2 is built in the new meaning given to J, namely $J = J_0^2$. We then get the pay-off:

Theorem 3.3. *For $G = \mathrm{SL}_n(\mathbf{C})$,*

$$\begin{aligned}(\mathrm{Pol}^G_{A^+})_*(\omega) &= J^{1/2}\omega_{B,\mathfrak{a}} \circ J^{1/2} - J^{-1/2}\omega_{B,\mathfrak{a}}(J^{1/2}) \\ &= J^{-1/2}\omega_{B,\mathfrak{a}} \circ J^{1/2} - \langle \rho, \rho \rangle.\end{aligned}$$

Proof. The first equation comes from the same formalism which proved Chapter VII, Theorem 4.5. The second equation follows by the identity of Corollary 1.7, with $2\rho_0$ now denoted by ρ.

XII, §4. SPHERICAL FUNCTIONS AND INVERSION

The definition of spherical function and a whole bunch of formulas were given in axiomatized form in Chapter III, so they apply directly to $\mathrm{SL}_n(\mathbf{C})$. In particular, spherical functions are defined by the integral which looks the same as in Chapter III, that is

$$\varphi_{G,\zeta}(a) = \int_K (ka)_A^{\zeta+\rho}\, dk.$$

Warning. here $K = \mathbf{U}$ is the complex unitary group (compact as it should be), and

$$\rho = 2\rho_0.$$

To avoid the subscript G, and to distinguish from the spherical function on G_0, the literature sometimes uses a capital letter, writing $\Phi_\Lambda(a)$ for the spherical function on G, whenever both are considered simultaneously.

By Proposition 2.1, the Harish-Chandra images of Casimir on G resp. G_0 to $A = A_0$ look the same. Hence we have the same formulation of Chapter VIII, Theorem 1.1, on G as on G_0, namely:

Theorem 4.1. *On spherical functions $\varphi_{G,\zeta}$ with $\zeta \in \mathfrak{a}_{\mathbf{C}}^\vee$ we have*

$$\omega(\varphi_{G,\zeta}) = \mathrm{ev}(\omega, \varphi_{G,\zeta})\varphi_{G,\zeta}$$

with eigenvalue

$$\mathrm{ev}(\omega, \varphi_{G,\zeta}) = \langle\zeta, \zeta\rangle - \langle\rho, \rho\rangle = \mathrm{ev}(\mathbf{h}(\omega), \chi_\zeta).$$

The Harish-Chandra series are series on $A = A_0$ so the discussion of these series as solutions of differential equations (eigenfunctions of Casimir) is strictly unchanged. The uniqueness of the solution of equation **EIGEN 1** by a Harish-Chandra series formulated in Chapter VIII, Proposition 2.6, holds, and we can then get a much more explicit solution in the present complex case.

First, we have the theorem corresponding to Theorem 2.3 of Chapter VIII. The formulation is the same, but the J-function is now the square of the function on the real group, and so the square root which caused an infinite series to appear is killed. Thus one has:

Theorem 4.2. *Let $G = \mathrm{SL}_n(\mathbf{C})$, and $J = J_G = J_0^2$. For ζ generic, let $F_\zeta^{\mathfrak{a}} = J_{\mathfrak{a}}^{-1/2}e^\zeta$. Then $F_\zeta^{\mathfrak{a}}$ is the unique Harish-Chandra series solution of the differential equation*

$$(\mathrm{Pol}_{A^+}^G)_*(\omega)F_\zeta^\# = (\langle\zeta, \zeta\rangle - \langle\rho, \rho\rangle)F_\zeta^\#,$$

with leading coefficient $f_0(\zeta) = 1$.

For ζ generic one has Theorem 3.1 of Chapter VIII, giving the linear independence of the functions $\{F_{w\zeta}\}_{w\in W}$. Theorem 6.2 of Chapter VI can be proved in the same way so that it is valid on G, with the map $\mathbf{p}_A$ being the polar projection from G to A. Then Corollary 3.2 of Chapter VIII is valid as before, telling us as before that the functions $\{F_{w\zeta}\}_{w\in W}$ form a basis of the space of (real) analytic functions on A^+ which are eigenfunctions of $(\mathrm{Pol}_{A^+}^G)_*(\mathrm{IDO}(G)^K)$ with the same eigencharacter as φ_ζ. Then the expression for the spherical functions as Harish-Chandra series collapses to exponential polynomials as follows.

Let

$$\Pi_+^\vee(\zeta) = \prod_{\alpha\in\mathcal{R}(\mathrm{n})} \langle\alpha, \zeta\rangle.$$

We put the $\vee$ sign to indicate that we now view Π_+ as polynomial function on $\mathfrak{a}_{\mathbf{C}}^{\vee}$. We shall deal with the functions

$$\Pi_+^{\vee}(\zeta)^{-1}\sum_{w\in W}(\det w)a^{w\zeta} \qquad \text{or} \qquad \Pi_+^{\vee}(\zeta)^{-1}\sum_{w\in W}(\det w)e^{w\zeta},$$

depending on whether one considers the functions on A or on $\mathfrak{a}$ via the logarithm.

Theorem 4.3. *Let $G = \mathrm{SL}_n(\mathbf{C})$. Then for $\zeta \in \mathfrak{a}_{\mathbf{C}}^{\vee}$, $\Pi_+^{\vee}(\zeta) \neq 0$ and $a \in A$,*

$$\varphi_{G,\zeta}(a) = \frac{\Pi_+^{\vee}(\rho)\sum_{w\in W}(\det w)a^{w\zeta}}{\Pi_+^{\vee}(\zeta)\sum_{w\in W}(\det w)a^{w\rho}}.$$

*The Harish-Chandra **c**-function is given by*

$$\mathbf{c}(\zeta) = \mathbf{c}_G(\zeta) = \Pi_+^{\vee}(\rho)/\Pi_+^{\vee}(\zeta).$$

Proof. In the present case, the function $J^{1/2}F_\zeta$ is just the exponential e^ζ. Hence

$$J^{1/2}\varphi_{G,\zeta}(a) = \sum_{w\in W}\mathbf{c}(w\zeta)a^{w\zeta} \qquad \text{for } a \in A^+.$$

Both sides are analytic functions on A. We can also pass to the corresponding functions on $\mathfrak{a}$ and express this relation as one between analytic functions on $\mathfrak{a}$. The spherical function $\varphi_{G,\zeta}$ is invariant under W, while $J^{1/2} = J_0$ is skew symmetric. By the linear independence of characters with all $w\zeta$ distinct ($w \in W$), it follows that the coefficients are skew symmetric, i.e., $\mathbf{c}(w\zeta) = (\det w)\mathbf{c}(\zeta)$ for generic ζ. From §3, (1), we can the rewrite the relation in the form

$$(1) \qquad J_0^{\mathfrak{a}}\varphi_{G,\zeta}^{\mathfrak{a}}(H) = \mathbf{c}(\zeta)\sum_{w\in W}(\det w)e^{w\zeta(H)} \qquad \text{for all } H \in \mathfrak{a}.$$

We view $\mathcal{D}(\Pi_+)$ as a differential operator on $\mathfrak{a}$, and apply it to (1). Applying $\mathcal{D}(\Pi_+)$ to the right side, using the fact that Π_+ is skew symmetric (Lemma 1.2), and evaluating at $H = 0$ gives $|W|\mathbf{c}(\zeta)\Pi_+(\zeta)$. Applying $\mathcal{D}(\Pi_+)$ to the left side, we use the product decomposition of $J_0^{\mathfrak{a}}$, and the Leibniz rule to differentiate the product $J_0^{\mathfrak{a}}\varphi_{G,\zeta}^{\mathfrak{a}}$. Unless all factors of degree 1 in $\mathcal{D}(\Pi_+)$ are applied to $J_0^{\mathfrak{a}}$, when we substitute $H = 0$ the corresponding term will be 0. On the other hand,

$$(\mathcal{D}(\Pi_+)J_0^{\mathfrak{a}})(0) = |W|\Pi_+^{\vee}(\rho).$$

Hence we get

$$\Pi_+^{\vee}(\rho) = \mathbf{c}(\zeta)\Pi_+^{\vee}(\zeta),$$

which proves the formula for the $\boldsymbol{c}$-function, and also concludes the proof of the theorem.

Note. The formula for the spherical function on $SL_n(\mathbf{C})$ is originally due to Gelfand–Naimark [GeN 52]. It is the analogue of "the" character formula of Weyl for compact groups. The formula in general for complex groups is due to Harish-Chandra [Har 58a], §14, and Berezin [Ber 57], who announced the results in two Doklady notes [Ber 56a,b]. We have closely followed [Har 58a], §14.

Remark. The identity of Theorem 1.4 shows that the ρ-part of the spherical function immediately implies the Important Inequality of Harish-Chandra stated in Chapter X, Theorem 2.8, and in fact gives an asymptotic estimate. It remains to be seen whether there is such an asymptotic estimate for the general real case. Note that the numerator of the expression in Theorem 4.3, i.e. the ζ-part, is regular at 0 and can thus be evaluated at $\zeta = 0$ because of the general analogue of Chapter V, Lemma 8.2, giving the precise denominator of the $\boldsymbol{c}$-function.

The inversion theory then proceeds essentially without change in its formulation. The **spherical transform** is defined by the same formula as before,

$$\mathbf{S}f(\zeta) = \int_G \varphi_G(\zeta, x) f(x)\, dx,$$

and of course the symbols should be indexed by G throughout, namely one should write

$$\mathbf{S}_G f(\zeta) = \int_G \varphi_G(\zeta, x) f(x)\, dx,$$

for $f \in C_c^\infty(K\backslash G/K)$.

The inversion measure and $\boldsymbol{c}$-functions should be indexed by G, and so they should be denoted by $\eta_G(\lambda)\, d\lambda$ and $\boldsymbol{c}_G(\zeta)$ respectively. However, we omit the subscript G, and we have

$$\eta(\lambda) = |\boldsymbol{c}^{-1}(\mathbf{i}\lambda)|^2,$$

where the $\boldsymbol{c}$-function is determined explicitly in Theorem 4.3, so $\boldsymbol{c}^{-1}$ is the polynomial function $\Pi_+^\vee$ up to a constant factor $\Pi_+^\vee(\rho)^{-1}$.

Since $A = A_0$, $\mathfrak{a} = \mathfrak{a}_0$ the symbols referring to $\mathfrak{a}$ or A alone have the same meaning in the real or complex case. We can therefore define the notion of the measures dH and $d\lambda$ being **Fourier normalized** to have the same meaning as before. For the **polar normalization**, the condition is still written in the form

$$\int_G f(x)\, dx = \int_{\mathfrak{a}^+} f_{\mathfrak{a}}(H) J_{\mathfrak{a}}(H)\, dH \qquad \text{for } f \in C_c^\infty(K\backslash G/K),$$

The Harish-Chandra **transpose spherical transform** ${}^t\mathbf{S}_\eta F$ is defined by the same integral as before, namely

$$({}^t\mathbf{S}_\eta F)(x) = \int_{\mathfrak{a}^\vee} F(\mathbf{i}\lambda)\varphi_G(\mathbf{i}\lambda, x)\eta(\lambda)\, d\lambda.$$

Theorem 6.1 of Chapter VIII is valid with the identical formulation on $SL_n(\mathbf{C})$:

Theorem 4.4. *Assume the measures dx, dH, $d\lambda$ are polar Fourier normalized. Then*

$$\mathbf{S}^-: C_c^\infty(K\backslash G/K) \to \mathrm{PW}(\mathbf{i}\mathfrak{a}^\vee)^W$$

is a linear isomorphism, and also an L^2-isometry, extending to an isometry

$$\mathbf{S}^-: L^2(K\backslash G/K, dx) \to L^2(\mathbf{i}\mathfrak{a}^\vee, \eta(\lambda)d\lambda)^W.$$

The inverse is given by the Harish-Chandra transpose $|W|^{-1}{}^t\mathbf{S}_\eta$.

One can carry out the proofs explicitly on $SL_n(\mathbf{C})$ just as well as on $SL_n(\mathbf{R})$, or one can follow the arguments in the general case following Rosenberg and Anker's improvements on preceding work.

Finally Chapters X and XI extending the inversion to the Harish-Chandra Schwartz spaces go through. The statements are the same, with the meaning of the symbols being relative to whichever group G is under consideration, for instance $G = SL_n(\mathbf{C})$, as exemplified above. In case of need, we refer to the theorems of those two chapters directly without restating them. We give an example in the next section, bringing to the fore some special collapsing features of the heat kernel in the complex case.

XII, §5. THE HEAT KERNEL

The collapse of the spherical function to an exponential polynomial causes the heat kernel to collapse also, to a form which is entirely similar to the heat kernel on the real line or in ordinary euclidean space. This was first shown in Gangolli's pioneer work [Gan 68], which we again follow in this section.

We let $G = SL_n(\mathbf{C})$, and all the usual invariants are those associated with G, for instance the spherical function $\varphi = \varphi_G$, the Jacobian $J = J_0^2$, the $\boldsymbol{c}$-function $\boldsymbol{c} = \boldsymbol{c}_G$ such that $\boldsymbol{c}^{-1}$ is a polynomial, etc. The scalar product B on $\mathfrak{g} = \mathrm{Lie}(G)$ is normalized as in the previous sections, so its restriction to $\mathfrak{g}_0$ is the chosen form on $\mathfrak{g}_0$, in our case the trace form. We write

$$\zeta^2 = \langle \zeta, \zeta \rangle = B(\zeta, \zeta)$$

as before. As in Chapter X, §7, we let

$$E_t(\zeta) = e^{(\zeta^2-\rho^2)t} \quad \text{for } \zeta \in \mathfrak{a}_{\mathbf{C}}^\vee \text{ and } t > 0.$$

We also let the G-**gaussian function** be the inverse spherical transform

$$(1) \qquad g_t(x) = \mathbf{S}^{-1}E_t(x) = |W|^{-1}\int_{\mathfrak{a}^\vee} E_t(\mathrm{i}\lambda)\varphi(\mathrm{i}\lambda, x)|c^{-1}(\mathrm{i}\lambda)|^2 d\lambda.$$

The Haar measures dH, $d\lambda$, dx are assumed polar Fourier normalized so that inversion holds without any extra constant factor.

Just as in the real case of Chapter X, Theorem 7.1, by using Theorem 4.1 we see that *$g(t, x)$ satisfies the heat equation*, that is

$$(-\omega_x + \partial_t)g(t, x) = 0.$$

The next theorem gives a formula from which it is immediate that the function $\mathbf{K}_t(x, y) = g_t(y^{-1}x)$ is the heat kernel. The number $\pi = 3.14\ldots$ will occur, as will the product

$$\Pi_+ = \prod_{\alpha\in\mathcal{R}(\mathfrak{n})} \alpha,$$

which we capitalize to make the distinction. Note that this product can be viewed as a function on $\mathfrak{a}^\vee$ and also on $\mathfrak{a}$, namely for $\lambda \in \mathfrak{a}^\vee$ and $H \in \mathfrak{a}$,

$$\Pi_+^\vee(\lambda) = \prod_{\alpha\in\mathcal{R}(\mathfrak{n})} \langle\alpha, \lambda\rangle \quad \text{and} \quad \Pi_+(H) = \prod_{\alpha\in\mathcal{R}(\mathfrak{n})} \alpha(H).$$

Theorem 5.1. *Let $N = \dim G/K = r + 2|\mathcal{R}(\mathfrak{n})|$, with*

$$r = \dim \mathfrak{a} = \dim \mathfrak{a}^\vee.$$

Thus $N = n^2 - 1$ for $\mathrm{SL}_n(\mathbf{C})$. *Then for $a \in A^+$,*

$$g_t(a) = \frac{1}{(4\pi t)^{N/2}} e^{-|\log a|^2/4t} e^{-\rho^2 t} F(a),$$

where F is the function defined by

$$F(a) = \prod_{\alpha\in\mathcal{R}(\mathfrak{n})} F_\alpha(a) \quad \textit{and} \quad F_\alpha(a) = C_\alpha \frac{\alpha(\log a)}{\sinh(\alpha(\log a))},$$

with the constant

$$C_\alpha = \frac{\pi}{\langle\alpha, \rho\rangle} \quad \textit{and} \quad \rho = \frac{1}{2}\sum_{\alpha\in\mathcal{R}(\mathfrak{n})} m_\alpha \alpha.$$

Before proving the theorem, we make some comments on the shape of the expression. Indeed, g_t is a perturbation of the euclidean formula for the heat kernel. In euclidian space, $\rho = 0$, and the function F is the constant 1. On G, the function F depends only on the A-variable. By K-bi-invariance, we could without loss of generality give the value only on elements $a \in A^+$.

Note further that $|\log a|$ is invariant under W, and so is the factor $F(a)$, as one verifies first for reflections, then for arbitrary elements of W which are products of reflections. Thus the expression given for $g_t(a)$ is determined by its values on A^+.

Proof. We substitute the values found in Theorem 4.3 for the spherical function and the $\mathbf{c}$-function, in formula (1), defining g_t. We note that for $\zeta = \mathbf{i}\lambda$ and λ real, we have

$$\mathbf{c}(-\mathbf{i}\lambda) = \overline{\mathbf{c}(\mathbf{i}\lambda)},$$

which is trivially verifiable in the expression of $\mathbf{c}$ as a polynomial. Then we find for $a \in A^+$:

$$\begin{aligned}(2)\quad J_0(a)g_t(a) &= |W|^{-1}\sum_{w\in W}\int_{\mathfrak{a}^\vee} e^{-(\lambda^2+\rho^2)t}a^{w\mathbf{i}\lambda}(\det w)\frac{\Pi_+^\vee(-\mathbf{i}\lambda)}{\Pi_+^\vee(\rho)}\,d\lambda \\ &= |W|^{-1}\sum_{w\in W}\int_{\mathfrak{a}^\vee} e^{-(\lambda^2+\rho^2)t}a^{\mathbf{i}w\lambda}\mathbf{c}(-\mathbf{i}w\lambda)^{-1}\,d\lambda \\ &= \frac{e^{-\rho^2 t}}{\Pi_+^\vee(\rho)}\int_{\mathfrak{a}^\vee} e^{-\lambda^2 t}a^{-\mathbf{i}\lambda}\Pi_+^\vee(\mathbf{i}\lambda)\,d\lambda,\end{aligned}$$

after a change of variables $\lambda \mapsto w^{-1}\lambda$ and $\lambda \mapsto -\lambda$. We are therefore reduced to the computation of the integral remaining in (2), which is a purely euclidean integral and is just calculus in several variables.

By definition,

$$(3)\qquad \Pi_+^\vee(\mathbf{i}\lambda) = \prod_{\alpha\in\mathcal{R}(n)} \mathbf{i}\lambda(H_\alpha)$$

where H_α represents α with respect to the scalar product. On the other hand, by the standard evaluation of calculus, putting $H = \log a$, we have the value

$$(4)\qquad \int_{\mathfrak{a}^\vee} e^{(\mathbf{i}\lambda)^2 t - \mathbf{i}\lambda(H)}\,d\lambda = (4\pi t)^{-r/2}e^{-H^2/4t},$$

where $r = \dim \mathfrak{a}^\vee$ and $H^2 = B(H, H) = \langle H, H\rangle$ is our euclidian scalar product. So the computation amounts to apply the usual rule for commuting polynomials and differential operators, which we tabulated in Chapter IV, §7, **M 3**. We get:

$$(5)\qquad \int_{\mathfrak{a}^\vee} e^{(i\lambda)^2 t - i\lambda(H)} \Pi_+^\vee(i\lambda)\, d\lambda = (4\pi t)^{-r/2} \prod_{\alpha\in\mathcal{R}(n)} (-D_\alpha) e^{-H^2/4t},$$

where $D_\alpha = \mathcal{D}(H_\alpha)$ is the differential operator associated with the vector H_α. For $\lambda \in \mathfrak{a}^\vee$ we have

$$D_\lambda e^{-H^2/4t} = -\frac{1}{2t}\lambda(H) e^{-H^2/4t}.$$

Hence

$$(6)\qquad \prod_{\alpha\in\mathcal{R}(n)} (-D_\alpha) e^{-H^2/4t} = (2t)^{-|\mathcal{R}(n)|} \prod_{\alpha\in\mathcal{R}(n)} \alpha(H) e^{-H^2/4t}.$$

Putting together (2), (5) and (6) yields the formula of Theorem 5.1, and concludes the proof.

In Chapter X, §7, we gave the three conditions **DIR 1**, **DIR 2**, **DIR 3** defining a Dirac family, and stated that the heat kernel satisfies these conditions. We proved **DIR 2** (total integral 1) for $\mathrm{SL}_n(\mathbf{R})$, and it follows in a similar manner for $\mathrm{SL}_n(\mathbf{C})$, using Theorem 2.3 of Chapter XI. We did not prove **DIR 2** nor **DIR 3**. We are now in a position to do so in the complex case.

Corollary 5.2. *The family of functions $\{g_t\}$ satisfies the conditions* **DIR 1, DIR 2, DIR 3** *in the complex case.*

Proof. For **DIR 1**, i.e. the positivity, we may limit ourselves to $a \in A^+$, in which case the condition $g_t > 0$ is immediate since each term in the product expression for g_t is positive. The condition **DIR 2** follows as in the real case. For **DIR 3**, we have to prove that given $c > 0$,

$$\lim_{t\to 0} \int_{\|x\|\geqq c} g_t(x)\, dx = 0.$$

The proof is essentially at the same calculus level as the usual proof for the heat kernel in euclidean space. Using polar coordinates, and without loss of generality, restricting ourselves to $a \in A^+$ so $H = \log a \in \mathfrak{a}^+$, we have the estimate

$$\int_{\|a\|\geqq c} g_t(x)\, dx \ll \int_{\substack{|H|\leqq c\\ H\in\mathfrak{a}^+}} g_t(\exp H) J_{\mathfrak{a}}(H)\, dH$$

$$\ll \int_{|H|\geqq c} \frac{1}{(4\pi t)^{N/2}} e^{-|H|^2/4t} e^{-\rho^2 t} \prod_\alpha \frac{\alpha(H)}{\sinh \alpha(H)} e^{2\rho(H)}\, dH.$$

For $|H| \geqq c$ the dominant term is the term with square exponential decay, which tends to 0 square exponentially. Since $\rho(H) \ll |H|$, the only possible competition is the term tending to infinity exponentially, but the square exponential term wins and the whole expression tends to 0 as $t \to 0$ thus proving the corollary.

Helgason has suggested to us that the heat kernel may be constructed by the same method he uses for the wave kernel in [Hel 94], Chapter V. He pulls back $J_{\mathfrak{p}}^{1/2} g_t$ to $\mathfrak{p}$, where the heat equation on G/K becomes the usual euclidean heat equation (translated by ρ^2) on $\mathfrak{p}$. Here $J_{\mathfrak{p}}$ is the Jacobian of the exponential map from $\mathfrak{p}$ to $\mathbf{P} \approx G/K$, namely

$$J_{\mathfrak{p}}(X) = \det\left(\left(\frac{\sinh[X]}{[X]}\right)_{\mathfrak{p}}\right), \qquad X \in \mathfrak{p},$$

where $[X]$ is the regular representation of X in the Lie algebra $\mathfrak{g}$, and the index $\mathfrak{p}$ signifies restriction to $\mathfrak{p}$. Cf. Mostow [Mos 53], Lemma 1, and [Hel 84], p. 273. In this presentation, Helgason then uses the Laplacian rather than the Casimir operator, and introduces more differential geometry in the background. Note that $J_{\mathfrak{p}}^{-1/2}$ restricted to A is the factor $F(a)$ that occurs in Gangolli's formula for the heat kernel in the complex case.

Estimates. Very good estimates for the heat kernel (and in some sense optimal) have been given by Anker and Ji [AnkJ 98] in general, following estimates on $SL_n(\mathbf{R})$ and $Pos_n(\mathbf{R})$ by Sawyer [Saw 92], [Saw 97]. Sawyer uses a new transform. Anker and Ji use the Harish-Chandra series to the hilt. The next section suggests another method, and we shall continue the present comments at the end.

XII, §6. THE FLENSTED-JENSEN DECOMPOSITION AND REDUCTION

Let

$$G = SL_n(\mathbf{C}) \qquad \text{and} \qquad G_0 = SL_n(\mathbf{R}).$$

Flensted-Jensen [FlJ 78] found a remarquable correspondence between the spherical inversion on those two groups, reducing the analysis on G_0 to that of G. He did it for a large class of Lie groups, but in 1978 was unable to do it for the usual class ("let G be a semisimple Lie group ..."). Recently, he has written that using some results of Delorme, he could do it now in general, but there is still no exposition of the general case in the literature. The 1978 class of groups of course includes $\mathrm{SL}_n(\mathbf{C})$ and $\mathrm{SL}_n(\mathbf{R})$. We shall state the results without proof. At the end, we make some more general comments.

Let $K_0^{\mathbf{C}}$ be the Lie subgroup of G with Lie algebra $\mathfrak{k}_0 + i\mathfrak{k}_0$. Then $K_0^{\mathbf{C}}$ is what some people call the **complex orthogonal group**, i.e. the group of linear automorphisms of the symmetric form on $\mathbf{C}^n$ given by the dot product of $\mathbf{C}^n$.

The basic item is a decomposition mixing features of a polar decomposition and Iwasawa decomposition.

Theorem 6.1 (Flensted-Jensen decomposition). *We have*

$$G = K_0^{\mathbf{C}}\,\mathrm{Cl}(A^+)K,$$

so in particular, $G = K_0^{\mathbf{C}}AK$, and $\mathrm{Cl}(A^+)$ is a fundamental domain for this decomposition.

Flensted-Jensen proves this theorem by using a differential geometric approach of Mostow [Mos 53]. The result should follow directly from the geometry of the linear action on $\mathbf{C}^n$. Using the Bernstein criterion as in the Appendix of Chapter VI, one obtains as corollary:

Theorem 6.2. *The restriction gives a linear isomorphism*

$$\mathrm{res}\colon C_c^\infty(K_0^{\mathbf{C}}\backslash G/K) \to C^\infty(A)^W,$$

and similarly with C_c^∞ and $C^0(=C)$.

Corollary 6.3. *There are unique isomorphisms $\mathrm{Pol}_{G_0}^G$ and $\mathrm{Pol}_G^{G_0}$ making the following diagram commutative:*

$$\begin{array}{ccc} C^\infty(K_0^{\mathbf{C}}\backslash G/K) & \overset{\mathrm{Pol}_{G_0}^G}{\underset{\mathrm{Pol}_G^{G_0}}{\rightleftarrows}} & C^\infty(K_0\backslash G_0/K_0) \\ & {}_{\mathrm{res}(6.2)}\searrow \quad \swarrow_{\mathrm{res}} & \\ & C^\infty(A)^W & \end{array}$$

and similarly with C^∞ replaced by C_c^∞ and C. The map on the left is the restriction isomorphism of Theorem 6.2, and the map on the right is the restriction to $A = A_0$.

The proof is immediate from Theorems 6.1, 6.2 and the polar decomposition of G_0, combined with the Bernstein theorem.

Note that

$$\mathrm{Pol}_{G_0}^G\colon C^\infty(K_0^{\mathbf{C}}\backslash G/K) \to C^\infty(K_0\backslash G_0/K_0)$$

is induced by the restriction of functions from G to G_0, while the other map is induced by the injection $G_0 \hookrightarrow G$, together with the appropriate bi-invariance conditions.

Next we describe the Flensted-Jensen transform relating the spherical transforms. We let $\mathrm{Av}_{K_0^{\mathbf{C}}}$ be the averaging integral over $K_0^{\mathbf{C}}$, so by definition,

$$\mathrm{Av}_{K_0^{\mathbf{C}}} \colon F(x) \mapsto \int_{K_0^{\mathbf{C}}} F(hx)\,dh,$$

for $F \in C_c(K\backslash G/K)$, so the integral is absolutely convergent. Of course, the integral extends to wider spaces of functions, to be discussed later. This is a matter of measure theory, and various continuity properties of the averaging map. Since $K_0^{\mathbf{C}}$ is not compact, the matter is not as trivial as when we averaged over a compact group (K or K_0). In any case, we get a linear map

$$\mathrm{Av}_{K_0^{\mathbf{C}}} \colon C_c^\infty(K\backslash G/K) \to C_c^\infty(K_0^{\mathbf{C}}\backslash G/K).$$

We may compose this map with $\mathrm{Pol}_{G_0}^G$, and define the **Flensted-Jensen transform** to be the composite,

$$\mathfrak{F}_{G_0}^G = \mathrm{Pol}_{G_0}^G \circ \mathrm{Av}_{K_0^{\mathbf{C}}}.$$

We write this composite in a full diagram:

$$C_c^\infty(K\backslash G/K) \xrightarrow{\mathrm{Av}_{K_0^{\mathbf{C}}}} C_c^\infty(K_0^{\mathbf{C}}\backslash G/K) \xrightarrow{\mathrm{Pol}_{G_0}^G} C_c^\infty(K_0\backslash G_0/K_0)$$

$$\mathfrak{F}_{G_0}^G$$

We also have the Flensted-Jensen transform going the other way, namely

$$C_c^\infty(K_0\backslash G_0/K_0) \xrightarrow{\mathrm{Pol}_G^{G_0}} C_c^\infty(K_0^{\mathbf{C}}\backslash G/K) \xrightarrow{\mathrm{Av}_K} C_c^\infty(K\backslash G/K)$$

$$\mathfrak{F}_{G_0}^G$$

so we write also

$$\mathfrak{F}_G^{G_0} = \mathrm{Av}_K \circ \mathrm{Pol}_G^{G_0}$$

on the spaces $C_c^\infty(K_0\backslash G_0/K_0)$ or $\mathrm{HCS}^{(1)}(K_0\backslash G_0/K_0)$. This relates the spherical functions.

Theorem 6.4. *Let φ_{G_0} and φ_G be the Harish-Chandra spherical functions on G_0 and G respectively. Let $\zeta \in \mathfrak{a}_{\mathbf{C}}^{\vee}$. Then*

$$\mathfrak{F}_G^{G_0}\varphi_{G_0,\zeta} = \varphi_{G,\zeta}.$$

Then the spherical transforms are similarly related as follows.

Theorem 6.5. *The spherical transforms on G_0 and G are related by the formula*

$$\mathbf{S}_{G_0} \circ \mathfrak{F}_{G_0}^{G} = \mathbf{S}_G$$

as linear maps on $C_c^\infty(K\backslash G/K)$ and also on $\mathrm{HCS}^{(1)}(K\backslash G/K)$.

Let $g_{G,t}$ and $g_{G_0,t}$ be the Gauss functions on G and G_0 respectively, i.e. the inverse spherical transforms of the Gauss function on $\mathfrak{a}_{\mathbf{C}}^{\vee}$, essentially the heat kernels as defined by Gangolli, so in the notation of §5,

$$\mathbf{S}_G g_{G,t} = E_t = \mathbf{S}_{G_0} g_{G_0,t}.$$

We get what amounts to a corollary of Theorem 6.4.

Theorem 6.6. *We have*

$$g_{G_0,t} = \mathfrak{F}_{G_0}^{G} g_{G,t}.$$

Proof. Immediate from Theorem 6.4.

Thus we see that the heat kernels on G and G_0 are related by the Flensted-Jensen transform. Several questions then arise concerning alternative developments of inversion theory and the heat kernel.

1. The Harish-Chandra series and/or the Flensted-Jensen decomposition. Overall, the Harish-Chandra spherical inversion as described in the present book is based on the Harish-Chandra series for spherical functions. An alternative approach would be to develop first the Flensted-Jensen decomposition and transform, get the simple formulas for the complex case, and then get the real case as a corollary. Monographs on specific groups such as SL_n as well as in the general case are needed.

2. The use of $\mathfrak{p}$. The Cartan decomposition $\mathfrak{g} = \mathfrak{k} + \mathfrak{p}$ has played a minor role in the development of the present book. The main use is in Lemma 1.2 of Chapter X, which is the only place we use the differential of the exponential map on $\mathfrak{p}$. Flensted-Jensen's proofs use $\mathfrak{p}$ repeatedly, especially combined with Mostow's differential geometric techniques with Lie triple systems [Mos 53]. Similarly, Helgason's proposed approach to the heat kernel is based on the pull back to $\mathfrak{p}$. One question is to what extent the greater use of $\mathfrak{p}$ via the Flensted-Jensen method would replace the pervasive use of the Harish-Chandra series, while yielding the same or better results. On the other hand, as of today, this series is still being used to prove that the series for the spherical function in the complex case collapses to an exponential polynomial. An analysis of the whole situation and the role of the Cartan decomposition remains to be worked out. What alternative approaches are there to Theorem 6.1, Theorem 6.4, and the inversion relation of Theorem 6.5? Is there a substantial simplification due to the special features of SL_n as distinguished from the most general case, so on SL_n one gets away with the "natural" representation on $\mathbf{C}^n$ or $\mathbf{R}^n$?

3. Differential geometry. Related to the above is the use of differential geometry for the proofs of basic theorems concerning inversion or the heat kernel. There is an obvious expository advantage in giving a treatment which does not use differential geometry, because the material can be made immediately accessible to many more people. On the other hand, there is no question that differential geometry ultimately plays its own central role. For one application which we had in mind, i.e. using the heat kernel to trigger our general program for constructing zeta functions, we are at first interested in minimizing complications which come from other branches of mathematics, simply to test how the heat kernel even in the simplest case of SL_n reacts with the methods of Dirichlet series, Bessel series and spectral theory. Cf. the Overview to the book. The present chapter makes it clear that $SL_n(\mathbf{R})$ was valuable as an introduction, but is already displaced by $SL_n(\mathbf{C})$ for certain applications. We see no end in sight for the need to have different expositions from different points of view for different purposes.

Bibliography

[Ank 87] J.-P. ANKER, La forme exacte de l'estimation fondamentale de Harish-Chandra, *C. R. Acad. Sci. Paris Sér. I* **305** (1987) pp. 371–374

[Ank 91] J.-P. ANKER, The spherical transform of rapidly decreasing functions. A simple proof of a characterization due to Harish-Chandra, Helgason, Trombi, and Varadarajan, *J. Funct. Anal.* **96** (1991) pp. 331–349

[AnkJ 98] J.-P. ANKER and L. JI, *Heat kernel and Green function estimates on noncompact symmetric spaces* I, II, Preprint, Institut Elie Cartan, Université Henri Poincaré Nancy, France, 1998

[Ben 83] T. BENGSTON, Bessel functions on $\mathcal{P}_n$, *Pacific J. Math.* **108** (1983) pp. 19–30

[Ber 56a] F. BEREZIN, Laplace operator on semisimple Lie groups, *Dokl. Akad. Nauk SSSR* **107** (1956) pp. 9–12

[Ber 56b] F. BEREZIN, Representation of complex semisimple Lie groups in Banach spaces, *Dokl. Akad. Nauk* **110** (1956) pp. 897–900

[Ber 57] F. BEREZIN, Laplace operator on semisimple Lie groups, *Trudy Moscov. Mah. Obshsc.* **6** (1957) pp. 371–463 (English translation, *Amer. Math. Soc. Transl.* **21** (1962) pp. 239–339)

[Bers 64] BERS ET AL., *PDE*, Proceedings of the Summer Seminar, Boulder, Colorado, Lectures by Bers and Schechter, Interscience, 1964, pp. 207–210

[Bha 60a] T. S. BHANU-MURTY, Plancherel's measure for the factor space $SL(n, \mathbf{R})/SO(n, \mathbf{R})$, *Soviet Math. Dokl.* **1** (1960) pp. 860–862

[Bha 60b] T. S. BHANU-MURTY, The asymptotic behavior of zonal spherical functions on the Siegel upper half plane, *Dokl. Akad. Nauk SSSR* **135** (1960) pp. 1027–1029

[Bor 97] A. BOREL, *Automorphic Forms on* $SL_2(\mathbf{R})$, Cambridge University Press, 1997

[Car 54/55] P. CARTIER, *Théorie des Caractères II*, Sém. Sophus Lie 1954–55, Exposé 19

[Cha 84] I. CHAVEL, *Eigenvalues in Riemannian Geometry*, Academic Press, 1984

[Che 55] C. CHEVALLEY, Invariants of finite groups generated by reflections, *Amer. J. Math.* **77** (1955) pp. 778–782

[Dad 82] J. DADOK, On the C^∞ Chevalley theorem, *Adv. in Math.* **44** (1982) pp. 121–131

[Del 86] P. DELORME, Formules limites et formules asymptotiques pour les multiplicités dans $L^2(\Gamma/G)$, *Duke Math. J.* **53** (1986) pp. 691–731

[DelF 91] P. DELORME and M. FLENSTED-JENSEN, Towards a Paley–Wiener theorem for semisimple symmetric spaces, *Acta Math.* **167** (1991) pp. 127–151

[EhM 55] L. EHRENPREIS and F. MAUNTNER, Some properties of Fourier transform on semisimple Lie groups, *Ann. of Math.* **61** (1955) pp. 406–554; II, III *Trans. Amer. Math. Soc.* **84** (1957) pp. 1–55

[FlJ 72] M. FLENSTED-JENSEN, Paley–Wiener type theorems for a differential operator connected with symmetric spaces, *Arch. Math.* **10** No. 1 (1972) pp. 143–162

[FlJ 78] M. FLENSTED-JENSEN, Spherical functions on semisimple Lie groups: A method of reduction to the complex case, *J. Funct. Anal.* **30** (1978) pp. 106–146

[Gan 68] R. GANGOLLI, Asymptotic behavior of spectra of compact quotients of certain symmetric spaces, *Acta Math.* **121** (1968) pp. 151–192

[Gan 71] R. GANGOLLI, On the Plancherel formula and the Paley–Wiener theorem for spherical functions on semisimple Lie groups, *Ann. of Math.* (2) **93** (1971) pp. 150–165

[GaV 88] R. GANGOLLI and V. VARADARAJAN, *Harmonic Analysis of Spherical Functions on Real Reductive Groups*, Ergebnisse der Math., Springer-Verlag, 1988

[Gar 60] L. GARDING, Vecteurs analytiques dans les représentations des groupes de Lie, *Bull. Soc. Math. France* **88** (1960) pp. 73–93

[Gel 50] I. M. GELFAND, Spherical functions on symmetric spaces, *Dokl. Akad. Nauk USSR* **70** (1950) pp. 5–8; *Amer. Math. Soc. Transl.* **37** (1964) pp. 39–44

[GeN 50] I. M. GELFAND and M. A. NAIMARK, Unitary representations of the classical groups, *Trudy Mat. Inst. Steklov* **36** (1950); German translation, Akademie-Verlag, 1957

[GeN 52] I. M. GELFAND and M. A. NAIMARK, Unitary representations of the unimodular group containing the identity representation of the unitary subgroup, *Trudy Moscow Mat. Obsc.* **1** (1952) pp. 423–475

[Gin 64] S. GINDIKIN, Analysis in homogeneous domains, *Russian Math. Surveys* **19** (1964) pp. 1–90

[GiK 62] S. GINDIKIN and F. I. KARPELEVIC, Plancherel measure for Riemannian symmetric spaces of non-positive curvature, *Dokl. Akad. Nauk SSSR* **145** (1962) pp. 252–255

[Gla 63] G. GLAESER, Fonctions composées différentiables, *Ann. of Math.* **77** (1963) pp. 193–209

[God 52a] R. GODEMENT, Une généralisation du théorème de la moyenne pour les fonctions harmoniques, *C. R. Acad. Sci. Paris Sér. I* **234** (1952) pp. 2137–2139

[God 52b] R. GODEMENT, A theory of spherical functions I, *Trans. Amer. Math. Soc.* **73** (1952) pp. 496–556

[God 57] R. GODEMENT, Introduction aux travaux de Selberg, *Séminaire Bourbaki*, 1957

[Har 49] HARISH-CHANDRA, On representations of Lie algebras, *Ann. of Math.* **250** (1949) pp. 900–915

[Har 53] HARISH-CHANDRA, Representations of semisimple Lie groups on Banach spaces I, *Trans. Amer. Math. Soc.* **38** (1953) pp. 185–243

[Har 54] HARISH-CHANDRA, The Plancherel formula for complex semisimple Lie groups, *Trans. Amer. Math. Soc.* **76** (1954) pp. 485–528

[Har 56] HARISH-CHANDRA, On a lemma of F. Bruhat, *J. Math. Pures Appl.* (9) **35** (1956) pp. 203–210

[Har 57a] HARISH-CHANDRA, Differential operators on a semisimple Lie algebra, *Amer. J. Math.* **79** (1957) pp. 87–120

[Har 57b] HARISH-CHANDRA, Fourier transform on a semisimple Lie algebra I, *Amer. J. Math.* **79** (1957) pp. 193–257

[Har 57c] HARISH-CHANDRA, Fourier transform on a semisimple Lie algebra II, *ibid.* pp. 653–686

[Har 57d] HARISH-CHANDRA, A formula for semisimple Lie groups, *Amer. J. Math.* **79** (1957) pp. 733–780

[Har 58a] HARISH-CHANDRA, Spherical functions on semisimple Lie groups I, *Amer. J. Math.* **79** (1958) pp. 241–310

[Har 58b] HARISH-CHANDRA, Spherical functions on a semisimple Lie group II, *Amer. J. Math.* **80** (1958) pp. 533–613

[Har 66] HARISH-CHANDRA, Discrete series for semisimple Lie groups II, *Acta Math.* **116** (1966) pp. 1–111

[Har 68] HARISH-CHANDRA, *Automorphic Forms on Semisimple Lie Groups*, Lecture Notes in Mathematics **62**, Springer-Verlag, 1968

[Hel 59] S. HELGASON, Differential operators on homogeneous spaces, *Acta Math.* **102** (1959) pp. 239–299

[Hel 62] S. HELGASON, *Differential Geometry and Symmetric Spaces*, Academic Press, 1962

[Hel 64] S. HELGASON, Fundamental solutions of invariant differential operators on symmetric spaces, *Amer. J. Math.* **86** (1964) pp. 565–601

[Hel 66] S. HELGASON, An analogue of the Paley–Wiener theorem for the Fourier transform on certain symmetric spaces, *Math. Ann.* **165** (1966) pp. 297–308

[Hel 68] S. HELGASON, *Lie groups and symmetric spaces*, Battelles Rencontres, pp. 1–71, Benjamin, 1968

[Hel 70] S. HELGASON, A duality for symmetric spaces with applications for group representations, *Adv. in Math.* **5** (1970) pp. 1–154

[Hel 72a] S. HELGASON, *Analysis on Lie Groups and Homogenous Spaces*, Conf. Board Math. Sci. Series 14, American Mathematical Society, 1972

[Hel 72b] S. HELGASON, A formula for the radial part of the Laplace Beltrami operator, *J. Differential Geom.* **6** (1973) pp. 411–419

[Hel 78] S. HELGASON, *Differential Geometry, Lie Groups, and Symmetric Spaces*, Academic Press, 1978

[Hel 84] S. HELGASON, *Groups and Geometric Analysis*, Academic Press, 1984

[Hel 94] S. HELGASON, *Geometric Analysis on Symmetric Spaces*, Mathematical Survey & Monograph 39, American Mathematical Society, 1994

[HelJ 69] S. HELGASON and K. JOHNSON, The bounded spherical functions on symmetric spaces, *Adv. in Math.* **3** (1969) pp. 586–593

[Hor 54a] A. HORN, Doubly stochastic matrices and the diagonal of a rotation matrix, *Amer. J. Math.* **76** (1954) pp. 620–630

[Hor 54b] A. HORN, On the eigenvalues of a matrix with prescribed singular values, *Proc. Amer. Math. Soc.* **5** (1954) pp. 4–7

[JoL 94] J. JORGENSON and S. LANG, *Explicit Formulas for Regularized Products and Series*, Springer Lecture Notes **1593**, 1994

[JoL 99] J. JORGENSON and S. LANG, Hilbert–Asai Eisenstein series, regularized products, and heat kernels, *Nagoya Math. J.* **153** (1999) pp. 155–188

[Jor 1880] C. JORDAN, Mémoire sur l'Equivalence des Formes, *J. École Polytech.* **XLVIII** (1880) pp. 112–150

[Kna 86] A. W. KNAPP, *Representation of semisimple Lie groups: An overview based on examples*, Princeton University Press, Princeton, N.J. 1986

[Kos 73] B. KOSTANT, On convexity, the Weyl group and the Iwasawa decomposition, *Ann. Sci. École Norm. Sup.* 4e série, **6** (1973) pp. 413–455

[Lan 56] S. LANG, Algebraic groups over finite fields, *Amer. J. Math.* **78** No. 3 (1956) pp. 555–563

[Lan 75/85] S. LANG, $SL_2(\mathbf{R})$, Addison-Wesley, 1975; Springer-Verlag, 1985

[Lan 87] S. LANG, *Linear Algebra*, Springer-Verlag, 1987

[Lan 93a] S. LANG, *Algebra*, Third Edition, Addison Wesley 1993

[Lan 93b] S. LANG, *Real and Functional Analysis*, Graduate Texts in Mathematics 142, Springer-Verlag, 1993

[Lan 95/99] S. LANG, *Differential and Riemannian Manifolds*, Springer-Verlag 1995, replaced by [Lan 99]

[Lan 99] S. LANG, *Fundamentals of Differential Geometry*, Springer-Verlag, 1999

[Len 71] A. LENARD, Generalization of the Golden–Thompson inequality, $\mathrm{Tr}\, e^A e^B \geqq \mathrm{Tr}\, e^{A+B}$, *Ind. Math. J.* **21** (1971) pp. 457–467

[Lgl 76] R. P. LANGLANDS, *On the Functional Equations Satisfied by Eisenstein Series*, Springer Lecture Notes **544**, 1976

[Maa 55] H. MAASS, Die Bestimmung der Dirichlet Reihen mit Grössencharakteren zu den Modulformen n-ten Grades, *J. Indian Math. Soc.* **19** (1955) pp. 1–23

[Maa 56] H. MAASS, Spherical functions and quadratic forms, *J. Indian Math. Soc.* **XX** (1956) pp. 117–162

[Maa 71] H. MAASS, *Siegel's Modular Forms and Dirichlet Series*, Lecture Notes in Mathematics **216**, Springer-Verlag, 1971

[Mos 53] D. MOSTOW, *Some New Decomposition Theorems for Semisimple Groups*, Mem. Amer. Math. Soc., 1953

[Na 68] R. NARASIMHAN, *Analysis on Real and Complex Manifolds*, North-Holland, 1968

[ProW 67] M. PROTTER and F. WEINBERGER, *Maximum Principles in Differential Equations*, Prentice-Hall, 1967

[Rag 72] M. S. RAGUNATHAN, *Discrete Subgroups of Lie Groups*, Ergebnisse der Math. **68** Springer-Verlag, 1972

[Ros 77] J. ROSENBERG, A quick proof of Harish-Chandra's Plancherel theorem for spherical functions on a semisimple Lie group, *Proc. Amer. Math. Soc.* **63** No. 1 (1977) pp. 143–149

[Saw 92] P. SAWYER, The heat equation on the space of positive definite matrices, *Can. J. Math.* **44** (3) (1992) pp. 624–651

[Saw 97] P. SAWYER, Estimates of the heat kernel on $SL_n(\mathbf{R})/SO(n)$, *Can. J. Math.* **49** (2) (1997) pp. 359–372

[Sel 56] A. SELBERG, Harmonic analysis and discontinuous groups, *J. Indian Math. Soc.* **XX** (1956) pp. 47–87

[Ste 64] R. STEINBERG, Differential equations invariant under finite reflection groups, *Trans. Amer. Math. Soc.* **112** (1964) pp. 392–400

[StW 71] E. M. STEIN and G. WEISS, *Introduction to Fourier Analysis on Euclidean Space*, Princeton University Press, 1971

[Ter 85] A. TERRAS, *Harmonic Analysis on Symmetric Spaces and Applications* I, Springer-Verlag, 1985

[Ter 88] A. TERRAS, *Harmonic Analysis on Symmetric Spaces and Applications* II, Springer-Verlag, 1988

[Tro 70] P. TROMBI, *Spherical transforms on symmetric spaces of rank one*, Thesis, University of Illinois, 1970

[TrV 71] P. TROMBI and V. VARADARAJAN, Spherical transforms on semisimple Lie groups, *Ann. of Math.* **94** (1971) pp. 246–303

[Tho 71] C. J. THOMPSON, Inequalities and partial orders on matrix spaces, *Ind. Math. J.* **21** No. 5 (1971) pp. 469–480

[Var 73] V. S. VARADARAJAN, The theory of characters and the discrete series for semisimple Lie groups, *AMS Proc. Symp. Pure Math.* **26** (1973) pp. 45–99

[Var 76] V. S. VARADARAJAN, *Harmonic Analysis on Real Reductive Groups*, Lecture Notes in Mathematics **576**, Springer-Verlag, 1976

[Wal 73] N. WALLACH, *Harmonic Analysis on Homogeneous Spaces*, Marcel Dekker, 1973

[Wal 88] N. WALLACH, *Real Reductive Groups* I, Academic Press, 1988

[War 72] G. WARNER, *Harmonic Analysis on Semisimple Lie Groups*, Vols. I, II, Springer-Verlag, 1972

[Yos 65/88] K. YOSIDA, *Functional Analysis*, Springer-Verlag, several editions from 1965 to 1988

Table of Notation

Unless otherwise specified, the items are defined on the indicated page. We also give a brief informal reminder of the definition here.

A	p. 2, group of diagonal matrices with positive components, determinant 1
A'	p. 88, subset of regular elements
$\mathcal{A}$	p. 21, set of simple $(\mathfrak{a}, \mathfrak{n})$-characters, basis of $\mathfrak{a}^\vee$
$\mathcal{A}'$	p. 17, dual basis of $\mathcal{A}$
$\mathfrak{a}$	p. 10, real space of diagonal matrices with trace 0
$\mathfrak{a}^+$	p. 18, set of $\mathcal{A}$-positive elements in $\mathfrak{a}$
$\mathfrak{a}^\vee$	p. 17, dual space of $\mathfrak{a}$
$\alpha_1, \ldots, \alpha_r$	p. 21, basis of $\mathfrak{a}$ consisting of the simple $(\mathfrak{a}, \mathfrak{n})$-characters
$\alpha'_1, \ldots, \alpha'_r$	p. 21, dual basis
α_{ij}	p. 81, $\alpha_{ij}(\mathrm{dia}(h_1, \ldots, h_n)) = h_i - h_j$
C_α	p. 247
$C_c(G, K)$	p. 60, continuous functions with compact support and K-conjugation invariant
$C_c^\infty(K\backslash G/K)$	p. 61, C_c^∞-functions on G which are K-bi-invariant
C_C	p. 306, continuous functions with support in C
$\mathbf{c}(g)$	p. 12, conjugation by a group element g
$\boldsymbol{c}_{\mathrm{Har}}$	p. 198, Harish-Chandra $\boldsymbol{c}$-function
$\mathrm{Cl}(\mathfrak{a}^+)$	p. 18, closure of $\mathfrak{a}^+$, so the set of $\mathcal{A}$-semipositive elements in $\mathfrak{a}$
$\mathrm{Cl}(A^+)$	p. 28, $\exp \mathrm{Cl}(\mathfrak{a}^+)$
$\mathrm{Co}(S)$	p. 25, convex closure of a set S

δ	pp. 6, 7, 8, Iwasawa character
Δ	p. 5, modular function on locally compact group
$\mathcal{D}(P)$	p. 38, differential operator associated with a polynomial P
E_{ij}	pp. 11, 81
E_t	p. 365, $E_t(\zeta) = e^{(\zeta^2-\rho^2)t}$
$\mathrm{ev}(D, f)$	p. 78, eigenvalue of D on f
Fu	p. 34, space of C^∞ functions
f^*	p. 118, $f^*(x) = \overline{f(x^{-1})}$
φ_χ	p. 80, spherical function
$\Phi(\zeta, H)$	p. 243, equal to $\varphi(\chi_\zeta, \exp H)$
Φ_{J_a}	p. 307
Φ_ν	p. 308
g_α	p. 247, 270
G_C	p. 172, equal to $K(\exp C)K$
G_R	p. 348, equal G_C with C the ball of radius R
gr	p. 40, graded algebra
g_t	p. 365, Gauss function on G, equal to $\mathbf{S}^{-1}E_t$
Har_W	p. 116, W-Harmonic polynomials
HCS	p. 346, Harish-Chandra Schwartz space
$\mathrm{HCS}^{(p)}$	p. 375, HCS with an L^p-condition
$\mathbf{H}f$	p. 60, Harish transform of f
$\mathbf{h}(D)$	p. 61, Harish-Chandra image of D
IDO	p. 36, invariant differential operators
Iw_A	p. 4, projection on A from $G = UAK$
J	pp. 92, 109, Jacobian determinant (absolute value)
$\mathcal{J}(\mathfrak{k})$	p. 49, equal to $\mathrm{IDO}(G)\tilde{\mathfrak{k}}$
$\mathcal{J}(\mathfrak{n}, \mathfrak{k})$	p. 54, $\tilde{\mathfrak{n}}\,\mathrm{IDO}(G) + \mathrm{IDO}(G)\tilde{\mathfrak{k}}$
χ_ζ	p. 77, $\chi_\zeta(a) = a^\zeta = e^{\zeta(\log a)}$
K	p. 1, unitary group (real)
$\lambda_1, \ldots, \lambda_r$	p. 21, dual basis of $\alpha_1, \ldots, \alpha_r$, so $\lambda_i = \alpha_i'$
M^G	p. 36, G-invariant elements of a set M on which G acts
$\mathbf{M}f$	pp. 97, 162, Mellin transform of f
$\mathfrak{n}$	pp. xii, 13, space of strictly upper triangular matrices $= \mathrm{Lie}(U)$
$\mathfrak{n}_\alpha$	p. 85, α-eigenspace of $\mathfrak{n}$ with $\alpha \neq 0$

$\mathbf{p}$	p. 220, polar coordinate map, $p(k_1, a, k_2) = k_1 a k_2$
p^s	p. 202, $p_1^{s_1} \cdots p_r^{s_r}$
p_d	p. 28, partial product $p_d(a) = a_1 \cdots a_d$
P^-	p. 68, $P^-(X) = P(-X)$
$\tilde{P}$	pp. 38, 39, same as $\mathcal{D}(P)$, differential operator associated to a polynomial
$\mathbf{p}_{A'}$	p. 251, polar projection
$\mathrm{Pol}(V)$	p. 10, polynomial functions on a vector space V
$\mathrm{Pos}_n(\mathbf{R})$	p. 1, space of symmetric real positive definite matrices
$(\mathrm{Pol}_{A^+})_*$	p. 252, polar projection on differential operators
PW	p. 160, Paley–Wiener functions
$\mathrm{PW}_{\mathcal{C}}$	p. 168, Paley–Wiener of exponential order $\leqq q^{\mathcal{C}}$
$\Pi_+, \Pi_+^\vee$	p. 215, polynomials, products over the $(\mathfrak{a}, \mathfrak{n})$-characters
$q^{\mathcal{C}}$	p. 167, quasi seminorm associated to a convex set $\mathcal{C}$ (max of a function on $\mathcal{C}$)
$\mathfrak{R}$	p. 248, ring generated by all g_α
$\mathcal{R}(\mathfrak{n})$	p. xii, the $(\mathfrak{a}, \mathfrak{n})$-characters
ρ	p. 27, one-half the sum of the $(\mathfrak{a}, \mathfrak{n})$-characters
$\mathbf{S}$	p. 80, spherical transform
$\mathcal{S}(\mathfrak{n})$	p. 284, set of simple $(\mathfrak{a}, \mathfrak{n})$-characters
S_α	p. 247
$\mathrm{Sch}(V)$	p. 348, Schwartz space of a real euclidean space
Sk	p. 12, skew-symmetric matrices or elements
$S(V)$	pp. 10, 37, symmetric algebra of V, so $\mathrm{Pol}(V^\vee)$
$\sigma(x)$	p. 31, $\lvert \log \mathrm{pol}_A(x) \rvert$
T_f	p. 153, integral scalar product with f
${}^tH$	pp. 65, 66, transpose of an operator H
${}^t\mathbf{\Phi}_\eta$	p. 299
${}^t\mathbf{S}_\eta$	p. 304
$\mathrm{Tub}_{R\rho}$	p. 374, equal to $\mathrm{Co}(WR\rho) + \mathbf{i}\mathfrak{a}^\vee$
$\tilde{v}$	pp. 38, 40, differential operator associated to a vector $v \in \mathrm{Lie}(G)$ same as $\mathcal{D}(v)$
W	pp. 10, 88, group of permutations of diagonal elements of a diagonal matrix, Weyl group
ω	p. 261, Casimir operator

Index

A

B

C

D

K

L

M

P

Q

R

S